Lecture Notes in Physics

Founding Editors

Wolf Beiglböck

Jürgen Ehlers

Klaus Hepp

Hans-Arwed Weidenmüller

Volume 1026

The series Lecture Notes in Physics (LNP), founded in 1969, reports new developments in physics research and teaching - quickly and informally, but with a high quality and the explicit aim to summarize and communicate current knowledge in an accessible way. Books published in this series are conceived as bridging material between advanced graduate textbooks and the forefront of research and to serve three purposes:

- to be a compact and modern up-to-date source of reference on a well-defined topic;
- to serve as an accessible introduction to the field to postgraduate students and non-specialist researchers from related areas;
- to be a source of advanced teaching material for specialized seminars, courses and schools.

Both monographs and multi-author volumes will be considered for publication. Edited volumes should however consist of a very limited number of contributions only. Proceedings will not be considered for LNP.

Volumes published in LNP are disseminated both in print and in electronic formats, the electronic archive being available at springerlink.com. The series content is indexed, abstracted and referenced by many abstracting and information services, bibliographic networks, subscription agencies, library networks, and consortia.

Proposals should be sent to a member of the Editorial Board, or directly to the responsible editor at Springer:

Dr Lisa Scalone
lisa.scalone@springernature.com

Henk N. W. Lekkerkerker · Remco Tuinier ·
Mark Vis

Colloids and the
Depletion Interaction

Second Edition

Henk N. W. Lekkerkerker
Van 't Hoff Laboratory
Utrecht University
Utrecht, The Netherlands

Remco Tuinier
Laboratory of Physical Chemistry
Eindhoven University of Technology
Eindhoven, The Netherlands

Mark Vis
Laboratory of Physical Chemistry
Eindhoven University of Technology
Eindhoven, The Netherlands

ISSN 0075-8450   ISSN 1616-6361   (electronic)
Lecture Notes in Physics
ISBN 978-3-031-52130-0   ISBN 978-3-031-52131-7   (eBook)
https://doi.org/10.1007/978-3-031-52131-7

This Springer imprint is published by the registered company Springer Nature Switzerland AG
The registered company address is: Gewerbestrasse 11, 6330 Cham, Switzerland

Paper in this product is recyclable.

*Dedicated to the memory of
S. Asakura (1927–2016),
F. Oosawa (1922–2019) and
A. Vrij (1932–2019)*

Preface to the Second Edition

Since 2011 when the First Edition of this book appeared, there has been substantial experimental and theoretical progress in the multidisciplinary field of colloids and the depletion interaction. This was made clear, for instance, by the special issues 'Self Assembly—Depletion Forces in Single Phase and Multi-Phase Complex Fluids' in *Curr. Opin. Colloid Interface Sci.* [1] and 'Depletion Forces and Asakura–Oosawa Theory' in *J. Chem. Phys.* [2]. This Second Edition, written with the invaluable contribution of Mark Vis as co-author, is an updated and enlarged version of the First. This updated edition includes recent developments in the field and selected references from the last decade. Further, the discussions on depletion interactions and resulting phase behaviour induced by colloidal spheres and rods have been extended with recent developments and are now presented separately in Chaps. 6 and 7. We have also significantly extended Chap. 8 on 'Phase behaviour of colloidal rods with depletants' with recent experimental and theoretical developments. This edition also contains several entirely new chapters. The first is a topic suggested by the late Kurt Binder [3]: 'The interface in demixed colloid–polymer dispersions', Chap. 5. The new Chaps. 9 and 10 concern the phase behaviour of colloidal platelets and colloidal cubes in the presence of depletants. Further, Chap. 11 focuses on various application areas of the concepts discussed in this book.

We want to express our gratitude to the chemists and physicists who have helped us with their constructive criticism and useful suggestions. Amongst these we mention Louise Bailey, Patrick Davidson, Geoff Maitland and Ben Widom. We are indebted to Max Martens, Max Schelling, Rik Wensink and Bert de With for commenting in detail on several chapters of the manuscript; Edgar Blokhuis for detailed comments on Chap. 5, Anja Kuhnhold for suggestions on Chaps. 7 and 8, Pavlik Lettinga for critical remarks on Chap. 8, Daniel de las Heras and Matthias Schmidt for useful comments on Chap. 9, and Frans Dekker, Laura Rossi and Janne-Mieke Meijer for suggestions on Chap. 10. We thank Dzina Kleschanok, Jasper Landman and Jolijn Nagelkerke for their work on depletion topics. We also acknowledge Evan Spruijt for useful suggestions on macromolecular crowding. We are most grateful to Elizabeth McKenzie for carefully proofreading this Second Edition, making excellent suggestions to improve the content and structure and particularly for helping us with our English.

When writing Chaps. 6 to 10, the work in the Ph.D. theses of Frans Dekker, Álvaro González García, Joeri Opdam and Vincent Peters has been very useful. The expert reader will recognise some of their work in the new chapters. Discussions with them on depletion and its intricacies, as well as practical help, have enabled us to extend the First Edition.

We express our appreciation for the pleasant cooperation with Annelies Kersbergen and Angela Lahee of Springer Nature, and the encouragement to think of a Second Edition from Maria Bellantone. Finally, R. T. and M. V. thank the members of the Laboratory of Physical Chemistry at Eindhoven University of Technology for the stimulating environment. We were fortunate to have meticulous help and management support from Pleunie Smits and, in the preparation of some of the texts and figures of the manuscript, from Mieke Kröner.

H. N. W. L. wants to thank his wonderful wife Loes for her understanding and support during the writing of this book: so many beautiful days that we could have done something together were sacrificed to this Second Edition. R. T. thanks Mieke, Luuk and Tim for their continuous support. M. V. thanks Maartje, Janneke and Sietske for joyful distractions.

This work is published open access thanks to financial support from the Open Access Books program of the Dutch Research Council (NWO) and by the Library & Information Services Department of Eindhoven University of Technology.

Utrecht/Eindhoven   Henk N. W. Lekkerkerker
February 2024   Remco Tuinier
                                                                      Mark Vis

References

1. T. Zemb, P.A. Kralchevsky, Curr. Opin. Colloid Interface Sci. **20**, 1 (2015)
2. K. Miyazaki, K.S. Schweizer, D. Thirumalai, R. Tuinier, E. Zaccarelli, J. Chem. Phys. **156**, 080401 (2022)
3. K. Binder, Eur. Phys. J. E **38**, 73 (2015)

Preface to the First Edition

The physical properties of colloidal suspensions are strongly affected by the forces that act between the colloidal particles. Attempts to explain them in these terms go back to the beginning of the twentieth century. Important and extensively studied forces in colloidal systems are Van der Waals forces, electrostatic forces, steric forces due to attached polymers and magnetic forces. In the last few decades researchers have observed that the stability of colloidal particles is also affected by nonadsorbing polymers in solution. The origin of this interaction was first explained successfully in 1954 by S. Asakura and F. Oosawa. They used the concept that the free volume available to nonadsorbing polymers increases whenever two hard particles are sufficiently close to each other, such that their depletion zones overlap and the total depletion zone decreases.

However, a number of important applications were used in technology and medicine (long) before the depletion concept was introduced. For example, the clustering of red blood cells induced by serum proteins had already been detected at the end of the eighteenth century and forms the basis of the blood sedimentation test that is still in use today. Furthermore, creaming of colloidal particles to concentrate latex dispersions upon the addition of polysaccharides was first studied in the 1920s. The 1940s saw the start of using polysaccharides to isolate plant viruses. Systematic and fundamental investigations on the effect of depletion interactions in colloidal systems started with the work of B. Vincent in the UK, S. Hachisu in Japan, and A. Vrij in The Netherlands in the 1970s. Work on the depletion interaction gained momentum after W. B. Russel and co-workers in the US clarified the relationship between the range and depth of the depletion interaction and the topology of the phase diagram in the 1980s. Since then, the depletion field has evolved rapidly.

This book aims to provide a self-contained treatment of the depletion interaction and the resulting phase behaviour in colloidal dispersions. It is hoped that the book may be equally useful to both junior and senior undergraduate students in physical chemistry, chemical and mechanical engineering, biophysicists or soft condensed matter physics. At the same time, we hope that professional chemists and engineers dealing with colloidal suspensions may find it a useful reference book to gain an understanding of the implications of the depletion interaction for handling suspensions.

In order to keep the size of the book within bounds a description of the interface between demixed phases has not been included and the discussion of phase transition kinetics is rather brief. We emphasise that we do not claim the references quoted to be a complete list. If the reader prefers to, they can read the book at three levels. For a general idea of depletion interactions and their implications in science and technology it is recommended to study Chapter 1. At the second level[1] one can study Sections 2.1 and 2.2, Chapter 3 and Sections 4.1, 4.2, 4.4 and 8.1 to 8.6. This material could be used for an 8-hour senior undergraduate or junior graduate course in physical chemistry or soft matter physics. The third level covers the complete text of this monograph.

Many people have stimulated us to write this book. Initially, we had hoped to write it with Dirk Aarts. His enthusiasm for the book project helped us greatly during the early stages but he was unable to reserve enough time for the book after his start in Oxford. We are indebted to him, to Jeroen van Duijneveldt and Gerard Fleer for commenting in detail on drafts of several chapters of the manuscript. It may well be that remaining errors and unclear aspects can be traced back to where we foolishly disagreed with them.

We were fortunate to have meticulous help in the preparation of texts and figures of the manuscript from Mieke Kröner, while the illustrations benefited from the advice of Jeannette Kröner.

R.T. wishes to thank Jan Dhont and the Soft Matter group at Forschungszentrum Jülich for their support during the initial stages leading to this book. The members of the Colloids & Interfaces group at DSM Research, Leon Bremer, Harm Langermans, Leo Vleugels, Benjamin Voogt, Jef Bisscheroux, and Feng Li, are acknowledged for the pleasant and stimulating interactions. Peter Jansens and Jeroen Kluytmans of DSM Research are thanked for supporting R.T. to finish the book. Collaborations with Martien Cohen Stuart, Tai-Hsi Fan, Kees de Kruif, Peter Schurtenberger, Takashi Taniguchi, and Agienus Vrij contributed to the evolution of this book.

H.N.W.L. wishes to thank the staff members and his PhD students and Postdocs at the Van 't Hoff laboratory, with whom he had the privilege to work in the period 1985–2010. He benefited from a long-term collaboration with Marc Baus, Louise Bailey, Mike Cates, Bob Evans, Seth Fraden, Daan Frenkel, Jean-Pierre Hansen, Joseph Indekeu, Geoff Maitland, Theo Odijk, Roberto Piazza, Wilson Poon, Peter Pusey, Bill Russel, Patrick Warren and Ben Widom. H.N.W.L. would like to thank the Royal Netherlands Academy of Arts and Sciences for the appointment as Academy Professor for the period 2006–2011, which made it possible to write this book.

---

[1] Updated for this Second Edition.

Finally, we express our appreciation for the encouragement from and pleasant cooperation with Maria Bellantone, Mieke van der Fluit and Liesbeth Mol of Springer Science + Business Media.

Geleen/Utrecht   Remco Tuinier
December 2010   Henk N. W. Lekkerkerker

Contents

Symbols

We refer to the section number where the symbol was first used. An asterisks (*) refers to the fact that this symbol is only used in a single Chapter.

Greek

α	Free volume fraction (3.3)
α_p	Static polarizability (1.2)
β	Scaling factor in the concentration dependence of the depletion thickness (4.3)
ß	Stretched exponential exponent (4.4)*
χ	Partition coefficient of polymers between a slit and the bulk (2.2)
δ	Depletion layer thickness (1.2)
δ_b	Brush thickness (1.2)
Δ	Range of attraction (1.2)
Δ	Lattice spacing (9.2)*
$\overline{\Delta}$	$= \Delta/D$ Relative lattice spacing with respect to D (9.2)*
η	Viscosity (5.3)
ε	Strength of attraction (1.2)
ε_0	Dielectric permittivity in vacuum (1.2)
ε_r	Relative dielectric permittivity (1.2)
ϕ	Colloid volume fraction (2.3)
ϕ^*	Reduced packing fraction (9.2)*
ϕ_c	Volume fraction of superballs (10.2)
ϕ_{cp}	Volume fraction at close packing (3.2)
ϕ_d	Relative depletant concentration (3.3)
ϕ_m	Melting volume fraction (3.2)*
ϕ_p	Relative polymer concentration (1.2)
φ	Volume fraction of polymer segments (1.2)
φ^*	Volume fraction of polymer segments at overlap (1.2)
γ	Angle between rods (7.2); only in Chaps. 7 and 8 and Chap. 9
γ	Scaling exponent for the characteristic polymer size (4.3)*
γ	Interfacial tension (1.3), only used in Chaps. 1 and 5 and Appendix A
Γ	Adsorbed amount (2.1)

Γ	Aspect ratio $L/D + 1$ (7.2)
$\Gamma_E(x)$	The Euler Gamma function (10.2)[*]
κ	Variational parameter in trial functions $f(\theta)$ (8.2); only in Chaps. 8 and 9
λ	Relative interparticle distance $(h - \sigma)/\sigma$ (2.3)[*]
λ	Scaling parameter for the size of a particle (3.3)
λ	Lagrange multiplier (8.2)[*]
λ_m	Wavelength in a medium (2.6)[*]
λ_B	Bjerrum length (1.2)[*]
λ_D	Debye screening length (1.2)
Λ	De Broglie wavelength (3.2)
Λ	Characteristic length scale (4.4)[*]
μ	Chemical potential (2.1)
μ^{ex}	Excess chemical potential (8.2)
$\tilde{\mu}$	Dimensionless chemical potential μ/kT (3.2)
ν	Overlap volume (2.3)[*]
ν	Scaling exponent for the characteristic polymer size (4.3)[*]
ν	Scaling parameter for the size of a rod (7.2)
θ	Polar angle (2.1)
θ_s	Scattering angle (2.6)
Π	Osmotic pressure of platelets (9.4)[*]
$\tilde{\Pi}$	Reduced osmotic pressure of platelets $\Pi D^3/kT$ (9.4)[*]
ρ	Mass density (1.2)
Δ_ρ	Density difference (1.2)
ρ	Packing parameter (8.2); only in Chaps. 8 and 9
$\tilde{\rho}$	Reduced linear platelet density (9.2)[*]
ϱ	Penetration depth of an evanescent wave (2.6)[*]
ρ^*	Relative platelet density (9.2)[*]
σ	Diameter of a penetrable hard sphere (2.1)
σ_{max}	Maximum standard deviation (3.2)[*]
σ_a	Diameter of an atom (1.2)[*]
σ_b	Brush anchor density (1.2)[*]
σ_c	Surface charge density (1.2)[*]
ς	Asphericity (10.2)[*]
τ_B	Brownian time scale (4.4)[*]
ω	Angular velocity (5.1)[*]
Ω	Grand potential (3.3)
$\overline{\Omega}$	Approximate grand potential (Appendix)[*]
$\tilde{\Omega}$	Normalised grand potential $\Omega v_0/kT V$ (3.3)
$\boldsymbol{\Omega}$	Solid angle (7.2)
Ξ	Grand canonical partition function (Appendix)[*]
ξ	Correlation length (4.3)
ψ	Electrostatic potential (8.2)
ψ_0	Surface potential (1.2)[*]
ζ	Constant in combination rule for depletion thickness (4.3)[*]

| ζ | Proportionality constant of the outer part of the double layer potential around a charged rod (8.2)[*] |

Latin

a	Parameter in expression for α (3.3)
A	Hamaker constant (1.2)[*]
A	Area (2.2)
A'	Parameter to describe effective rod diameter of a charged rod (8.2)[*]
A_i	ith osmotic virial coefficient between polymer chains (4.3)
b	Parameter in expression for α (3.3)
b	Segment length (2.2)
b	Excluded volume between two rods (8.2)[*]
B	Amplitude of the double layer repulsion (1.2)
B_i	ith osmotic virial coefficient between colloids (1.3)
c	Concentration (2.2)
c	Parameter in expression for α (3.3)
$c(\tilde{r})$	Direct correlation function (5.2)[*]
c_c	Surface integrated mean curvature of a superball (10.4)[*]
$\tilde{c}_c$	Normalised surface integrated mean curvature c_c/R_{el} of a superball (10.4)[*]
C	Coefficient in the dispersion attraction (1.2)[*]
C_i	Curvature terms for the depletion thickness (4.3)[*]
D	Rod/platelet diameter (2.4 and 2.5)
d	Scaling exponent describing the scattering of aggregates (4.4)[*]
d	Diameter of a colloidal sphere (1.3)
d_f	Fractal dimension describing the internal structure of aggregates (4.4)[*]
D	Droplet diameter (5.1)[*]
D_s	Self-diffusion coefficient (4.4)[*]
e	The elementary charge (1.2)
$E(x)$	Complete elliptic integral of the second kind (9.2)[*]
E_i	Internal electric field (1.2)[*]
E_e	External electric field (1.2)[*]
$\overline{E}$	Electronic excitation energy (1.2)[*]
f	Function for the distance dependence (2.2)
$f(\Omega)$	(or: $f(\theta)$) orientational distribution function (7.2); only in Chaps. 7 to 9
$f(m)$	Superball parameter (10.1)[*]
$\mathfrak{f}$	Superball shape description function (10.2)[*]
F	Helmholtz energy (2.1)
$\tilde{F}$	Dimensionless Helmholtz energy Fv_0/kTV (3.3)
$\tilde{f}$	Scattering scaling function (4.4)[*]
g	Gravitational acceleration (1.1)
G_P	Parsons-Lee scaling factor (9.2)
h	Closest distance between the surfaces of two particles (1.2)

h	Contact height (5.1)
H	Distance between spherical surfaces (2.1)*
H_{sample}	Sample height (9.4)*
I	Ionic strength (1.2)
I	Scattered intensity (2.6)
k	Boltzmann's constant (3.1)
k_E	Euler's constant (8.2.1)
kT	Thermal energy (1.1)
K	Force between plates per unit area (2.1)
K_s	Force between two spheres (2.1)
K_{sp}	Force between a sphere and a flat plate (2.1)
K_0	Modified Bessel function of the second kind of order 0 (8.2)*
ℓ	Characteristic length scale of depletion agent (2.5)*
ℓ_c	Capillary length (5.1)*
ℓ_{sed}	Sedimentation length (1.1)
L	Rod length (2.4)
L	Droplet length (5.1)*
L	Thickness of a platelet (2.5)*
L	Characteristic length (5.3)*
L_T	Thermal length (5.3)
L_0	Typical droplet coalescence length scale (5.3)*
m	Superball shape parameter (10.1)
$\tilde{m}$	Is the second moment of the direct correlation function (5.2)
M	Number of monomers or chain length (1.2)
M_p	Polymer or particle molar mass (1.2)
M_x	Molar mass of polymer 'x' (1.2)
m^*	Buoyant mass (1.1)
n	Number density (1.2)
n_b	Bulk number density (2.1)
n_d	Number density of depletants (3.3)
n_i	Number density between plates (2.2)
n_o	Contact number density outside plates (2.3)
n_s	Salt number density (1.2)
$\tilde{n}$	Reduced platelet density (9.4)
N	Number of particles (2.1)
$\tilde{N}$	Normalisation constant (8.2)
N_{Av}	Avogadro's number (1.2)
p	Probability (11)
p_c	Number of elementary charges e on a surface area $2\pi\lambda_D\lambda_B$ (1.2)
P	(osmotic) pressure (1.2)
P	Form factor (2.6)*
P_e	Peclet number (4.4)*
$\tilde{P}$	Dimensionless pressure $P\nu/kT$ (3.2)
P_i	Osmotic pressure inside plates (2.1)
P_o	Osmotic pressure outside plates (2.1)

q	Size ratio of depletant over colloid size (1.2)
q_e	Number of charges (1.2)[*]
Q	Canonical partition function (Appendix)[*]
Q	Scattering wave vector (2.6)
Q_m	Scattering wave vector where the scattering intensity I passes through a maximum (4.4)
$\mathfrak{Q}$	Parameter in the EOS of a fluid of superballs (10.2)[*]
r	Center-to-center distance between two spheres (2.1)
r_{cp}	Distance between the centres of a superball and its nearest neighbours at close packing (10.2)[*]
R	Radius of a sphere (1.1)
Re	Reynolds number (5.3)
R_c	Radius of curvature (5.1)[*]
R_d	Effective depletion radius hard-sphere radius (2.1)
R_{eff}	Effective hard-sphere radius (4.1)
R_{ell}	Superball edge length (10.1)[*]
$\sqrt{\langle R^2 \rangle}$	End-to-end distance (2.2)
$\langle ... \rangle$	Ensemble-averaged value (2.2)
R_g	Radius of gyration of a polymer chain (1.2)
$\mathfrak{R}$	Parameter in the EOS of a fluid of superballs (10.2)[*]
s_c	Surface area of a superball (10.4)[*]
$\tilde{s}_c$	Normalised surface area s_c/R_{el}^2 of a superball (10.4)[*]
$\mathfrak{s}$	Orientational entropy parameter (8.2)
S	Nematic order parameter (8.2); only in Chaps. 8 and 9
S	Structure factor (2.6)
t	Time (7.2)
$\hat{t}$	Diffusion time (4.4)
t_0	Typical droplet coalescence time scale (5.3)[*]
T	Temperature (1.1)
$\mathfrak{T}$	Parameter in the EOS of a fluid of superballs (10.2)[*]
U_c	Interaction energy between colloids (Appendix)[*]
U_{cd}	Interaction between colloids and depletants (Appendix)[*]
$\mathbf{v}$	Excluded volume of a polymer segment (4.3)
v	Velocity (5.3)
v_d	Volume of a depletant (3.3)
v_{excl}	Excluded volume (7.2)
$\tilde{v}_{excl}$	Normalised excluded volume (9.2)
v_0	Colloidal particle volume (1.3)
v_p	Polymer coil volume (1.2)
v_R	Volume of a rod (8.2)
v_{rod}	Volume of a spherocylinder (7.2)
v_s	Volume of a monomer (segment) (1.2)
v^*	Free volume (3.2)
V	Volume (2.2)

V_{ov}	Overlap volume between depletion layers (1.2)
$\langle V_{free} \rangle$	Ensemble-averaged free volume for the depletants (3.3)
$\langle V_{free} \rangle_0$	Undistorted ensemble-averaged free volume (3.3)
W	Interaction potential (1.2)
W_s	Interaction potential between two spheres (2.1)
W_{sp}	Interaction potential between a sphere and a plate (2.1)
W	Particle insertion work (3.3)
x	Distance from the surface (2.2)
x^R	Property x in the reservoir (3.1)
y	$\phi/(1 - \phi)$ (8.3)
z	Partition function of one ideal chain (2.2)
z	Valency of charged surface groups (1.2)[*]
z	Normalised distance from the surface (2.2)
z	Height of interface (5.1)[*]
z	Position or normalised position (from a surface/interface)(2.1)
z_p	Polymer fugacity (9.3)[*]
Z	Partition function (2.2)[*]
Z	Linear charge density (8.2)[*]

Introduction

1.1 Colloids

According to IUPAC [1], the term colloidal refers to 'a state of subdivision, implying that the molecules or polymolecular particles dispersed in a medium have at least in one direction a dimension roughly between 1 nm and 1 μm, or that in a system discontinuities are found at distances of that order'. This means that colloidal particles are submicrometre sized substances dispersed in a medium that can be a liquid or a gas [2–10]. This implies that colloids are much bigger than 'normal' molecules (though they may be comparable in size to macromolecules). The lower limit of the length scale for a colloidal particle is close to a nm. The medium of low molecular mass substances in a colloidal suspension can often be regarded as 'background' with respect to the colloidal size range and, in that case, this medium may be approximated as a continuum.

From a physics point of view, colloidal particles are characterised by observable Brownian motion, originating from a thermal energy of order of kT for each colloidal particle. Particles in a solvent are considered to be Brownian if sedimentation can be neglected with respect to thermal motion. This means that the sedimentation length, the ratio of thermal energy and gravity force should be larger than the colloid radius. The sedimentation length is defined as [11]:

$$\ell_{\text{sed}} = \frac{kT}{m^*g}.$$

(1.1)

Here, the buoyant mass is given by $m^* = (4\pi/3)\Delta\rho R^3$ for a spherical colloid with radius R, where $\Delta\rho$ is the density difference between particle and solvent. Hence, the upper colloidal size corresponds to the condition where $\ell_{\text{sed}} \approx R$. For $\Delta\rho = 100$ kg/m^3, this implies an upper limit of the radius of about 1 μm at 300 K.

© The Author(s) 2024
H. N. W. Lekkerkerker, R. Tuinier, M. Vis, *Colloids and the Depletion Interaction*,
Lecture Notes in Physics 1026, https://doi.org/10.1007/978-3-031-52131-7_1

Perrin [12] studied dispersed resin colloids and detected Brownian motion as a visible manifestation of thermal fluctuations, verifying Einstein's theoretical results [13]. The height distribution of the resin colloids in the field of gravity was shown to obey Boltzmann's law for the sedimentation equilibrium. The picture emerged that colloids behave as big atoms in many respects. Later, Onsager [14, 15] and McMillan and Mayer [16] laid down a statistical mechanics foundation for the colloid–atom analogy. They pointed out that the degrees of freedom of the solvent molecules in a colloidal dispersion can be integrated out, implying the solvent can be considered as 'background'. The resulting description only involves colloidal particles interacting through an *effective* potential, the potential of mean force, that accounts for the presence of the solvent.

Often, the (rotationally averaged) interactions between small spherical atoms and some molecules can be reasonably described using the Lennard-Jones interaction [17] (see Ref. [18] for an in-depth critical discussion). For many of these systems, the phase diagrams scaled by the critical values of temperature, pressure and molar volume appear similar as well. The fact that the thermodynamic properties of all simple gases exhibit basic similarities is expressed by the law of corresponding states of Van der Waals. A statistical mechanical derivation of this law was provided by Pitzer [19].

Just as the pressure of an atomic gas is affected by the interaction between the atoms, the physical properties of a colloidal dispersion depend on the potential of mean force between colloidal particles. An extended law of corresponding states has been conjectured [20], stating that knowledge of the potential of mean force between spherical colloidal particles enables prediction of the phase diagram (topology). Therefore, one may expect similarities between the phase diagrams of atomic and colloidal systems.

Apart from such similarities, there are also distinct differences between atoms and colloidal particles. In contrast to pair interactions between atoms, interactions between colloidal particles can be tuned by choosing particle type or solvent, by supplementing additives such as electrolytes, polymers or other colloidal particles, or by modifying the particle surface. Since the 1970s it has gradually become clear that adding small particles or polymers that do not adsorb onto the colloids opens up a wide variety of possibilities for tuning the phase behaviour of colloidal dispersions. The interactions mediated by such nonadsorbing species and the resulting phase behaviour are at the core of this book.

The science of colloids is important for applications ranging from drug delivery and dairying to coating technology and energy storage materials. Colloidal dispersions can be found in a wide range of environments and products. Industrial examples include emulsions (mayonnaise), foams (shaving cream), surfactant solutions (shampoo) or polymer latex dispersions (paint). Long-term stability of a colloidal dispersion is often desired, for example, in storage of paint [21] or food [22]; and this is regularly achieved by adjusting the particle surface chemically or via adsorption.

Unconsciously, humankind has long held an interest in colloidal stability. For example, carbon is the oldest ink material known, and its use for writing in Egypt can be dated back to 3400 BCE [23]. The carbon used for making ink was soot in most cases. By mixing it with gum arabic and water, soot was made into ink. Without understanding the underlying principles, the Egyptians effectively used the principle of stabilising dispersions by adsorbed macromolecules [3,24]. This is nowadays recognised as an example of polymeric stabilisation (see Sect. 1.2.4). In this manner the Egyptians succeeded in engineering the soot particles, such that they can remain suspended for an indefinite period.

An example where the instability of colloids (clay) plays a role in nature is delta formation. Deltas [25] are formed due to precipitation of colloidal (clay) particles carried by the river as its flow meets the sea (or ocean), where the fresh river water mixes with salty sea water. The delta formation process had already been described by Barton [26] in 1918 before a clear understanding of the role of salt on colloidal stability had been established.

Milk is a natural colloidal dispersion that contains casein micelles—self-assembled protein associates with a diameter of about 200 nm [27]. The casein micelles are protected against flocculation by an assembly of dense 'hairs' (often called a 'brush') at their surfaces. Polymer brushes can thus provide steric stabilisation of colloids. For millennia man has used the fact that milk flocculates and gels when it is acidified, as in yogurt production. Below pH = 5 macroscopic flocculation of the casein micelles in milk is observed [28]. This means that the interactions between casein micelles change from repulsive to attractive. The explanation is that acidification leads to collapse of the casein brushes [29]. In cheese-making the steric stabilisation is removed by enzymes that induce gelation into cheese curd.

Modest solvent composition changes can also affect the state of a colloidal dispersion. A charge-stabilised dispersion of polymer latex particles or gold colloids may flocculate irreversibly upon adding salt, while ion removal through dialysis may turn the dispersion into an ordered structure that exhibits Bragg reflection [30]. Obviously, the physical state of a colloidal dispersion is a function of the interactions between the colloidal particles.

In foods, paints and biological systems such as living cells, colloids and polymers are often present simultaneously. When the polymers do not adsorb onto the colloidal particles the result is a so-called depletion layer: a zone near the particle surface which contains a lower polymer concentration than the bulk of the solution. As we shall see, overlap of depletion layers leads to an attractive depletion interaction between the colloidal particles. The term depletion derives from Latin meaning 'emptied out'. The verb 'plere' is 'to fill' [31]. Thus a 'pletion' force is due to accumulation of some substance between two colloids. The reversal, a 'depletion' force, is due to the expulsion of material. Feigin and Napper [32] were probably the first to introduce the term depletion.

Mixing colloids with polymers or other colloids can lead to phase transitions or aggregation resulting in, for instance, gelation, crystallisation, glass transition, flocculation, or fluid–fluid demixing of the dispersion. Figure 1.1 illustrates a colloid–polymer mixture and its tendency to phase separate into a phase enriched in colloids and a phase concentrated in polymers due to the attraction mediated by nonadsorbing polymer chains.

Fig. 1.1 Representation of a
colloid–polymer mixture.
Top: the system just after
mixing. *Bottom*: the
dispersion becoming
inhomogeneous after mixing
the colloidal spheres with a
sufficiently high amount of
nonadsorbing polymers

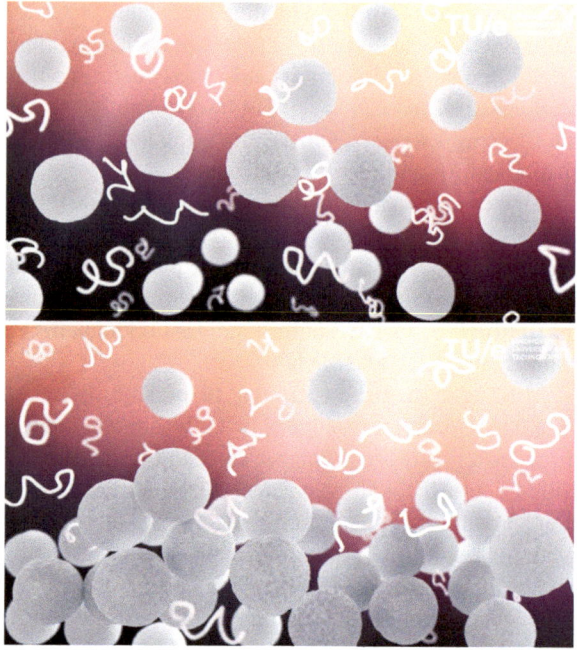

The type of instability depends on the range and strength of the particle interactions involved. The knowledge gained over the last decades on depletion effects in mixtures of colloidal particles and polymers is of great interest for designers of new products. Insight into the factors determining the stability of mixtures changes product development from trial-and-error towards knowledge-driven innovation. This book serves as a guide to help understand what happens when colloids are mixed with polymers or other colloids.

This chapter gives an introduction to colloidal interactions (including the depletion force in a historical context) and provides examples of the manifestations of depletion effects. First, we start with a brief overview of colloidal interactions in Sect. 1.2, including the basic concept of the depletion interaction. In Sect. 1.3, we outline the effects of unbalanced forces, addressing depletion forces in colloidal dispersions from a historical perspective, and including an overview of selected literature. Finally, a brief outline of the other chapters of this book are given in Sect. 1.4.

1.2   Colloidal Interactions

The basic understanding of colloidal interactions [10] commenced in the 1940s. Derjaguin and Landau [33] in the former USSR, and Verwey and Overbeek [34] in The Netherlands pointed out that, in a dispersion of charged colloids in an electrolyte solution, the Van der Waals attraction between two colloidal particles is opposed by a repulsion that originates from electrical double layers. This foundation for the stability of colloids is known as the DLVO theory and has been remarkably successful in explaining the results of a vast number and broad range of experiments, including direct force measurements [35]. Polymers, either depleted from or adsorbed or anchored to colloidal surfaces, also turned out to strongly influence colloidal interactions; these were not considered by DLVO.

In Sects. 1.2.1 and 1.2.2 we shall first consider Van der Waals and double layer interactions (the two contributions to the DLVO potential (Sect. 1.2.3), and then discuss (polymeric) steric stabilisation by end-attached polymers in Sect. 1.2.4. Finally, the depletion interaction will be addressed in Sect. 1.2.5.

1.2.1 Van der Waals Attraction

The attractive interaction between two colloidal particles is due to London–Van der Waals attraction between their constituent atoms or molecules. For two atoms at a centre-to-centre distance r apart, the attraction has the form:

$$W(r) = -\frac{C}{r^6} \tag{1.2}$$

(see Chap. 13 and Table 13.3 in Ref. [35]). According to London theory [36], C is given by the (approximate) expression

$$C = \frac{3}{4}\overline{E}\left(\frac{\alpha_p}{4\pi\varepsilon_0}\right)^2. \tag{1.3}$$

Here, $\overline{E}$ is a typical (average) electronic excitation energy, α_p is the static polarisability and ε_0 is the vacuum permittivity.

Hamaker [37] calculated the London–Van der Waals attraction between two colloidal particles by summation of all atomic/molecular Van der Waals interactions between these two colloidal particles. For two colloidal spheres with radius R (Fig. 1.2) and closest distance between the spheres h, the resulting Van der Waals attraction without intervening medium reads (see for instance Ref. [38])

$$W_{\text{VdW}}(h) = -\frac{A}{6}f(h/R), \tag{1.4}$$

with

$$f(h/R) = \frac{2R^2}{h^2 + 4Rh} + \frac{2R^2}{h^2 + 4Rh + 4R^2} + \ln\left(\frac{h^2 + 4Rh}{h^2 + 4Rh + 4R^2}\right),$$

and

$$A = C\pi^2 n^2, \tag{1.5}$$

where n is the number density of the atoms/molecules in the colloidal particles.

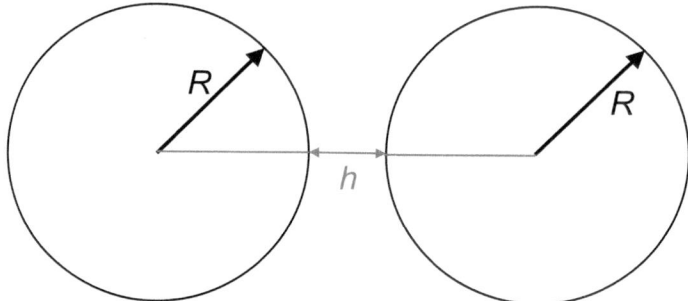

Fig. 1.2 Two colloidal spheres with radii R at closest separation distance h

To make an estimate of the quantity A, which is known as the Hamaker constant we follow Israelachvilli [35] and use $C = 10^{-77}$ J·m^6 and a number density $n = 2 \cdot 10^{28}$ m^{-3} (corresponding to a diameter of 0.4 nm of the atoms/molecules in the colloidal particles). We then find

$$A = \pi^2 \cdot 10^{-77} \cdot (3 \cdot 10^{28})^2 \simeq 10^{-19} \text{ J}. \tag{1.6}$$

Values for the Hamaker constants for different materials range between $(0.4 - 5) \cdot 10^{-19}$ J and can be found, for instance, in Table 13.2 in Ref. [35]. With an intervening medium between the colloidal particles the Hamaker constants are (significantly) lower than without intervening medium. From the values presented in Table 13.3 of [35] it turns out that the reduction may be as much as a factor of 2 or 3.

From Eq. (1.4) it follows that the Van der Waals attraction is very strong at short interparticle separations. For small h,

$$W_{\text{VdW}}(h) \simeq -\frac{AR}{12h}. \tag{1.7}$$

To stabilise a colloidal dispersion, a significant repulsion that prevents the particles getting too close and aggregating irreversibly is needed. The Van der Waals interaction is shown schematically in Fig. 1.3 as the lower dashed curve.

Fig. 1.3 Common contributions to the interaction potential $W(h)$ between colloidal particles as a function of separation distance h: typical double layer repulsion between charged colloidal spheres (top), Van der Waals attraction (bottom) and their sum (solid curve), which is the DLVO interaction potential

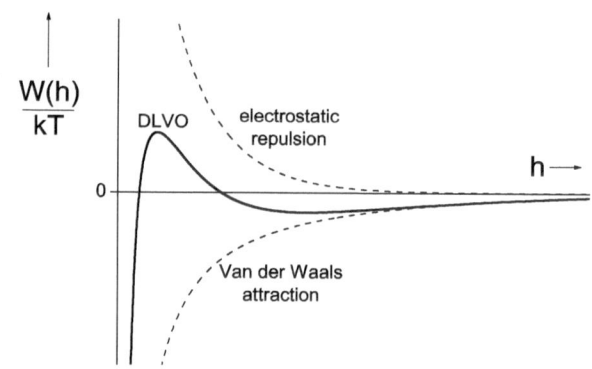

1.2.2 Double Layer Interaction

A charged colloid is surrounded by a solution with an inhomogeneous distribution of ions. Co-ions (with the same charge as the colloids) are repelled from the colloid surface, whereas counterions (with opposite charge) accumulate at the surface. Far from the colloidal surface the concentrations of the two ion types attain a constant averaged value. The inhomogeneous layer is termed 'double layer', and its width depends on the ion concentration in the bulk solution: adding more ions screens the charges on the colloidal surfaces.

When two double layers overlap, a repulsive pair potential develops, which leads to a repulsive pressure. Dispersed like-charged colloids hence repel each other upon approach due to screened-Coulomb or double layer repulsion. The length scale over which this force is operational is set by the Debye screening length λ_D which, for a simple 1–1 salt, reads

$$\lambda_D = \sqrt{\frac{1}{8\pi\lambda_B n_s}}, \tag{1.8}$$

where n_s is the salt number density and λ_B is the Bjerrum length,

$$\lambda_B = \frac{e^2}{4\pi\varepsilon_0\varepsilon_r kT}, \tag{1.9}$$

with e the elementary charge and ε_r the relative dielectric constant (≈ 80 in water). The Bjerrum length is the distance between two elementary charges at which their interaction equals kT. In water at room temperature its value is ≈ 0.7 nm. For the Debye length we can then use the expression $\lambda_D = 0.3/\sqrt{c_s}$ [35], with the salt concentration c_s in mol/L and λ_D in nm.

The interparticle separation dependence of double layer repulsion is approximately exponential for a thin double layer ($\lambda_D \ll R$) [34]

$$W_{DR}(h) = B\frac{R}{\lambda_B}\exp(-h/\lambda_D), \tag{1.10}$$

which shows that the range of the screened double layer repulsion is λ_D, which depends on the salt concentration. The double layer interaction between two like-charged colloidal particles is represented in Fig. 1.3 (upper dashed curve).

The quantity B can be expressed in terms of the surface charge density σ_c of the interacting colloids [34]

$$\frac{B}{kT} = \frac{8p_c^2}{(1+q_c)^2}, \tag{1.11}$$

where $p_c = 2\pi\lambda_D\lambda_B |\sigma_c/e|$ and $q_c = \sqrt{1+p_c^2}$, with p_c the number of elementary charges e on a surface area $2\pi\lambda_D\lambda_B$. Given the fact that $|\sigma_c|$ varies roughly between

0.1 and 2 $e \cdot nm^{-2}$, the value of p_c ranges from 0.1 to 100 and thus B has a typical value of 0.1–$8\,kT$. The quantity B can also be expressed as a function of the surface potential ψ_0 [34]:

$$\frac{B}{kT} = 8\left[\tanh\left(\frac{e\psi_0}{4kT}\right)\right]^2. \qquad (1.12)$$

The surface potential of a charged colloidal particle typically varies from 10 to 100 mV, leading to B values in the same range as given above.

1.2.3 DLVO Interaction

By assuming additivity of the interactions, the total DLVO potential is simply given by

$$W_{\mathrm{DLVO}} = W_{\mathrm{VdW}} + W_{\mathrm{DR}}. \qquad (1.13)$$

In Fig. 1.3 the DLVO interaction potential W_{DLVO} is represented alongside its two contributions. If the maximum of W_{DLVO} is sufficiently high (larger than a few kT), flocculation is prevented. Flocculation does occur when the particles get very close together and reach the so-called primary minimum. This minimum is usually deep enough for irreversible flocculation.

For a given Van der Waals attraction and particle size the DLVO potential depends on the ionic strength. The DLVO potential is qualitatively represented in Fig. 1.4, from (i) towards (iv) by increasing the salt concentration. At low salt concentration (i) the double layer repulsion dominates, the maximum of W_{DLVO} exceeds several kT, and a stable colloidal dispersion is expected. In situation (ii) the salt concentration is larger but there is still a local maximum that may be significant, preventing the particles from irreversibly sticking into the primary minimum. A shallow secondary minimum now manifests itself at large interparticle distances. If this local minimum is sufficiently deep (i.e. for large particles), weak flocculation can take place. Such weakly flocculated aggregates can be redispersed by shaking or by lowering the salt content. Adding still more salt (iii, iv) leads to irreversible aggregation: the Van der Waals attraction gets dominant and the colloidal dispersion will be unstable. DLVO

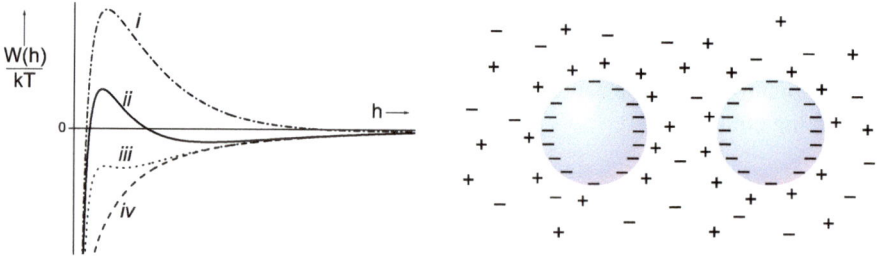

Fig. 1.4 Illustrative DLVO pair interactions (*left*) between two charged colloidal spheres (*right*) in an electrolyte solution as a function of increasing salt content from (i) $\rightarrow$ (iv)

theory is capable of accurately describing early stage aggregation of dilute charged colloidal spheres for $\lambda_D \gtrsim 3\,\text{nm}$ [39].

Using the surface force apparatus, Israelachvili and Adams [40] measured a repulsive force between surfaces in aqueous solution at short separations that could not be interpreted in terms of DLVO theory. This interaction is due to hydration forces caused by the ordering of water molecules. Its range is very short, typically below 2 nm. For a discussion on the limitations of DLVO theory and possible improvements see Ref. [41].

In the above descriptions we concentrated on situations where a polar background solvent was implicitly assumed. In apolar solvents double layer repulsion is difficult to achieve because dissociation (which leads to charged surface groups) is less likely to occur, and it then becomes essential to stabilise colloids with polymers. In the first decades after the establishment of the DLVO theory, most papers on forces between colloidal particles focused on Van der Waals and double layer interactions. Forces of other origin, such as polymeric steric stabilisation [24], depletion [42], or effects of a critical solvent mixture [43], gained interest at a later stage.

1.2.4 Influence of Attached Polymers

Colloidal dispersions can be very well stabilised by attaching polymers to the particle surfaces [24]. Here, we consider polymer chains that are in a 'good solvent'. This means that the chains are swollen and repel each other. As two colloidal particles protected with attached polymers approach each other, the local osmotic pressure increases dramatically due to mutual steric hindrance of the polymer chains on the particles. This competition between the chains for the same volume leads to a repulsive interaction, as was realised by Fischer [44].

Polymers can be attached to surfaces as, for instance, mushrooms, brushes or adsorbed chains (Fig. 1.5). In the case of mushrooms and brushes, the (nonadsorbing) chains are chemically bound to the surface by one chain end. When polymers adsorb at a surface many segments stick and densely pack at the interface. Attached polymers can contribute to a (significant) repulsive interaction between the particles. Upon

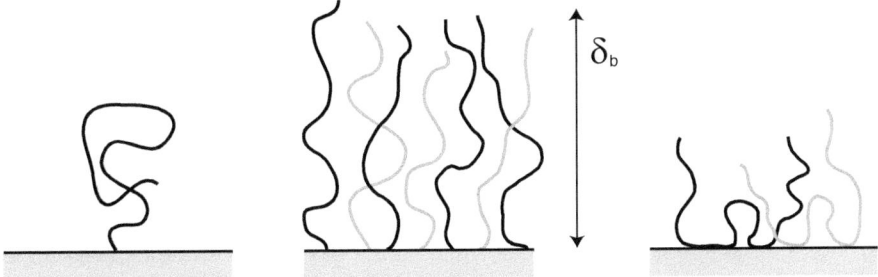

Fig. 1.5 Polymers attached at a surface: a mushroom (*left*), a brush (*middle*) and a layer of adsorbed polymer (*right*)

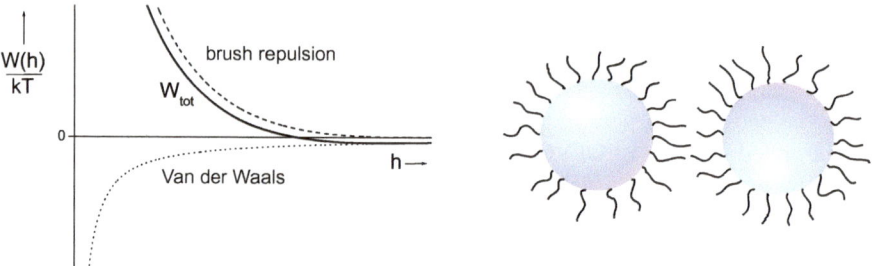

Fig. 1.6 The influence of a brush repulsion on the interaction potential $W(h)$ between two spheres with Van der Waals attraction

overlap of the attached polymers the osmotic pressure between the surfaces strongly increases which leads to a repulsive interaction between the particles.

For polymer brushes—chains that are anchored to the surface by an end segment with a high anchor density—the chains are highly stretched. The Helmholtz (free) energy of interaction between brushes consists of two terms: an osmotic repulsion contribution and a stretching factor. The Alexander–De Gennes theory [45–47] considers the repulsive interaction of overlapping brushes of thickness δ_b (Fig. 1.5) in a good solvent. This thickness scales as $M\sigma_b^{1/3}$, where M is the chain length and σ_b the anchor density. For $h \leq 2\delta_b$ the pressure P between two parallel plates with anchored brushes at separation h reads:

$$\frac{P(h)}{kT\sigma_b^{3/2}} \approx \left(\frac{2\delta_b}{h}\right)^{9/4} - \left(\frac{h}{2\delta_b}\right)^{3/4}. \tag{1.14}$$

The first positive term on the right-hand side represents the osmotic repulsion between the brushes, and the second negative term originates from the elastic energy gain upon retraction of chains (less stretching). The repulsion dominates the interaction for $h < \delta_b$. As will become clear in Sect. 2.1, the pressure yields the interaction potential between two plates, from which the interaction between two spheres can also be derived.

Figure 1.6 is a qualitative representation of the effect of adding a polymer brush to the interaction between two (uncharged) colloidal spheres subject to Van der Waals attraction. Commonly, one assumes the total interaction is the sum of all pair interactions:

$$W_{\text{tot}} = \sum_i W_i. \tag{1.15}$$

So, in Fig. 1.6 the total interaction potential is $W_{\text{tot}} = W_{\text{VdW}} + W_{\text{brush}}$. Without the anchored polymer chains the particles would coagulate spontaneously since the Van der Waals attraction is very strong at small values of h. However, upon adding the polymer brush repulsion, the total interaction (solid curve) is repulsive for a wide h-range, with no significant attraction left.

The Van der Waals attraction can be reduced by choosing a solvent (mixture) that allows for refractive index matching of colloid and solvent. For model studies

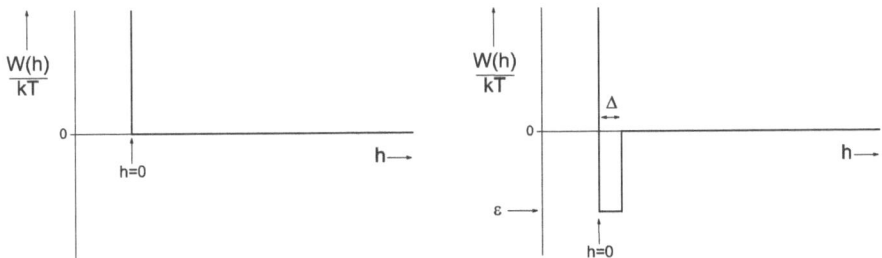

Fig. 1.7 Hard sphere (*left*) and square well or adhesive hard sphere (*right*) interaction

where one desires hard sphere-like particles, refractive index matching is combined with attaching short hairs (a thin brush; $\delta_b \ll R$) to the colloidal particles. This leads to absence of effective attractions and only a short-ranged repulsion between the particles, i.e. we now have a system of hard spheres (imagine submicrometer sized billiard balls). The pair interaction may then be approximated as

$$W(h) = \begin{cases} \infty & h \le 0, \\ 0 & h > 0, \end{cases} \tag{1.16}$$

the hard sphere interaction plotted in Fig. 1.7 (left panel). In the next subsection we consider the effect of adding nonadsorbing polymers to such a hard sphere dispersion.

When the medium is a poor solvent for the attached polymers a rather different situation is encountered. The polymer chains then tend to assume collapsed configurations in order to minimise contact with solvent molecules and the polymer segments prefer to interact with each other. This results in (short ranged) attraction between colloidal particles covered with polymer chains in a poor solvent (see Sect. 5.5 in Ref. [48]). The interaction of such sticky spheres (billiard balls with a thin layer of honey [49]) is often described in a simple manner using the (square well or) adhesive hard sphere interaction (see right panel in Fig. 1.7),

$$W(h) = \begin{cases} \infty & h \le 0, \\ -\varepsilon & 0 < h \le \Delta, \\ 0 & h > \Delta, \end{cases} \tag{1.17}$$

where Δ is the range of the attraction set by the thickness of the polymer layers and ε is the strength of attraction upon overlap of the polymer layers. For $\Delta \ll R$ the sticky sphere model of Baxter [50] can be employed providing simple expressions for the second osmotic virial coefficient and the equation of state. When the attractions are sufficiently strong, phase separation or aggregation occurs [51–54].

1.2.5 Depletion Interaction

Consider a room in a restaurant on two different occasions, as sketched in Fig. 1.8. On regular evenings the staff arranges the tables in a typical dinner set-up. Sometimes

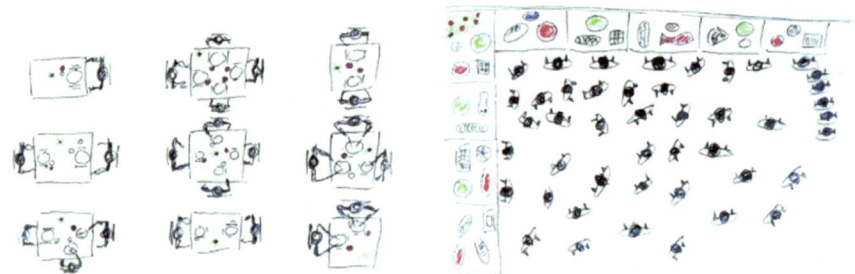

Fig. 1.8 *Left*: dinner set-up in the restaurant on a quiet evening. *Right*: buffet set-up in the same restaurant after 'phase separation'. Drawings by D. Frenkel (personal communication)

the room is booked for a cocktail party with many people present. In such a busy cocktail party the tables are laden with drinks and snacks and the configuration of the tables is rather different. Obviously, when the number of visitors exceeds a certain value, it is more efficient to push the tables close to each other and towards the wall in order to gain more translational freedom for the visitors.

The 'phase separation' in Fig. 1.8 is driven by entropy only. The apparent attraction between the tables originates from purely repulsive people–people, people–table and table–table interactions: the visitors do not wish to be too close to each other (and can still fetch a drink from a table). It is, just like depletion, an example of what prof. Vrij [55,56] referred to as 'attraction through repulsion'. Below we explain the origin of the depletion effect, first by considering colloidal hard spheres in a solution of nonadsorbing polymer.

Suppose colloidal spheres are mixed with nonadsorbing polymers. The loss of configurational entropy of the polymer chains in the region near the surface results in negative adsorption. Hence the colloidal particles are surrounded by depletion layers: zones in which the polymer concentration is lower than in the bulk. The mechanism that is responsible for the depletion attraction originates from the presence of these depletion layers.

Consider the depiction of a few colloidal spheres in a polymer solution shown in Fig. 1.9. Effective depletion layers are indicated by the (dashed) circles around the spheres. When the depletion layers overlap (lower two spheres) the volume available for the polymer chains increases. It follows that the free energy of the polymers is minimised by states in which the colloidal spheres are close together. The effect of this is just as if there were an attractive force between the spheres even though the direct colloid–colloid and colloid–polymer interactions are both repulsive [42]. For small depletant concentrations the attraction equals the product of the osmotic pressure and the overlap volume, indicated by the hatched region between the lower spheres in Fig. 1.9. The model illustrated above first became clear in the 1950s through the work of Asakura and Oosawa [57,58], and gained full attention only once Vincent et al. [59,60] and Vrij [42] started systematic experimental and theoretical work on colloid–polymer mixtures.

Consider two colloidal spheres each with radius R, each surrounded by a depletion layer with thickness δ. In that case the depletion potential can be calculated from the

Fig. 1.9 Colloidal spheres in a polymer solution with nonadsorbing polymers. The depletion layers are indicated by the short dashes. When there is no overlap of depletion layers (upper two spheres) the osmotic pressure on the spheres due to the polymers is isotropic. For overlapping depletion layers (lower two spheres) the osmotic pressure on the spheres is unbalanced; the excess pressure is indicated by the arrows

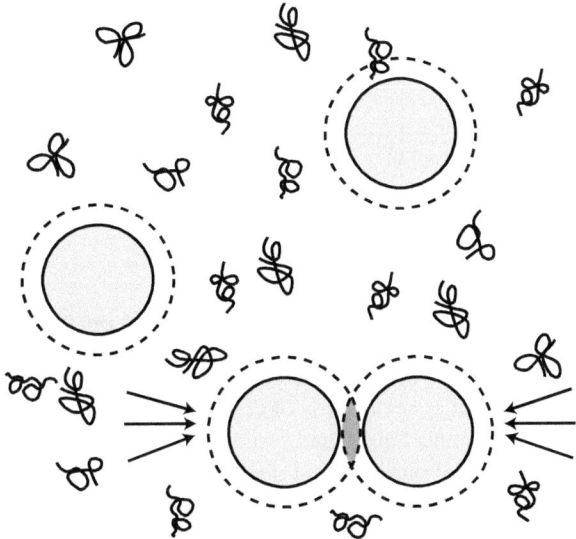

product of $P = n_b kT$, the (ideal) osmotic pressure of depletants with bulk number density n_b, multiplied by V_{ov}, the overlap volume of the depletion layers. Hence, the Asakura–Oosawa–Vrij (AOV) depletion potential equals [42,57,58]:

$$W_{dep}(h) = \begin{cases} \infty & h < 0, \\ -PV_{ov}(h) & 0 \leq h \leq 2\delta, \\ 0 & h \geq 2\delta, \end{cases} \quad (1.18)$$

with overlap volume $V_{ov}(h)$,

$$V_{ov}(h) = \frac{\pi}{6}(2\delta - h)^2(3R + 2\delta + h/2). \quad (1.19)$$

This simple expression is often used for the depletion interaction and will be derived in more detail in Chap. 2.

The AOV interaction potential $W_{dep}(h)$ between two hard spheres in a solution containing free polymers is plotted in Fig. 1.10. The minimum value of the potential W_{dep} is achieved when the particles touch ($h = 0$).

We note that in the original paper of Asakura and Oosawa [57], where Eq. (1.18) was first derived, the polymers were regarded as pure (infinitely dilute) hard spheres. Vrij [42,61] achieved the same result by describing the polymer chains as penetrable hard spheres (PHSs) (see Sect. 2.1). Inspection of Eqs. (1.18) and (1.19) reveals that the *range* of the depletion attraction is determined by the size 2δ of the depletant, whereas the *strength* of the attraction increases with the osmotic pressure and subsequently, with the depletant concentration. Depletion effects offer the possibility to independently modify the range and the strength of attraction between colloids. In dilute polymer solutions, the depletion thickness δ is close to the polymer's radius of gyration R_g (Sect. 2.2.1).

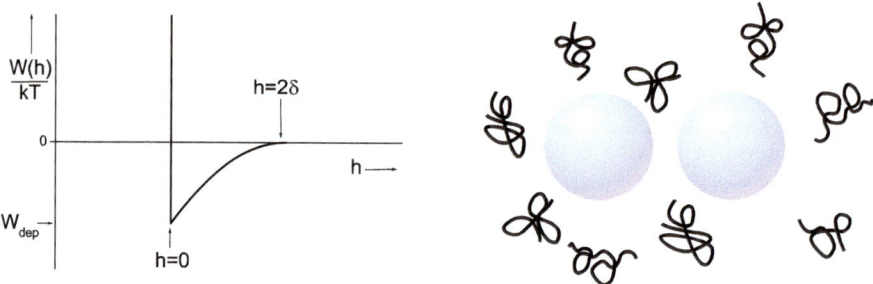

Fig. 1.10 The depletion interaction $W(h)$ between two hard spheres

In a mixture of hard spheres and depletants, a phase transition occurs upon exceeding a certain concentration of colloidal spheres and/or depletants. This is the subject of Chaps. 3 and 4 and 6 and 7 in this book. This is extended to mixtures of anisotropic hard colloidal particles and depletants in Chaps. 8 to 10.

A key parameter in describing the phase stability of colloid–polymer mixtures is the size ratio q,

$$q = \frac{R_g}{R}. \tag{1.20}$$

Throughout, colloid–polymer mixtures are described in terms of the volume fraction of colloids ϕ and the *relative* polymer concentration:

$$\phi_p = \frac{n_b}{n_b^*} = \frac{\varphi}{\varphi^*}, \tag{1.21}$$

which is unity at the (polymer coil) overlap concentration and can be regarded as the 'volume fraction' of polymer coils (and exceeds unity in the semidilute concentration regime). Here, n_b is the bulk polymer number density and n_b^* is its value at which the polymer coils overlap. In terms of the volume fraction of polymer segments φ ($0 \le \varphi \le 1$), one then uses $\phi_p = \varphi/\varphi^*$, with φ^* representing the segment volume fraction where the chains start to overlap:

$$\varphi^* = \frac{M v_s}{v_p}, \tag{1.22}$$

where M is the number of monomers per chain, v_s is the monomer (segment) volume, and $v_p = (4\pi/3) R_g^3$ the coil volume. The overlap number density n_b^* follows as $n_b^* = 3/(4\pi R_g^3)$.

It has actually become standard practice to normalise polymer concentrations in this way and use φ/φ^* (or n_b/n_b^*) as the parameter for 'polymer concentration'. In terms of the more practically accessible concentration c_p (in, e.g. kg/m^3 or g/L), the polymer coil overlap concentration is expressed as

$$c_p^* = \frac{3 M_p}{4\pi R_g^3 N_{Av}}, \tag{1.23}$$

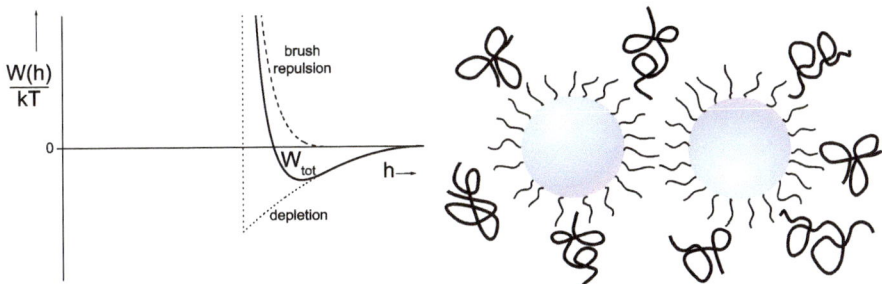

Fig. 1.11 The total interaction potential between two spheres covered with polymer brushes in a good solvent. The solution contains nonadsorbing polymer chains

where M_p is the polymer's molar mass and N_{Av} is Avogadro's number. Note that $c_p^* = c_p/\phi_p$.

Exercise 1.1. Show that, when using the approximation $\delta = R_g$, the attractive part of Eq. (1.18) for ideal depletants can be written in normalised quantities as

$$\frac{W_{dep}(h)}{kT} = -\frac{\phi_p}{q^3}\left(q - \frac{h}{2R}\right)^2\left(\frac{3}{2} + q + \frac{h}{4R}\right).$$

Hint: For ideal depletants the pressure P in Eq. (1.18) can be rewritten as $Pv_p/kT = \phi_p$ by using $\phi_p = n_b v_p$.

Figure 1.11 shows the influence of a combined depletion attraction and a brush repulsion on the total interaction. The presence of brushes reduces the attraction and the minimum value of the attraction is found at $h > 0$ [62].

The fact that depletion forces enable the range and strength of attraction to be varied independently is helpful for studying fundamental properties of liquids, as well as crystallisation and gelation phenomena, using colloidal systems instead of low molar mass substances. Another advantage of colloid–polymer mixtures is that colloids can be investigated using microscopy. Aarts et al. [63] could even detect capillary waves at the colloidal gas–liquid interface. Observations of wetting phenomena can also be studied at the particle level [64,65].

1.3   Historical Overview on Depletion

Depletion in colloidal dispersions is a central theme in this book. As we saw in Sect. 1.2.5 depletion effects in colloidal dispersions are caused by an unbalanced force. From a physics point of view, the depletion force between colloidal particles due to nonadsorbing polymer chains or small particles has common features with any other unbalanced force, whether of a colloidal nature or not. Before we focus

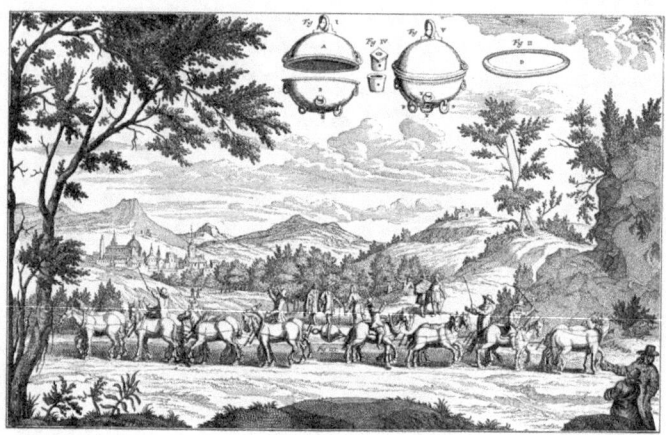

Die Magdeburger Halbfugeln auf dem Reichstag zu Regensburg, 1654.

Fig. 1.12 Caspar Schott's (1602–1666) illustration of the vacuum force demonstration by two teams of horses attempting to separate the hemispheres

on depletion effects on a mesoscopic level, we first give two classical examples of unbalanced forces.

1.3.1 Early Interest in Unbalanced Forces

1.3.1.1 The von Guericke Force
Halfway through the 17th century a series of remarkable experiments were performed, initiated by Otto von Guericke. One took place in 1657 at the court of King Friedrich Wilhelm III of Brandenburg in Berlin, Germany. Two hollow copper hemispheres, each with a diameter of 51 cm, were joined together and air pumped out to create a partial vacuum. A team of horses was then harnessed to each hemisphere (Fig. 1.12). The teams, each pulling with a force of about 1500 N each, could not pull the two joined hemispheres apart, demonstrating the tremendous force of air pressure. This proved the existence of the nothing we now call a vacuum.

Exercise 1.2. Show that the force on the *hemispheres* due to air pressure is one order of magnitude larger than the force that can be produced by 24 horses.

This experiment was the brainchild of scientist, inventor and politician Otto von Gericke (later spelled von Guericke), who lived between 1602 and 1686. The vacuum pump he used was invented by himself in 1650. His book on vacuum [66] reminds one of the difficulties in understanding vacuum at the time. Von Guericke abandoned established views and developed an independent vision on vacuum [67].

The result of this experiment showed that the surrounding air molecules push against the *Magdeburg* hemispheres (von Guericke was the mayor of Magdeburg). While it appears that there is an attractive force that pulls the hemispheres together, the vacuum in fact results from unbalanced repulsive forces.

Exercise 1.3. Explain why the osmotic pressure of the polymer solution in a colloid–polymer mixture plays a similar role to air pressure in von Guericke's experiment.

1.3.1.2 Le Sage's Gravitation Theory

In 1690, Nicolas de Fatio (and later in 1748 Georges-Luis Le Sage) proposed a mechanical theory for the explanation of both Newton's gravitational force and cohesive forces in materials [68]. This theory assumes the existence of 'ultramundane corpuscles'. Streams of such corpuscles are thought to impact on all materials from all directions. Now, if two bodies of materials are close to one another they can partially shield each other from the incoming 'ultramundane corpuscles'; the bodies will be struck by fewer corpuscles from the side of the other body. This mutual shielding was then supposed to push the bodies together due to the unbalanced force of the colliding corpuscles. This line of reasoning was first formulated by de Fatio in a letter to Huygens [69], but Newton also had contact with de Fatio on this matter. While Huygens, Newton and Leibniz were interested, they never accepted de Fatio's explanation as the driving force for gravity.

At a later stage Le Sage published a similar, more refined version of the theory [70]. He was in contact with some of the greatest physicists and mathematicians of his time, including Euler and Bernouilli, who found the theory rather speculative. Inconsistencies in the theory were later revealed by, for instance, Laplace, Lord Kelvin, Lorentz and, for didactic reasons, by Feynman [71] more recently. Occasionally, there is still interest in Le Sage's theory [72]. In an interesting paper, Rowlinson [68] drew attention to the fact that the work of Le Sage has a remarkable similarity to the depletion force.

1.3.2 Experimental Observations on Depletion Before the 1950s

Long before Asakura and Oosawa rationalised the attractive interaction caused by depletants, the effects of depletion were already noted in various areas of specialisation. In this overview, we first give examples of such studies and try to interpret them with our current knowledge of depletion forces. Subsequently, we discuss several studies that were performed after the work of Asakura and Oosawa, often in light of the theoretical progress that is being made over the last decades especially. Although it is nearly impossible to cover all developments within the area of deple-

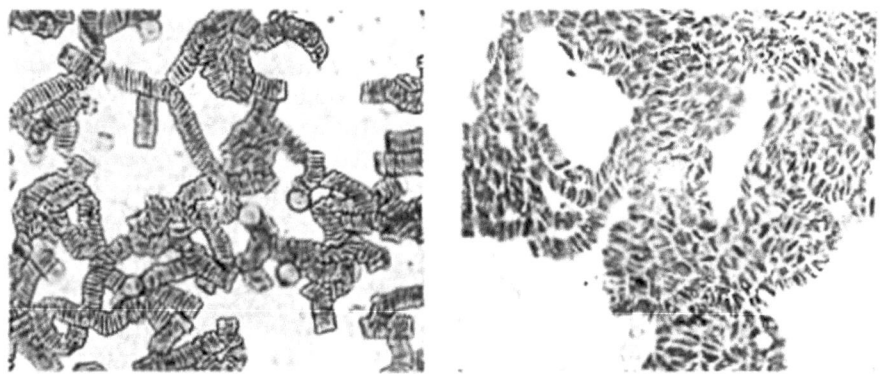

Fig. 1.13 Red blood cells in the blood of a pneumonia patient (*left*), in which rouleaux formation took place (strong aggregation), compared to weak aggregation in healthy blood (*right*) [73].
Reprinted with permission from Ref. [73]. Copyright 1929 APS

tion phenomena in physics and chemistry, we aim to give the reader a broad overview
here.

1.3.2.1 Clustering of Red Blood Cells

Red blood cells (RBCs) are biconcave particles and their detailed shape and size
depend on the RBC type. The human RBC may be considered a disc with a diameter
$D = 6.6\,\mu$m and a thickness of $L = 2\,\mu$m, its volume thus being of the order of 10^2
μm^3. The RBCs occupy about 40–50 vol% of our blood.

Exercise 1.4. Demonstrate that stacking all red blood cells in a human being
(having about 5 L of blood) in a single column provides an RBC cylinder with
a height that is of the order of the earth's circumference.

By the 18th century it was already known that RBCs tend to cluster, preferably
with their flat sides facing each other, like a stack of coins [73]. These structures are
commonly denoted as 'rouleaux'. In the blood of healthy human beings the tendency
of RBCs to aggregate is low. Aggregation is found to be enhanced in pregnancy or a
wide range of illnesses, giving rather pronounced rouleaux (Fig. 1.13). An impressive
review on RBC clustering was written by Fåhraeus [73]. Thysegen [74] provides
another historical review.

Enhanced RBC aggregation can be detected, for instance, by measuring the sedimentation rate. The sedimentation rate varies between 1–3 mm per hour for healthy
blood and up to 100 mm per hour in the case of severe illnesses. The blood sedimentation test, based on monitoring the aggregation of red blood cells, became a
standard method for detecting illnesses. The relationship between pathological con-

dition, RBC aggregation and enhanced sedimentation rate has been known for at least two centuries, as described in Refs. [73–76].

Fåhraeus [73,75] related enhanced aggregation of RBCs longer and stronger rouleaux to the concentration of the blood serum proteins fibrinogen, globulin and albumin. The tendency to promote aggregation depends on the type of protein. Rouleaux formation is most sensitive to increased serum concentrations of fibrinogen (molar mass 340 kg/mol) compared to β- and γ-globulins (90 and 156 kg/mol, respectively). The globulins in turn lead to RBC aggregation at lower protein concentrations than albumin proteins (69 kg/mol). Further, it has been shown that adding several types of macromolecules also promotes rouleaux formation [77]. Asakura and Oosawa [58] suggested that RBC aggregation might be caused by depletion forces between the RBCs induced by serum proteins. This is in line with the finding that the sedimentation rate is more sensitive to larger serum proteins.

Some authors interpret rouleaux formation as being caused by bridging of RBCs by serum proteins. There is, however, no evidence for protein adsorption onto RBCs. A study on rouleaux formation in mixtures of human RBCs ($D = 6.6$ μm) and rabbit RBCs ($D = 7.8$ μm) resulted in rouleaux structures that consisted (mainly) of only a *single* type of RBC [78]. This can be explained by a depletion effect (the overlap volume, hence entropy, is maximised if similar RBCs stack onto each other). However, the formation of mixed aggregates is expected if bridging were to occur; and since this is not observed, there is little support for the bridging hypothesis [79]. The general picture is that red blood cells tend to cluster at elevated concentrations of the blood serum proteins, which act as depletants [80,81].

1.3.2.2 Demixing of Biopolymers in Solution

Another manifestation of segregative interactions leading to demixing was reported by the microbiologist Beijerinck [82] who tried to mix gelatin (denatured protein coil) with starch (polysaccharide) in aqueous solution in order to prepare new Petri dish growth media for bacteria. He reported that these biopolymers could not be mixed; emulsion droplets appeared instead. With current knowledge [83,84] this can be regarded as an early detection of depletion-induced demixing. Tolstoguzov, Grinberg and co-workers extensively studied many mixtures of polysaccharides and proteins and concluded that such mixtures tend to segregate [85–87], unless there are specific interactions such as opposite charges. They further found that adding salt decreases the miscibility region in protein/polysaccharide mixtures [85]. It is obvious that, as well as pure depletion forces, double layer interactions play a role in such mixtures. The separate liquid phases in demixed protein–polysaccharide mixtures can sometimes be characterised by a sharp liquid–liquid interface. The interfacial tension between the coexisting phases in protein–polysaccharide mixtures has been determined and is of $\mathcal{O}(\mu N/m)$ [88,89], in agreement with Eq. 1.24 below.

1.3.2.3 Creaming of Particles in Latex and of Emulsion Droplets

In the beginning of the 20th century, large scale production of latex for rubber and paint production commenced. The term 'latex' is nowadays identified with a

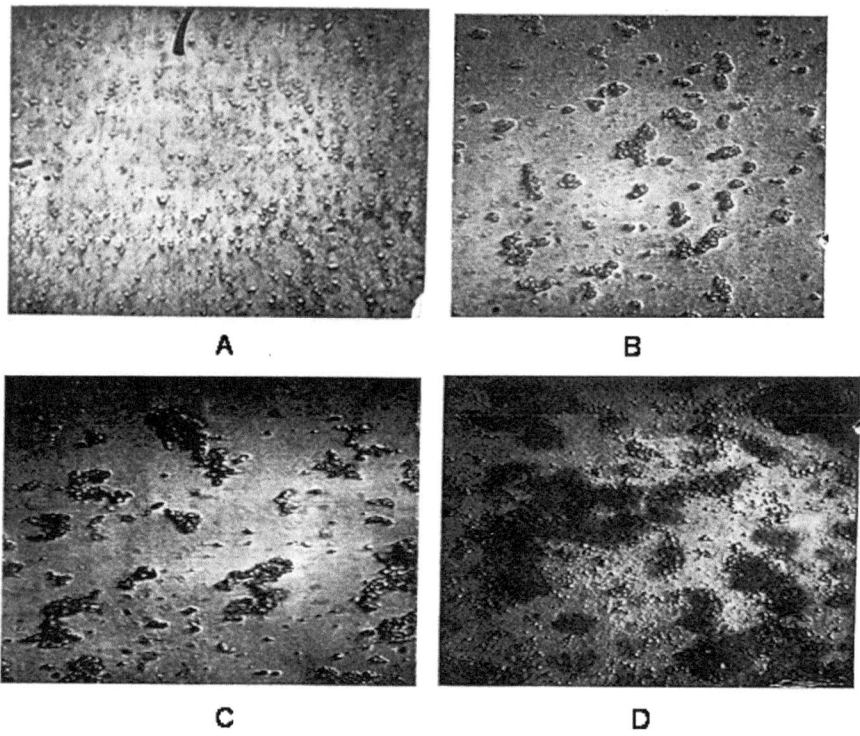

Fig. 1.14 Optical micrographs of a rubber latex dispersion [91]. **a** 1% suspension without polymer.
b 2 min after addition of 0.2% polysaccharide. **c** 10 min after addition of 0.2% polysaccharide. **d**
Image of the creaming layer. Size of the images are about 130 by 100 μm. Reprinted from Ref. [91]

stable dispersion of polymeric particles in an aqueous medium. In order to lower
transport costs there was a significant interest in concentrating the polymeric latex.
Centrifugation is highly energy consuming, and thus expensive.

Traube [90] showed that adding plant and seaweed polysaccharides led to a phase
separation between an extremely dilute and a very concentrated phase. Since the
particles are lighter than the solvent, the concentrated phase (with volume fraction
$0.5 \leq \phi \leq 0.8$) floats on top. The lower phase is clear and hardly contains particles.
Baker [91] and Vester [92] systematically investigated the mechanism that leads to
what they called (enhanced) creaming.

In Fig. 1.14 we show microscopy images of the latex dispersion investigated by
Baker [91]. The images are for a 1% latex dispersion, first without added polymer (A).
Images B and C were taken respectively 2 and 10 min after adding 0.3% of polymer
(the polysaccharide tragon seed gum). After polymer addition, Baker reports an
immediate deceleration of Brownian motion adjacent to particle aggregation. After
about 10 min particle aggregation discontinues and the aggregates start creaming. The
entire creaming process takes about 1 day. Image D was taken in the cream layer.
Upon diluting the cream layer to 1% of latex particles Brownian motion restarts,
suggesting that the flocs segregated into individual particles again. From the work

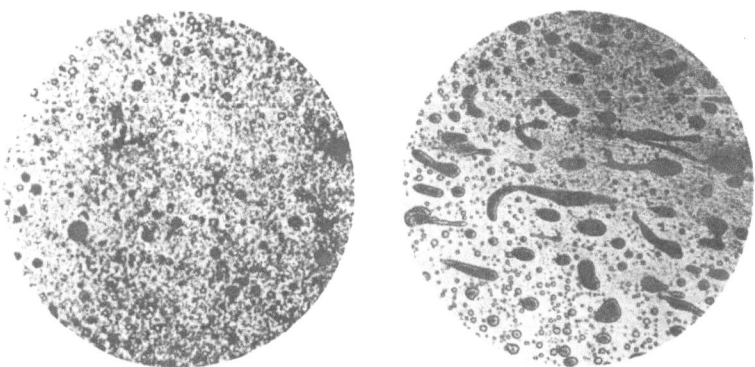

Fig. 1.15 Optical micrographs of demixed dispersions of emulsion droplets (black) in a polymer solution before (*left*) and after (*right*) pressing the microscopy slide. Micrograph diameters correspond to 610 µm. Reprinted with permission from Ref. [92]. Copyright 1938 Springer

of Baker [91] it can thus be concluded that the particles aggregate reversibly; upon dilution the latex particles can be resuspended. This suggests that bridging, which can also cause creaming [93], is not the driving force for enhanced creaming.

Vester [92] reviewed ways to optimise the creaming speed of lattices by using non-adsorbing polymer chains as depletants. He found that polymer addition can also lead to formation of short-lived emulsion droplets with diameters of $\mathcal{O}(10\text{--}100\,\mu m)$ that are enriched in latex, while the continuous phase is dilute in latex particles. Nowadays, this is interpreted as a colloidal gas–liquid phase coexistence. A microscopy image of the resulting emulsion is given in Fig. 1.15 (left panel). The droplets deform very easily upon confining the emulsion (right panel Fig. 1.15). This must imply that the interfacial tension γ is very low, say $\ll 1$ mN/m. Indeed, we do expect a small interfacial tension. The order of magnitude of a surface tension can be estimated from

$$\gamma \approx \frac{kT}{d^2}, \tag{1.24}$$

where d is the particle diameter [94]. For molecular systems this yields values for γ of at least a few mN/m up to hundreds mN/m, which is indeed approximately the range of measured surface tensions. For particles in the colloidal size domain with, for instance, $d \approx 30$ nm, an interfacial tension of only 1 µN/m is expected at room temperature. Indeed, this is the order of magnitude of the ultra-low interfacial tension measured in demixed colloid–polymer mixtures that are in colloidal gas–liquid equilibrium (see Chap. 5).

Cockbain [95] found that creaming of oil droplets in a surfactant-stabilised oil-in-water emulsion is enhanced when the surfactant concentration exceeds the critical micelle concentration. This phenomenon was left unexplained at the time, but thirty years later Fairhurst et al. [96] made a connection with depletion interaction theories and suggested that the micelles play a similar role to nonadsorbing polymers or small colloidal particles in acting as depletants (see Chap. 6).

1.3.2.4 Precipitation and Isolation of Viruses

Cohen [97] demonstrated that adding less than a percent of heparin to solutions of rod-like viruses results in the precipitation of the virus particles. The isolated precipitate phase consists of 'paracrystals'. The connection Cohen [97] makes with the work of Bernal and Farkuchen [98] suggests the phase appears liquid crystalline. This allows viruses to be isolated and concentrated [99,100]. A microscopy image of clusters of tobacco mosaic virus (TMV) particles in a dispersion with 0.5 wt% of heparin [97] is shown in Fig. 1.16. In Chap. 8 we consider depletion effects in colloidal rod dispersions.

1.3.3 1950–1969

The work of Fumio Oosawa (1922–2019) played a crucial role in the development of the insights into the depletion interaction. After finishing his education in physics in 1944 [102], Oosawa specialised in statistical mechanics because he wanted to do some 'unorthodox physics' [103]. Oosawa [101], who in the early 1950s was a young Associate Professor at Nagoya University in Japan, organised a winter symposium in Nagoya and invited a multidisciplinary group of Japanese scholars, mainly active in biology. He asked the participants to present work on phenomena in biological systems where statistical physics could be helpful to understand certain mechanisms. During the meeting the 'aggregation' of particles under the influence of macromolecules was a re-occurring theme [103]. It was observed, for instance, in suspensions of red blood cells [77], bacterial cells, soil powder and gum latex particles, as explained by Professor Oosawa during a symposium in 2014 [104]. During the winter symposium in 1953, professor Tachibana commented that these

Fig. 1.16 Light microscopy image of TMV paracrystals upon adding heparin [97]. Size of image: 80 by 60 μm. Reprinted from Ref. [97] under the terms of CC-BY-4.0

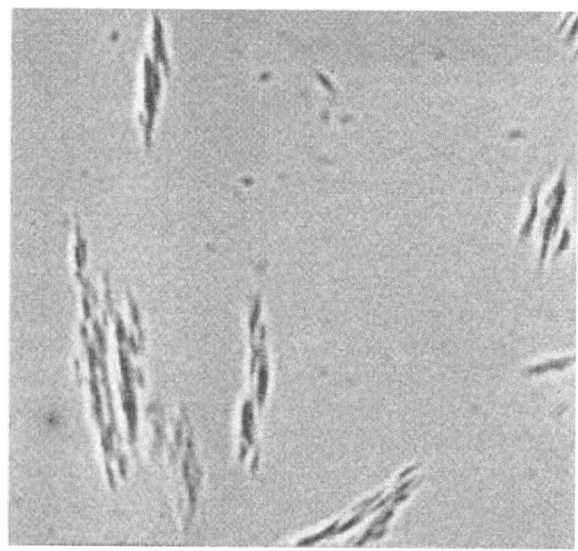

Fig. 1.17 Oosawa (*left*) and Asakura (*right*) at Nagoya University in the 1960s [101]. Photograph courtesy of the family of Professor Fumio Oosawa

similar phenomena might originate from the same physical principle [103]. This inspired Oosawa to start work with Sho Asakura, then a graduate student, on the influence of polymers on the interaction between particles.

Soon after, later Noble Prize (1974) winner P.J. Flory met Oosawa at a conference in Tokyo, organised by professor Yukawa. At this conference Oosawa invited professor Flory to come to Nagoya University [101]. During this visit Asakura and Oosawa (Fig. 1.17) reported unpublished theoretical results on two particles immersed in a solution containing nonadsorbing polymer chains, showing the chains impose an effective *attractive* interaction between the particles. The very positive response of Flory, at that time Associate Editor of *J. Chem. Phys.*, resulted in submission of this work and lead to a seminal paper, in which Asakura and Oosawa [57] presented a derivation of the interaction between two particles immersed in a solution of other nonadsorbing species. The theory of Asakura and Oosawa is the first theoretical prediction of a depletion force.

The effective attraction is a purely entropic effect; the indirect attraction originates from purely repulsive interactions. In this short paper Askura and Oosawa [57] considered four cases. The first three concern the interaction between parallel flat hard plates mediated by dilute (i) hard spheres, (ii) thin rods and (iii) ideal polymer chains. The segments of ideal polymer chains have no excluded volume; segments do not feel other segments. This will be explained in more detail in Sect. 2.2. Further, they also considered the interaction between hard spheres induced by small dilute hard spheres.

Not long after the publication of the work of Asakura and Oosawa, Sieglaff [105] demonstrated that a depletion-induced phase transition may occur upon adding polystyrene to a dispersion of microgel spheres in toluene. This demonstrated that the attractive depletion force is sufficiently strong to induce a phase separation. Sieglaff rationalised his findings in terms of the theory of Asakura and Oosawa. It took several years before subsequent work was done. This study of Sieglaff was later extended by Clarke and Vincent [106].

1.3.4 1970–1982

Early systematic experimental studies with respect to phase stability for colloid–polymer mixtures were performed by Vincent and co-workers [60, 107, 108]. At ICI and in Bristol, they concentrated on mixtures of colloidal spheres (latex particles) and nonadsorbing polymers such as polyethylene oxide (PEO). The work started when Vincent worked at ICI in Slough (1970–1972), where he investigated the origin of the flocculation of pigment particles in paint dispersions with F. Waite. In the papers of Vincent et al. [107, 109–111] a lot of attention is given to properly qualifying the demixing phenomena in colloid–polymer mixtures. These experiments were ahead of a full theoretical understanding of the phase behaviour of colloid–polymer mixtures. One of the systems studied was polystyrene spheres with terminally attached PEO brushes dispersed in mixtures of free PEO and water for a wide range of concentrations. The spherical particles were stable in both pure water and pure PEO melts. However, in mixed solutions of PEO and water (for instance, 50% water and 50% of PEO) a 'slow flocculation' of the particles was observed. The maximum flocculation rate was measured and was found to shift to lower PEO concentrations upon increasing the molar mass. This restabilisation at very high polymer concentrations (reported in a series of papers [107, 109, 110, 112]) was also found in a dispersion of grafted silica spheres mixed with polydimethyl siloxane (PDMS) polymer chains. Only polymer melts that are sufficiently liquid-like allow systems at such high polymer concentrations to be studied because other polymers would be too viscous for a proper analysis of the phase behaviour.

In the same period, Hachisu et al. [114] investigated aqueous dispersions of negatively charged polystyrene latex particles that undergo a colloidal fluid-to-solid phase transition upon lowering the salt concentration using dialysis or increasing the particle concentration. Under conditions where the latex dispersion (particles with $R = 170$ nm) is not ordered (fluid-like), Kose and Hachisu [113] added sodium polyacrylate to polystyrene latex particles (both components are negatively charged) and observed crystallisation of the colloidal spheres (Fig. 1.18). Since polymers and particles repel each other, the crystallisation process is probably induced by depletion interaction, although the authors themselves did not explicitly mention depletion. They do suggest that the ordering is due to 'some attractive force'. When the polymer concentration is increased, crystallisation occurs faster (Fig. 1.18, bottom panel).

Theoretical work on depletion interactions and their effects on macroscopic properties such as phase stability commenced along various routes. First, Vrij [42] studied the polymer-mediated depletion interaction between hard spheres. He described the nonadsorbing polymers as PHSs (see Sects. 1.2.5 and 2.1). Vrij [42] referred to the work of Vester [92], Li-In-On et al. [60] and preliminary experiments at the

Fig. 1.18 Optical micrographs showing mixtures of monodisperse polystyrene latex particles with 185 mg/L sodium polyacrylate polymers after (*top*) 25 min and (*middle*) 55 min; and (*bottom*) with 370 mg/L sodium polyacrylate after 25 min. Reprinted with permission from Ref. [113]. Copyright 1976 Elsevier

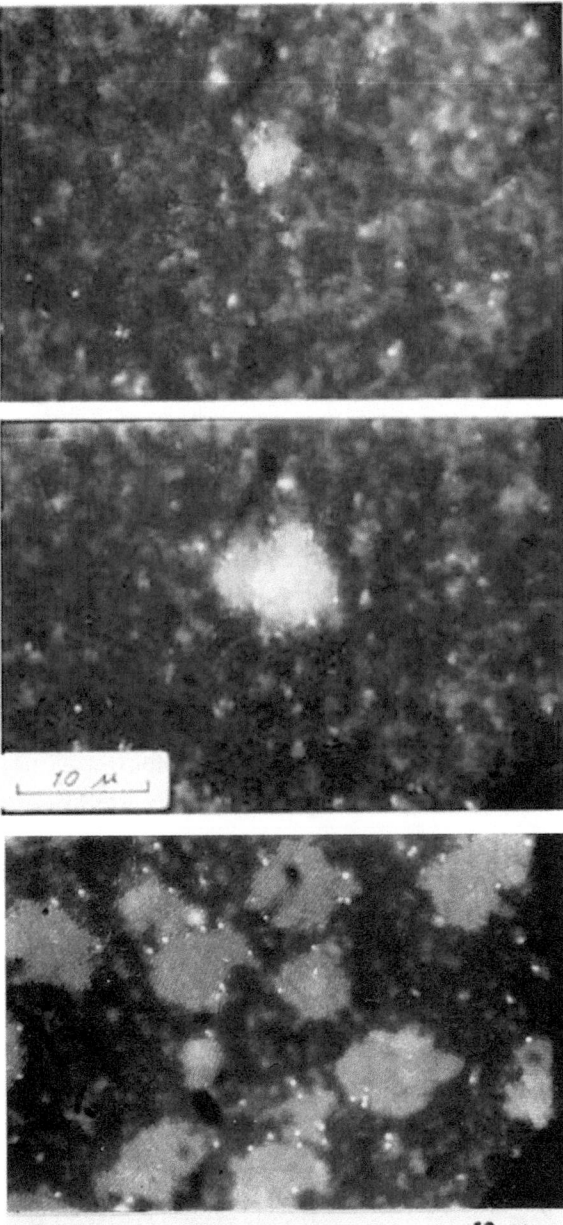

Van 't Hoff Laboratory on micro-emulsion droplets mixed with free polymer [42] for experimental evidence of depletion effects.

Progress on the quantification of the depletion layer thickness was triggered by 1991 Noble Prize winner De Gennes. In his seminal book [115], De Gennes derived an expression for the density profile of a semidilute polymer solution near a hard wall and demonstrated that the depletion thickness equals the correlation length, the length scale over which the segments are correlated. In dilute polymer solutions (below coil overlap) it is close to the radius of gyration of the polymer chains. However, in semidilute solutions (above overlap) the correlation length becomes independent of the chain length and is a (decreasing) function of the polymer concentration. Hence, also the depletion thickness decreases with polymer concentration in the semidilute regime. De Gennes considered the depletion contact potential between two colloidal hard spheres in a semidilute polymer solution in a good solvent. For this case, where the only relevant length scales are the sphere radius R and the correlation length ξ, he derived the following scaling relation for the minimum of the interaction potential [116]:

$$\frac{W_{\text{dep}}(h=0)}{kT} \cong \begin{cases} -\frac{R}{\xi} & R \gg \xi, \\ -\left(\frac{R}{\xi}\right)^{4/3} & R \ll \xi, \end{cases} \tag{1.25}$$

with an unknown prefactor $\mathcal{O}(1)$.

Exercise 1.5. What is expected with respect to colloidal stability of large $(R \gg \xi)$ and tiny $(R \ll \xi)$ colloidal spheres in a semidilute polymer solution?

Depletion effects have been studied using mean-field methods since the end of the 1970s. Insights into polymer physics have increased tremendously through the development of mean-field theories. The advantage of these theories is that they simultaneously include excluded volume interactions and give insights into details of polymer configurations near interfaces. A detailed analytical mean-field treatment for depletion interaction was made by Joanny et al. [117] who calculated the polymer segment concentration profile between two plates in the semidilute regime, in agreement with De Gennes' scaling prediction discussed above.

Using a Flory–Huggins-like mean-field model, Feigin and Napper [32] calculated the free energy of interaction between two flat plates mediated by nonadsorbing polymers and noted that a repulsive barrier is present for polymer concentrations in the concentrated regime. The potential at plate contact is, however, still attractive. The authors suggested that if the repulsive barrier is large enough this might lead to so-called depletion stabilisation; a colloidal dispersion is destabilised at low polymer concentrations but restabilised at high concentrations. A conceivable intuitive explanation is kinetic: at high polymer concentrations it is hard to push polymer chains out of the gap between two particles. The bulk osmotic pressure is very high in a concentrated polymer solution. The polymer chains between the particles thus need to be transported towards a very steep osmotic pressure gradient.

Scheutjens and Fleer [118, 119] developed a numerical self-consistent field (SCF) method that enables the calculation of equilibrium concentration profiles near interfaces. This SCF method was applied to depletion effects in [120], showing that the depletion layer thickness is close to R_g at low polymer concentrations but decreases with increasing polymer concentration in the semidilute regime. In the concentrated regime, very close to the melt concentration, the polymer concentration between two parallel plates oscillates around the bulk polymer concentration. This finding is supported by Monte Carlo computer simulations of Broukhno et al. [121] The interaction potential between the plates was also calculated by Scheutjens and Fleer [120] using SCF. For dilute polymer solutions the range of the potential is close to $2R_g$ and the depth of the potential increases with increasing solvent quality. When the volume fraction of polymer segments in the system is 0.1 (a very high polymer concentration, in practice) a weak repulsive part appears in the interaction potential, as was also found by Feigin and Napper [32]. This repulsion appears at lower concentrations for better solvent quality [24, 120].

A direct link between theoretical and experimental work on depletion-induced phase separation of a colloidal dispersion due to nonadsorbing polymers was made by De Hek and Vrij [61, 122]. They mixed sterically stabilised silica dispersions with polystyrene in cyclohexane and measured the limiting polymer concentration (phase separation threshold). Commonly, one uses the binodal or spinodal as the experimental phase boundary. A binodal denotes the condition (compositions, temperature) in which two or more distinct phases coexist (see Chap. 3). A tie-line connects two binodal points. A spinodal corresponds to the boundary of absolute instability of a system to decomposition. At or beyond the spinodal boundary infinitesimally small fluctuations in composition will lead to phase separation.

De Hek and Vrij [61] used the pair potential of Eq. 1.18 to estimate the stability of colloidal spheres in a polymer solution by calculating the second osmotic virial coefficient B_2:

$$B_2 = 2\pi \int_0^\infty r^2 (1 - \exp[-W(r)/kT]) dr, \tag{1.26}$$

where we used the centre-to-centre distance r between the spheres, which equals $2R + h$. A simple argument was used to estimate the spinodal [61]. For colloid or polymer concentrations [123] exceeding the spinodal, phase separation occurs spontaneously. Therefore, at the spinodal

$$\frac{dP}{d\phi} = 0. \tag{1.27}$$

The virial expansion for the osmotic pressure P of a colloidal dispersion reads

$$\frac{Pv_0}{kT} = \phi + B_2^* \phi^2 + \cdots . \tag{1.28}$$

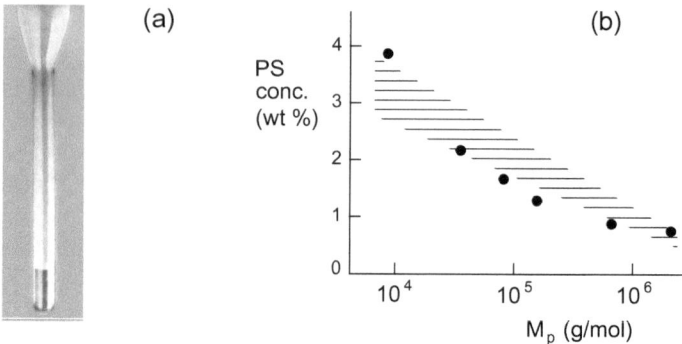

Fig. 1.19 **a** Photograph of a test tube containing a phase-separated mixture of polystyrene polymer chains (molar mass: 32.4 kg/mol) and sterically stabilised silica spheres ($R = 21$ nm) in cyclohexane. Initial concentrations: 1 wt % of silica spheres and 2.5 g/L polystyrene. The concentration below which no phase separation was found was 17 g/L. The two demixed phases are separated by a sharp interface. Reprinted with permission from Ref. [122]. Copyright 1979 Elsevier. **b** State diagram of 1 wt % silica spheres ($R = 46$ nm) in cyclohexane mixed with polystyrene polymer chains varying in molar mass M_p [61]. The limiting polystyrene concentrations below which no phase separation occurred are indicated as the filled circles. Hatched region: theoretical limits between which the spinodal curve is situated. Reprinted with permission from Ref. [61] Copyright 1981 Elsevier

with $B_2^* = B_2/v_0$. Here, v_0 is the volume of the colloidal sphere. In the limit of low ϕ, Eqs. (1.27) and (1.28) provide

$$1 + 2B_2^* \phi^{\mathrm{sp}} = 0. \tag{1.29}$$

This relates the polymer activity (which determines B_2) to the colloid volume fraction ϕ^{sp} at the spinodal. De Hek and Vrij [61] were able to give a good description of the phase line of mixtures of polystyrene chains and small volume fractions of (hard sphere like) octadecyl silica spheres dispersed in cyclohexane [122].

Figure 1.19 depicts results obtained by de Hek and Vrij [61, 122] on a mixture of octadecyl silica spheres and polystyrene polymers in cyclohexane. Both separated phases are fluid. The limiting polymer concentration below which no phase separation occurs in a solution containing a given amount of silica is plotted versus the molar mass of the added polymer polystyrene. It was found that less polymer is required to induce a phase separation when the molar mass is larger. This experimental trend can be predicted by using the spinodal condition Eq. (1.29), but this is only a semi-quantitative test because, in fact, formally, one should compare the stability curve with the binodal, as this denotes the compositions of the coexisting phases in a demixed dispersion. The spinodal is, however, not too far off from the binodal and probably gives a good and simple estimate. For the smallest molar mass, the separated phase was gel-like instead of a fluid. A statistical mechanics calculation of ideal polymer chains between two walls by De Hek and Vrij [61] demonstrated that the range of attraction between two flat parallel plates due to ideal polymer chains is close to $2.25 R_g$, implying a depletion thickness at each plate of about R_g.

A static light scattering (SLS) contrast variation study on elucidating the negative adsorption of polystyrene chains next to a silica sphere in cyclohexane solutions was

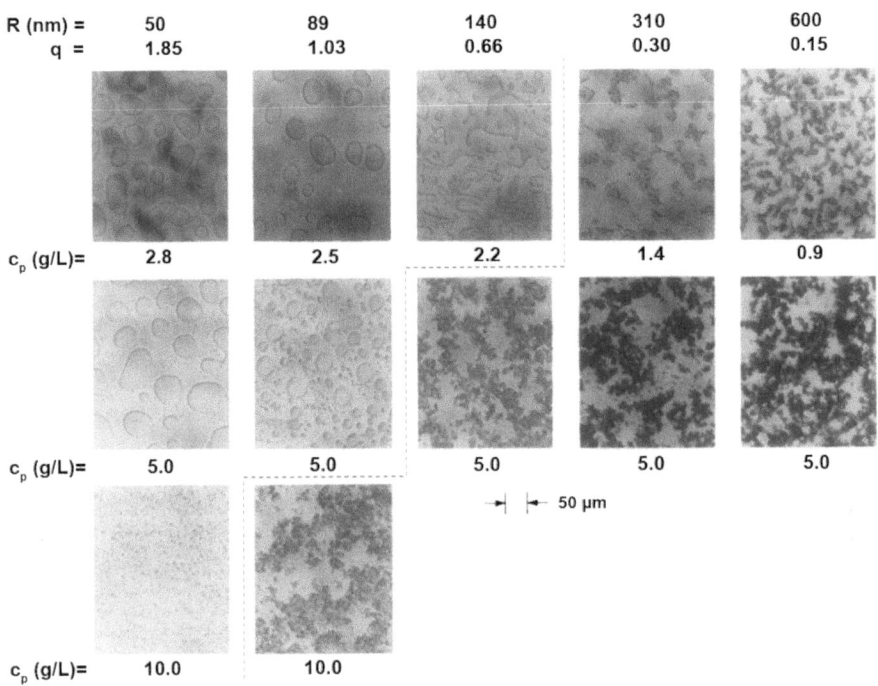

Fig. 1.20 Optical micrographs of latex colloids mixed with HEC by Sperry [125–127]. Reprinted with permission from Ref. [127]. Copyright 1984 Elsevier

described in another paper by de Hek and Vrij [124]. This negative adsorption can be converted to a depletion thickness, which is approximately the radius of gyration of polystyrene in cyclohexane. The second virial coefficient of the silica particles could be determined from the SLS experiments, and its value was shown to become negative when a sufficient amount of nonadsorbing polystyrene is added, which implies attraction between the spheres.

By mixing aqueous hydroxyethylcellulose (HEC) with latex, Sperry and co-workers [125–127] observed phase separation and made a study on the effect of the structure of the colloid-rich phase as a function of the colloid–polymer size ratio $q = R_g/R$. The micrographs in Fig. 1.20 of phase separating mixtures demonstrate how the morphology of the segregating systems varies upon changing q and polymer concentration. Unstable systems at large q and not too high polymer concentrations are characterised by smooth interfaces, implying colloidal gas–liquid coexistence. For small q, demixed systems are characterised by irregular interfaces that indicate (colloidal) fluid–solid coexistence. This suggests that the width of the region where a colloidal liquid is found in colloid–polymer mixtures is limited. We return to this issue in Sect. 4.3. Irregular interfaces are also detected for $q > 1/3$ when the polymer concentration is substantially increased.

1.3.5   1983–1999

The work of Sperry inspired Gast, Hall and Russel to develop a theory which might explain the experimental phenomena. Gast et al. [128] used thermodynamic perturbation theory (TPT) [129] to derive the free energy of a mixture of colloidal particles and polymers (described as penetrable hard spheres, PHSs), based on pair-wise additivity of the interactions between the colloids. This is an approach which is based upon a perturbation of the free energy of a pure colloidal dispersion due to depletion forces, with Eq. (1.18) as input. Using equations of state for the hard sphere fluid and the FCC crystal structure as references, they calculated the phase behaviour from the (perturbed) free energy. This made it possible to assign the nature (i.e. colloidal gas, liquid or solid) of the coexisting phases as a function of the size ratio q, the concentration of the polymers, and the volume fraction of colloids. For small values of q, say, $q = R_g/R < 0.3$, increasing the polymer concentration broadens the hard sphere fluid–solid coexistence region; a (stable) colloidal fluid–solid coexistence is expected if the polymer chains are significantly smaller than the colloidal spheres (low q). Inside the unstable regions a (metastable) colloidal gas–liquid branch is located. For intermediate values of q, the gas–liquid coexistence curve crosses the fluid–solid curve; and for large q-values, mainly gas–liquid coexistence is found for $\phi < 0.49$, where ϕ is the volume fraction of colloids [123]. The results are in agreement with the findings of Sperry [125–127].

Exercise 1.6. Use the Gibbs phase rule and derive how many coexisting phases a system can assume when it consists of two components. For a discussion see Ref. [130].

Experimentally, Gast, Russel and Hall [131] later verified the predicted types of phase coexistence regions for a model colloid–polymer system. Colloid–polymer phase diagrams [123] are commonly plotted in terms of the volume fraction of colloids ϕ and the *relative* polymer concentration ϕ_p, defined in Eq. 1.21. In both the descriptions by De Hek and Vrij and by Gast, Russel and Hall, the depletion thickness δ was assumed to be equal to the radius of gyration of the polymers. This assumption can be rationalised by calculating the density profile of polymer chains at a surface. This was done by Lépine and Caillé [132], who solved the Edwards equation for ideal polymer chains near a reflective, attractive and repulsive surface. Eisenriegler [133] also calculated the density profile of nonadsorbing ideal chains near a flat surface and from this density profile it follows that $\delta/R_g = 2/\sqrt{\pi} \approx 1.13$ [134] (see Sect. 2.2). This agrees with an earlier result derived by Casassa and Tagami [135] using the end segment distribution at a nonadsorbing flat surface. Later, it was shown [136] that the depletion layer thickness is independent of the reference point (for instance, centre of mass, middle segment and end segments) used to describe the depletion density profile.

Experimental work on the determination of the depletion layer thickness commenced in this period, although these are indirect measurements. The depletion thickness δ of polystyrene in ethyl acetate at a nonadsorbing glass plate was measured using an evanescent wave technique by Allain et al. [137]. The value found for δ was indeed close to the radius of gyration of the polymer. Ausserré et al. [138] measured the depletion thickness of xanthan (a polysaccharide) in water at a quartz wall below and above the polymer overlap concentration. In dilute solutions below overlap, δ was close to the radius of gyration of xanthan, whereas in the semidilute regime (i.e. above overlap ($\phi_p > 1$)) it decreases as $\delta \sim \phi_p^{-0.8}$. This is in accordance with what is expected theoretically (see Sects. 4.3.1 and 4.3.2). Pashley and Ninham [139] succeeded in measuring the depletion potential between mica plates (as induced by CTAB micelles) using the surface force apparatus.

The polymer density profile of nonadsorbing ideal chains next to a hard sphere for arbitrary size ratio q was first calculated by Taniguchi et al. [140] and later independently by Eisenriegler et al. [141] Eisenriegler also considered the pair interaction between two colloidal hard spheres for $R_g \ll R$ [142] and for $R_g \gg R$ [143], as well as the interaction between a sphere and a flat wall due to ideal chains [144]. Depletion of excluded volume polymer chains at a wall and near a sphere was considered by Hanke et al. [145] One of their results is that the ratio δ/R_g at a flat plate, which is 1.13 for ideal chains [133,134], is slightly smaller for excluded volume chains (1.07). The precise value for the depletion thickness is important. From Eq. (1.19) it follows that V_{ov} scales with δ^2 for large colloidal spheres ($R \gg \delta$) and increases even more strongly for larger depletion zones.

Inspired by the work of De Gennes [115,116], fundamental work commenced on colloid–polymer mixtures in which the polymers are relatively large compared to the colloids. This regime is relevant for mixtures of polymer or polysaccharides mixed with proteins and is often denoted as the *protein limit*. The opposite case (small q) is known as the *colloid limit*. We distinguish three regimes in colloid–polymer mixtures (Fig. 1.21): small q (also termed the 'colloid limit') of $q \lesssim 0.5$, 'equal sized' ($0.5 \lesssim q \lesssim 2$) and the large q regime (also termed the 'protein limit') of $q \gtrsim 2$.

Odijk [146–149] published a series of papers devoted to the protein limit $\xi \gg R$; he considered semidilute polymer solutions where the correlation length ξ scales as $\phi_p^{-3/4}$. He first calculated the density profile of a small colloid in a semidilute polymer solution with $\xi \gg R$ and found a very simple shape of the density profile that is independent of ξ and only depends on R [146]. By considering the second virial coefficient between a large polymer and a tiny colloid, he concluded that phase separation is not expected in this case. This was confirmed later by Eisenriegler [150], who from renormalisation group theory found that the second osmotic virial coefficient of small colloidal spheres, B_2, only marginally decreases with increasing polymer concentration up to the coil overlap concentration above which it increases. Odijk [147] also considered many-body effects by involving void–void correlations and statistical geometrical approaches [151]. He concluded that the depletion-induced interaction between small colloids due to large semidilute polymers levels off to a maximum attraction near a volume fraction $\phi \sim 0.3$. To mimic proteins, Odijk [149]

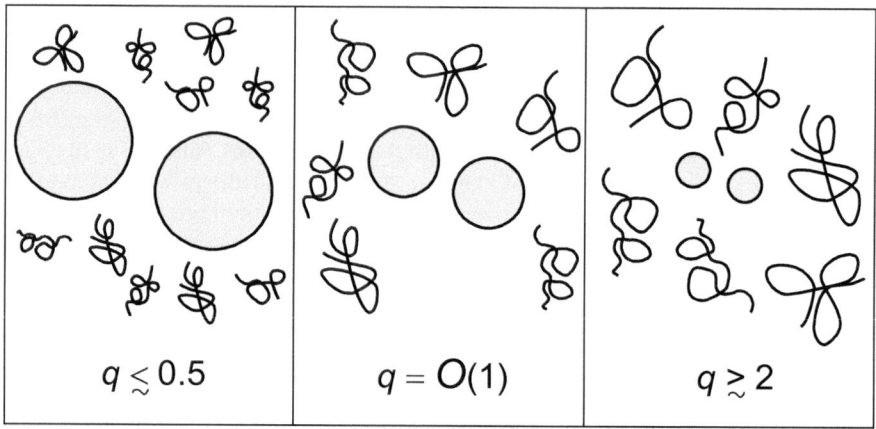

Fig. 1.21 The size ratios in colloid–polymer mixtures in different regimes. *Left*: the 'colloid limit' of relatively small polymer chains. *Middle*: the equal size regime. *Right*: the 'protein limit' regime of relatively large polymer chains

and Eisenriegler [152–155] extended the approach of polymer depletion and small colloidal spheres to colloidal particles with ellipsoidal shape.

A semi-grand canonical treatment for the phase behaviour of colloidal spheres with nonadsorbing polymers was proposed by Lekkerkerker [156], who developed 'free volume theory' (also called 'osmotic equilibrium theory', see Chap. 3). The main difference with TPT [128] is that free volume theory (FVT) accounts for polymer partitioning between the phases and for multiple overlap of depletion layers, hence avoiding the assumption of pair-wise additivity, which becomes inaccurate for relatively thick depletion layers. These effects are incorporated through scaled particle theory (see, for instance, [151] and references therein). The resulting free volume theory (FVT) phase diagrams calculated by Lekkerkerker et al. [157] revealed that for $q < 0.3$ coexisting fluid–solid phases are predicted, whereas a gas–liquid coexistence is found for $q > 0.3$ at low colloid volume fraction, as was predicted by TPT.

A coexisting *three*-phase colloidal gas–liquid–solid region (not present in TPT phase diagrams) was predicted by FVT for $q > 0.3$ and gained much attention. Experimental work [158, 159] demonstrated that this three-phase region indeed exists. Both Leal-Calderon [158] and Ilett et al. [159] measured phase diagrams of colloid–polymer mixtures as a function of the size ratio q. The topology of the phase diagrams corresponds well to FVT predictions, as long as q is below 0.6 (see Chap. 4).

As another example of a three-phase system, photographs of dispersions containing 16 vol% polystyrene latex spheres (with a diameter of 67 nm), published by Faers and Luckham [160], are reproduced in Fig. 1.22. The numbers shown represent the concentration (in wt%) of the polysaccharide hydroxyethylcellulose (HEC). In the dispersion with 0.3 wt% of HEC three phases coexist. From top to bottom colloidal gas, liquid and solid phases can be recognised. The rigidity of the solid–liquid interface is demonstrated in the lower photographs where the tubes are tilted. The

Fig. 1.22 Photograph of a polystyrene latex dispersion (16 vol %) in 10 mM NaCl at pH 7 with (as indicated in wt %) added hydroxyethylcellulose (HEC) studied by Faers and Luckham [160]. In the lower photograph the tubes are tilted, demonstrating the difference between rigid colloidal solid–liquid and fluid colloidal gas–liquid interfaces for the three-phase coexistence at 0.3 wt % HEC. Reprinted with permission from Ref. [160]. Copyright 1997 American Chemical Society

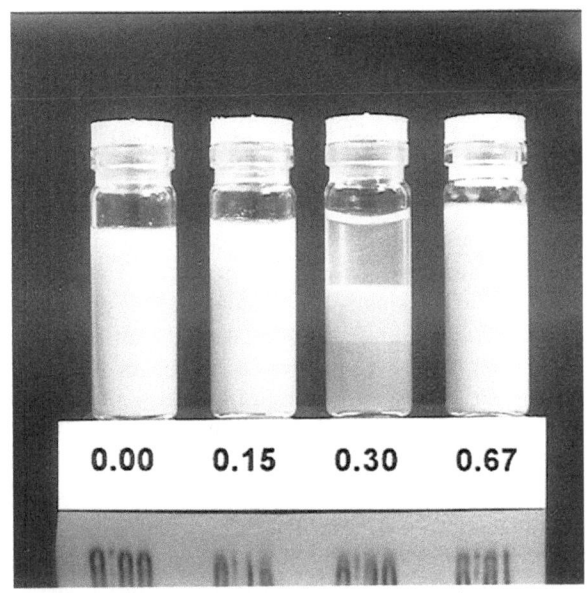

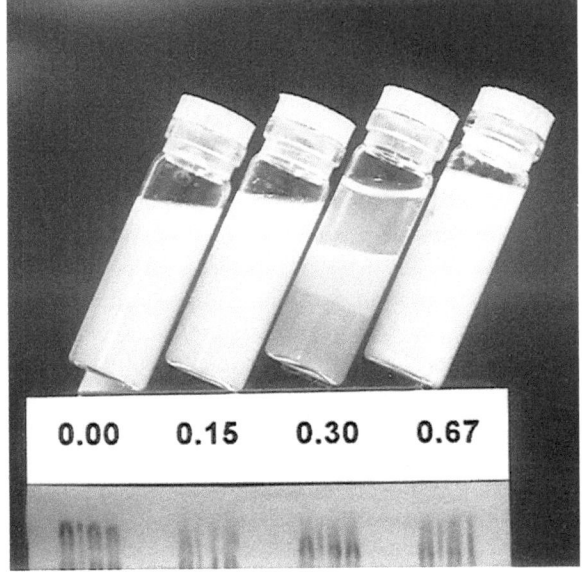

gas–liquid interface flows upon tilting the sample, for the gas–solid interface this is not the case. Using the theory of Lekkerkerker et al. [157] it is also possible to calculate the tie-lines along which the system demixes, enabling a comparison of the theory with experimental phase boundaries. The theory describes the experimental phase diagrams rather accurately [159] for small q. FVT for colloidal spheres mixed with PHSs was tested by Meijer and Frenkel [161]. Their Monte Carlo computer simulation results on a dispersion of spheres immersed in a solution of ideal polymer chains showed that the agreement with the osmotic equilibrium theory of Lekkerkerker et al. [157] is very good for small values of q.

Faers and Luckham [160] also studied the effect of the amount of polymer grafted onto the colloid surfaces. Decreasing the amount of grafted polymer increased the phase separation concentration of polymers at fixed colloid concentration, demonstrating that it is worthwhile to investigate the effect of the presence of brushes in combination with nonadsorbing polymers.

Polymers are often added to oil-in-water emulsions in order to impose a certain emulsion viscosity. However, this may lead to instability problems, as is known in food emulsions [22, 162]. Bibette et al. [163–165] were the first to quantitatively relate phase transitions in emulsions due to nonadsorbing polymers to depletion-induced forces. They showed that it is possible to size fractionate an emulsion with a depletion-induced phase transition. An interesting aspect of (micro) emulsion droplets is that they are not hard spheres, as assumed in FVT [157]. Several groups [42, 166–168] studied the phase behaviour of droplets in a micro-emulsion mixed with nonadsorbing polymers. The phase behaviour could be explained by describing the micro-emulsion itself as a collection of sticky hard spheres rather than pure hard spheres. The colloid–polymer mixture is then described as a mixture of sticky spheres mixed with nonadsorbing polymers [166, 167]. The phase behaviour for the colloid limit has been studied extensively by, for instance, Meller and Stavans [169] for emulsions. The FVT of Lekkerkerker et al. [157] was found to agree well with these experimental studies. The B_2-approach of Vrij [42, 61] could explain the phase line measured for an aqueous mixture of casein micelles and nonadsorbing exocellular polysaccharides [170]. However, the polymer is often larger (protein limit) or has a similar size to the spherical droplets in polymer/micro-emulsion mixtures. Then phase transitions occur near or above the polymer overlap concentrations. Obviously, the assumption $\delta = R_g$ is then no longer correct. For a proper description of the phase behaviour in this case, the effect of interactions between the polymers must be taken into account: more accurate descriptions of the depletion thickness and osmotic pressure as a function of the polymer concentration are needed.

Free volume theory is an approximate approach; therefore, theoreticians worked on a more formal way of accounting for the influence of depletants on the properties of colloidal mixtures. The exact procedure of integrating out the degrees of freedom [171] was applied to a binary hard sphere mixture by Dijkstra, van Roij and Evans [172]. They presented a method to derive an expression for the effective grand potential for the large hard spheres by formally integrating out the degrees of freedom of the small spheres.

For mixtures of hard spheres and PHSs as depletants, integrating out the depletant was laid out by Dijkstra, Brader and Evans [173], who formally derived the semi-grand canonical ensemble mixtures of hard spheres and PHSs and made a connection to FVT. They showed that for small $q < 0.154$ only one and two body terms are needed. Although for larger q three and higher body terms are needed, integrating out is formally still possible and can be done numerically [174].

Depletion potentials were measured indirectly using scattering techniques [124, 175, 176] and methods such as atomic force microscopy [177–179] and total internal reflection microscopy [180–182] (Sect. 2.6). Work using the surface force apparatus was also extended (see, for instance, [183–185]). The structure factor of dispersed colloidal particles is sensitive to the details of the effective pair interactions. In colloidal dispersions the influence of added nonadsorbing polymers on the colloid structure factor was measured using neutron scattering by making the polymer chains invisible [175]. A characteristic feature of the structure factor is the upswing at small wave vectors (see Sect. 2.6.4). Depletion effects were also quantified by measuring the spin-spin nuclear resonance time. Cosgrove et al. [186] performed such a study using a dispersion of silica with added sodium polystyrene sulfonate (NaPSS). The resonance time could be related to the depletion thickness, which decreased with increasing concentration of NaPSS.

When a colloid–polymer mixture phase separates into a colloid-rich and polymer-rich phase, an interface appears in between. For a colloidal gas–liquid interface it is possible to measure the interfacial tension using a number of techniques. The value of the interfacial tension [187] is interesting since it is related to phase separation kinetics (see Sect. 4.4). The spinning drop method was successfully used in the past to determine the interfacial tensions in demixed colloid–polymer mixtures [188, 189], yielding tensions with values of a few μN/m, corroborating the relation between the interfacial tension expressed in Eq. (1.24). The order of magnitude of the data of De Hoog and Lekkerkerker [189] were comparable with the theoretical results of Vrij [187], Van der Schoot [190] and of Brader and Evans [191]. From the results of Chen et al. [192], it follows that the interfacial tension increases with the distance from the critical point, in agreement with scaling theory [94]. By analysis of the break-up of an elongated droplet in a centrifugal field De Hoog and Lekkerkerker [193] demonstrated that the value of the measured interfacial tension was independent of the method used. Overall, it can be concluded that the colloidal and the 'molecular' gas–liquid interface behave similarly. The difference is that the interfacial tension between a colloidal liquid and gas is ultra-low.

Systematic experimental studies were made on various aspects of colloid–polymer mixtures. The phase behaviour of hard sphere binary asymmetric mixtures gained attention from theoretical [194, 195] and experimental [196, 197] points of view (see Chap. 6). Detailed investigations were published on the phase behaviour [158, 159] of well-defined colloid–polymer mixtures using hard sphere-like colloids mixed with rather monodisperse flexible polymer chains. Studies also appeared on the role of depletion effects on the dynamics of colloid–polymer mixtures, such as the diffusion of colloids [198] or the rheology of colloidal dispersions [199, 200] in solutions containing nonadsorbing polymers.

Specific effects (such as the presence of polymer brushes [62, 160]) affect depletion phenomena, and studies on these themes were also initiated in the 1990s. The same holds for the influence of charges. Many theories and depletion studies with model systems are based on hard sphere like colloidal particles. In practice, many stable dispersions containing spherical colloids consist of particles that are not 'pseudo-hard', but can be characterised by a pair potential containing an additional soft repulsive tail. An example is a stable dispersion of charged colloids in a polar solvent [201]. Here, double layer interactions provide a soft repulsive interaction between the particles (Sect. 1.2.2).

When charged colloids are dispersed in an aqueous salt solution in the presence of neutral depletion agents, adjusting the salt concentration influences the stability of the dispersion [85, 202, 203]. Grinberg and Tolstoguzov [87] presented generalised phase diagrams of (globular) proteins mixed with neutral nonadsorbing polysaccharides in aqueous salt solutions. The miscibility or compatibility was shown to increase when the ionic strength of the solvent was lowered. Patel and Russel [204] studied the phase behaviour of mixtures of charged colloidal polystyrene latex spheres and dextran as (neutral) polymer chains, and reported a significant shift of the gas–liquid binodal curve towards higher polymer concentrations when compared to predictions for neutral polymer chains mixed with hard spheres.

An early theoretical study on polyelectrolytes as nonadsorbing polymers was made by Böhmer et al. [205], who used the self-consistent field method of Scheutjens and Fleer [24, 118–120]. For high salt concentrations, the polymer concentration dependence of the depletion layer thickness matches with that of an uncharged polymer in solution. Below a salt concentration of 1 mol/l, the depletion layer thickness starts to decrease with increasing polyelectrolyte concentration at lower polymer concentration. At low salt concentrations a significant repulsive barrier in the potential between two uncharged parallel flat plates was found.

Walz and Sharma [206] proposed a force balance theory on the Derjaguin approximation level for the interaction between two spheres (regarded as hard spheres) dispersed in a solvent containing charged macromolecules. The magnitude of the interaction potential at contact increases as the Debye length increases or if the charge density on the large colloidal spheres (same sign as the 'macromolecules') increases. The range of the interaction potential also increases as the Debye length increases. At higher concentrations of the small particles a repulsive barrier in the interaction potential curve appears for sufficiently large size ratio of small and large colloid and sufficient Debye lengths. This might lead to the 'depletion stabilisation' that was also discussed for colloid–polymer mixtures by Napper [207]. In the model of Walz and Sharma [206], however, the polymers are modelled as charged hard spheres. It is therefore questionable as to whether this method applies to colloid–polymer mixtures, for which the polymer–colloid repulsion is soft.

Odijk [148] incorporated the effect of (like) charges on both polymer and colloid in theory [146] for two small colloidal spheres immersed in a polyelectrolyte solution. He related the effective depletion radius for small charged spheres, immersed in a solution with like-charged polyelectrolytes to the Debye length, the effective number of charges on the protein, the hard sphere radius and the Kuhn length [208]. When the effective depletion radius becomes larger than the correlation length of the polymer solution, phase separation due to depletion is expected.

Theoretical work was done on the influence of polydispersity on the depletion interaction and phase behaviour of colloid–polymer mixtures. Sear and Frenkel [209] investigated the phase behaviour of a colloid–polymer mixture by treating the polymers as PHSs using a distribution of polydisperse PHSs. Their calculations demonstrated that phase separation leads to size fractionation of the PHSs. FVT was extended to model polydispersity by replacing the monodisperse polymers with bidisperse polymers by Warren [210]. Warren found that polydispersity enhances the tendency to phase separate when a bidisperse polymer mixture is compared to a monodisperse mixture having identical number-averaged molar masses. It followed that the location of the binodals of the colloid–bidisperse polymer mixture is almost identical to that of a colloid–monodisperse polymer mixture when the weight-averaged molar mass of the bidisperse mixture is taken as the monodisperse molar mass.

1.3.6 2000–2022

Further progress was made on measuring depletion forces directly with high precision using a wide range of techniques [181,182,211,212]. Using modern advances in microscopy techniques [213] or total internal reflection microscopy, it is possible to measure depletion forces [214] and analyse, for instance, the radial distribution function [215] (see Sect. 2.6). Confocal microscopy allows the potential of mean force between colloids in colloid–polymer mixtures to be measured via the radial distribution function, as explored by Royall et al. [213]. These techniques make it possible to directly test theoretical concepts at the level of the effective depletion-mediated pair interaction between colloidal particles.

Advances were also made using theoretical methods and computer simulations. Until the end of the 1990s most theoretical approaches were based on describing polymer chains as ideal or as PHSs. At the turn of the last century especially, a wealth of different approaches was proposed to describe colloid–polymer mixtures in which interactions between polymer segments were accounted for. Essential was the progress made in Monte Carlo computer simulation studies on depletion effects [216–235] to test such theories. Below we first discuss some examples of theoretical developments.

1.3.6.1 Theoretical Equilibrium Approaches

(G)FVT
Despite the success of FVT in predicting the phase diagram of colloid–polymer mixtures for the colloid limit (small q) (semi-)quantitatively, in the protein limit (large q) the FVT predictions were far less convincing: it mainly provides useful qualitative information for large q. Quantitative deviations appear for $q > 0.5$ when comparing FVT with Monte Carlo computer simulations [220,221], theory [236] and

experiment [159,168,189,198,237–241] with realistic polymers. In short, classical FVT predicts binodal curves at too-small polymer concentrations for large q (see Sect. 4.1).

Under conditions where the polymer chains are much larger than the colloidal particles, such as in dispersions of proteins [242–245], tiny colloids [246–248] or micro-emulsion droplets [168,240] mixed with large polymers (or polysaccharides), instability occurs at rather high polymer concentration. In such situations it does not suffice to stick to the classical Asakura–Oosawa–Vrij description. Van der Schoot [249] showed that polymer collapse can take place when adding small colloids to a polymer solution. He derived an expression for the free energy of a polymer solution in a good solvent in the presence of small colloidal spheres and showed that adding colloids decreases the conformational entropy of a polymer chain. Effectively, adding spheres thus turns the solvent quality from good to poor. As a consequence, a polymer chain is expected to collapse above a certain colloid concentration. This effect originates from the mutual exclusion of polymer segments and colloidal particles. Experimental work confirms shrinkage of a polymer chain caused by adding nanospheres [246–248]. Computer simulations of large polymer chains in a system with random small obstacles by Wu et al. [250] are in line with polymer shrinkage due to added nanospheres: the size of the polymers was found to decrease when small particles are added.

To better describe some of the phenomena mentioned above, FVT has also been extended to incorporate the effects of interactions between the polymer chains [251–254] (see Sect. 4.3). This generalised free volume theory (GFVT) includes the correct dependencies for the depletion thickness and osmotic pressure on the polymer concentration for interacting chains [255], and gives a good description of colloid–polymer phase diagrams of model systems up to large q [254]. GFVT is in (semi)quantitative agreement with experiments and computer simulations [253,254] for a wide range of q values.

PRISM

Integral equation methods [17] are widely employed to understand structure and thermodynamics in atomic, colloidal and small molecule fluids, and have been generalised to treat macromolecular materials in the 1990s. Shaw and Thirumalai [256] applied the reference interaction site model (RISM) [257,258] to the case of colloids and combined it with the Edwards model for polymers to explain depletion stabilisation effects [32,259,260]: at high polymer concentrations, repulsive contributions to the pair interactions appear. RISM was later extended to the polymer reference interaction site model (PRISM), a continuous space liquid state approach that allows computation of the equilibrium properties of polymeric systems [261].

The PRISM integral equation approach has been generalised to explicitly treat polymers and their conformational degrees of freedom in order to arrive at microscopic equilibrium theories of the thermodynamics of colloid–polymer dispersions [262,263]. PRISM enables one to account for the role of particle size and polymer concentration to quantify the structure and phase stability of dilute and semidilute dispersions [264] and melts [265]. Although results on the structural and microscopic properties of colloid–polymer mixtures [262,263,266] heavily rely on the accuracy of approximate closure relations it can provide an accurate description of, for instance, the second osmotic virial coefficient of proteins with added nonadsorbing polymer chains [242].

Predicting complete phase diagrams (including binodals) using PRISM still requires extreme computational effort. PRISM can be used to predict fluid–fluid (colloidal gas–liquid) spinodal curves, which are close to binodals near the critical point. PRISM equilibrium predictions agree well with experiments [267]. The measured binodals of Ramakrishnan et al. [268] agree well with spinodals computed with PRISM.

GCM

Several liquid state theories have been developed that are based on *effective* potentials [269] from which the thermodynamic properties of many-body systems can be computed. Louis and Bolhuis et al. [216, 270–274] developed a Gaussian core model (GCM) to describe interacting polymer chains. In this model, the polymer chains are replaced with spherical particles with a soft repulsion between them. This model enables structure, depletion interactions and the full phase behaviour to be studied (in combination with Monte Carlo computer simulations). Basically, the theory is a liquid state approach. On the level of the depletion interaction mediated by interacting polymers, Louis and Bolhuis showed that their GCM agrees very well with Monte Carlo computer simulations [270, 274], except for a slight oscillation in the density profile in the case of Gaussian cores due to their 'particle' character. In the colloid limit GCM predictions for the phase diagram of colloid–polymer mixtures are similar to those for free volume theory [220]. For larger q-values FVT predicts phase separation at slightly smaller polymer concentrations compared to the GCM, although the trends are very similar.

DFT

Classical density functional theory (DFT) [275] is a formal procedure that can be used to quantify thermodynamic properties of fluids. To apply DFT, approximations must be made. Fundamental measure theory [276, 277] is an accurate approximation for hard sphere mixtures and permits the study of the interactions in and structure and phase behaviour of, e.g. colloidal systems [278], colloid–polymer mixtures [191, 279, 280] and star polymer plus linear polymer mixtures [281].

Within DFT the polymers are commonly treated as PHSs. Oversteegen and Roth [282] discussed the close analogy and discrepancies between FVT and fundamental measure theory. For asymmetric additive hard sphere mixtures DFT can be exploited to study the influence of the degree of repulsive interaction (the 'additivity') between the small spheres and the interaction between the large spheres. From self-avoiding walk computer simulations, it follows that the degree of additivity of excluded volume polymers is very small [283]. DFT also allows the colloidal gas–liquid interface of a demixed dispersion to be studied [191, 284]. This made it possible to evaluate, for instance, the interfacial tension. For a review on the possibilities of DFT for studying colloid–colloid and colloid–polymer mixtures, we refer to Ref. [285]. These fundamental DFT studies [286] helped to quantify the effective interactions and microstructural effects [278] in hard particle mixtures.

Mortazavifar and Oettel [287] proposed a DFT for the Asakura–Oosawa model of colloid–polymer mixtures, describing both fluid and crystal phases. They find good agreement with available computer simulation studies. Also, they showed that the phase diagram is fairly sensitive to the specific (non-ideality) approximations within DFT.

1.3.6.2 Further Insights into the Large Polymer Chain Regime

The equilibrium phase diagram for $q \lesssim 0.3$ is much simpler than for larger q [158–160,253,288,289]. For small q, the only effect of adding polymer chains to the pure hard sphere dispersion is the widening of the fluid–solid coexistence region. A gas–liquid phase transition occurs at larger polymer concentrations above the fluid–solid phase line and is therefore metastable (see Sect. 3.3.4). It is only above a certain range of attraction that the (colloidal) gas–liquid phase transition shifts to polymer concentrations below the fluid–solid coexistence curve. For a specific q value close to 1/3 the fluid–fluid critical point hits the fluid–solid coexistence curve. This critical point is known as the *critical endpoint* and denotes the boundary of stable gas–liquid phase coexistence. It is rather insensitive to the shape of the interaction potential used [290].

In the protein limit ($q \gtrsim 2$) the phase behaviour is dominated by the gas–liquid phase transition at low colloid volume fractions ϕ. Colloidal gas–liquid coexistence concentrations have been determined using Monte Carlo simulations by Bolhuis, Meijer and Louis [221]. They studied mixtures of hard spheres and self-avoiding walk polymer chains consisting of segments with hard sphere interactions. For three q-values the phase coexistence data are shown in Fig. 1.23. Phase transitions then take place near and above the polymer overlap concentration ($\phi_p \geq 1$). In such cases, a more detailed description of the physics of polymer solutions is required in order to describe depletion forces and the resulting phase transitions. Rotenberg et al. [291] extended TPT to incorporate interactions between the polymer chains. This shifts the binodal curves for larger q to higher relative polymer concentrations, as was also predicted using GFVT [251,254]. This is explained in some detail in Sect. 4.3. Mahynski et al. [292] performed more detailed simulations for large q. They found that the polymer osmotic pressures at the binodals collapse onto a single curve for various size ratios, even when far from the critical point. They also studied details of the structure of the mixture and of the potential of mean force between the small colloidal spheres in a solution of long polymer chains with excluded volume interaction [229].

Most of the models for depletion of polymer chains are based upon the assumption that the polymer segments are always much smaller than the colloidal spheres. Depletion of freely jointed chains near a spherical colloid can also be considered for arbitrary size of the segment, showing that depletion effects get weaker as the segments become longer (for a fixed value of the polymer's radius of gyration) [293,294]. This is in agreement with work of Paricaud, Varga and Jackson [295] who used Wertheim perturbation theory.

Fig. 1.23 Monte Carlo computer simulation results for the gas–liquid coexistences of hard spheres mixed with excluded volume polymers for $q = 3.86$ (O), 5.58 ($\times$) and 7.78 ($\blacklozenge$), redrawn from Bolhuis et al. [221]. The binodal curves are drawn to guide the eye

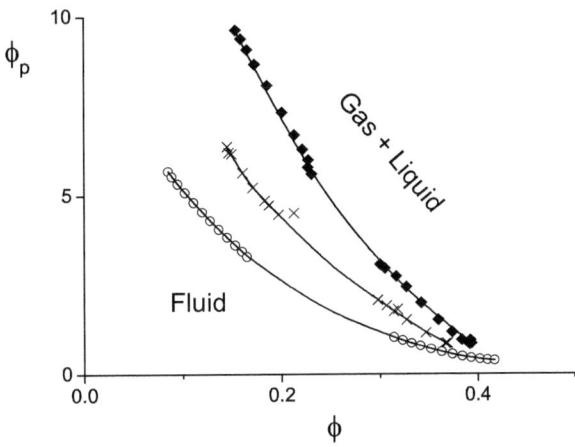

For large q many-body interactions become increasingly relevant. In the case of colloid–polymer mixtures, this means that multiple overlap of depletion zones play an important role in that high q regime. It is interesting to verify whether the effects of direct attractive forces can be compared to indirect depletion forces. For several types of direct attractions it has been shown that at the fluid–fluid (or gas–liquid) critical point, the $B_2^* \approx -6$ criterion [296] holds. However, in the case of a colloid–polymer mixture B_2^* strongly depends on the sphere/depletant size ratio [297], highlighting the fundamental difference between the depletion-mediated interaction and direct attractions (see also Santos et al. [298]).

1.3.6.3 Structure of Colloid–polymer Mixtures

As mentioned, scattering techniques are very useful to indirectly measure the ensemble-averaged structure of colloidal and colloid–polymer mixtures, and here we mention a few further examples of the progress made. Mutch et al. [241] measured the structure factor of colloids in a polymer solution in the protein limit. Muratov et al. [299] presented a revised form of the Percus–Yevick approach to describe the scattering of colloid–polymer mixtures with short-range depletion attraction. An extensive small-angle neutron scattering (SANS) and small-angle X-ray scattering (SAXS) study on the static and dynamic properties of silica spheres in semidilute solutions of high molar mass polystyrene in 2-butanone was performed by Poling-Skutvik et al. [300]. Their investigations revealed physical particle–polymer coupling on short length scales and long-ranged particle interactions, as well as sub-diffusive particle dynamics.

Kumar et al. [301] used SANS to study the influence that adding nonadsorbing polymers to charged silica spheres in an aqueous salt solution has on the structure factor. They found interesting re-entrant phase behaviour, which is reflected in the measured structure factors at small wave vectors. Peláez-Fernández et al. [302] used static light scattering (SLS) to study the effect of nonadsorbing polyelectrolytes on a dispersion of like-charged colloids. The experimental results for the colloid–colloid structure factor revealed that the main structure factor peak moves to higher wave vectors as the polyelectrolyte concentration increases. The authors interpret this in terms of an electrostatically enhanced depletion attraction. Later, Mehan et al. [303] showed that SANS enables measurement of the structure in multi-component (like-charged) systems.

The influence of the nonadsorbing polymers' branching on the structure of mixtures of hard sphere-like colloids with star polymers (with varying number of arms but constant radius of gyration) was investigated by Stellbrink et al. [304]. They measured partial structure factors in mixtures of star polymers with colloids using SANS. The relative distance to the demixing transition was reflected in a change of the structure factor at small wave vectors.

Spin-echo SANS (SESANS) is a SANS technique that was developed to probe length scales from 10 nm up to several tens of micrometres [305]. SESANS detects the polarisation of the neutron beam after scattering. Van Gruijthijsen et al. [306] and Washington et al. [307] used SESANS to quantify and describe polymer depletion-mediated structural effects in the dispersions, and showed that they can be interpreted using depletion forces. Van Gruijthijsen et al. [306] also compared SESANS to SAXS measurements conducted on the same experimental system, and found that similar structural information can be obtained. While SAXS has the advantage that it also provides form factors, SESANS can be applied more easily to study larger colloidal particles.

1.3.6.4 Polydispersity Effects

The particles and polymers in any real experiment have a finite polydispersity. The influence of polydispersity on depletion interaction and phase behaviour was investigated by extending existing approaches. Goulding and Hansen [308] computed the interaction potential between two spheres in a polydisperse bath of PHSs (polydisperse PHS model). When the polydispersity is characterised by a standard deviation of up to 30%, there is hardly an effect on the somewhat increased range and slightly deeper potential between the hard spheres. Above 30% polydispersity the effects become more significant. The original Asakura–Oosawa theory for two parallel plates immersed in a solution of nonadsorbing ideal polymer chains [57] could be extended to involve polymer size polydispersity [309], and would still provide analytical expressions for the interaction between the plates. This work was extended towards the interaction between two spheres in a solution of polydisperse ideal chains. It followed that the influence of polydispersity on the interaction is rather weak [309]. Even a polydispersity of 70% (standard deviation) only increases the attraction by less than 20%.

The phase behaviour of mixtures of monodisperse hard spheres and polydisperse ideal polymers has been investigated using original FVT [310]. At fixed mean polymer size, polydispersity favours gas–liquid coexistence and delays the onset of fluid–solid separation. On the other hand, systems with different size polydispersity but the same mass-averaged polymer chain length have nearly polydispersity-independent phase diagrams. The influence of polymer polydispersity on the colloidal gas–liquid phase coexistence of *interacting* polymers with spherical colloids is a complicated issue that has only been investigated using TPT by Paricaud et al. [311].

Nguyen et al. [312] derived an exact analytic expression for the many-body depletion interactions between the colloidal particles in the limit of long nonadsorbing polydisperse polymer chains. They also showed that depletion interactions in such systems can be described using mean-field theory.

The effect of particle polydispersity on the phase behaviour of mixtures of polydisperse hard spheres and ideal polymers has also been explored [313], also based on original FVT. Even modest polydispersities ($< 10\%$) can significantly change the phase diagram topology by introducing a host of new, multiphasic equilibria involving multiple solid phases. In practice, such multiphasic equilibria may show up as kinetic effects, preventing the system from reaching equilibrium. The nonequilibrium behaviour observed at higher polymer and particle concentrations may partly be due to this effect. Colloidal gas–liquid phase separation is, however, less sensitive to polydispersity [310].

1.3.6.5 Interfaces in Demixed Colloid–Polymer Mixtures

Since the 2000s the interface between coexisting phases has gained much more attention, as it was realised that it has special properties. For instance, more insight was gained on the ultra-low interfacial tension at the colloidal gas–liquid interface in demixed dispersions containing colloids and polymers. It became clear that this ultra-low interfacial tension affects the relevant characteristic length- and timescales [314]. The capillary length [315] decreases down to the order of micrometres, while the thermal length can become of the order of (sub)micrometres. This is special because other length scales (such as particle sizes) get bigger. The typical interface velocity in such systems is just a few micrometres per second. Inertial terms only become important at large length- and timescales. By means of confocal scanning laser microscopy, Aarts et al. [64] studied the influence of the ultra-low interfacial tension on wetting of colloid–polymer mixtures on a solid surface and on capillary waves [63] at the interface of a demixed colloid–polymer dispersion [316]. Studies on the bending rigidity of the colloidal gas–liquid interface in a demixed colloid–polymer dispersion have also been performed [317]. Interface physics in colloid–polymer mixtures has received ample attention. See, for instance, [213,231,318–322]. For more details, see Chap. 5.

1.3.6.6 Nonequilibrium Phenomena in Colloid–Polymer Mixtures

The influence of depletion effects on nonequilibrium phenomena in multi-component mixtures [323,324] gained increasing interest from both theoreticians and experimentalists [325]. PRISM enabled calculations of the structural correlations, allowing the microscopic evaluation of slow colloid dynamics. Interest focused on arrested states of colloid–polymer mixtures: upon adding a significant amount of depletants the mixtures tend to assume space-spanning structures of aggregated colloidal particles; hence, gelation or glass formation occurs (see Sect. 4.4 or the reviews [325,326]).

The structural relaxation time, formation of glasses and gels, nonlinear rheology and delayed gel collapse [327–329] were predicted quantitatively and compared to experimental results [330,331]. Experimental studies on the rheology of (gel) networks made of dispersions [332,333] and emulsions [334–336] at high concentrations of added nonadsorbing polymers also gained interest. Also the transition between gelation and glass formation was studied [337]. Wu et al. [338] experimentally studied the colloidal particle dynamics in colloid–polymer mixtures using polymers with different architectures: linear, subgranular cross-linked and branched microgels.

The use of relatively small polymeric depletants induced a short-range attractive interaction with a controlled strength. This enabled various research groups to study the glass transition and gelation of dense colloidal dispersions. In 2000, mode-coupling theory (MCT) was applied to predict the slow dynamics of the glass transition in colloid–polymer mixtures [339]. It revealed re-entrance of the repulsive-to-attractive glass transitions. The theoretical predictions were soon verified experimentally [323,340].

Further progress was stimulated by computer simulation studies on the influence that depletion attraction has on the structural and dynamic behaviour of colloids. These include phenomena such as the onset of attractive and repulsive glasses and the occurrence of re-entrant melting when the range of the depletion attraction is very small [341,342]. The idea to control the 'stickiness' of the interaction by changing the polymer concentration was also studied at lower colloid concentrations providing insights on colloidal gels [343,344]. These studies gave an explanation of the route from the glass transition at high densities to gelation of colloidal particles by varying the concentration of nonadsorbing polymers. Percolation of rod-like colloids through nonadsorbing polymers also gained interest [226,345].

The interplay of the attractive glass/gel line with phase separation [346] was also studied. A combination of computer simulations and confocal microscopy experiments showed that the gel line intersects the binodal at high densities, giving rise to a so-called arrested phase separation [344,347] (see Sect. 4.4). Arrested states induced by nonadsorbing polymer chains have also been studied extensively in the context of hard sphere/star polymer [348] or star/star mixtures [349]. The introduction of soft interactions is found to enrich the phenomenology of glass transitions and the interplay between the two species [350] compared to binary mixtures of hard spheres [351].

1.3.6.7 Towards Complexity in Colloid–Depletant Mixtures

Interest in depletion phenomena began to broaden after the turn of the last century [352,353]. A few selected items are briefly discussed below.

Influence of solvent quality on colloid–polymer mixtures
Although many initial approaches ignored details of the solvent, it became clearer that solvent quality [354–358] can play an important and sometimes complex [359] role. Hence, it can be important to properly treat the non-ideality of polymer chains in solution. As an example, we mention a study by Taylor, Evans and Royall [360]

on the response to temperature of a well-known model colloid–polymer mixture. At room temperature they found that the critical value of the second virial coefficient for the colloidal gas–liquid phase transition for colloidal spheres can be described using the AOV concept. They could also accurately predict the onset of gelation observed experimentally. Upon cooling the system, the depletion attraction between colloids is reduced because the polymer radius of gyration decreases as the Θ-temperature is approached. Paradoxically, this raises the 'effective' temperature and leads to 'melting' of the colloidal gels.

Depletion effects mediated by complex polymer mixtures
Work on less conventional depletants (other than, for instance, simple polymers or hard spheres) started to appear. Preisler et al. [361] performed SCF calculations and Monte Carlo computer simulations to analyse the depletion profiles of star-like and H-shaped polymers in a good solvent at a wall. The influence of polymer chain stiffness was also studied and it turned out that, at a fixed coil size, stiffer chains decrease the depletion thickness [293,294,362] for dilute polymer solutions. In the case of semidilute polymer solutions the depletion thickness goes through a maximum as a function of chain stiffness [294]. Lim and Denton [363] demonstrated that polymer shape distributions influence the resulting depletion-induced interaction potentials between colloidal particles. Depletion of ring polymers in solution next to hard nonadsorbing walls was also studied [364,365]. The authors found more pronounced structuring of rings at a nonadsorbing hard surface as compared to linear chains. This structuring strongly affects the shape of the depletion potential between two hard walls.

Depletion forces between colloidal particles in a binary polymer blend were studied by Chervanyov [366], who showed that the relative contributions to the range and strength of the effective depletion attraction strongly depend on the mass fractions of the polymer species and their chain length ratio. Interactions between colloidal particles embedded in a polymer network were considered by Di Michele, Zaccone and Eiser [367]. They presented a theoretical framework to quantify the attractive interactions between the particles mediated by such a polymer network. These predictions agreed with Monte Carlo simulations performed by the authors.

Depletion effects mediated by complex colloids
The depletion forces between colloidal hard spheres mediated by self-assembling patchy particles were studied by García, Gnan and Zaccarelli [368]. They found that the depletion interaction is completely attractive and oscillatory. This may be relevant for understanding the behaviour of complex mixtures in crowded environments, or for targeted self-assembly aimed at building desired superstructures. For an overview of the use of complex depletants such as rods, platelets and ellipsoids, see the review by Briscoe [369].

Surfactant micelles can also induce depletion attraction between colloidal particles [370]. Additionally, use can be made of the temperature dependence of the shape of self-assembled surfactants. Gratale et al. [371] showed that the strength and range of the depletion interaction can be tuned via shape anisotropy of the surfactant micelles. Depletion effects also can be encountered in mixtures of self-assembling block copolymers in a selective solvent under the influence of nonadsorbing polymers [372,373]. This work showed that nonadsorbing polymers also induce attraction between block copolymer micelles.

A promising class of depletants involves a self-assembling medium forming either supramolecular chains [374–376] or clusters [377]. Also interesting are depletants in a solvent in the vicinity of a critical point [378,379], which provide a connection between depletion interactions and so-called critical Casimir forces [43,380] in colloidal dispersions. A solution of depletants near a gas–liquid critical point was also studied numerically, showing how critical solvent mixture forces can be used to effectively manipulate colloidal aggregation [381] and percolation [382]. For a surfactant solution near a fluid–fluid critical point, depletion forces can merge into critical Casimir forces [352].

Systems such as colloidal spheres with multi-component depletants [383–387] and dispersions of star polymers (soft colloids) and linear polymers [388,389] received interest, as well as mixtures of different types of star polymers [390]. Also, microgel particles as depletants were studied [391]. A full understanding of such more complex mixtures is a topic of future research.

Non-hard colloids with depletants
Accounting for mixtures of colloidal spheres with a hard-core attractive or repulsive Yukawa interactions was studied within FVT [392]. It appeared that additional direct repulsive interactions between the colloids increased the single-phase stability region, whereas additional attractions reduce the stability regions. Rovigatti et al. [393] investigated the influence of brushes anchored onto the colloidal particles in colloid–polymer mixtures.

Deposition of colloids via depletion forces
Linse and Wennerström [394] reported an interesting theoretical and simulation study on a mixture of particles interacting with each other and with a flat wall through a square well attractive potential. They found an interval of attraction strengths over which surface adsorption of the particles is significant, while bulk instability through nucleation remains negligible solely due to geometrical effects. In hindsight, this effect was already demonstrated experimentally by Dinsmore et al. [395] using mixtures of small and large colloidal spheres at a wall (see Sect. 6.3). Ouhajji et al. [396] realised the deposition of the silica spheres onto a glass plate mediated by adding PDMS. For mixtures of indented colloids, Ashton et al. [397] found conditions at which the particles crystallised at a wall by adding nonadsorbing polymers.

The depletion–adsorption transition
In 2015, Feng et al. [398] published intriguing results on the temperature-dependent phase stability of an aqueous mixture of relatively large silica spheres mixed with PEO polymer chains. They found that a reversible transition from depletion-induced crystallisation to adsorption-mediated bridging flocculation occurs by changing the temperature. See also the study by Kwon et al. [399] on the temperature dependence in a similar system. It becomes clear that the classical depletion description in which it is assumed that the polymer concentration vanishes at the surface of a colloidal particle does not always suffice. Ouhajji et al. [396] studied dispersions of silica

spheres in cyclohexane containing nonadsorbing PDMS polymers. They could only interpret the phase diagram by considering weak depletion effects [400], which is consistent with earlier force measurements on this system by Wijting, Besseling and Cohen Stuart [65]. The so-called depletion–adsorption transition (DAT) was theoretically studied in some detail on the level of pair interactions [401–403] and complete phase behaviour [402]. The predicted non-monotonic DAT temperature dependence [401] was confirmed experimentally [404]. Chen et al. [405] and Fantoni et al. [406] proposed models in which the DAT arises from the interactions with the solvent molecules.

Charged colloid–polymer mixtures
Insights into mixtures of charged colloids and nonadsorbing (charged) polymers also developed further. Studies of aqueous mixtures of proteins and nonadsorbing polymers such as polyethylene glycol (PEG) or (uncharged) polysaccharides yielded some interesting observations. Finet and Tardieu [202] studied the stability of solutions of the lens protein α-crystallin. Adding an excess of salt to this system does not destabilise the protein dispersion. It follows that the effective attractions between the proteins are absent or are very weak in the case of screened charges. Adding PEG, however, induces significant attractions [202], and results in a shift of the fluid–fluid (in the protein field also termed liquid–liquid) phase transition to higher temperatures [407]. Adding excess salt *and* PEG induces instant phase separation [202]. A similar synergistic effect of salt and PEG was found in aqueous solutions of (spherical) brome mosaic virus particles [203] and lysozyme [408]. Adding PEG also influences protein crystallisation (see Sect. 11.2). Royall, Aarts and Tanaka [409] studied the influence of double layer repulsion on depletion forces using confocal microscopy. In conclusion, the trend found in experimental studies on mixtures of charged 'colloids' and neutral polymers is that the miscibility is, as expected, increased upon decreasing the salt concentration, i.e. increasing the range of the double layer repulsion.

There still are few theoretical studies on mixtures of colloids with a screened-Coulomb repulsion mixed with neutral or charged polymer chains. Ferreira et al. [410] made a PRISM analysis for mixtures of charged colloids and polyelectrolytes up to the level of the pair interaction and computed gas–liquid spinodal curves from the colloid–colloid structure factor. Denton and Schmidt [411] proposed a simple theory yielding the colloidal gas–liquid binodal curve for charged spheres mixed with free neutral polymer chains, described as PHSs. Fortini et al. [222] extended free volume theory to account for a short-ranged soft repulsion between the spherical colloids, allowing a description of the full phase diagrams. They also performed Monte Carlo simulations, and the theory was found to agree quite well with the extended free volume theory. It was found that the colloidal fluid–solid coexistence is especially sensitive to the screened-Coulomb repulsion.

The work of Fortini et al. [222] was later extended towards highly screened, charged spheres mixed with *interacting* polymers [412]. Zhou, van Duijneveldt and Vincent [413] have shown that generalised free volume theory (GFVT), including short-ranged soft repulsion, is capable of quantitatively describing the depletion-induced phase separation in mixtures of charged silica particles and nonadsorbing

Exercise 1.7. What happens to the miscibility region of a stable colloidal fluid with added nonadsorbing polymers upon adding a screened double layer repulsion between the spheres?

---

polystyrene polymer chains in dimethylformamide (Θ-solvent conditions). They varied both the range of the double layer repulsion and the size ratio q.

Stradner et al. [414] and Sedgwick et al. [415] considered mixtures of charged spherical colloids with a long-ranged double layer repulsion mixed with very short polymer chains that induce a short-ranged depletion attraction. In such systems small equilibrium clusters are formed that can be described theoretically [416] or using Molecular Dynamics computer simulations [417]. The finite cluster size is a result of a competition between short-ranged depletion attraction and long-ranged repulsion.

Some aspects that could be relevant have not yet been incorporated into the theory for the phase behaviour. A first issue is the effect of gradients in permittivity. Croze and Cates [418] demonstrated that even the depletion zones caused by neutral polymers are affected by charged surfaces. The electrical field present between like-charged surfaces polarise the neutral polymer chains because of their (usually) low permittivity. Curtis and Lue [419] also showed dielectric discontinuities can be quite relevant for colloidal dispersions with added depletants and electrolytes in solution. These effects can enhance polymer depletion and increase the screening of double layer interactions.

The situation gets more complicated when the free polymers are (like-)charged as well [420]. Work of Israelachvili, Pincus and others [421] revealed that the addition of free polyelectrolyte mainly decreases the effective Debye length in aqueous salt solutions, leading to a decrease in the double layer repulsion. Grillo et al. [422] made an interesting study on aqueous mixtures of Pluronic F127 surfactants mixed with hyaluronic acid polyelectrolytes in the semidilute concentration regime. They found that the surfactant micelles and polyelectrolytes were homogeneously distributed in salt free solutions. By increasing the ionic strength the micelles start to cluster and the self-assembly is explained by depletion forces. The salt type is found to play an important role.

The interactions between charged particles mediated by like-charged polyelectrolytes were measured by Moazzami-Gudarzi et al. [423] (see also the review by Scarratt et al. [424]). The crystallisation of charged colloidal spheres mixed with like-charged polyelectrolytes was studied by Ioka et al. [425]. They found that the polyelectrolytes induced depletion forces but simultaneously cause screening of double layer forces. Experimental work was also done on binary asymmetric mixtures of charged colloids by Toyotama et al. [426], revealing a eutectic point. Colloidal probe atomic force microscopy measurements between large charged colloidal spheres mediated by small charged colloidal spheres were performed by Ludwig and von Klitzing et al. [427–429]. They observed an oscillatory force and analysed its wavelength in detail. The case of nonadsorbing polyelectrolytes near uncharged surfaces and the relation to the Donnan potential was explored theoretically [430].

In summary, it seems that at high salt concentrations like charges on polymers and colloids do not seem to strongly affect the depletion-induced attraction between colloids. At low ionic strength, however, the situation becomes quite complicated and detailed theories that enable a computation of the stability of such systems still have to be developed. For charged multi-component colloidal mixtures a rich phase behaviour is found. It is clear that there is much work left to be done on the role of (charged) depletants in (charged) systems before we obtain a complete picture.

1.3.6.8 Depletion Effects and Its Relevance for Biology

Nonadsorbing polymers or colloids can also be used to concentrate bacteria. This is very useful for water treatments, where one attempts to achieve the formation of bioflocs of bacteria. Schwarz-Linek et al. [431] studied the addition of nonadsorbing polymers on mixtures of *Escherichia coli* bacteria and found that concentration of these bacteria is possible in this way. Sun et al. [432] used rod-like nanoparticles to induce phase separation of suspensions containing *Pseudomanas aeruginosa* bacteria. They concluded that rod-shaped nanoparticles are very effective at inducing phase separation of suspensions containing bacteria. Phase transitions of dispersions containing rod-like viruses mediated by the addition of nonadsorbing polymers are discussed in Sect. 8.5.1 and other parts of Chap. 8.

Depletion effects play a role in protein dispersions [243,244] similar to that in colloidal suspensions (as was made clear above) and can lead to protein aggregation and phase separation [433] in living matter. As summarised by Sapir and Harries [434,435], excluded volume effects are thought to be of importance in explaining several intracellular processes [436,437] and the appearance of membraneless organelles (MLOs). Hence, the resulting depletion effects that are operational are suggested to mediate several types of biological processes such as endocytose [438], microtubule bundling [439], protein dynamics [440–442], transcription and self-organisation of the molecules of life [443–445]. Crowding of the subcellular environment by macromolecules is supposed to mediate conformational switches between active states of RNA [446], can influence the conformations of DNA [447–449], and may induce structural transitions in protein-like polymers [450,451]. Quantifying the effects of entropic forces in biology remains a virgin area for additional research. See also the brief illustration on macromolecular crowding in Sect. 11.1.

1.3.6.9 Anisotropic Colloids and Depletion Effects

In this overview on the history of depletion in colloidal dispersions we have mostly focused on mixtures of colloidal spheres and nonadsorbing polymers, which also received much attention. At the beginning of the 21st century, colloid synthesis evolved to such a degree [452–456] that it became possible to make colloidal particles of a wide range of shapes [457–461]. This, and the fact that anisotropic shapes occur in nature, has triggered experimental, theoretical and computer simulation studies on mixtures of non-spherical colloids in the presence of nonadsorbing polymers, and on binary colloidal mixtures containing anisotropic particles. Hence, insights have been obtained into the phase behaviour of mixtures containing, for instance, colloidal

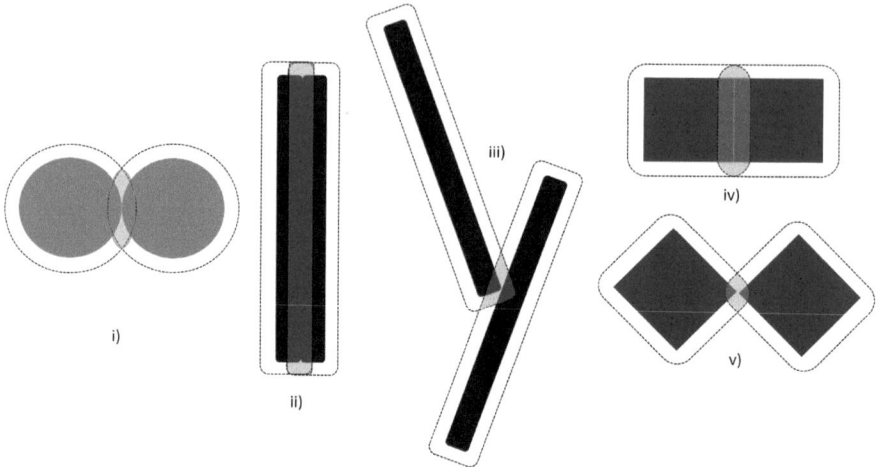

Fig. 1.24 Illustration of the depletion zones around and overlap volumes between (i) spheres, (ii) and (iii) rods, and (iv) and (v) cubes with fixed particle volume, mediated by depletants that induce a fixed depletion thickness (indicated by the dashes). Reprinted with permission from Ref. [488]. Copyright 2017 Elsevier

rods [130,462–469], platelets [469–477], dendrimers [478], rocks (colloidal particles with an irregular surface) [479], boards [480], 'golf-balls' [481], ellipsoids [482] and cubes [483–485] with added polymers.

In Chaps. 8–10, the focus is on the phase behaviour of anisotropic colloidal particles and the influence of nonadsorbing polymers. An interesting feature of non-spherical hard colloidal particles [486] is that they can exhibit directionality purely based on (entropic) excluded volume interactions [452,487], because flat faces tend to align. The interplay between orientational and excluded volume entropy enables (multiple) liquid-crystalline phases to occur. The addition of depletants can mediate *entropic patchiness* of anisotropic colloidal particles [488].

In Fig. 1.24, overlap volumes are indicated for a few colloidal particles of different shapes. For isotropic spheres, the overlap volume reaches a maximum value as the spheres touch, see (i). For anisotropic particles, the overlap volume depends on their orientation. When comparing the overlap volumes in Fig. 1.24(ii) and (iii) it becomes clear that the overlap volume is maximised for rods as the particles align with their largest surface areas close to each other. Hence, aligned rod configurations are induced by the depletion effect. For cubes the overlap volume is maximised when two (flat) edges are aligned (see Fig. 1.24(iv) and (v)). This illustrates that particles tend to align due to the addition of depletants. As we shall see, the variation of particle shape, and the strength and range of depletion attraction yield a wide variety of self-assembled structures.

Besides studying the effects of nonadsorbing polymers as depletants, it is also of interest to treat colloids themselves as depletants when added to a dispersion of larger or different colloids (see Chaps. 6 and 7). Studies of depletion effects in colloidal dispersions reveal that mixing more complex particle shapes leads to increasingly exotic phase behaviour. Examples include mixtures of rods and platelets [489–492], rods and spheres [467,492–497] (see Chap. 7), platelets and spheres [498–508], rods and cubes [509] and bidisperse platelet mixtures [510–513]. Xie et al. [514] showed that mixing colloidal rods with surfactant micelles can lead to phase states in which the rods assume orientationally ordered nematic and smectic-like membrane superstructures.

Van der Schoot [515,516] theoretically considered the interactions between hard spheres immersed in a dispersion of hard rods in the nematic phase state, and also studied their self-assembly [516]. Further, nonequilibrium phenomena are also quite relevant in such systems [226,517]. Aggregation was found to occur in dispersions of spherical colloidal particles and worm-like micelles, leading to transient gels [518].

Depletion effects in dispersions of more complex shapes have also attracted attention. Krüger et al. [519] derived expressions for the depletion force between two arbitrarily shaped large convex colloidal particles immersed in a suspension of small spherical particles. Damasceno et al. [520] studied the thermodynamic self-assembly of a family of truncated tetrahedra, and reported several atomic crystal isostructures as the polyhedron shape varies from tetrahedral to octahedral. The self-assembled crystal structures can be understood as a tendency for polyhedra to maximise face-to-face alignment, which can be generalised as directional entropic forces. Interestingly, the self-assembled structures differ from the densest packing.

Although we cover several examples of mixtures of anisotropic particles mixed with (mainly polymeric) depletants in Chaps. 8 and 10, it is noted that binary mixtures of anisotropic colloids have also been investigated. Nakato et al. [521] studied pure titanate platelets ($D = 7.1$ μm, $L = 0.75$ nm). They observed an isotropic–nematic phase transition at very small volume fractions $\phi_I = 3.7 \cdot 10^{-5}$ and $\phi_N = 1.9 \cdot 10^{-3}$. The large width of the transition region indicates that the platelets are quite polydisperse. Upon adding small laponite platelets ($D = 30$ nm, $L = 1$ nm), Nakato et al. [512] observed both biphasic I–N *and* triphasic N_1–N_2–I equilibria. This triphasic region is observed for extremely small laponite volume fractions between $1.6 \cdot 10^{-7}$ and $3.6 \cdot 10^{-6}$. For higher laponite concentrations the depletion-induced attraction between the large platelets becomes so strong that the dispersion flocculates.

Free volume theory for binary platelets [513] confirms the observed equilibrium phase transitions. The calculations reveal that the biphasic regime (without added laponite) lies between titanate volume fractions $\phi_I = 5.8 \cdot 10^{-4}$ and $\phi_N = 8.4 \cdot 10^{-4}$, so in between the observed experimental values. The triphasic triangle region lies between laponite volume fractions of $4 \cdot 10^{-7}$ and $2 \cdot 10^{-6}$, close to the experimental volume fractions.

1.4 Outline

In this chapter we provided an introduction to colloidal interactions, a historical perspective on early observations, and a qualitative understanding of the basic depletion effects. Further, an overview was provided of the important developments in the field of the depletion interaction and the resulting phase behaviour of colloidal dispersions.

In Chap. 2 we address the fundamentals of depletion interactions, including pair potentials and the effects of anisotropic depletants. The focus will be on small depletant concentrations which allow simple treatments using both the force method and the adsorption method to arrive at depletion potentials. The basics of phase behaviour in colloidal dispersions with added depletants are set out in Chap. 3. This is followed in Chap. 4 by extending the model to also include a more detailed description of the polymer physics involved so that it can be applied to mixtures of spherical colloids and polymers. Experimental phase diagrams of well-defined colloid–polymer mixtures are discussed and compared to theories for colloid–polymer mixtures. Phase separation kinetics and nonequilibrium states in colloid–polymer mixtures are treated as well. Chapter 5 concerns the properties of the interface that appears between coexisting colloidal gas and colloidal liquid phases, induced by nonadsorbing polymers. Chapter 6 deals with the phase behaviour of binary colloidal sphere mixtures in the absence of nonadsorbing polymer; and we will discuss the effect of adding small rod-like colloids to a suspension with colloidal spheres in Chap. 7.

Rod-like colloids are considered in Chap. 8, first without polymer: the physics of the isotropic to nematic phase transition is discussed in some detail, followed by a treatment of charged rods. Next, polymer-induced depletion effects for rod-like colloids are discussed. At the end of the chapter it is shown how highly ordered phases (smectic and solid-like) can be treated, and the richness of the phase behaviour of rod–polymer mixtures is revealed. In Chap. 9 mixtures of platelets and depletants are discussed, and mixtures of cube-like colloids with added nonadsorbing polymers are discussed in Chap. 10. These anisotropic colloids have gained increasing attention in more recent years. Treatments of the equilibrium phase states of pure hard platelets involve the isotropic, nematic and columnar phase states, and lead to intriguing phase behaviour. Cube-like colloids are dispersions containing particles with superball-like shapes. Adding nonadsorbing polymers to such systems can promote the formation of ordered, simple, cubic crystalline structures. Throughout, the concepts will be illustrated by experimental and computer simulation results.

In Chap. 11 we highlight manifestations of depletion effects in more complex systems, in particular in biology and technology. This book ends with an Epilogue (Chap. 12), with reflections and an outlook on the possible areas where extensions of the current knowledge of depletion phenomena are needed.

References

1. Everett, D.H.: Pure Appl. Chem. **31**, 577 (1972)
2. Everett, D.H.: Basic Principles of Colloid Science. Royal Society of Chemistry, London(1988)
3. Russel, W.B., Saville, D.A., Schowalter, W.R.: Colloidal Dispersions. Cambridge University Press (1989)
4. Evans, D.F., Wennerström, H.: The Colloidal Domain: Where Physics, Chemistry, Biology, and Technology Meet, 2nd edn. Wiley-VCH, New York (1999)
5. Hunter, R.J.: Foundations of Colloid Science, 2nd edn. Oxford University Press, Oxford(2001)
6. Jones, R.A.L.: Soft Condensed Matter. Oxford University Press, Oxford (2002)
7. Lyklema, J.: Fundamentals in Colloid and Interface Science, vol. 1–5. Elsevier, Amsterdam (1991–2005)

8. Barrat, J.L., Hansen, J.P.: Basic Concepts for Simple and Complex Liquids. Cambridge University Press, Cambridge, U.K. (2010)
9. Doi, M.: Soft Matter Physics. Oxford University Press, Oxford (2013)
10. Butt, H.J., Kappl, M.: Surface and Interfacial Forces, 2nd edn. Wiley-VCH, New York (2018)
11. Philipse, A.P.: Brownian Motion: Elements of Colloid Dynamics. Undergraduate Lecture Notes in Physics. Springer, Berlin (2018)
12. Perrin, J.: Ann. de Chem. et de Phys. **18**, 5 (1909)
13. Einstein, A.: Ann. Phys. **17**, 549 (1905)
14. Onsager, L.: Chem. Rev. **13**, 73 (1933)
15. Onsager, L.: Ann. NY. Acad. Sci. **51**, 627 (1949)
16. McMillan, W.G., Mayer, J.E.: J. Chem. Phys. **13**, 276 (1945)
17. Hansen, J.P., McDonald, I.R.: Theory of Simple Liquids, 2nd edn. Academic Press, San Diego, CA, USA (1986)
18. Maitland, G.C., Rigby, M., Smith, E.B., Wakeham, W.A.: Intermolecular Forces: Their Origin and Determination. Clarendon Press, Oxford (1981)
19. Pitzer, K.S.: J. Chem. Phys. **7**, 583 (1939)
20. Noro, M.G., Frenkel, D.: J. Chem. Phys. **113**, 2941 (2000)
21. de With, G.: Polymer Coatings. Wiley, New York (2018)
22. Walstra, P.: Physical Chemistry of Foods. Marcel Decker, New York (2003)
23. Lucas, A., Harris, J.: Ancient Egyptian Materials and Industries. Dover, London (1999)
24. Fleer, G.J., Cohen Stuart, M.A., Scheutjens, J.M.H.M., Cosgrove, T., Vincent, B.: Polymers at Interfaces. Chapman and Hall, New York (1993)
25. The Greek historian Herodotus coined the term delta for the landform where the Nile river flows into the Mediterranean Sea; the sediment deposited at the river's mouth has the shape of the upper-case Greek letter Δ
26. Barton, E.C.: Geograph. J. **51**, 100 (1918)
27. de Kruif, C.G., Holt, C.: Advanced Dairy Chemistry Proteins, P.L.H.M.P.F. Fox (ed.) vol. 1, Chap. 5, pp. 233–276. Kluwer Academic, Plenum, New York (2002)
28. Walstra, P., Jenness, R.: Dairy Chemistry and Physics. Wiley, New York (1984)
29. de Kruif, C.G., Zhulina, E.B.: Colloids Surf. A **117**, 151 (1996)
30. Kose, A., Ozka, M., Takano, K., Kobayashi, Y., Hachisu, S.: J. Colloid Interface Sci. **44**, 330 (1973)
31. Soanes, C., Stevenson, A.: Oxford Dictionary Of English. Oxford University Press, New York (2005)
32. Feigin, R.I., Napper, D.H.: J. Colloid Interface Sci. **75**, 525 (1980)
33. Derjaguin, B.V., Landau, L.: Acta Physicochimica USSR **14**, 633 (1941)
34. Verwey, E.J.W., Overbeek, J.T.: Theory of the Stability of Lyophobic Colloids. Elsevier, Amsterdam (1948)
35. Israelachvili, J.N.: Intermolecular and Surface Forces, 3rd edn. Academic Press, Amsterdam (2011)
36. London, F.: Trans. Faraday Soc. **33**, 8b–26 (1937)
37. Hamaker, H.C.: Physica **4**, 1058 (1937)
38. Lyklema, J.: Fundamentals in Colloid and Interface Science, vol. 1. Elsevier, Amsterdam (1991)
39. Behrens, S.H., Christel, D.I., Emmerzael, R., Schurtenberger, P., Borkovec, M.: Langmuir **16**, 2566 (2000)
40. Israelachvili, J.N., Adams, G.E.: J. Chem. Soc. Faraday Trans. **74**, 975 (1978)
41. Ninham, B.W.: Adv. Colloid Interface Sci. **83**, 1 (1999)
42. Vrij, A.: Pure Appl. Chem. **48**, 471 (1976)
43. Hertlein, C., Helden, L., Gambassi, A., Dietrich, S., Bechinger, C.: Nature **451**, 172 (2008)
44. Fischer, E.W.: Kolloid Z. **160**, 120 (1958)
45. Alexander, S.J.: J. Phys. France **38**, 983 (1977)
46. De Gennes, P.G., Acad, C.R.: Sc. Paris ser. B **300**, 839 (1985)
47. De Gennes, P.G.: Adv. Colloid Interface Sci. **27**, 189 (1987)

48. Vrij, A., Tuinier, R.: Fundamentals in Colloid and Interface Science, Lyklema, J. (ed.), vol. 4, Chap. 5. Elsevier, Amsterdam (2005)
49. de Kruif, C.G.: Personal Communication
50. Baxter, R.J.: J. Chem. Phys. **49**, 2770 (1968)
51. Rouw, P.W., Vrij, A., de Kruif, C.G.: Colloids Surf. **31**, 299 (1988)
52. Rouw, P.W., de Kruif, C.G.: J. Chem. Phys. **88**, 7799 (1988)
53. Rouw, P.W., Vrij, A., de Kruif, C.G.: Prog. Colloid Polym. Sci. **76**, 1 (1988)
54. Pelssers, E.G.M., Cohen Stuart, M.A., Fleer, G.J.: Chem. Soc. Faraday Trans. **86**, 1355 (1990)
55. Vrij, A.: Personal Communication
56. Philipse, A.P., Lekkerkerker, H.N.W., Vroege, G.J., Tuinier, R.: J. Colloid Interface Sci. **543**, 352 (2019)
57. Asakura, S., Oosawa, F.: J. Chem. Phys. **22**, 1255 (1954)
58. Asakura, S., Oosawa, F.: J. Pol. Sci. **33**, 183 (1958)
59. Long, J.A., Osmond, D.W.J., Vincent, B.: J. Colloid Interface Sci. **42**, 545 (1973)
60. Li-In-On, R., Vincent, B., Waite, F.A.: ACS Symp. Ser. **9**, 165 (1974)
61. De Hek, H., Vrij, A.: J. Colloid Interface Sci. **84**, 409 (1981)
62. Wijmans, C.M., Zhulina, E.B., Fleer, G.J.: Macromolecules **27**, 3238 (1994)
63. Aarts, D.G.A.L., Schmidt, M., Lekkerkerker, H.N.W.: Science **304**, 847 (2004)
64. Aarts, D.G.A.L., van der Wiel, J.H., Lekkerkerker, H.N.W., Phys, J.: Condens. Matter **15**, S245 (2003)
65. Wijting, W.K., Besseling, N.A.M., Cohen Stuart, M.A.: Phys. Rev. Lett. **90**, 196101 (2003)
66. Von Guericke, O.: Experimenta Nova Magdeburgica de Vacuo Spatio. Waesberge, Amsterdam (1672)
67. Mach, E.: The Science of Mechanics. The Open Court Publishing Company, Illinois, USA (1960)
68. Rowlinson, J.: Notes Rec. R. Soc. Lond. **57**, 35 (2003)
69. Fatio de Dullier, N.: Oeuvres Completes de Christiaan Huygens. The Hague **9**, 381 (1888–1950)
70. Le Sage, G.L.: Lettre á une académicien de Dijon. Mercure de France **153** (1756)
71. Feynman, R.P.: The Character of Physical Law. MIT, Cambridge (1967)
72. Edwards, M.R. (ed.): Pushing Gravity: New Perspectives on Le Sage's Theory for Gravitation. C. Roy Keys Inc, Montreal (2002)
73. Fåhraeus, R.: Physiol. Rev. **9**, 241 (1929)
74. Thysegen, J.E.: Acta Med. Scand. Suppl. **134**, 1 (1942)
75. Fåhraeus, R.: Acta. Med. Scand. **55**, 1 (1921)
76. Bedell, S.E., Booker, B.T.: Am. J. Med. **78**, 1001 (1985)
77. Inokuchi, K.: Bull. Chem. Soc. Jpn. **24**, 78–82 (1951)
78. Forsdyke, D.R., Ford, P.M.: Biochem. J. **214**, 257 (1983)
79. Janzen, J., Brooks, D.E.: Clin. Hemorheol. **9**, 695 (1989)
80. Neu, B., Meiselman, H.J.: Biophys. J. **83**, 2482 (2002)
81. Nehring, A., Shendruk, T.N., de Haan, H.W.: Soft Matter **14**, 8160 (2018)
82. Beijerinck, M.W.: Zentr. Bakteriol. Paras. Infektionskr. **2**, 697 (1896)
83. Doublier, J.L., Garnier, C., Renard, C., Sanchez, C.: Curr. Opin. Colloid Interface Sci. **5**, 184 (2000)
84. Goh, K.K., Teo, A., Sarkar, A., Singh, H.: Milk Proteins, Boland, M., Singh, H. (eds.) 3rd edn, pp. 499–535. Academic Press (2020)
85. Tolstoguzov, V.B.: Food Hydrocolloids **4**, 429 (1991)
86. Tolstoguzov, V.B.: Food Hydrocoll. **11**, 181 (1997)
87. Grinberg, V.Y., Tolstoguzov, V.B.: Food Hydrocoll. **11**, 145 (1997)
88. Tolstoguzov, V.B., Mzhel'sky, A.L., Gulov, V.Y.: Colloid Polym. Sci. **252**, 124 (1974)
89. Scholten, E., Tuinier, R., Tromp, R.H., Lekkerkerker, H.N.W.: Langmuir **18**, 2234 (2002)
90. Traube, J.: Gummi Zeitung **39**, 434 (1925)
91. Baker, H.C.: Trans. Inst. Rubber Ind. **13**, 70 (1937)
92. Vester, C.F.: Kolloid Z. **84**, 63 (1938)

93. Dickinson, E.: Food Hydrocoll. **17**, 25 (2003)
94. Rowlinson, J.S., Widom, B.: Molecular Theory of Capillarity. Clarendon Press, Oxford (1982)
95. Cockbain, E.G.: Trans. Faraday Soc. **48**, 185 (1952)
96. Fairhurst, D., Aronson, M., Ohm, M.L., Goddard, E.D.: Colloids Surf. **7**, 153 (1983)
97. Cohen, S.S.: J. Biol. Chem. **144**, 353 (1942)
98. Bernal, J.D., Fankuchen, I.: J. Gen. Physiol. **25**, 111 (1941)
99. Cohen, S.S.: Proc. Soc. Exp. Biol. Med. **48**, 163 (1941)
100. Leberman, R.: Virology **30**, 341 (1966)
101. Oosawa, F.: Hyo-Hyo Rakugaku. Autobiography, Nagoya (2005)
102. Kurihara, K., Vincent, B.: J. Chem. Phys. **154**, 220401 (2021)
103. Oosawa, F.: J. Chem. Phys. **155**, 084104 (2021)
104. Nagoya Symposium on Depletion Forces: Celebrating the 60th Anniversary of the Asakura-Oosawa Theory, Nagoya, Japan, 14–15 Mar 2014
105. Sieglaff, C.: J. Polym. Sci. **41**, 319 (1959)
106. Clarke, J., Vincent, B.: J. Chem. Soc. Faraday Trans. I **77**, 1831 (1981)
107. Cowell, C., Li-In-On, R., Vincent, B., Waite, F.A.: J. Chem. Soc. Faraday Trans. **74**, 337 (1978)
108. Vincent, B., Luckham, P.F., Waite, F.A.: J. Colloid Interface Sci. **73**, 508 (1980)
109. Vincent, B., Edwards, J., Emmett, S., Jones, A.: Colloids Surf. **17**, 261 (1986)
110. Vincent, B.: Colloids Surf. **24**, 269 (1987)
111. Vincent, B., Edwards, J., Emmett, S., Croot, R.: Colloids Surf. **31**, 267 (1988)
112. Emmett, S., Vincent, B.: Phase Trans. **21**, 197 (1990)
113. Kose, A., Hachisu, S.: J. Colloid Interface Sci. **55**, 487 (1976)
114. Hachisu, S., Kose, A., Kobayashi, Y.: J. Colloid Interface Sci. **55**, 499 (1976)
115. De Gennes, P.G.: Scaling Concepts in Polymer Physics. Cornell University Press, Ithaca (1979)
116. De Gennes, P.G., Acad, C.R.: Sc. Paris Ser. B **288**, 203 (1979)
117. Joanny, J.F., Leibler, L., De Gennes, P.G., Polymer Sci.: Polym. Phys. **17**, 1073 (1979)
118. Scheutjens, J.M.H.M., Fleer, G.J.: J. Phys. Chem. **83**, 1619 (1979)
119. Scheutjens, J.M.H.M., Fleer, G.J.: J. Phys. Chem. **84**, 178 (1980)
120. Scheutjens, J.M.H.M., Fleer, G.J.: Adv. Colloid Interface Sci. **16**, 361 (1982)
121. Broukhno, A., Jönsson, B., Åkesson, T., Vorontsov-Velyaminov, P.N.: J. Chem. Phys. **113**, 5493 (2000)
122. De Hek, H., Vrij, A.: J. Colloid Interface Sci. **70**, 592 (1979)
123. As dimensionless concentration variable ϕ is used throughout. In case of hard colloidal particles the quantity ϕ is the volume fraction. For polymers and penetrable hard spheres ϕ refers to the relative concentration with respect to overlap (see (1.21))
124. De Hek, H., Vrij, A.: J. Colloid Interface Sci. **88**, 258 (1982)
125. Sperry, P.R., Hopfenberg, H.B., Thomas, N.L.: J. Colloid Interface Sci. **82**, 62 (1981)
126. Sperry, P.R.: J. Colloid Interface Sci. **87**, 375 (1982)
127. Sperry, P.R.: J. Colloid Interface Sci. **99**, 97 (1984)
128. Gast, A.P., Hall, C.K., Russel, W.B.: J. Colloid Interface Sci. **96**, 251 (1983)
129. Barker, J.A., Henderson, D.: Rev. Mod. Phys. **48**, 587 (1976)
130. Peters, V.F.D., Vis, M., González García, Á., Wensink, H.H., Tuinier, R.: Phys. Rev. Lett. **125**, 127803 (2020)
131. Gast, A.P., Russel, W.B., Hall, C.K.: J. Colloid Interface Sci. **109**, 161 (1986)
132. Lépine, Y., Caillé, A.: Can. J. Phys. **56**, 403 (1978)
133. Eisenriegler, E.: J. Chem. Phys. **79**, 1052 (1983)
134. Tuinier, R., Vliegenthart, G.A., Lekkerkerker, H.N.W.: J. Chem. Phys. **113**, 10768 (2000)
135. Casassa, E.F., Tagami, Y.: Macromolecules **2**, 14 (1969)
136. Wang, Y., Hansen, F.Y., Peters, G.H., Hassager, O.: J. Chem. Phys. **129**, 074904 (2008)
137. Allain, C., Ausserré, D., Rondelez, F.: Phys. Rev. Lett. **49**, 1694 (1982)
138. Ausserré, D., Hervet, H., Rondelez, F.: Macromolecules **19**, 85 (1986)
139. Pashley, R.M., Ninham, B.W.: J. Phys. Chem. **91**, 2902 (1987)

140. Taniguchi, T., Kawakatsu, T., Kawasaki, K.: Slow Dynamics in Condensed Matter–Proceedings of the 1st Tohwa University International Symposium, AIP Conference Proceedings, Kawasaki, K., Kawakatsu, T., Tokuyama, M. (eds.), vol. 256, p. 503. AIP, New York, N.Y. (1992)
141. Eisenriegler, E., Hanke, A., Dietrich, S.: Phys. Rev. E **54**, 1134 (1996)
142. Eisenriegler, E.: Phys. Rev. E **55**, 3116 (1997)
143. Eisenriegler, E.: J. Phys. D: Condens. Matter **12**, A227 (2000)
144. Bringer, A., Eisenriegler, E., Schlesener, F., Hanke, A.: Eur. Phys. J. B **11**, 101 (1999)
145. Hanke, A., Eisenriegler, E., Dietrich, S.: Phys. Rev. E **59**, 6853 (1999)
146. Odijk, T.: Macromolecules **29**, 1842 (1996)
147. Odijk, T.: J. Chem. Phys. **106**, 3402 (1997)
148. Odijk, T.: Langmuir **13**, 3579 (1997)
149. Odijk, T.: Biophys. J. **79**, 2314 (2000)
150. Eisenriegler, E.: J. Chem. Phys. **113**, 5091 (2000)
151. Reiss, H.: J. Phys. Chem. **96**, 4736 (1992)
152. Eisenriegler, E., Bringer, A.: J. Phys.: Condens. Matter **17**, 1711 (2005)
153. Eisenriegler, E.: J. Chem. Phys. **124**, 144912 (2006)
154. Eisenriegler, E.: J. Chem. Phys. **125**, 204903 (2006)
155. Eisenriegler, E., Bringer, A.: J. Chem. Phys. **127**, 034904 (2007)
156. Lekkerkerker, H.N.W.: Colloids Surf. **51**, 419 (1990)
157. Lekkerkerker, H.N.W., Poon, W.C.K., Pusey, P.N., Stroobants, A., Warren, P.B.: Europhys. Lett. **20**, 559 (1992)
158. Leal-Calderon, F., Bibette, J., Biais, J.: Europhys. Lett. **23**, 653 (1993)
159. Ilett, S.M., Orrock, A., Poon, W.C.K., Pusey, P.N.: Phys. Rev. E **51**, 1344 (1995)
160. Faers, M.A., Luckham, P.F.: Langmuir **13**, 2922 (1997)
161. Meijer, E.J., Frenkel, D.: J. Chem. Phys. **100**, 6873 (1994)
162. Dickinson, E.: Soft Matter **2**, 642 (2006)
163. Bibette, J., Roux, D., Nallet, F.: Phys. Rev. Lett. **65**, 2470 (1990)
164. Bibette, J.: J. Colloid Interface Sci. **147**, 474 (1992)
165. Bibette, J., Roux, D., Pouligny, B.: J. Phys. II **2**, 401 (1992)
166. Snowden, M.J., Williams, P.A., Garvey, M.J., Robb, I.D.: J. Colloid Interface Sci. **166**, 160 (1994)
167. Xia, K.Q., Zhang, Y.B., Tong, P., Wu, C.: Phys. Rev. E **55**, 5792 (1997)
168. Lynch, I., Cornen, S., Piculell, L.: J. Phys. Chem. B **108**, 5443 (2004)
169. Meller, A., Stavans, J.: Langmuir **12**, 301 (1996)
170. Tuinier, R., de Kruif, C.G.: J. Colloid Interface Sci. **218**, 201 (1999)
171. van Kampen, N.G., Oppenheim, I.: Physica **138A**, 231 (1986)
172. Dijkstra, M., Van Roij, R., Evans, R.: Phys. Rev. E **59**, 5744 (1999)
173. Dijkstra, M., Brader, J.M., Evans, R., Phys, J.: Condens. Matter **11**, 10079 (1999)
174. Dijkstra, M., van Roij, R.: Phys. Rev. Lett. **89**, 208303 (2002)
175. Ye, X., Narayanan, T., Tong, P., Huang, J.S., Lin, M.Y., Carvalho, B.L., Fetters, L.J.: Phys. Rev. E **54**, 6500 (1996)
176. Horner, K.D., Topper, M., Ballauff, M.: Langmuir **13**, 551 (1997)
177. Milling, A.J., Biggs, S.: J. Colloid Interface Sci. **170**, 604 (1995)
178. Milling, A.J., Vincent, B.: Chem. Soc., Faraday Trans. **93**, 3179 (1997)
179. Biggs, S., Burns, J.L., Yan, Y., Jameson, G.J., Jenkins, P.: Langmuir **16**, 9242 (2000)
180. Sober, D.L., Walz, J.Y.: Langmuir **11**, 2352 (1995)
181. Sharma, A., Walz, J.Y.: J. Chem. Soc. Faraday Trans. **92**, 4997 (1996)
182. Rudhardt, D., Bechinger, C., Leiderer, P.: Phys. Rev. Lett. **81**, 1330 (1998)
183. Richetti, P., Kékicheff, P.: Phys. Rev. Lett. **68**, 1951 (1992)
184. Chu, X.L., Nikolov, A.D., Wasan, D.T.: J. Chem. Phys. **103**, 6653 (1995)
185. Chu, X.L., Nikolov, A., Wasan, D.T.: J. Chem. Phys. **105**, 4892 (1996)
186. Cosgrove, T., Obey, T.M., Ryan, K.: Colloids Surf. **651**, 1 (1992)
187. Vrij, A.: Phys. A **235**, 120 (1997)

188. Vliegenthart, G.A., Lekkerkerker, H.N.W.: Prog. Colloid Polym. Sci. **105**, 27 (1997)
189. de Hoog, E.H.A., Lekkerkerker, H.N.W.: J. Phys. Chem. B **103**, 5274 (1999)
190. van der Schoot, P.: J. Phys. Chem. B **103**, 8804 (1999)
191. Brader, J.M., Evans, R.: Europhys. Lett. **49**, 678 (2000)
192. Chen, B.H., Payandeh, B., Robert, M.: Phys. Rev. E **62**, 2369 (2000)
193. de Hoog, E.H.A., Lekkerkerker, H.N.W.: J. Phys. Chem. B **105**, 11636 (2001)
194. Biben, T., Hansen, J.P.: Phys. Rev. Lett. **66**, 2215 (1991)
195. Lekkerkerker, H.N.W., Stroobants, A.: Phys. A **195**, 387 (1993)
196. Van Duijneveldt, J.S., Heinen, A.W., Lekkerkerker, H.N.W.: Europhys. Lett. **21**, 369 (1993)
197. Imhof, A., Dhont, J.K.G.: Phys. Rev. Lett. **75**, 1662 (1995)
198. Bodnár, I., Dhont, J.K.G., Lekkerkerker, H.N.W.: J. Phys. Chem. **100**, 19614 (1994)
199. Wolthers, W., van den Ende, H., Duits, M., Mellema, J.: J. Rheol. **1996**, 55 (1996)
200. Bodnár, I., Dhont, J.K.G.: Phys. Rev. Lett. **77**, 5304 (1996)
201. Monovoukas, Y., Gast, A.P.: J. Colloid Interface Sci. **128**, 533 (1989)
202. Finet, S., Tardieu, A.: J. Cryst. Growth **232**, 40 (2001)
203. Casselyn, M., Perez, J., Tardieu, A., Vachette, P., Witz, J., Delacroix, H.: Acta Cryst. D **57**, 1799 (2001)
204. Patel, P.D., Russel, W.B.: J. Colloid Interface Sci. **131**, 192 (1989)
205. Böhmer, M.R., Evers, O.A., Scheutjens, J.M.H.M.: Macromolecules **23**, 2288 (1990)
206. Walz, J.Y., Sharma, A.: J. Colloid Interface Sci. **168**, 485 (1994)
207. Napper, D.H.: Polymeric Stabilization of Colloidal Dispersions. Academic Press, San Diego, CA, USA (1983)
208. Rubinstein, M., Colby, R.: Polymer Physics. Oxford University Press, New York (2003)
209. Sear, R., Frenkel, D.: Phys. Rev. E **55**, 1677 (1997)
210. Warren, P.B.: Langmuir **13**, 4388 (1997)
211. Verma, R., Crocker, J.C., Lubensky, T.C., Ya, G.: Macromolecules **33**, 177 (2000)
212. Kleshchanok, D., Tuinier, R., Lang, P.R.: Langmuir **22**, 9121 (2007)
213. Royall, C.P., Louis, A.A., Tanaka, H.: J. Chem. Phys. **127**, 044507 (2007)
214. Xing, X., Hua, L., Ngai, T.: Curr. Opin. Colloid Interface Sci. **20**, 54 (2015)
215. Edwards, T.D., Bevan, M.A.: Macromolecules **45**, 585 (2012)
216. Louis, A.A., Bolhuis, P.G., Hansen, J.P., Meijer, E.J.: Phys. Rev. Lett. **85**, 2522 (2000)
217. Dijkstra, M.: Curr. Opin. Colloid Interface Sci. **4**, 372 (2001)
218. Louis, A.A., Allahyarov, E., Löwen, H., Roth, R.: Phys. Rev. E **65**, 061407 (2002)
219. Moncho-Jordá, A., Louis, A.A., Bolhuis, P.G., Roth, R.: J. Phys.: Condens. Matter **15**, S3429 (2003)
220. Bolhuis, P.G., Louis, A.A., Hansen, J.P.: Phys. Rev. Lett. **89**, 128302 (2002)
221. Bolhuis, P.G., Meijer, E.J., Louis, A.A.: Phys. Rev. Lett. **90**, 068304 (2003)
222. Fortini, A., Dijkstra, M., Tuinier, R.: J. Phys.: Condens. Matter **17**, 7783 (2005)
223. Tanaka, H., Araki, T.: Chem. Engin. Sci. **61**, 2108 (2006)
224. Chou, C.Y., Vo, T., Panagiotopoulos, A., Robert, M.: Phys. A **369**, 275 (2006)
225. Dijkstra, M., van Roij, R., Roth, R., Fortini, A.: Phys. Rev. E **73**, 041409 (2006)
226. Schilling, T., Jungblut, S., Miller, M.A.: Phys. Rev. Lett. **98**, 108303 (2007)
227. Jungblut, S., Tuinier, R., Binder, K., Schilling, T.: J. Chem. Phys. **127**, 244909 (2007)
228. Fortini, A., Bolhuis, P.G., Dijkstra, M.: J. Chem. Phys. **128**, 024904 (2008)
229. Mahynski, N.A., Irick, B., Panagiotopoulos, A.Z.: Phys. Rev. E **87**, 022309 (2013)
230. Nikoubashman, A., Mahynski, N., Pirayandeh, A., Panagiotopoulos, A.: J. Chem. Phys. **140**, 094903 (2014)
231. Jover, J., Galindo, A., Jackson, G., Müller, E.A., Haslam, A.J.: Mol. Phys. **113**, 2608 (2015)
232. Wu, L., Malijevský, A., Jackson, G., Müller, E.A., Avendano, C.: J. Chem. Phys. **143**, 044906 (2015)
233. Howard, M.P., Nikoubashman, A., Panagiotopoulos, A.Z.: Langmuir **33**, 11390 (2017)
234. Wu, L., Malijevský, A., Avendano, C., Müller, E.A., Jackson, G.: J. Chem. Phys. **148**, 164701 (2018)

235. Barcelos, E.I., Khani, S., Boromand, A., Vieira, L.F., Lee, J.A., Peet, J., Naccache, M.F., Maia, J.: Comp. Phys. Comm. **258**, 107618 (2021)
236. Surve, M., Pryamitsyn, V., Ganesan, V.: J. Chem. Phys. **122**, 154901 (2005)
237. Y. Hennequin, M. Evens, C.M. Quezada Angulo, J.S. Van Duijneveldt, J. Chem. Phys. **123**, 054906 (2005)
238. Zhang, Z.X., van Duijneveldt, J.S.: J. Chem. Phys. **124**, 154910 (2006)
239. Zhang, Z.X., van Duijneveldt, J.S.: Langmuir **22**, 63 (2006)
240. Van Duijneveldt, J.S., Mutch, K., Eastoe, J.: Soft Matter **3**, 155 (2007)
241. Mutch, K.J., van Duijneveldt, J.S., Eastoe, J., Grillo, I., Heenan, R.K.: Langmuir **25**, 3944 (2009)
242. Kulkarni, A.M., Chatterjee, A.P., Schweizer, K.S., Zukoski, C.F.: Phys. Rev. Lett. **83**, 4554 (1999)
243. Kulkarni, A.M., Chatterjee, A.P., Schweizer, K.S., Zukoski, C.F.: J. Chem. Phys. **113**, 9863 (2000)
244. Tuinier, R., Dhont, J.K.G., de Kruif, C.G.: Langmuir **16**, 1497 (2000)
245. Tuinier, R., Brûlet, A.: Biomacromolecules **4**, 28 (2003)
246. Kramer, T., Scholz, S., Maskros, M., Huber, K.: J. Colloid Interface Sci. **279**, 447 (2004)
247. Kramer, T., Schweins, R., Huber, K.: J. Chem. Phys. **123**, 014903 (2005)
248. Kramer, T., Schweins, R., Huber, K.: Macromolecules **38**, 9783 (2005)
249. van der Schoot, P.: Macromolecules **31**, 4635 (1998)
250. Wu, D., Hui, K., Chandler, D.: J. Chem. Phys. **96**, 835 (1991)
251. Aarts, D.G.A.L., Tuinier, R., Lekkerkerker, H.N.W., Phys, J.: Condens. Matter **14**, 7551 (2002)
252. Fleer, G.J., Tuinier, R.: Phys. Rev. E **76**, 041802 (2007)
253. Tuinier, R., Smith, P.A., Poon, W.C.K., Egelhaaf, S.U., Aarts, D.G.A.L., Lekkerkerker, H.N.W., Fleer, G.J.: Europhys. Lett. **82**, 68002 (2008)
254. Fleer, G.J., Tuinier, R.: Adv. Colloid Interface Sci. **143**, 1 (2008)
255. Fleer, G.J., Skvortsov, A.M., Tuinier, R.: Macromol. Theory Sim. **16**, 531 (2007)
256. Shaw, M.R., Thirumalai, D.: Phys. Rev. A **44**, R4797 (1991)
257. Chandler, D., Singh, Y., Richardson, D.: J. Chem. Phys. **81**, 1975 (1984)
258. Curro, J.G., Schweizer, K.S.: Macromolecules **20**, 1928 (1987)
259. van der Gucht, J., Besseling, N.A.M., van Male, J., Cohen Stuart, M.A.: J. Chem. Phys. **112**, 2886 (2000)
260. Semenov, A.N., Shvets, A.A.: Soft Matter **11**, 8863–8878 (2015)
261. Schweizer, K.S., Curro, J.G.: Adv. Polym. Sci. **116**, 319 (1994)
262. Fuchs, M., Schweizer, K.S.: Europhys. Lett. **51**, 621 (2000)
263. Fuchs, M., Schweizer, K.S., Phys, J.: Condens. Matter **14**, R239 (2002)
264. Chen, Y.L., Schweizer, K.S., Fuchs, M.: J. Chem. Phys. **118**, 3880 (2003)
265. Banarjee, D., Schweizer, K.S.: J. Chem. Phys. **142**, 214903 (2015)
266. Fuchs, M., Schweizer, K.S.: Phys. Rev. E **64**, 021514 (2001)
267. Shah, A., Ramakrishnan, S., Chen, Y.L., Schweizer, K.S., Zukoski, C.F.: J. Phys.: Condens. Matter **15**, 4751 (2003)
268. Ramakrishnan, S., Fuchs, M., Schweizer, K.S., Zukoski, C.F.: J. Chem. Phys. **116**, 2201 (2002)
269. Likos, C.N.: Phys. Rep. **348**, 267 (2001)
270. Bolhuis, P.G., Louis, A.A., Hansen, J.P., Meijer, E.J.: J. Chem. Phys. **114**, 4296 (2001)
271. Bolhuis, P.G., Louis, A.A., Hansen, J.P.: Phys. Rev. E **6402**, 021801 (2001)
272. Bolhuis, P.G., Louis, A.A.: Macromolecules **35**, 1860 (2002)
273. Louis, A.A., Bolhuis, P.G., Meijer, E.J., Hansen, J.P.: J. Chem. Phys. **116**, 10547 (2002)
274. Louis, A.A., Bolhuis, P.G., Meijer, E.J., Hansen, J.P.: J. Chem. Phys. **117**, 1893 (2002)
275. Evans, R.: Adv. Phys. **28**, 143 (1979)
276. Rosenfeld, Y.: Phys. Rev. Lett. **63**, 980 (1989)
277. Rosenfeld, Y.: J. Chem. Phys. **98**, 8126 (1993)
278. Roth, R., Evans, R., Dietrich, S.: Phys. Rev. E **62**, 62 (2000)
279. Schmidt, M., Löwen, H., Brader, J.M., Evans, R.: Phys. Rev. Lett. **85**, 1934 (2000)
280. Schmidt, M., Löwen, H., Brader, J.M., Evans, R.: J. Phys.: Condens. Matter **14**, 9353 (2002)

281. Dzubiella, J., Likos, C.N., Löwen, H.: J. Chem. Phys. **116**, 9518 (2002)
282. Oversteegen, S.M., Roth, R.: J. Chem. Phys. **122**, 214502 (2005)
283. Roth, R., Evans, R.: Europhys. Lett. **53**, 271 (2001)
284. Brader, J.M., Evans, R., Schmidt, M., Löwen, H.: J. Phys.: Condens. Matter **14**, L1 (2002)
285. Brader, J.M., Evans, R., Schmidt, M.: Mol. Phys. **101**, 3349 (2003)
286. Evans, R., Oettel, M., Roth, R., Kahl, G.: J. Phys.: Condens. Matter **28**, 240401 (2016)
287. Mortazavifar, M., Oettel, M.: J. Phys.: Condens. Matter **28**, 244018 (2016)
288. Poon, W.C.K., Renth, F., Evans, R.M.L., Fairhurst, D.J., Cates, M.E., Pusey, P.N.: Phys. Rev. Lett. **83**, 1239 (1999)
289. Moussaïd, A., Poon, W.C.K., Pusey, P.N., Soliva, M.F.: Phys. Rev. Lett. **82**, 225 (1999)
290. Fleer, G.J., Tuinier, R.: Phys. A **379**, 52 (2007)
291. Rotenberg, B., Dzubiella, J., Hansen, J.P., Louis, A.A.: Mol. Phys. **102**, 1 (2004)
292. Mahynski, N.A., Lafitte, T., Panagiotopoulos, A.Z.: Phys. Rev. E **85**, 051402 (2012)
293. Tuinier, R.: Eur. Phys. J. E **10**, 123 (2003)
294. Martens, C.M., van Leuken, S.H.M., Opdam, J., Vis, M., Tuinier, R.: Phys. Chem. Chem. Phys. **24**, 3618 (2022)
295. Paricaud, P., Varga, S., Jackson, G.J.: J. Chem. Phys. **118**, 8525 (2003)
296. Vliegenthart, G.A., Lekkerkerker, H.N.W.: J. Chem. Phys. **112**, 5364 (2000)
297. Tuinier, R., Feenstra, M.: Langmuir **30**, 13121 (2014)
298. Santos, A., López de Haro, M., Fiumara, G., Saija, F.: J. Chem. Phys. **142**, 224903 (2015)
299. Muratov, A., Moussaïd, A., Narayanan, T., Kats, E.I.: J. Chem. Phys. **131**, 054902 (2009)
300. Poling-Skutvik, R., Mongcopa, K.I.S., Faraone, A., Narayanan, S., Conrad, J.C., Krishnamoorti, R.: Macromolecules **49**, 6568 (2016)
301. Kumar, S., Ray, D., Aswal, V.K., Kohlbrecher, J.: Phys. Rev. E **90**, 042316 (2014)
302. Peláez-Fernández, M., Moncho-Jordá, A., Callejas-Fernández, J.: J. Chem. Phys. **134**, 054905 (2011)
303. Mehan, S., Chinchalikar, A.J., Kumar, S., Aswal, V.K., Schweins, R.: Langmuir **29**, 11290 (2013)
304. Stellbrink, J., Allgaier, J., Richter, D., Moussaïd, A., Schofield, A., Poon, W., Pusey, P., Lindner, P., Dzubiella, J., Likos, C., Löwen, H.: Appl. Phys. A **74**, 355 (2002)
305. Rekveldt, M.T., Plomp, J., G, B.W., Kraan, W.H., Grigoriev, S., Blaauw, M.: Rev. Sci. Instrum. **76**, 033901 (2005)
306. van Gruijthuijsen, K., Bouwman, W.G., Schurtenberger, P., Stradner, A.: Europhys. Lett. **106**, 28002 (2014)
307. Washington, A.L., Li, X., Schofield, A.B., Hong, K., Fitzsimmons, M.R., Dalgliesh, R., Pynn, R.: Soft Matter **10**, 3016 (2014)
308. Goulding, D., Hansen, J.P.: Mol. Phys. **99**, 865 (2001)
309. Tuinier, R., Petukhov, A.V.: Macromol. Theory Simul. **11**, 975 (2002)
310. Fasolo, M., Sollich, P.: J. Phys.: Condens. Matter **17**, 797 (2005)
311. Paricaud, P., Varga, S., Cummings, P.T., Jackson, G.J.: Chem. Phys. Lett. **398**, 489 (2004)
312. Nguyen, H.S., Forsman, J., Woodward, C.E.: Soft Matter **14**, 6921 (2018)
313. Fasolo, M., Sollich, P.: J. Chem. Phys. **122**, 074904 (2005)
314. Aarts, D.G.A.L., Dullens, R.P.A., Lekkerkerker, H.N.W.: New J. Phys. **7**, 40 (2005)
315. Aarts, D.G.A.L.: J. Phys. Chem. B **109**, 7407 (2005)
316. Aarts, D.G.A.L.: Soft Matter **3**, 19 (2007)
317. Kuipers, J., Blokhuis, E.M.: J. Colloid Interface Sci. **315**, 270 (2007)
318. Brader, J.M., Evans, R.R., Schmidt, M.M., Löwen, H.: J. Phys.: Condens. Matter **14**, L1 (2001)
319. Moncho-Jordá, A., Dzubiella, J., Hansen, J.P., Louis, A.A.: J. Phys. Chem. B **109**, 6640 (2005)
320. Lekkerkerker, H.N.W., de Villeneuve, V.W.A., de Folter, J.W.J., Schmidt, M., Hennequin, Y., Bonn, D., Indekeu, J.O., Aarts, D.G.A.L.: Eur. Phys. J. B **64**, 341 (2008)
321. Binder, K., Virnau, P., Statt, A.: J. Chem. Phys. **141**, 559 (2014)
322. Vis, M., Brouwer, K.J.H., González García, Á., Petukhov, A.V., Konovalov, O., Tuinier, R.: J. Phys. Chem. Lett. **11**, 8372 (2020)

323. Pham, K.N., Puertas, A.M., Bergenholtz, J., Egelhaaf, S.U., Moussaïd, A., Pusey, P.N., Schofield, A.B., Cates, M.E., Fuchs, M., Poon, W.C.K.: Science **296**, 104 (2002)
324. Tuinier, R., Dhont, J.K.G., Fan, T.H.: Europhys. Lett. **75**, 929 (2006)
325. Anderson, V.J., Lekkerkerker, H.N.W.: Nature **416**, 811 (2002)
326. Verhaegh, N.A.M., Lekkerkerker, H.N.W.: The Physics of Complex Systems, Proceedings of the International School of Physics "Enrico Fermi" Course CXXXIV, Mallamace, F., Stanley, H. (eds.) IOS Press, Amsterdam (1997)
327. Chen, Y.L., Schweizer, K.S.: J. Chem. Phys. **120**, 7212 (2004)
328. Chen, Y.L., Kobelev, V., Schweizer, K.S.: Phys. Rev. E **71**, 041405 (2005)
329. Kobelev, V., Schweizer, K.S.: J. Chem. Phys. **123**, 164903 (2005)
330. Shah, A., Ramakrishnan, S., Chen, Y.L., Schweizer, K.S., Zukoski, C.F.: J. Chem. Phys. **119**, 8747 (2003)
331. Ramakrishnan, S., Chen, Y.L., Schweizer, K.S., Zukoski, C.F.: Phys. Rev. E **70**, 040401 (R) (2004)
332. Laurati, M., Egelhaaf, S.U., Petekidis, G.: J. Rheol. **55**, 673 (2011)
333. Moghimi, E., Jacob, A.R., Koumakis, N., Petekidis, G.: Soft Matter **13**, 2371 (2017)
334. Berli, C.L., Quemada, D., Parker, A.: Colloids Surf. A **203**, 11 (2002)
335. Quemada, D., Berli, C.: Adv. Colloid Interface Sci. **98**, 51 (2002)
336. Berli, C.L., Quemada, D., Parker, A.: Colloids Surf. A **215**, 201 (2003)
337. Koumakis, N., Petekidis, G.: Soft Matter **7**, 2456 (2011)
338. Wu, Q., Higler, R., Kodger, T.E., van der Gucht, J., Appl, A.C.S.: Mater. Interfaces **12**, 42041 (2020)
339. Dawson, K., Foffi, G., Fuchs, M., Götze, W., Sciortino, F., Sperl, M., Tartaglia, P., Voigtmann, T., Zaccarelli, E.: Phys. Rev. E **63**, 011401 (2000)
340. Eckert, T., Bartsch, E.: Phys. Rev. Lett. **89**, 125701 (2002)
341. Puertas, A.M., Fuchs, M., Cates, M.E.: Phys. Rev. Lett. **88**, 098301 (2002)
342. Zaccarelli, E., Foffi, G., Dawson, K.A., Buldyrev, S., Sciortino, F., Tartaglia, P.: Phys. Rev. E **66**, 041402 (2002)
343. Zaccarelli, E.: J. Phys.: Condens. Matter **19**, 323101 (2007)
344. Lu, P.J., Zaccarelli, E., Ciulla, F., Schofield, A., Sciortino, F., Weitz, D.A.: Nature **453**, 499 (2008)
345. Surve, M., Pryamitsyn, V., Ganesan, V.: Macromolecules **40**, 344 (2007)
346. Cardinaux, F., Gibaud, T., Stradner, A., Schurtenberger, P.: Phys. Rev. Lett. **99**, 118301 (2007)
347. Verhaegh, N.A.M., Asnaghi, D., Lekkerkerker, H.N.W., Giglio, M., Cipelletti, L.: Phys. A **242**, 104 (1997)
348. Stiakakis, E., Vlassopoulos, D., Likos, C.N., Roovers, J., Meier, G.: Phys. Rev. Lett. **89**, 208302 (2002)
349. Mayer, C., Sciortino, F., Likos, C.N., Tartaglia, P., Löwen, H., Zaccarelli, E.: Macromolecules **42**, 423 (2009)
350. Mayer, C., Zaccarelli, E., Stiakakis, E., Likos, C., Sciortino, F., Munam, A., Gauthier, M., Hadjichristidis, N., Iatrou, H., Tartaglia, P., Löwen, H., Vlassopoulos, D.: Nat. Mat. **7**, 780 (2008)
351. Voigtmann, T.: Europhys. Lett. **96**, 36006 (2011)
352. Buzzaccaro, S.: Il Nuovo Cim. C **34**, 109 (2011)
353. Piazza, R., Buzzaccaro, S., Parola, A., Colombo, J.: J. Phys.: Condens. Matter **23**, 194114 (2011)
354. Fleer, G.J., Skvortsov, A.M., Tuinier, R.: Macromolecules **36**, 7857 (2003)
355. D'Adamo, G., Pelissetto, A., Pierleoni, C.: Mol. Phys. **111**, 3372 (2013)
356. D'Adamo, G., Pelissetto, A., Pierleoni, C.: J. Chem. Phys. **141**, 024902 (2014)
357. Denton, A.R., Davis, W.J.: J. Chem. Phys. **155**, 084904 (2021)
358. Nakamura, Y., Yoshimori, A., Akiyama, R.: J. Chem. Phys. **154**, 084501 (2021)
359. Mukherji, D., Marques, C.M., Stuehn, T., Kremer, K.: Nat. Comm. **8**, 1374 (2017)
360. Taylor, S.L., Evans, R., Royall, C.P.: J. Phys.: Condens. Matter **24**, 464128 (2012)

361. Preisler, Z., Kosovan, P., Kuldova, J., Uhlik, F., Limpouchova, Z., Prochazka, K., Leermakers, F.A.M.: Soft Matter **7**, 10258 (2011)
362. Ganesan, V., Khounlavong, L., Pryamitsyn, V.: Phys. Rev. E **78**, 051804 (2008)
363. Lim, W.K., Denton, A.R.: Soft Matter **12**, 2247 (2016)
364. Usatenko, Z., Halun, J., Kuterba, P.: Condens. Matter Phys. **19**, 43602 (2016)
365. Chubak, I., Likos, C.N., Egorov, S.A.: J. Phys. Chem. B **125**, 4910 (2021)
366. Chervanyov, A.I.: Phys. Rev. E **97**, 032508 (2018)
367. Michele, L.D., Zaccone, A., Eiser, E.: Proc. Natl. Acad. Sci. **109**, 10187 (2012)
368. García, N.A., Gnan, N., Zaccarelli, E.: Soft Matter **13**, 6051 (2017)
369. Briscoe, W.H.: Curr. Opin. Colloid Interface Sci. **20**, 46 (2015)
370. Ray, D., Aswal, V.K., Kohlbrecher, J.: J. Appl. Phys. **117**, 164310 (2015)
371. Gratale, M.D., Still, T., Matyas, C., Davidson, Z.S., Lobel, S., Collings, P.J., Yodh, A.G.: Phys. Rev. E **93**, 050601 (2016)
372. Abbas, S., Lodge, T.P.: Phys. Rev. Lett. **99**, 137802 (2007)
373. González García, Á., Ianiro, A., Beljon, R., Leermakers, F.A.M., Tuinier, R.: Soft Matter **16**, 1560 (2020)
374. Knoben, W., Besseling, N.A.M., Cohen Stuart, M.A.: Phys. Rev. Lett. **97**, 068301 (2006)
375. Peters, V.F.D., Vis, M., Tuinier, R.: J. Pol. Sci. **59**, 1175 (2021)
376. Peters, V.F.D., Tuinier, R., Vis, M.: J. Colloid Interface Sci. **608**, 644 (2022)
377. Gnan, N., Zaccarelli, E., Sciortino, F.: Nature Commun. **5**, 1 (2014)
378. Buzzaccaro, S., Colombo, J., Parola, A., Piazza, R.: Phys. Rev. Lett. **105**, 198301 (2010)
379. Gnan, N., Zaccarelli, E., Tartaglia, P., Sciortino, F.: Soft Matter **8**(6), 1991 (2012)
380. Bonn, D., Otwinowski, J., Sacanna, S., Guo, H., Wegdam, G., Schall, P.: Phys. Rev. Lett. **103**, 156101 (2009)
381. Gnan, N., Zaccarelli, E., Tartaglia, P., Sciortino, F.: Soft Matter **8**, 1991 (2012)
382. Gnan, N., Zaccarelli, E., Sciortino, F.: Nat. Commun. **5**, 3267 (2014)
383. Sun, Z., Kang, Y., Kang, Y.: J. Phys. Chem. B **118**, 11826 (2014)
384. Ji, S., Walz, J.Y.: Curr. Opin. Colloid Interface Sci. **20**, 39 (2015)
385. Park, N., Conrad, J.C.: Soft Matter **13**, 2781 (2017)
386. Zhang, I., Pinchaipat, R., Wilding, N.B., Faers, M.A., Bartlett, P., Evans, R., Royall, C.P.: J. Chem. Phys. **148**(18), 184902 (2018)
387. Lele, B.J., Tilton, R.D.: Langmuir **35**, 15937 (2019)
388. Stiakakis, E., Petekidis, G., Vlassopoulos, D., Likos, C.N., Iatrou, H., Hadjichristidis, N., Roovers, J.: Europhys. Lett. **72**(4), 664 (2005)
389. Camargo, M., Likos, C.N.: Phys. Rev. Lett. **104**, 078301 (2010)
390. Marzi, D., Capone, B., Marakis, J., Merola, M.C., Truzzolillo, D., Cipelletti, L., Moingeon, F., Gauthier, M., Vlassopoulos, D., Likos, C.N., Camargo, M.: Soft Matter **11**, 8296 (2015)
391. Bayliss, K., van Duijneveldt, J.S., Faers, M.A., Vermeer, A.W.P.: Soft Matter **7**, 10345 (2011)
392. González García, Á., Tuinier, R.: Phys. Rev. E **94**, 062607 (2016)
393. Rovigatti, L., Gnan, N., Parola, A., Zaccarelli, E.: Soft Matter **11**, 692 (2015)
394. Linse, P., Wennerström, H.: Soft Matter **8**, 2486 (2012)
395. Dinsmore, A.D., Warren, P.B., Poon, W.C.K., Yodh, A.G.: Europhys. Lett. **40**, 337 (1997)
396. Ouhajji, S., Nylander, T., Piculell, L., Tuinier, R., Linse, P., Philipse, A.P.: Soft Matter **12**, 3963 (2016)
397. Ashton, D.J., Ivell, S.J., Dullens, R.P.A., Jack, R.L., Wilding, N.B., Aarts, D.G.A.L.: Soft Matter **11**, 6089 (2015)
398. Feng, L., Laderman, B., Sacanna, S., Chaikin, P.: Nature Mater. **14**, 61 (2015)
399. Kwon, N.K., Park, C.S., Lee, C.H., Kim, Y.S., Zukoski, C.F., Kim, S.Y.: Macromolecules **49**, 2307 (2016)
400. Tuinier, R., Ouhajji, S., Linse, P.: Soft Matter. **39**, 115 (2016)
401. Xie, F., Woodward, C.E., Forsman, J.: Soft Matter **12**, 658 (2016)
402. González García, Á., Nagelkerke, M.M., Tuinier, R., Vis, M.: Adv. Colloid Interface Sci. **275**, 102077 (2020)
403. Zhang, P.: Macromolecules **54**, 3790 (2021)

404. Haddadi, S., Skepö, M., Forsman, J.: ACS Nanosci. Au **1**, 69 (2021)
405. Chen, J., Kline, S.R., Liu, Y.: J. Chem. Phys. **142**, 084904 (2015)
406. Fantoni, R., Giacometti, A., Santos, A.: J. Chem. Phys. **142**, 224905 (2015)
407. Annunziata, O., Asherie, N., Lomakin, A., Pande, J., Ogun, O., Benedek, G.B.: Proc. Natl. Acad. Sci. **99**, 14165 (2002)
408. Kulkarni, A.M., Zukoski, C.F.: J. Cryst. Growth **232**, 156–164 (2001)
409. Royall, C.P., Aarts, D.G.A.L., Tanaka, H.: J. Phys.: Condens. Matter **17**, S3401 (2005)
410. Ferreira, P.G., Dymitrowska, M., Belloni, L.: J. Chem. Phys. **113**, 9849 (2000)
411. Denton, A.R., Schmidt, M.: J. Chem. Phys. **122**, 244911 (2005)
412. Gögelein, C., Tuinier, R.: Eur. Phys. J. E. **27**, 171 (2008)
413. Zhou, J., van Duijneveldt, J.S., Vincent, B.: Langmuir **26**, 9397 (2010)
414. Stradner, A., Sedgwick, H., Cardinaux, F., Poon, W.C.K., Egelhaaf, S.U.: Nature **432**, 492 (2004)
415. Sedgwick, H., Egelhaaf, S.U., Poon, W.C.K.: J. Phys.: Condens. Matter **16**, 4913 (2004)
416. Groenewold, J., Kegel, W.K.: J. Phys. Chem. B **66**, 066108 (2002)
417. Sciortino, F., Mossa, S., Zaccarelli, E., Tartaglia, P.: Phys. Rev. Lett. **93**, 055701 (2004)
418. Croze, O.A., Cates, M.E.: Langmuir **21**, 5627 (2005)
419. Curtis, R.A., Lue, L.: Curr. Opin. Colloid Interface Sci. **20**, 19 (2015)
420. Pryamitsyn, V., Ganesan, V.: Macromolecules **47**, 6095 (2014)
421. Tadmor, R., Hernandez-Zapata, E., Chen, N., Pincus, P., Israelachvili, J.N.: Macromolecules **35**, 2380 (2002)
422. Grillo, I., Morfin, I., Combet, J.: J. Colloid Interface Sci. **561**, 426 (2020)
423. Moazzami-Gudarzi, M., Maroni, P., Borkovec, M., Trefalt, G.: Soft Matter **13**, 3284 (2017)
424. Scarratt, L.R., Trefalt, G., Borkovec, M.: Curr. Opin. Colloid Interface Sci. **55**, 101482 (2021)
425. Ioka, M., Toyotama, A., Yamaguchi, M., Nozawa, J., Uda, S., Okuzono, T., Yoshimura, M., Yamanaka, J.: J. Chem. Phys. **154**, 234901 (2021)
426. Toyotama, A., Okuzono, T., Yamanaka, J.: Sci. Rep. **6**, 23292 (2016)
427. Ludwig, M., Witt, M.U., von Klitzing, R.: Adv. Colloid Interface Sci. **269**, 270 (2019)
428. Ludwig, M., von Klitzing, R.: Current Opin. Colloid. Interface Sci. **47**, 137 (2020)
429. Ludwig, M., von Klitzing, R.: Phys. Chem. Chem. Phys. **23**, 1325 (2021)
430. Landman, J., Schelling, M.P.M., Tuinier, R., Vis, M.: J. Chem. Phys. **154**, 164904 (2021)
431. Schwarz-Linek, J., Dorken, G., Winkler, A., Wilson, L.G., Pham, N.T., French, C.E., Schilling, T., Poon, W.C.K.: Europhys. Lett. **89**, 68003 (2010)
432. Sun, X., Danumah, C., Liu, Y., Boluk, Y.: Chem. Engin. J. **198–199**, 476 (2012)
433. Mitchison, T.J.: Mol. Biol. Cell **30**, 173 (2019)
434. Sapir, L., Harries, D.: J. Phys. Chem. Lett. **5**, 1061 (2014)
435. Sapir, L., Harries, D.: Bunsen-Mag. **19**, 152 (2017)
436. Minton, A.P.: Curr. Opin. Struct. Biol. **10**, 34 (2000)
437. Cheung, M.S., Klimov, D., Thirumalai, D.: Proc. Natl. Acad. Sci. **102**, 4753 (2005)
438. Zeng, Y., Liu, Y., Chen, Y., Mao, W., Hu, L., Mao, Z., Song, X., Xu, H.: J. Biol. Syst. **23**, 49 (2015)
439. Hilitski, F., Ward, A.R., Cajamarca, L., Hagan, M.F., Grason, G.M., Dogic, Z.: Phys. Rev. Lett. **114**, 138102 (2015)
440. Zöttl, A., Yeomans, J.M.: J. Phys.: Condens. Matter **31**, 234001 (2019)
441. Zosel, F., Soranno, A., Buholzer, K.J., Nettels, D., Schuler, B.: Proc. Natl. Acad. Sci. **117**, 13480 (2020)
442. König, I., Soranno, A., Nettels, D., Schuler, B.: Angew. Chem. **60**, 10724 (2021)
443. Hancock, R.: J. Struct. Biol. **2004**, 281–290 (2004)
444. Hancock, R.: New Models of the Cell Nucleus: Crowding, Entropic Forces, Phase Separation, and Fractals, Hancock, R., Jeon, K.W. (eds.) International Review of Cell and Molecular Biology, vol. 307, pp. 15–26. Academic Press (2014)
445. Hancock, R.: Front. Phys. **2**, 53 (2014)
446. Denesyuk, N., Thirumalai, D.: J. Am. Chem. Soc. **133**, 11858 (2011)
447. Odijk, T.: Biophys. Chem. **73**, 23 (1998)

448. Marenduzzo, D., Finan, K., Cook, P.R.: J. Cell. Biol. **175**, 681–686 (2006)
449. Kang, H., Toan, N.M., Hyeon, H., Thirumalai, D.: J. Am. Chem. Soc. **137**, 10970 (2015)
450. Snir, Y., Kamien, R.: Science **307**, 1067 (2005)
451. Kudlay, A., Cheung, M., Thirumalai, D.: Phys. Rev. Lett. **102**, 118101 (2009)
452. Glotzer, S.C., Solomon, M.J.: Nat. Mat. **6**, 557–562 (2007)
453. Sacanna, S., Pine, D.J.: Current Opin. Colloid. Interface Sci. **16**, 96 (2011)
454. Gong, Z., Hueckel, T., Yi, G.R., Sacanna, S.: Nature **550**, 234 (2017)
455. Li, W., Palis, H., Mérindol, R., Majimel, J., Ravaine, S., Duguet, E.: Chem. Soc. Rev. **49**, 1955 (2020)
456. Hueckel, T., Hocky, G.M., Sacanna, S.: Nat. Rev. Mater. **6**, 1053 (2021)
457. Manoharan, V.N., Elsesser, M.T., Pine, D.J.: Science **301**, 483–487 (2003)
458. Kraft, D.J., Vlug, W.S., van Kats, C.M., van Blaaderen, A., Imhof, A., Kegel, W.K.: J. Am. Chem. Soc. **131**, 1182 (2009)
459. Meng, G., Arkus, N., Brenner, M.P., Manoharan, V.N.: Science **327**, 560–563 (2010)
460. Wang, Y., Wang, Y., Breed, D., Manoharan, V.N., Feng, L., Hollingsworth, A.D., Weck, M., Pine, D.J.: Nat. Mat. **491**, 51–55 (2012)
461. Meijer, J.M., Rossi, L.: Soft Matter **17**, 2354 (2021)
462. Lekkerkerker, H.N.W., Stroobants, A.: Il Nuovo Cimento D **16**, 949 (1994)
463. Buitenhuis, J., Donselaar, L.N., Buining, P.A., Stroobants, A., Lekkerkerker, H.N.W.: J. Colloid Interface Sci. **175**, 46 (1995)
464. Bolhuis, P.G., Stroobants, A., Frenkel, D., Lekkerkerker, H.N.W.: J. Chem. Phys. **107**, 1551 (1997)
465. Dogic, Z., Purdy, K.R., Grelet, E., Adams, M., Fraden, S.: Phys. Rev. E. **69**, 051702 (2004)
466. Dogic, Z., Fraden, S.: Curr. Opin. Colloid Interface Sci. **11**, 47 (2006)
467. Dogic, Z., Fraden, S.: Soft Matter: Complex Colloidal Suspensions, Gompper, G., Schick, M. (eds.), vol. 2, Chap. 1, pp. 1–86. Wiley, Ltd (2006)
468. Savenko, S.V., Dijkstra, M.: J. Chem. Phys. **124** (2006)
469. Peters, V.F.D., González García, Á., Wensink, H.H., Vis, M., Tuinier, R.: Langmuir **37**, 11582 (2021)
470. Van der Kooij, F.M., Vogel, M., Lekkerkerker, H.N.W.: Phys. Rev. E **62**, 5397 (2000)
471. Zhang, S.D., Reynolds, P.A., van Duijneveldt, J.S.: J. Chem. Phys. **117**, 9947 (2002)
472. Zhang, S.D., Reynolds, P.A., van Duijneveldt, J.: Mol. Phys. **100**, 3041 (2002)
473. Wensink, H.H., Lekkerkerker, H.N.W.: Europhys. Lett. **66**, 125 (2004)
474. Luan, S., Li, W., Liu, S., Sun, D.: Langmuir **25**, 6394 (2009)
475. de las Heras, D., Schmidt, M.: J. Phys.: Condens. Matter **27**, 194115 (2015)
476. González García, Á., Tuinier, R., Maring, J.V., Opdam, J., Wensink, H.H., Lekkerkerker, H.N.W.: Mol. Phys. **116**, 2757 (2018)
477. Zhao, C., Wang, G., Takarada, T., Liang, X., Komiyama, M., Maeda, M.: Colloids Surf. A **568**, 216 (2019)
478. Wengenmayr, M., Dockhorn, R., Sommer, J.U.: Macromolecules **52**, 2616 (2019)
479. Rice, R., Roth, R., Royall, C.P.: Soft Matter **8**, 1163 (2012)
480. Belli, S., Dijkstra, M., van Roij, R.: J. Phys.: Condens. Matter **24**, 284128 (2012)
481. Watanabe, K., Tajima, Y., Shimura, T., Ishii, H., Nagao, D.: J. Colloid Interface Sci. **534**, 81 (2019)
482. Florea, D., Wyss, H.M.: J. Colloid Interface Sci. **416**, 30 (2014)
483. Rossi, L., Sacanna, S., Irvine, W.T.M., Chaikin, P.M., Pine, D.J., Philipse, A.P.: Soft Matter **7**, 4139 (2011)
484. Dekker, F., González García, Á., Philipse, A.P., Tuinier, R.: Eur. Phys. J. E **43**, 38 (2020)
485. Saez Cabezas, C.A., Sherman, Z.M., Howard, M.P., Dominguez, M.N., Cho, S.H., Ong, G.K., Green, A.M., Truskett, T.M., Milliron, D.J.: Nano Lett. **20**, 4007 (2020)
486. van der Schoot, P.: Molecular Theory of Nematic (and other) Liquid Crystals, Springer, Berlin Heidelberg New York (2022)
487. van Anders, G., Klotsa, D., Ahmed, N.K., Engel, M., Glotzer, S.C.: Proc. Natl. Acad. Sci. **111**, E4812 (2014)

488. Petukhov, A.V., Tuinier, R., Vroege, G.J.: Current Opin. Colloid Interface Sci. **30**, 54 (2017)
489. van der Kooij, F.M., Lekkerkerker, H.N.W.: Phys. Rev. Lett. **84**, 781 (2000)
490. Wensink, H.H., Vroege, G.J., Lekkerkerker, H.N.W.: J. Chem. Phys. **115**, 7319 (2001)
491. Woolston, P., van Duijneveldt, J.S.: Langmuir **31**, 9290 (2015)
492. Opdam, J., Peters, V.F.D., Wensink, H.H., Tuinier, R.: J. Phys. Chem. Lett. **14**, 199 (2023)
493. Adams, M., Dogic, Z., Keller, S.L., Fraden, S.: Nature **393**, 349 (1998)
494. Guu, D., Dhont, J.K.G., Vliegenthart, G.A., Lettinga, M.P.: J. Phys.: Condens. Matter **24**, 464101 (2012)
495. Holovko, M.F., Hvozd, M.V.: Condens. Matter Phys. **20**, 1 (2017)
496. Opdam, J., Guu, D., Schelling, M.P.M., Aarts, D.G.A.L., Tuinier, R., Lettinga, M.P.: J. Chem. Phys. **154**, 204906 (2021)
497. Opdam, J., Gandhi, P., Kuhnhold, A., Schilling, T., Tuinier, R.: Phys. Chem. Chem. Phys. **24**, 11820 (2022)
498. Harnau, L., Dietrich, S.: Phys. Rev. E **71**, 011504 (2005)
499. Oversteegen, S.M., Vonk, C., Wijnhoven, J.E.G.J., Lekkerkerker, H.N.W.: Phys. Rev. E **71**, 041406 (2005)
500. Kleshchanok, D., Petukhov, A.V., Holmqvist, P., Byelov, D.V., Lekkerkerker, H.N.W.: Langmuir **26**, 13614 (2010)
501. Kleshchanok, D., Meijer, J.M., Petukhov, A.V., Portale, G., Lekkerkerker, H.N.W.: Soft Matter **7**, 2832 (2011)
502. de las Heras, D., Doshi, N., Cosgrove, T., Phipps, J., Gittins, D.I., Van Duijneveldt, J.S., Schmidt, M.: Sci. Rep. **2**, 789 (2012)
503. Kleshchanok, D., Meijer, J.M., Petukhov, A.V., Portale, G., Lekkerkerker, H.N.W.: Soft Matter **8**, 191 (2012)
504. Chen, M., Li, H., Chen, Y., Mejia, A.F., Wang, X., Cheng, Z.: Soft Matter **11**, 5775 (2015)
505. Doshi, N., Cinacchi, G., Van Duijneveldt, J., Cosgrove, T., Prescott, S., Grillo, I., Phipps, J., Gittins, D.: J. Phys.: Condens. Matter **23**, 194109 (2011)
506. Aliabadi, R., Moradi, M., Varga, S.: J. Chem. Phys. **144**, 074902 (2016)
507. Chen, M., He, M., Lin, P., Chen, Y., Cheng, Z.: Soft Matter **13**, 4457 (2017)
508. Eckert, T., Schmidt, M., de las Heras, D.: Phys. Rev. Res. **4**, 013189 (2022)
509. Park, K., Koerner, H., Vaia, R.A.: Nano Lett. **10**, 1433 (2010)
510. Wensink, H.H., Vroege, G.J., Lekkerkerker, H.N.W.: J. Phys. Chem B **105**, 10610 (2001)
511. Sun, D., Sue, H.J., Cheng, Z., Martínez-Ratón, Y., Velasco, E.: Phys. Rev. E **80**, 041704 (2009)
512. Nakato, T., Yamashita, Y., Mouri, E., Kuroda, K.: Soft Matter **10**, 3161 (2014)
513. Lekkerkerker, H.N.W., Tuinier, R., Wensink, H.H.: Mol. Phys. **113**, 2666 (2015)
514. Xie, Y., Li, Y., Wei, G., Liu, Q., Mundoor, H., Chen, Z., Smalyukh, I.I.: Nanoscale **10**, 4218 (2018)
515. van der Schoot, P.: J. Chem. Phys. **112**, 9132 (2000)
516. van der Schoot, P.: J. Chem. Phys. **117**, 3537 (2002)
517. Landman, J., Paineau, E., Davidson, P., Bihannic, I., Michot, L.J., Philippe, A.M., Petukhov, A.V., Lekkerkerker, H.N.W.: J. Phys. Chem. B **118**, 4913 (2014)
518. Petekidis, G., Galloway, L.A., Egelhaaf, S.U., Cates, M.E., Poon, W.C.K.: Langmuir **18**, 4248 (2002)
519. Krüger, S., Mögel, H.J., Wahab, M., Schiller, P.: Langmuir **27**, 646 (2011)
520. Damasceno, P.F., Engel, M., Glotzer, S.C.: ACS Nano **6**, 609 (2012)
521. Nakato, T., Yamashita, Y., Kuroda, K.: Thin Solid Films **495**, 24 (2006)

Depletion Interaction

2

In this chapter, we consider the depletion interaction between two flat plates and between two spherical colloidal particles for different depletants (polymers, small colloidal spheres, rods and plates). First of all, we focus on the depletion interaction due to a somewhat hypothetical model depletant—the penetrable hard sphere (PHS)—to mimic a (ideal) polymer molecule. This model, implicitly introduced by Asakura and Oosawa [1] and considered in detail by Vrij [2], is characterised by the fact that the spheres freely overlap each other but act as hard spheres with diameter σ when interacting with a wall or a colloidal particle. The thermodynamic properties of a system of hard spheres with added PHSs have been considered by Widom and Rowlinson [3], and provided much of the inspiration for the theory of phase behaviour developed in Chap. 3.

The depletion potential is a *potential of mean force* and, following Onsager [4,5], the system is considered at a given chemical potential of the solvent (and other solution components). As such, the relevant pressure exerted by the depletants is the osmotic pressure. Two methods are used to derive interactions between particles mediated by different types of depletants: the adsorption method and the force method.

2.1 Depletion Interaction Due to Penetrable Hard Spheres

2.1.1 Depletion Interaction Between Two Flat Plates

Interaction Potential Between Two Flat Plates Using the Force Method

The force per unit area $K(h)$ between two parallel plates separated by a distance h, is the difference between the osmotic pressure P_i inside the plates and the outside pressure P_o

© The Author(s) 2024
H. N. W. Lekkerkerker, R. Tuinier, M. Vis, *Colloids and the Depletion Interaction*,
Lecture Notes in Physics 1026, https://doi.org/10.1007/978-3-031-52131-7_2

$$K = P_i - P_o. \tag{2.1}$$

Since the PHSs behave thermodynamically ideally, the osmotic pressure outside the plates is given by the van 't Hoff law

$$P_o = n_b kT,$$

where n_b is the bulk number density of the PHS. When the plate separation h (see Fig. 2.1) is equal to or larger than the diameter σ of the PHSs the osmotic pressure inside the plates is the same as outside:

$$P_i = P_o = n_b kT.$$

On the other hand, when the plate separation is less than the diameter of the PHSs, no particles can enter the gap and

$$P_i = 0.$$

This means that

$$K(h) = \begin{cases} -n_b kT & h < \sigma, \\ 0 & h \geq \sigma. \end{cases} \tag{2.2}$$

This is the classical result derived by Asakura and Oosawa [1].

Since $K = -dW/dh$, integration from ∞ to h yields the interaction potential $W(h)$ per unit area between the plates

$$W(h) = \begin{cases} -n_b kT(\sigma - h) & h < \sigma, \\ 0 & h \geq \sigma. \end{cases} \tag{2.3}$$

Fig. 2.1 Schematic picture of two parallel flat plates in the presence of penetrable hard spheres (dashed circles)

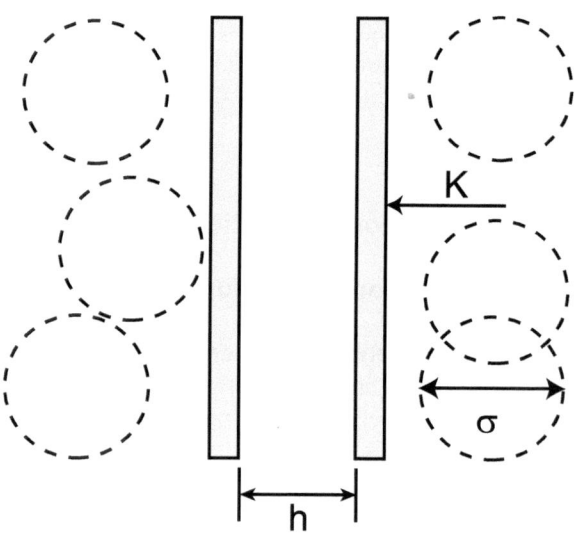

Interaction Potential Between Two Flat Plates Using the Extended Gibbs Adsorption Equation

An alternative and insightful way to obtain the interaction potential is from the extended Gibbs adsorption equation [6–8]. The natural thermodynamic potential to describe the system depicted in Fig. 2.2 is the grand potential $\Omega(T, V, \mu, h)$

$$\Omega = F - \mu N, \tag{2.4}$$

where $F = F(T, V, N, h)$ is the Helmholtz (free) energy, N the number of PHSs in the system and μ their chemical potential. At constant temperature and volume, we have $dF = \mu dN - K A dh$, so $d\Omega$ is given by

$$d\Omega = -K A dh - N d\mu, \tag{2.5}$$

where K is again the force per unit area between the plates and A is the area of the plates.

From cross-differentiating Eq. (2.5), we obtain

$$\left(\frac{\partial K}{\partial \mu}\right)_h = \frac{1}{A}\left(\frac{\partial N}{\partial h}\right)_\mu. \tag{2.6}$$

Combining this with

$$K = -\left(\frac{\partial W}{\partial h}\right)_\mu, \tag{2.7}$$

we obtain

$$-\left(\frac{\partial}{\partial h}\left(\frac{\partial W}{\partial \mu}\right)_h\right)_\mu = \frac{1}{A}\left(\frac{\partial N}{\partial h}\right)_\mu. \tag{2.8}$$

Fig. 2.2 Illustration of the (depletion) force between two plates in the system of interest (I) at a given chemical potential of the depletant in the reservoir (II). The system is connected to the reservoir through a hypothetical membrane (M) that allows permeation of depletant

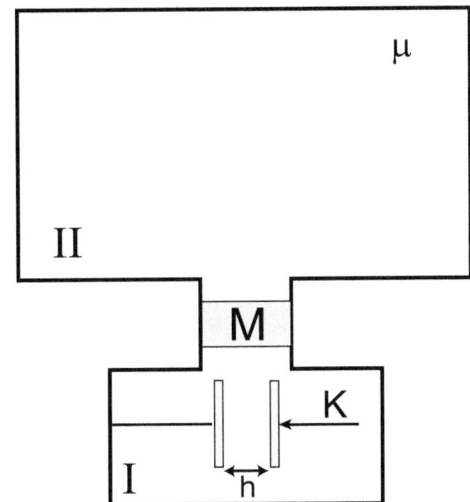

Since the depletion potential W vanishes at infinite separation for all values of the chemical potential μ of the depletion agent, integration over h gives

$$-\left(\frac{\partial W}{\partial \mu}\right)_h = \frac{N(h) - N(\infty)}{A}, \tag{2.9}$$

where $N(h)$ is the number of PHSs in the system when the plates are at separation h and $N(\infty)$ is that at infinite separation. The right-hand side of Eq. (2.9) can be conveniently written in terms of the surface adsorption

$$-\left(\frac{\partial W}{\partial \mu}\right)_h = \Gamma(h) - \Gamma(\infty), \tag{2.10}$$

where [9]

$$\Gamma(h) = \int_0^h [n(x) - n_b]dx, \tag{2.11}$$

and

$$\Gamma(\infty) = 2\Gamma_{\text{single wall}}$$
$$= 2\int_0^\infty [n(x) - n_b]dx. \tag{2.12}$$

Note that, in Eq. (2.12), $n(x)$ refers to the PHS concentration profile near a single wall, whereas, in Eq. (2.11), $n(x)$ is the profile between two walls. Equation (2.10) is the extension of the Gibbs adsorption equation for a single surface to the case of two surfaces at finite separation [6–8]. Integration of Eq. (2.10) gives

$$W(h) = -\int_{-\infty}^\mu [\Gamma(h) - \Gamma(\infty)]d\mu. \tag{2.13}$$

The zone which is excluded for depletants is termed the depletion layer thickness or depletion thickness δ. When the surface of a spherical depletant particle touches a hard object it cannot get closer. Therefore, the centre of the spherical depletant is at least one radius away from this hard object. Hence, the depletion thickness then equals the particle radius. The depletion thickness of PHSs is therefore $\sigma/2$, and $A[\Gamma(h) - \Gamma(\infty)]$ equals the overlap volume $A(\sigma - h)$ times n_b (see Fig. 2.3), so that

$$\Gamma(h) - \Gamma(\infty) = \begin{cases} n_b(\sigma - h) & h < \sigma, \\ 0 & h \geq \sigma. \end{cases} \tag{2.14}$$

The chemical potential of the PHSs is

$$\mu = kT \ln n_b. \tag{2.15}$$

Fig. 2.3 The overlap volume
(hatched area) of depletion
layers due to PHSs between
two parallel flat plates equals
$A(\sigma - h)$

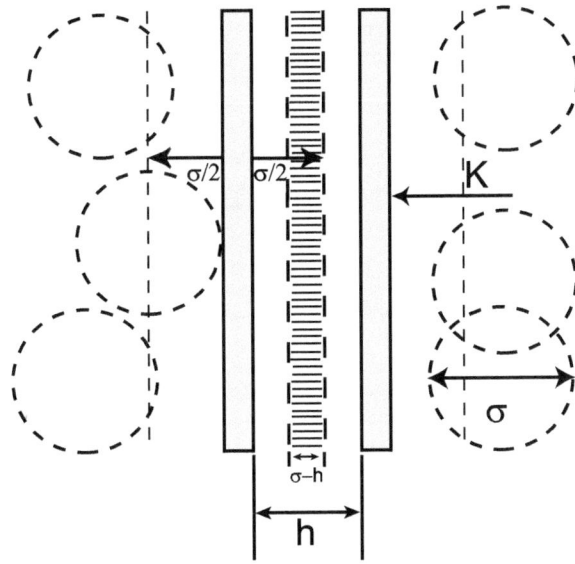

Inserting Eqs. (2.14) and (2.15) into Eq. (2.13) yields (again) the interaction potential given by Eq. (2.3). The conceptual advantage of the calculation with the extended Gibbs adsorption equation is that it provides a direct link between the depletion of particles with the depletion potential, which is frequently highly illuminating. The method also offers advantages to obtain physically motivated approximate expressions for the depletion interaction where an exact calculation is not possible.

2.1.2   Depletion Interaction Between Two Spheres

Interaction Potential Between Two Spheres Using the Force Method
When the depletion zones with thickness $\sigma/2$ around spherical colloidal particles with radius R start to overlap, i.e., when the distance r between the centres of the colloidal particles is smaller than $2R + \sigma = 2R_\mathrm{d}$, a net force arises between the colloidal particles. For convenience, as was used by Odijk [10], we define an effective depletion radius R_d as

$$R_\mathrm{d} = R + \sigma/2. \tag{2.16}$$

The (attractive) force originates from an uncompensated (osmotic) pressure due to the depletion of PHSs from the gap between the colloidal particles. This is depicted in Fig. 2.4, from which we immediately deduce that the uncompensated pressure acts on the surface between $\theta = 0$ and $\theta_0 = \arccos(r/2R_\mathrm{d})$.

For obvious symmetry reasons, only the component along the line connecting the centres of the colloidal spheres contributes to the total force. For the angle θ, this component is $P \cos \theta$, where the pressure is $P = n_\mathrm{b} kT$. The surface on which this

Fig. 2.4 Two hard spheres in the presence of penetrable hard spheres (PHSs) as depletants. The PHSs impose an unbalanced pressure P between the hard spheres resulting in an attractive force between them. The overlap volume of depletion layers (hatched) has the shape of a lens with width $\sigma - h$ and height $2R_d \sin\theta_0$

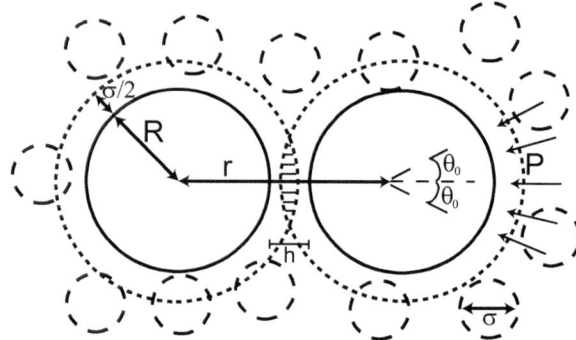

force acts between θ and $\theta + d\theta$ equals $2\pi R_d^2 \sin\theta \, d\theta$. The total force between the colloidal spheres is obtained by integration over θ from 0 to θ_0

$$\frac{K_s(r)}{n_b kT} = -2\pi (R + \sigma/2)^2 \int_0^{\theta_0} \sin\theta \cos\theta \, d\theta$$

$$= \begin{cases} -\pi R_d^2 [1 - (r/2R_d)^2] & 2R \leq r < 2R_d \\ 0 & r \geq 2R_d. \end{cases} \tag{2.17}$$

This result was also obtained by Asakura and Oosawa [1]. The minus sign in the right-hand side of Eq. (2.17) implies that the force is attractive.

Exercise 2.1. Show that Eq. (2.17) can also be written as the pressure multiplied by the area of the overlap of the depletion zones (see Fig. 2.4).

The depletion potential is now obtained by integration of the depletion force (Eq. (2.17))

$$W_s(r) = \int_r^{2R_d} K_s(r') dr' \tag{2.18}$$

$$= \begin{cases} -n_b kT V_{ov}(r) & 2R \leq r < 2R_d \\ 0 & r \geq 2R_d \end{cases} \tag{2.19}$$

with

$$V_{ov}(r) = \frac{4\pi}{3} R_d^3 \left[1 - \frac{3}{4} \frac{r}{R_d} + \frac{1}{16} \left(\frac{r}{R_d} \right)^3 \right] \tag{2.20a}$$

$$= \frac{2\pi}{3} (R_d - r/2)^2 (2R_d + r/2) \tag{2.20b}$$

$$V_{ov}(h) = \frac{\pi}{6} (\sigma - h)^2 (3R + \sigma + h/2) \tag{2.20c}$$

This result of Eq. (2.20a), in which r is the variable, was first obtained by Vrij [2]. Equation (2.20c), where the variable is $h = r - 2R$, was already given without explicit derivation in Eq. (1.19). Both Eqs. (2.20a) and (2.20c) are frequently used in the literature. Equation (2.20b) provides an intermediate step between the two forms. Note that $W_s(r)$ in Eq. (2.19) is equal to pressure times the overlap volume V_{ov}. The reason for this simple form will become clearer after consideration of the interaction between two spheres using the extended Gibbs equation. In the limit that $\sigma/2 \ll R$, the force Eq. (2.17) and potential Eq. (2.19) between the spheres take very simple forms:

$$\frac{K_s(h)}{n_b kT} = -\pi R(\sigma - h) \tag{2.21}$$

and

$$\frac{W_s(h)}{n_b kT} = -\pi R \frac{1}{2}(\sigma - h)^2, \tag{2.22}$$

for separations $h = r - 2R$ smaller than σ.

Exercise 2.2. Derive Eqs. (2.21) and (2.22) from Eqs. (2.17) and (2.19).

Interaction Potential Between Two Spheres From the Extended Gibbs Adsorption Equation

Applying exactly the same line of reasoning as for the derivation of the extended Gibbs adsorption equation for two flat plates (see Eq. (2.9)), we now obtain

$$-\left(\frac{\partial W_s}{\partial \mu}\right) = N(r) - N(\infty), \tag{2.23}$$

where $N(r)$ is the number of PHSs in the system when the colloidal spheres are at centre-to-centre separation r and $N(\infty)$ that at infinite separation. Clearly, the difference between $N(r)$ and $N(\infty)$ is caused by the overlap of the depletion zones

$$N(r) - N(\infty) = \begin{cases} n_b V_{ov}(r) & 2R \leq r < 2R_d \\ 0 & r \geq 2R_d \end{cases} \tag{2.24}$$

with V_{ov} defined in Eq. (2.20). Integration of Eq. (2.23) with Eqs. (2.24) and (2.15) immediately leads to the interaction potential of Eq. (2.19). This route to the interaction potential makes clear why the overlap volume of the depletion zones appears.

2.1.3   Depletion Interaction Between a Sphere and a Plate

The force method and the extended Gibbs adsorption equation can also be applied to obtain the depletion interaction between a sphere and a flat plate. For the Gibbs adsorption route, we (again) use

$$-\left(\frac{\partial W_{sp}}{\partial \mu}\right) = N(h) - N(\infty), \qquad (2.25)$$

where $N(h)$ is now the number of PHSs in the system when the colloidal sphere is at a separation h from the plate, and $N(\infty)$ is that at infinite separation. Again, the difference between $N(h)$ and $N(\infty)$ is caused by the overlap of the depletion zones, now of the sphere and of the plate (see Fig. 2.5)

$$\frac{N(h) - N(\infty)}{n_b} = V_{ov}(h) \qquad (2.26)$$

$$= \begin{cases} \frac{1}{3}\pi(\sigma - h)^2(3R + \frac{\sigma}{2} + h) & 0 \le h < \sigma \\ 0 & h \ge \sigma. \end{cases} \qquad (2.27)$$

Integration of Eq. (2.25) now leads to

$$\frac{W_{sp}(h)}{n_b kT} = \begin{cases} -\frac{1}{3}\pi(\sigma - h)^2 \left(3R + \frac{\sigma}{2} + h\right) & 0 \le h < \sigma \\ 0 & h \ge \sigma. \end{cases} \qquad (2.28)$$

For $R \gg \sigma$, Eq. (2.28) simplifies to

$$\frac{W_{sp}(h)}{n_b kT} = -\pi R(\sigma - h)^2 \qquad 0 \le h < \sigma, \qquad (2.29)$$

which, obviously, is twice Eq. (2.22).

Fig. 2.5 Illustration of the overlap volume (hatched) of depletion layers between a hard wall and a hard sphere

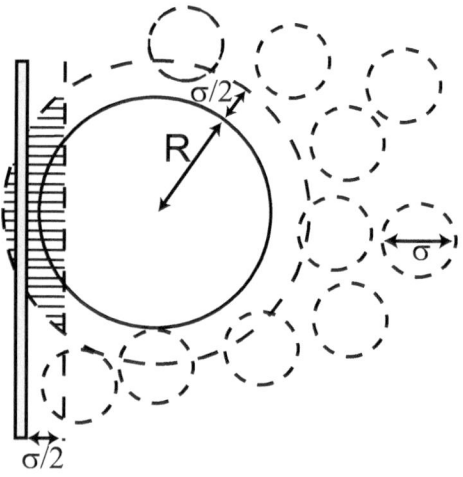

2.1.4 Derjaguin Approximation

Some of the above results also follow directly from the so-called Derjaguin approximation. Derjaguin [11] showed that, for any type of interaction, there exists a simple (approximate) relation for the force between curved objects and the interaction potential between two flat plates. In the Derjaguin approximation, the spherical surface is replaced by a collection of flat rings. Consider two spheres with radius R at a centre-to-centre distance $r = 2R + h$. The distance H between the sphere surfaces at a distance z from the line joining the centres is $H = h + 2\Delta$, where $(R - \Delta)^2 + z^2 = R^2$ (Fig. 2.6a). When the range of interaction is short it is sufficient to consider only small values of h/R or z/R (Fig. 2.6b). For $z \ll R$, we can write to a good approximation $\Delta = z^2/2R$. Hence, $H = h + z^2/R$ and thus, $dH = (2z/R)dz$. The interaction between two spheres can now be written as the sum (integral) of the interactions of flat rings with radius z and surface $2\pi z dz$ at a distance H from each other (Fig. 2.6b). Assuming that the interaction is sufficiently short-ranged, the contribution of rings with high values of H may be neglected, and thus the integration may be extended to $z = \infty$. For the interaction energy between two spheres, we obtain

$$W_s(h) = \int_0^\infty W(H)2\pi z dz$$
$$= \pi R \int_h^\infty W(H)dH, \tag{2.30}$$

or

$$W_s(h) = \pi R \int_h^\infty W(h')dh', \tag{2.31}$$

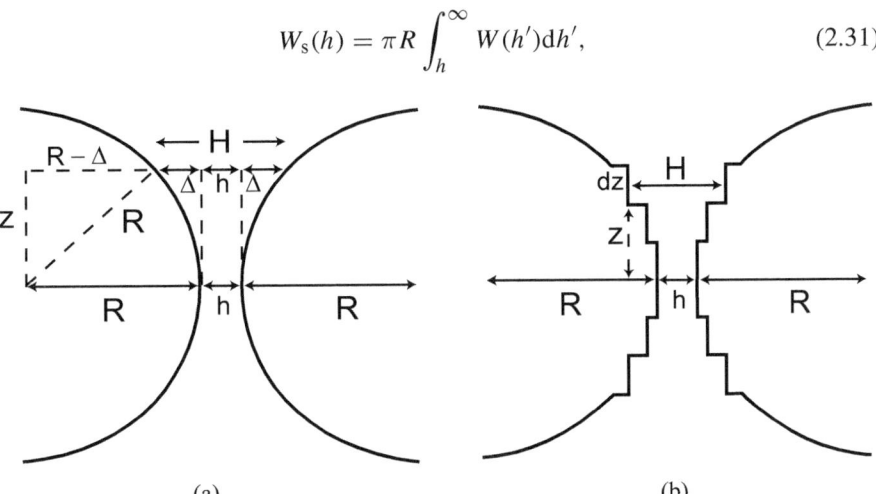

$$(a) \qquad\qquad\qquad (b)$$

Fig. 2.6 a Relevant length scales for describing the interaction force between two big spheres and **b** the Derjaguin approximation of **a**

and hence

$$K_s(h) = -\frac{\partial W_s(h)}{\partial h} = \pi R W(h). \tag{2.32}$$

Here, $W(h)$ is the interaction potential (energy per unit area) between two flat plates at distance h. Clearly this approximate relationship between the force for spheres and the interaction potential for plates is more accurate the larger the radius of the spheres is compared to the range of the interaction. In this chapter, we shall frequently use this Derjaguin approximation. It is a useful tool that is very accurate under the right conditions (see above), but one has to be careful and be aware of its limitations.

With respect to the depletion interaction, the Derjaguin approximation becomes accurate when considering a depletion agent that is small compared to the radius of the colloidal spheres. As an example, we apply the Derjaguin approximation to the depletion interaction potential between plates $W(h)$ given by Eq. (2.3). It follows that insertion of Eq. (2.3) into Eq. (2.31) correctly yields Eq. (2.22) as a result of the depletion interaction between the two spheres. Further using the Derjaguin approximation Eq. (2.32) with Eq. (2.3) for $W(h)$ matches with the full result of Eq. (2.21) for $\sigma/2 \ll R$.

One can also apply the Derjaguin approximation to the interaction between a sphere and a flat plate. One then obtains

$$K_{sp}(h) = 2\pi R W(h). \tag{2.33}$$

This is a useful relationship as it allows one to obtain the interaction potential between two parallel plates from the measured force between a sphere and a wall (see Sect. 2.6).

Exercise 2.3. Derive equation Eq. (2.33).

From Eq. (2.33) it follows that

$$W_{sp}(h) = 2\pi R \int_h^\infty W(h')\mathrm{d}h'. \tag{2.34}$$

For the case of the PHS as depletion agent, this leads to

$$W_{sp}(h) = \begin{cases} -n_b kT \pi R(\sigma - h)^2 & 0 \le h < \sigma, \\ 0 & h \ge \sigma, \end{cases} \tag{2.35}$$

for the interaction between a sphere and a plate. Its derivation is now much simpler than that of Eq. (2.29).

2.2 Depletion Interaction Due to Ideal Polymers

2.2.1 Depletion Interaction Between Two Flat Plates

Interaction Potential Between Two Flat Plates Using the Force Method
The simplest model to describe polymers is the ideal-chain model. For books on polymer physics where all the relevant background material can be found, see Refs. [12–21]. In this model, the polymer consists of M subunits, each with a fixed bond length b, and their orientation is completely independent of the orientation and positions of the previous monomers, even to the extent that two different monomers can occupy the same position in space: there is no excluded volume. This model plays the same role in polymer physics as an ideal gas in molecular physics. It allows the polymer chain to be described as a (Gaussian) random walk of M steps, as depicted in Fig. 2.7.

The average value $\langle \mathbf{R} \rangle$ of the end-to-end vector $\mathbf{R}$ joining one end of the polymer to the other is zero, as 'negative' steps have the same probability as 'positive' ones. Mathematically, the probability of the end-to-end vector being $\mathbf{R}$ is the same as it being $-\mathbf{R}$ so that, for symmetry reasons, the two contributions cancel on average. A straightforward calculation (see any of the references [12–15, 18, 20, 21]) shows that $\langle \mathbf{R}^2 \rangle$, the average of the square of $\mathbf{R}$, is given by

$$\langle \mathbf{R}^2 \rangle = Mb^2. \tag{2.36}$$

This quantity is a measure of the size of the polymer chain. We see that the size of the ideal polymer chain, of the order $b\sqrt{M}$, is much smaller than the total unfolded contour length bM of the polymer.

Another commonly used and convenient quantity to describe the size of a polymer is the radius of gyration R_g, the root-mean-square of the average monomer position from the centre of mass [21], which is

$$R_g^2 = \frac{1}{6}Mb^2. \tag{2.37}$$

The result $R_g = b\sqrt{M/6}$ holds for sufficiently long Gaussian chains in the bulk solution.

Fig. 2.7 Sketch of a random walk chain consisting of monomers with length b. For any given walk, the end-to-end vector $\mathbf{R} = \sum r_i$

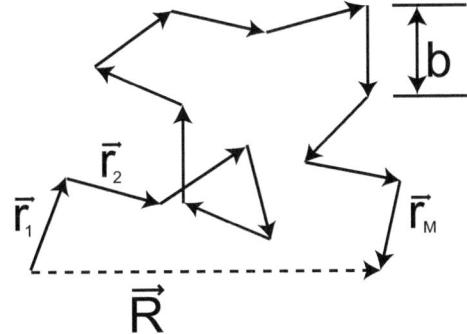

Fig. 2.8 An ideal chain
confined between two
parallel flat plates

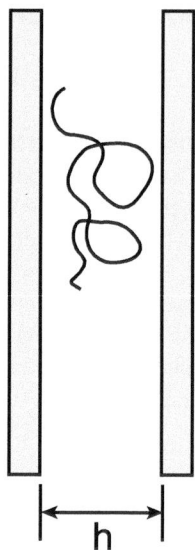

We now consider an ideal Gaussian chain confined between two (large) flat plates
with area A at a plate separation h (Fig. 2.8). For the computation of segment density
profiles in polymer solutions near interfaces, one can use the fact that there is a close
analogy between the diffusion of a Brownian particle and the flight of a random walk
[22, 23]. A diffusion-like equation can be derived to evaluate the partition function
of polymer chains. Given the boundary condition, this 'diffusion' equation can be
solved. The partition function $z(h)$ of one confined chain is given by [1, 15, 24–26]:

$$z(h) = V\chi(h), \tag{2.38}$$

where $V = Ah$ is the volume of the system and $\chi(h)$ the partition coefficient

$$\chi(h) = \frac{8}{\pi^2} \sum_{n=1,3,5,\ldots} \frac{1}{n^2} \exp\left(-\frac{n^2\pi^2 R_g^2}{h^2}\right). \tag{2.39}$$

Note that, since

$$\sum_{n=1,3,5,\ldots} \frac{1}{n^2} = \frac{\pi^2}{8}, \tag{2.40}$$

clearly

$$0 \le \chi(h) \le 1. \tag{2.41}$$

Since the ideal chains do not interact, the partition function for N confined chains
can be written as

$$Z(h) = \frac{z(h)^N}{N!}. \tag{2.43}$$

Exercise 2.4. Show that Eq. (2.39) can be approximated as [27]:

$$\chi(h) = \begin{cases} \frac{8}{\pi^2} e^{-\pi^2 R_g^2/h^2} & h \lesssim \frac{8R_g}{\sqrt{\pi}}, \\ 1 - \frac{4R_g}{h\sqrt{\pi}} & h \gtrsim \frac{8R_g}{\sqrt{\pi}}. \end{cases} \qquad (2.42)$$

Hint: for the large h limit, first write $\chi(x) \simeq \chi(x=0) + x \left(\frac{d\chi}{dx}\right)_{x=0}$ with $x = R_g/h$. Make use of Eq. (2.40) and introduce a new dummy variable using $n = 2j + 1$ with $j = 0, 1, 2, 3, \ldots, \infty$ and replace the summation by integration in the derivative.

The Helmholtz energy is given by

$$F(h) = -kT \ln Z(h); \qquad (2.44)$$

hence, the result

$$F(h) = NkT \left[\ln\left(\frac{N}{V}\right) - 1 - \ln \chi(h) \right] \qquad (2.45)$$

is obtained after insertion of Eq. (2.43). This free energy can be written as

$$F(h) = F_{\text{unconfined}} - T \Delta S(h), \qquad (2.46)$$

where $\Delta S(h)$ is the entropy of confinement:

$$\Delta S(h) = N \Delta s(h) = Nk \ln \chi(h). \qquad (2.47)$$

From Eq. (2.41) it follows that the confinement entropy is negative, as expected, because confinement leads to a decrease of the entropy. From the free energy given in Eq. (2.45), we obtain

$$P_i = -\left(\frac{\partial F}{\partial V}\right) = n_i kT \left[1 + \frac{h}{\chi} \frac{\partial \chi}{\partial h} \right] \qquad (2.48)$$

for the pressure of the chains inside the plates, where $n_i = (N/V)_i$ is the number density of the ideal chains between the plates. The first term $n_i kT$ corresponds to the van 't Hoff law. Likewise, the pressure of the ideal chains outside the plates is given by

$$P_o = n_b kT, \qquad (2.49)$$

Exercise 2.5. Derive Eq. (2.50) by using the equality of the chemical potential of the ideal chains inside and outside of the plates $\mu_i = \mu_o$.

---

where n_b is the bulk number density of the polymer chains. Using Einstein's fluctuation theory [28, 29] it immediately follows that

$$n_i = n_b e^{\Delta s/k}$$
$$= n_b \chi. \tag{2.50}$$

Combining Eqs. (2.1) and (2.48)–(2.50), we find

$$K(h) = P_i - P_o$$
$$= -n_b k T \left[1 - \chi - h \frac{\partial \chi}{\partial h} \right] \tag{2.51}$$

for the force per unit area $K(h)$ between the plates. The attraction due to the osmotic effect is given by $-n_b k T (1 - \chi)$, whereas $n_b k T h \partial \chi / \partial h$ is the repulsive contribution due to confinement of polymer chains by the plates. Also this result was first derived by Asakura and Oosawa [1]. Integration of Eq. (2.51) yields the interaction potential per unit area $W(h)$ between the plates [30]

$$W(h) = -n_b k T \left[\frac{4 R_g}{\sqrt{\pi}} - h + h \chi(h) \right]. \tag{2.52}$$

Here, we have used

$$\lim_{h \to \infty} [h - h \chi(h)] = \frac{4 R_g}{\sqrt{\pi}} \tag{2.53}$$

according to Eq. (2.42). Comparing Eq. (2.52) with Eq. (2.3)—the interaction potential between flat plates due to PHSs—we find that the contact potentials ($h = 0$) match if we take $\sigma = 4 R_g / \sqrt{\pi} = 2.26 R_g$. The two potentials are plotted in Fig. 2.9. For small h, where $\chi(h)$ is negligible, the two potentials coincide: this is in the region $0 < h < 3 R_g/2$. For $h > 2 R_g$, the two potentials deviate because the discontinuous behaviour of Eq. (2.3) is replaced by the smooth crossover of Eq. (2.52). In the transition region, ideal polymers have a longer range of attraction than PHSs. Eisenriegler [31] has shown that Eq. (2.52) is identical to Eq. (2.3) (with $\sigma = 4 R_g / \sqrt{\pi}$) up to and including terms of order h^4.

Fig. 2.9 Depletion potential $W(h)$ as a function of inter-plate distance h between two parallel flat plates caused by nonadsorbing ideal chains (solid curve). The dotted lines give $W(h)$ according to Eq. (2.3) with $\sigma = 4R_g/\sqrt{\pi}$

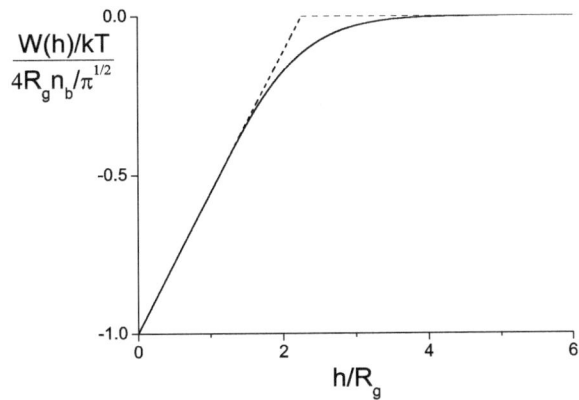

Interaction Potential Between Two Flat Plates From the Extended Gibbs Equation

From Eqs. (2.11) and (2.50) it follows that

$$\Gamma(h) = [n_i(h) - n_b]\, h = n_b h [\chi(h) - 1] \tag{2.54}$$

and hence, in view of Eq. (2.53)

$$\Gamma(\infty) = -n_b \frac{4R_g}{\sqrt{\pi}}. \tag{2.55}$$

Substituting Eqs. (2.54) and (2.55) in Eq. (2.13), and using the fact that ideal chains show ideal thermodynamic behaviour, i.e.,

$$\mu = kT \ln n_b, \tag{2.56}$$

we again obtain the result given by Eq. (2.52) for the interaction potential per unit area $W(h)$ between two plates. While the above thermodynamic route to the calculation of the adsorption is very efficient (as thermodynamics always is!), it is instructive (and useful for future reference) to consider the calculation of Γ starting from the polymer segment concentration profile $\varphi(x)$ near a single flat plate (with bulk concentration φ_b) and between two flat plates. Eisenriegler [32] (and later Marques and Joanny [33]) calculated the polymer concentration near one flat plate for ideal Gaussian ($M \gg 1$) chains and found the following expression for the relative polymer segment concentration $f(x) = \varphi(x)/\varphi_b$:

$$f(x) = 2\psi(z) - \psi(2z), \tag{2.57}$$

with

$$\psi(z) = \operatorname{erf}(z) + \frac{2z}{\sqrt{\pi}} e^{-z^2} - 2z^2 \operatorname{erfc}(z), \tag{2.58}$$

where z is defined as $x/(2R_g)$ and x is the distance from the surface. The (Gauss) error function $\mathrm{erf}(y)$ is defined as

$$\mathrm{erf}(y) = \frac{2}{\sqrt{\pi}} \int_0^y e^{-t^2} dt, \tag{2.59}$$

and the complimentary error function $\mathrm{erfc}(y) = 1 - \mathrm{erf}(y)$.

One can characterise the negative adsorption by the depletion layer thickness δ, which is defined as

$$\delta = \int_0^\infty dx (1 - f(x)). \tag{2.60}$$

For the case of ideal polymer chains near a flat plate with the profile Eq. (2.57), we find

$$\delta = \frac{2R_g}{\sqrt{\pi}}. \tag{2.61}$$

This is in full agreement with Eq. (2.55) as $\Gamma(\infty) = 2\Gamma_{\text{single wall}} = -n_b 2\delta$.

Exercise 2.6. Derive $\delta = 2R_g/\sqrt{\pi}$ for ideal chains starting from its definition in Eq. (2.60) and by using the profile Eq. (2.57).

A simple approximation [30,34] for the rather involved Eq. (2.57) is

$$f(x) = \tanh^2\left(\frac{x}{\delta}\right), \tag{2.62}$$

with δ given by Eq. (2.61).

Figure 2.10 depicts the concentration profile of an ideal polymer near a flat wall and its replacement by a step profile with width $\delta = 2R_g/\sqrt{\pi}$ (dashed). The simple approximation of Eq. (2.62) reproduces the exact result to within an accuracy of 1%.

Exercise 2.7. Show that the profile $f(x) = \tanh^2(x/\delta)$ has a depletion thickness δ.

For the concentration profile between two flat plates separated by a distance h, the following product function approximation has been proposed [30]

$$\varphi(x) = f(x)f(h - x). \tag{2.63}$$

In Eq. (2.63), $\varphi(x)$ is the polymer segment concentration between the plates and $f(x)$ and $f(h - x)$ are the individual one plate profiles given by Eq. (2.57) or, more simply

Fig. 2.10 Relative segment concentration of ideal chain segments from Eq. (2.57) as a function of the distance from a flat plate (solid curve). Dashed lines represent the step function profile and the dotted curve is the approximation of Eq. (2.62)

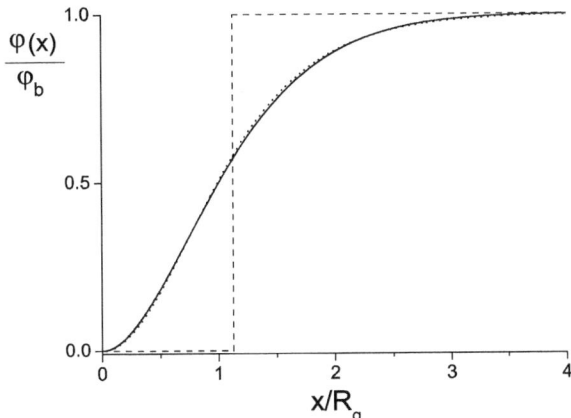

Fig. 2.11 Segment concentrations between two flat plates. Monte Carlo simulations with ideal chains of 100 segments for $h/R_g = 24.5$ (O), 5.88 (■) and 3.42 (△) are compared with Eq. (2.63) as solid curves

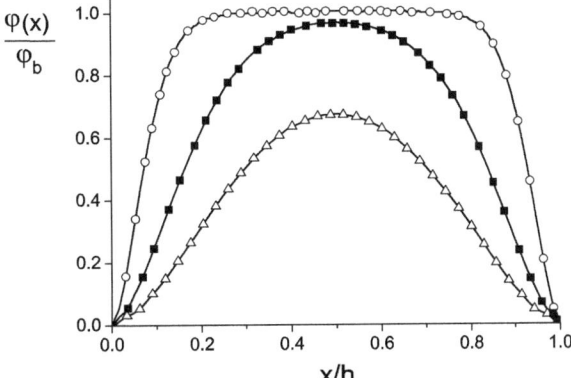

by Eq. (2.62). The concentration near a single plate, say plate 1, can be expressed by a Boltzmann factor such as $f(x) = \exp[-W_{\text{wall}}(x)/kT]$, where $W_{\text{wall}}(x)$ is the free energy giving rise to the profile. For the second plate located at a distance h, we can then write $f(h - x) = c(h - x)/n_b M = \exp[-W_{\text{wall}}(h - x)/kT]$.

Subsequently, the product function Eq. (2.63) follows from the superposition approximation:

$$W_{\text{wall, tot}}(x) = W_{\text{wall}}(x) + W_{\text{wall}}(h - x), \tag{2.64}$$

which is expected to work well for sufficiently large h/R_g. This is indeed supported by computer simulations [30], see the comparison in Fig. 2.11. For $h/R_g < 3$, the product function overestimates the segment concentration between the plates. In such narrow slits, the configurations of an ideal chain are then affected by both walls, which is not accounted for by the superposition approximation. While the relative deviation of the production function is largest for small plate separation h, for these distances $\Gamma(h) \to 0$ and, hence, the absolute error is small. The resulting adsorption is plotted in Fig. 2.12 (dashed curve). The exact result from Eq. (2.54) (with Eq. (2.39) for $\chi(h)$) is plotted as the solid curve. We therefore conclude that, overall, the product

Fig. 2.12 (Negative) adsorption of ideal chains between two walls as a function of the distance between the walls. The exact result is represented by the solid curve. The product function approximation is represented by the dashed curve

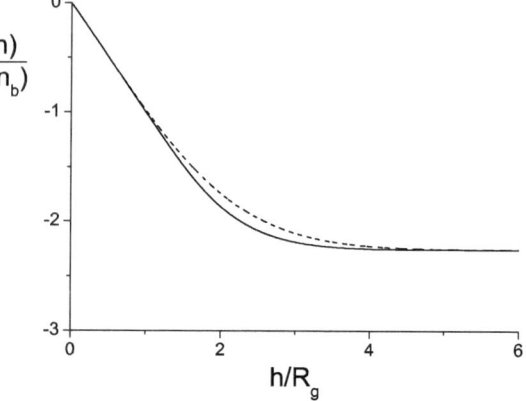

function gives a good prediction for the adsorption, and we will use it to calculate the depletion interaction between two spheres due to ideal polymers by using the extended Gibbs adsorption equation.

2.2.2 Interaction Between Two Spheres

Interaction Between Two Spheres From the Derjaguin Approximation
Using the Derjaguin approximation Eq. (2.31), the interaction potential between two spheres $W_s(h)$ can be obtained from the interaction potential per unit area $W(h)$ between two flat plates (Eq. (2.52)). The result for $q = 0.01$ is plotted in Fig. 2.13. Eisenriegler [31] obtained the following analytical expression:

$$W_s(h) = -n_b kTRR_g^2 \left(4\pi \ln 2 - 4\sqrt{\pi}\frac{h}{R_g} + \frac{\pi}{2}\frac{h^2}{R_g^2} \right) \qquad (2.65)$$

for small values of h valid up to and including terms of order h^4. This equation matches the numerical results for $R/R_g = 100$ presented in Fig. 2.13 very closely for $h < (3/2)R_g$, see Refs. [30,31].

Comparing Eq. (2.22) for PHSs and Eq. (2.65) for ideal chains reveals that we match the contact potentials for

$$\sigma = \sqrt{8\ln(2)}R_g. \qquad (2.66)$$

The result $\sigma = 2.35R_g$ agrees closely with the value $\sigma = 4R_g/\sqrt{\pi} = 2.26R_g$ for flat plates (within 5%). Hence, in the limit $R \gg R_g$ ideal polymers behave almost as PHSs with a diameter $\sigma \approx 2R_g$, just as for ideal chains between flat plates. In the next section, we will see that this picture changes when $R \lesssim R_g$.

Fig. 2.13 Interaction potential between two big hard spheres as a function of the closest distance between the surfaces of the spheres

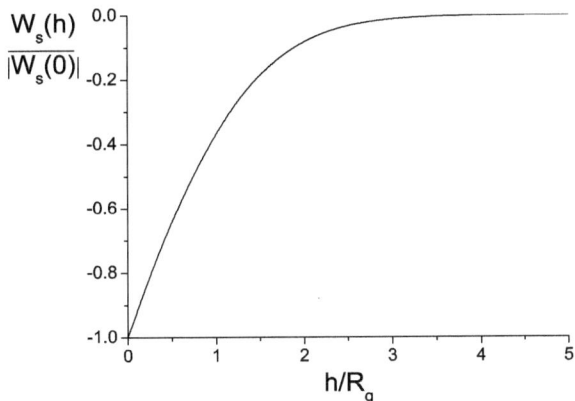

Interaction Potential Between Two Spheres From the Extended Gibbs Equation

The limitation of the Derjaguin approximation is that it only provides reliable results for $R \geq R_g$. To obtain results for the interaction potential between spheres for arbitrary $q = R_g/R$, we use the extended Gibbs adsorption equation. Taniguchi et al. [35] and, independently, Eisenriegler et al. [36] found the concentration profile of Gaussian ideal polymer chains around a single hard sphere with radius R, which reads

$$f_s(x) = \frac{\left(\frac{x}{R}\right)^2 + 2\left(\frac{x}{R}\right)\psi(z) + f(x)}{\left(\frac{x}{R} + 1\right)^2}, \tag{2.67}$$

where z again equals $x/2R_g$ and x is now the distance from the surface of the sphere. The functions $f(x)$ and $\psi(z)$ are defined in Eqs. (2.57) and (2.58). A simpler, yet accurate, form of Eq. (2.67) was first presented by Fleer et al. [34]:

$$f_s(x) = \left(\frac{\frac{x}{R} + \tanh(x/\delta)}{\frac{x}{R} + 1}\right)^2. \tag{2.68}$$

For various ratios of $q = R_g/R$, we plotted the profiles $f_s(x)$ in Fig. 2.14. For $R \ll R_g$, the Odijk [37] result

$$f_s(x) = \left(\frac{x}{x + R}\right)^2 \tag{2.69}$$

is recovered, which is independent of the polymer length scale.

For large hard sphere radii ($q = 0.1$), we see that the sphere profile approaches that of a flat plate. However, for $R_g/R = 1$ the depletion layer thickness becomes significantly smaller than $2R_g/\sqrt{\pi}$ and it further decreases with increasing q.

Fig. 2.14 Relative ideal chain segment concentrations at a wall and at a sphere for $q = 0.1$, $q = 1$ and $q = 10$ according to Eq. (2.67). With increasing q the profile shifts closer to the surface

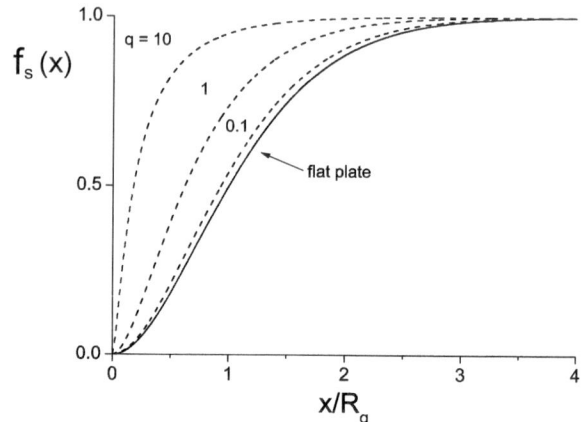

Exercise 2.8. (a) Show that, in the limit $R \gg R_g$, the expression in Eq. (2.67) for the profile around spheres becomes equal to Eq. (2.57) for the profile at a flat plate.
(b) Give a physical argument as to why the concentration profile shifts towards the particle surface when R_g/R increases.

Starting from Eq. (2.67), we can obtain an analytical expression for the depletion thickness around a sphere δ_s, which is now defined by

$$\frac{4\pi}{3}\left[(R + \delta_s)^3 - R^3\right] = \int_0^\infty 4\pi(R + x)^2 \left(1 - f_s(x)\right) dx. \qquad (2.70)$$

After carrying out the integration of the right-hand side of Eq. (2.70), we obtain [38,39]

$$\frac{\delta_s}{R_g} = \left[\left(1 + \frac{6q}{\sqrt{\pi}} + 3q^2\right)^{1/3} - 1\right]/q. \qquad (2.71)$$

Note that, in the limit $q \to 0$, Eq. (2.71) yields, as expected, the flat plate result $\delta_s/R_g = 2/\sqrt{\pi}$. The result in Eq. (2.71) holds for Gaussian ideal chains, implying that the segment size b is smaller than all other length scales, R_g and R. For freely jointed ideal chains, the depletion thickness also depends on the size ratio b/R for $R \lesssim 50b$ [40]. The q-dependence of the depletion thickness of Eq. (2.71) is plotted in Fig. 2.15.

Such curvature dependence of the depletion thickness of ideal polymer chains around a sphere also exists for ideal polymer chains around a spherocylinder ('rod'), which is relevant for Chap. 8. The depletion thickness of an ideal polymer around a cylinder requires a numerical calculation (see Ref. [41]). For practical purposes, an

Fig. 2.15 Depletion thickness (normalised with R_g) of ideal chains at a sphere (dashed curve) and a spherocylinder (solid curve) as a function of the size ratio $q = R_g/R$ (sphere) or $q = 2R_g/D$ (spherocylinder). Data points are exact numerical results. Full curve follows Eq. (2.72). For comparison, the flat wall case is given as the dotted line

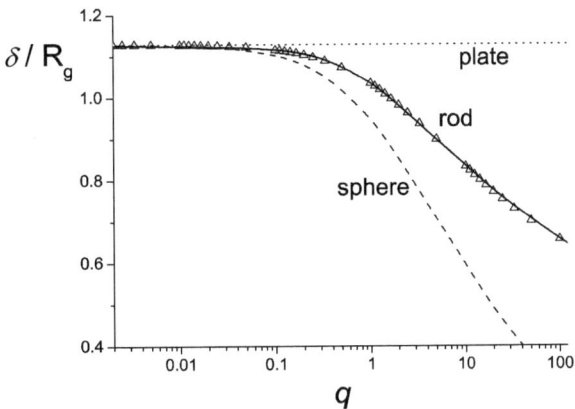

Exercise 2.9. (a) Show that, in the limit $R \gg R_g$ (and hence $R \gg \delta_s$), Eq. (2.71) for δ_s reduces to the flat plate result (Eq. (2.61)) for δ. The result for the relative depletion thickness δ/R_g as a function of the size ratio $q = R_g/R$ is plotted in Fig. 2.15.
(b) Carry out the integration in Eq. (2.70) and show that the result for δ is given by Eq. (2.71).

---

empirical expression was given that describes the numerical data with an accuracy within a per cent for $q = 2R_g/D$ values up to 100 [41],

$$\frac{\delta}{R_g} = \frac{\left(1 + \frac{4}{\sqrt{\pi}}q - k_1 q^{1.6} + k_2 q^{1.77}\right)^{1/2} - 1}{q}, \tag{2.72}$$

with the constants $k_1 = 0.62133$ and $k_2 = 1.50338$. In Fig. 2.15 the depletion thickness around a cylinder as a function of q is plotted and compared to the flat wall and sphere result. The data points are the numerical results, the curve follows Eq. (2.72). It follows that the depletion thickness around a cylinder (rod) is of the order of the polymer's radius of gyration for $q < 10$.

In Sect. 2.2.1, it was shown that the product function, Eq. (2.63), describes the polymer concentration profile between two flat plates quite well. Here, we apply the product function ansatz to calculate the concentration profile around two spheres. We assume that the local polymer concentration $n_s(\mathbf{r})$ in every point P (see Fig. 2.16) outside the spheres is given by

$$\frac{n_s(\mathbf{r})}{n_b} = f_s(x_1) f_s(x_2), \tag{2.73}$$

where $f_s(x_i)$ is the polymer concentration profile around a sphere given by Eq. (2.67) with x_i denoting the closest distance to the surface of the sphere.

Fig. 2.16 The geometry of
two spheres separated by a
distance h

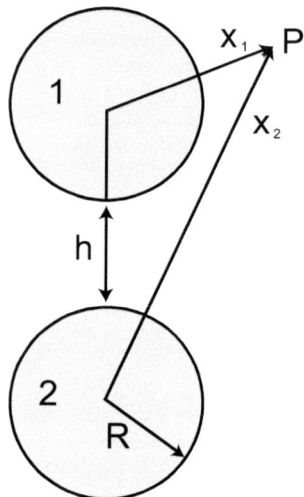

The interaction between two spheres can now be calculated from Eq. (2.23), which
for ideal chains becomes

$$\frac{W_s(h)}{kT} = N(\infty) - N(h), \tag{2.74}$$

where $N(h)$ and $N(\infty)$ are the number of polymer molecules in the system when the
colloidal particles are at a distance h (Fig. 2.16) and infinitely far apart, respectively.
The quantity $N(h)$ can be calculated numerically from

$$N(h) = \int d\mathbf{r} n_s(\mathbf{r}) \tag{2.75}$$

using the profile function Eq. (2.73), and obviously

$$N(\infty) = n_b \left(V - 2\frac{4\pi}{3}(R + \delta_s)^3 \right).$$

The result for $q = 0.01$ is plotted in Fig. 2.17a (dashed curve). We normalised the
interaction curve by dividing by the absolute value at contact. The depth of the interaction at contact can be computed numerically and we find $W_s(0)/n_b kT \approx -8.0 R R_g^2$.
It can be compared to the result that follows from Eq. (2.65): $W_s(0)/n_b kT =
-4\pi R R_g^2 \ln(2) \approx -8.7 R R_g^2$, which is close. A plot of the results for different values
of the size ratio q is shown in Fig. 2.17b.

We observe that the range of the interaction becomes smaller with decreasing
colloid radius R in agreement with the decrease of the depletion thickness δ_s with
decreasing colloid radius. In fact, by replacing $\sigma/2$ in Eq. (2.18) by δ_s (Eq. (2.71)), we
obtain interaction curves in good agreement with the results presented in Fig. 2.17b
obtained from the extended Gibbs adsorption equation using the product function
(Eq. (2.73)).

Fig. 2.17 Depletion potential between two spheres as a function of their closest distance. **a** Size ratio $R/R_g = 100$; dashed curve: Eq. (2.74) using the Derjaguin limit result Eq. (2.73); solid curve: Eq. (2.65). **b** Four different size ratios: $1/q = R/R_g = 10$ (solid), 3 (dashed), 1 (dotted-dashed) and 0.3 (dotted) calculated using Eq. (2.74)

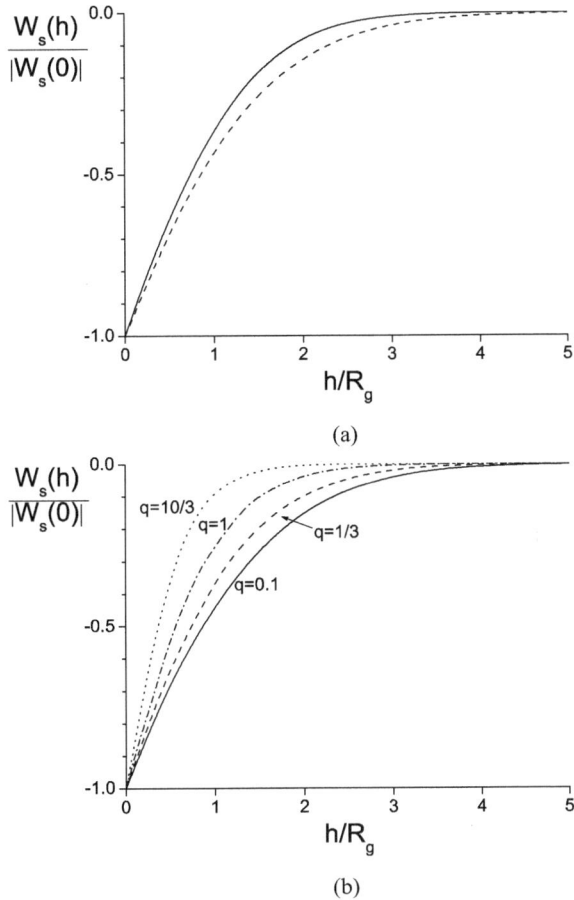

(a)

(b)

This brings us to the conclusion that, as far as the depletion interaction is concerned, ideal polymer chains can be replaced (to a good approximation) by PHSs with a diameter $\sigma = 2\delta_s$, where the depletion thickness δ_s now depends on the size ratio $q = R_g/R$. The ideal chain description sufficiently describes depletion effects in dilute polymer solutions. In Chap. 4, we shall see that, for polymers with excluded volume, the depletion thickness not only depends on the size ration q but also on the polymer concentration (see Refs. [39,42–44]). Also, the (osmotic) pressure is no longer given by the ideal (van 't Hoff) expression. Both features significantly affect depletion effects.

2.3   Depletion Interaction Due to Colloidal Hard Spheres

2.3.1   Concentration Profiles Near a Hard Wall and Between Two Hard Walls

We now consider the depletion interaction due to (small) colloidal hard spheres with diameter σ. At very low concentrations, where we may neglect the interaction between the spheres so that the system can regarded as thermodynamically ideal, the results for the depletion interaction are identical to those for PHSs. At higher concentrations—say, at volume fractions larger than a few per cent—the interactions between the spheres cannot be neglected. This has two important consequences for the depletion interaction. First of all, the pressure and chemical potential are no longer given by the ideal expressions. The corrections to ideal behaviour can be written in terms of the virial series (see textbooks on statistical thermodynamics, e.g., Hill [45] or Widom [46]):

$$\frac{P}{n_b kT} = 1 + B_2 n_b + \cdots \tag{2.76}$$

$$\frac{\mu}{kT} = \ln n_b + 2 B_2 n_b + \cdots \tag{2.77}$$

Now, n_b is the bulk concentration of (small) hard spheres. The quantity B_2 is the second osmotic virial coefficient

$$B_2 = \frac{2\pi \sigma^3}{3} = 4 v_0, \tag{2.78}$$

where $v_0 = \pi \sigma^3 / 6$ is the volume of a hard sphere. Secondly, the interactions between the particles among themselves and with a wall leads to a concentration profile near the wall. Obviously, in a layer at the wall with thickness $\sigma/2$ no centres of the hard spheres can penetrate. In the case of PHSs, the concentration takes on the bulk value n_b outside the depletion layer. However, in the case of hard spheres the interactions lead to an effective attraction between a sphere and the wall and the concentration profile at distance $x = \sigma/2 + y$ from the wall can be written as

$$n(x) = n_b \exp\left[-W_{\text{wall}}(x)/kT\right], \tag{2.79}$$

where $W_{\text{wall}}(x)$ is the effective interaction between the hard sphere and the wall. In fact, this is the potential of mean force between the sphere and the wall due to the other hard spheres. To first order in density, we can write

$$W_{\text{wall}}(x) = -n_b kT \nu(y), \tag{2.80}$$

where $\nu(y)$ is the overlap volume of the depletion zone around the sphere and the depletion layer of the wall depicted in Fig. 2.18,

$$\nu(y) = \frac{\pi}{3}\left(2\sigma^3 - 3\sigma^2 y + y^3\right) \qquad 0 \leq y \leq \sigma. \tag{2.81}$$

Fig. 2.18 Overlap volume (hatched) between a hard wall and a hard sphere

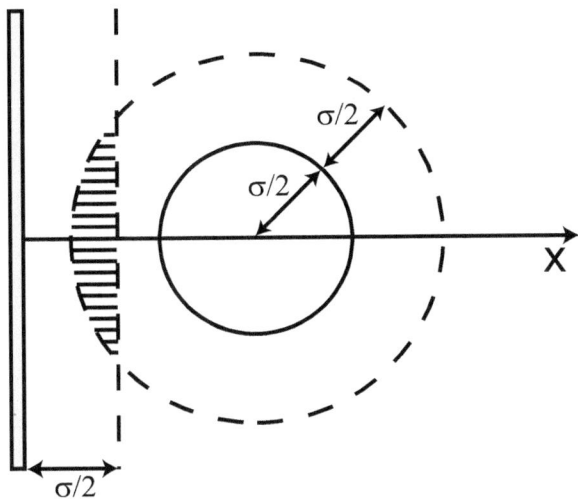

or

$$n_b\nu(y) = \phi\left(4 - 6\frac{y}{\sigma} + 2\frac{y^3}{\sigma^3}\right) \qquad 0 \le y \le \sigma. \qquad (2.82)$$

Here, $\phi = n_b\pi\sigma^3/6$ is the volume fraction of the (small) spheres [47].

The profile $n(x)$ can be obtained from Eqs. (2.79) to (2.81):

$$\frac{n(x)}{n_b} = \begin{cases} 0 & 0 \le x < \sigma/2, \\ 1 + n_b\nu(y) & \sigma/2 \le x \le 3\sigma/2. \end{cases} \qquad (2.83)$$

This profile of hard spheres at a single wall to order n_b^2 is depicted in Fig. 2.19.

We see that, in addition to the depletion layer, there is an 'accumulation' layer, where $n(x) > n_b$. The hard spheres located close to the depletion layer tend to 'push' one another into the layer next to the excluded depletion layer, and this leads to the accumulation. The concentration profile at a single wall to order n_b^3 was calculated by Fisher [48]. This accumulation layer has important consequences for the depletion interaction.

For the calculation of the depletion interaction due to hard spheres, we need the concentration profile between two confining walls. This problem was treated analytically by Glandt [49] and by Antonchenko et al. [50] using Monte Carlo computer simulations. Like for a single wall, we present the calculation of the concentration profile between two confining walls to order n_b^2. For $h < \sigma$, no spheres can penetrate between the walls and hence, the concentration is zero. For $\sigma \le h \le 3\sigma$, the depletion zone of a sphere may overlap simultaneously with the depletion zones of both walls (Fig. 2.20) and we can write

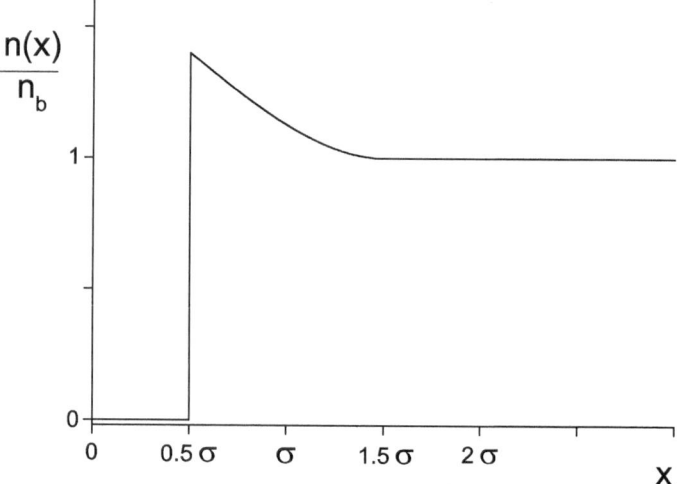

Fig. 2.19 Density profile $n(x)$ of hard spheres with $\phi = 0.1$ as a function of the distance from the wall x (Eq. (2.83))

Fig. 2.20 A sphere between two walls. Hatched areas indicate the overlap of the depletion zones of the hard spheres and the hard walls

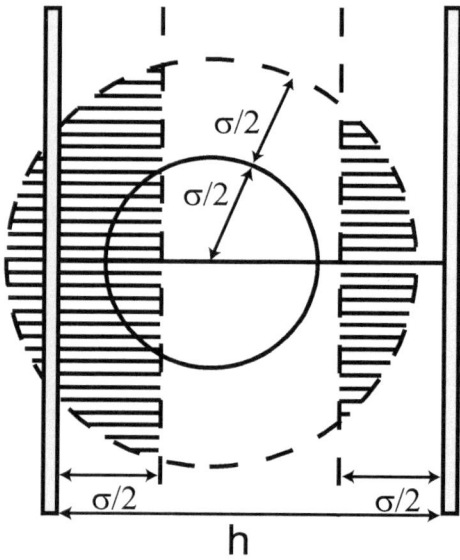

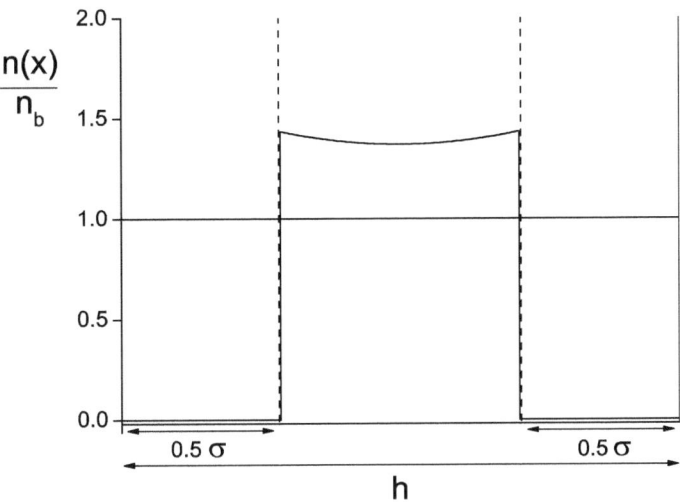

Fig. 2.21 Density profile of hard spheres between two hard walls for $h = 7\sigma/4$ and $\phi = 0.1$

$$\frac{n(x)}{n_b} = 1 + n_b\nu(y) + n_b\nu(h - \sigma - y) \qquad 0 \le y \le h - \sigma, \tag{2.84}$$

with $n_b\nu$ defined in Eq. (2.82).

The profile of hard spheres between two confining walls to order n_b^2 for $\sigma \le h \le 2\sigma$ is depicted in Fig. 2.21.

For $h > 3\sigma$, there is never simultaneous overlap of the depletion layer of a sphere with the depletion layers of the confining walls; as a result, there is no depletion-induced interaction.

2.3.2 Depletion Interaction Between Two Flat Plates

Interaction Potential Between Two Flat Plates From the Force Method

We follow the work of Mao et al. [51]. The same results as presented here were obtained earlier by Walz and Sharma [52] using a somewhat different method. The starting point for our treatment is a result by Henderson [53] that the force per unit area between two parallel hard plates immersed in a suspension of hard spheres is given by

$$K = P_i - P_o = kT(n_i - n_o), \tag{2.85}$$

where n_i and n_o are the *contact densities* of the hard spheres inside and outside the plates. The contact density is the ensemble-averaged density *at* the surface. Equation (2.85) can be explained as follows: the particle velocities are separable degrees of freedom, and therefore always obey the Maxwell–Boltzmann distribution. The force per unit area on a hard plate is therefore given rigorously by elementary kinetic theory as [54]

$$P = n^*kT, \qquad (2.86)$$

where n^* is the number density of particles at a distance corresponding to the point of impact (the position at which a particle hits the surface). This is, of course, the contact density. For P_o, $n^* = n_o$ and for P_i, $n^* = n_i$. This argument applies whenever there is a hard interaction between the particles and the plate [54]. The generality of Eq. (2.85) will also be exploited in Sects. 2.4 and 2.5, where we consider the depletion interaction due to hard rods and hard discs.

Up to order n_b^2, we find from Eq. (2.83) that

$$\frac{P_o}{kT} = n_o = n_b[1 + n_b \nu(0)] \qquad (2.87)$$

$$= n_b \left[1 + n_b \frac{2\pi\sigma^3}{3} \right], \qquad (2.88)$$

or, in terms of ϕ,

$$\frac{P_o v_0}{kT} = \phi + 4\phi^2. \qquad (2.89)$$

This is in agreement with the virial series in Eq. (2.76) using the second virial coefficient from Eq. (2.78). Between the plates we find

$$\frac{P_i}{kT} = \begin{cases} 0 & 0 \leq h < \sigma, \\ n_b \left[1 + n_b \nu(0) + n_b \nu(h-\sigma) \right] & \sigma \leq h < 2\sigma, \\ n_b[1 + n_b \nu(0)] & h \geq 2\sigma. \end{cases} \qquad (2.90)$$

Hence,

$$\frac{K(h)}{n_b kT} = \begin{cases} -1 - 4\phi & 0 \leq h < \sigma, \\ \phi \left[4 - 6\lambda + 2\lambda^3 \right] & \sigma \leq h < 2\sigma, \\ 0 & h \geq 2\sigma, \end{cases} \qquad (2.91)$$

where $\lambda = (h - \sigma)/\sigma$, which runs from 0 at $h = \sigma$ to 1 at $h = 2\sigma$.

The depletion force of Eq. (2.91) depicted in Fig. 2.22 jumps from negative (attractive) at $h = \sigma^-$ to positive (repulsive) at $h = \sigma^+$. The key idea behind the origin of the repulsive part of the depletion force is that, for small λ, the mutual repulsion of spheres is substantially reduced due to the fact that the excluded volumes of the spheres are hidden behind the depletion zones of the walls. In the limit $h = \sigma^+$, the spheres behave effectively thermodynamically ideal. To match the chemical potential (Eq. (2.77)) of the spheres in the bulk, the number density inside the gap must be

$$n_i = n_b \left[1 + 2B_2 n_b \right], \qquad (2.92)$$

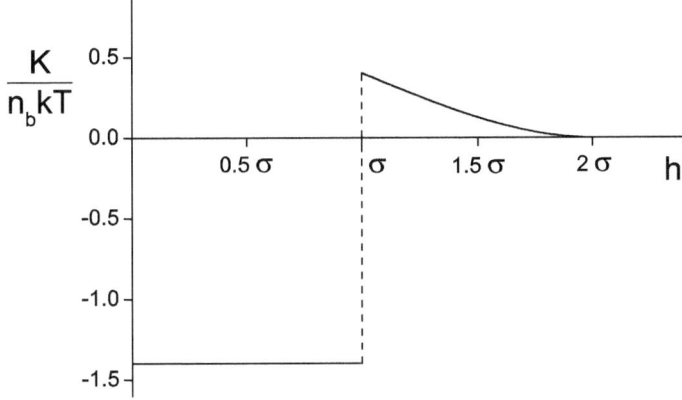

Fig. 2.22 Depletion force between two hard plates due to small hard spheres ($\phi = 0.1$), calculated using Eq. (2.91)

and hence, for $h = \sigma^+$

$$P_i = kTn_b[1 + 2B_2n_b],\tag{2.93}$$

giving rise to a maximum repulsive depletion force

$$K_{\max}(h = \sigma^+) = P_i - P_o$$
$$= 4kTn_b\phi.\tag{2.94}$$

Exercise 2.10. Derive Eq. (2.92).

Integrating the force, Eq. (2.91) yields the interaction potential per unit area $W(h)$ between the plates

$$\frac{W(h)}{kTn_b} = \begin{cases} \sigma\left(\lambda + \frac{3}{2}\phi + 4\lambda\phi\right) & 0 \leq h < \sigma, \\ \sigma\phi\left(\frac{3}{2} - 4\lambda + 3\lambda^2 - \frac{1}{2}\lambda^4\right) & \sigma \leq h < 2\sigma, \\ 0 & h \geq 2\sigma. \end{cases}\tag{2.95}$$

In Fig. 2.23, we present the interaction potential, which has a significant attraction at small separation distance h, but also has a repulsive part of the potential.

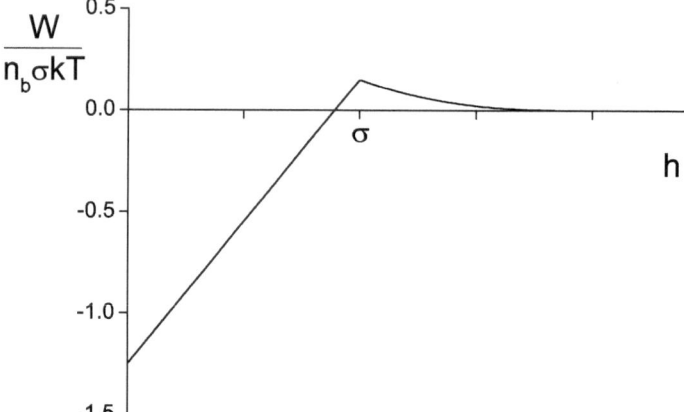

Fig. 2.23 Interaction potential between two hard plates due to small hard spheres ($\phi = 0.1$) from Eq. (2.91)

Exercise 2.11. (a) Explain why the force K has a discontinuity at $h = \sigma$, whereas the interaction potential W is continuous at that point.
(b) Why is the interaction potential still repulsive for h just below σ, while the force is attractive?

Interaction Potential Between Two Flat Plates From the Extended Gibbs Equation

From the depletion layer with thickness $\sigma/2$ and concentration profile of Eq. (2.83) it follows that the adsorption on a single plate is given by

$$
\Gamma_{\text{single wall}} = -\frac{\sigma}{2} n_b + n_b^2 \int_0^\sigma \nu(x)\,\mathrm{d}x
$$
$$
= -\frac{\sigma}{2} n_b + n_b^2 \frac{\pi}{4} \sigma^4. \tag{2.96}
$$

Hence,

$$
\Gamma(\infty) = 2\Gamma_{\text{single wall}}
$$
$$
= -\sigma n_b + n_b^2 \frac{\pi}{2} \sigma^4
$$
$$
= \sigma n_b (3\phi - 1). \tag{2.97}
$$

For two confining walls, it is clear that, for $h < \sigma$, no spheres can penetrate the gap between the walls. Hence,

$$
\Gamma(h) = -n_b h \qquad 0 \le h < \sigma. \tag{2.98}
$$

Using the concentration profile Eq. (2.84), we obtain the following for $\sigma \leq h < 2\sigma$:

$$\Gamma(h) = -\sigma n_{\mathrm{b}} + n_{\mathrm{b}}^2 \int_0^{h-\sigma} [\nu(x) + \nu(h - \sigma - x)] \, \mathrm{d}x$$

$$= -\sigma n_{\mathrm{b}} + \frac{2\pi}{3} n_{\mathrm{b}}^2 \sigma^4 \left[2\lambda - \frac{3}{2}\lambda^2 + \frac{1}{4}\lambda^4 \right] \tag{2.99}$$

or

$$\frac{\Gamma(h)}{n_{\mathrm{b}}\sigma} = \phi \left[8\lambda - 6\lambda^2 + \lambda^4 \right] - 1. \tag{2.100}$$

For $h \geq 2\sigma$,

$$\Gamma(h) = \Gamma(\infty). \tag{2.101}$$

Combining Eqs. (2.97)–(2.101) gives

$$\frac{\Gamma(h) - \Gamma(\infty)}{n_{\mathrm{b}}\sigma} = \begin{cases} 1 - \frac{h}{\sigma} - 3\phi & 0 \leq h < \sigma, \\ -\phi \left[3 - 8\lambda + 6\lambda^2 - \lambda^4 \right] & \sigma \leq h < 2\sigma, \\ 0 & h \geq 2\sigma. \end{cases} \tag{2.102}$$

Taking into account that the chemical potential is now given by Eq. (2.77), we obtain from Eq. (2.13)

$$W(h) = -kT \int_0^{n_{\mathrm{b}}} [\Gamma(h) - \Gamma(\infty)] \left[\frac{1}{n_{\mathrm{b}}} + 2B_2 \right] \mathrm{d}n_{\mathrm{b}}. \tag{2.103}$$

Substituting Eq. (2.102) in Eq. (2.103) after some algebra yields Eq. (2.95). Note that, in all cases considered so far (PHSs, polymers), the quantity $[\Gamma(h) - \Gamma(\infty)]$ was always positive (or zero) for all values for h. Here, we see that $[\Gamma(h) - \Gamma(\infty)]$ is negative for a certain range of h values due to accumulation effects in the concentration profiles. This leads to a positive interaction energy, as is clear from Eq. (2.103).

Such a repulsive contribution to the depletion interaction originates from excluded volume interactions between the depletants; in the case of ideal polymers and PHSs it is absent. Therefore, one might also expect accumulation effects when considering interacting polymers. From Monte Carlo simulation studies [55] and numerical self-consistent field computations [56,57], it follows that interacting polymers do contribute to repulsive depletion interactions but with a strength of the repulsion that is nearly imperceptible. Nonadsorbing polymers contribute to an extremely weak repulsion at high polymer concentration [55,57]. This repulsion is more pronounced when hard spheres are the depletants, as we have seen. However, polydispersity suppresses this repulsive interaction [58]. Hence, it is expected that the weak repulsion mediated by polymer chains as depletants is even more dampened due to polydispersity.

2.3.3 Depletion Interaction Between Two (Big) Spheres

Using the Derjaguin approximation (Eq. (2.31)), we obtain the interaction between two big spheres due to the small spheres by integration:

$$W_s(h) = \pi R \int_h^{2\sigma} W(h')dh'. \tag{2.104}$$

Using Eq. (2.95) for the interaction potential per unit area in Eq. (2.104), we obtain

$$\frac{W_s(h)}{kT} = \begin{cases} -\frac{R}{\sigma}\left[3\phi\lambda^2 - \phi^2\left(\frac{12}{5} - 9\lambda - 12\lambda^2\right)\right] & 0 \le h < \sigma, \\ \frac{R\phi^2}{5\sigma}\left[12 - 45\lambda + 60\lambda^2 - 30\lambda^3 + 3\lambda^5\right] & \sigma \le h < 2\sigma, \\ 0 & h \ge 2\sigma, \end{cases} \tag{2.105}$$

which has a positive maximum value of

$$\frac{W_{s,\max}}{kT} = \frac{12R}{5\sigma}\phi^2 \quad \text{at} \quad h = \sigma\left(1 - \frac{3}{2}\phi\right), \tag{2.106}$$

and a minimum value at contact

$$\frac{W_{s,\min}}{kT} = -3\frac{R}{\sigma}\left(\phi + \frac{1}{5}\phi^2\right). \tag{2.107}$$

In Fig. 2.24, we present the interaction potential between spheres (valid up to n_b^2 or, equivalently, up to ϕ^2). In [51], results are obtained for the interaction valid up to n_b^3, including a comparison with Monte Carlo computer simulation results of Biben et al. [59]. These results are reproduced in Fig. 2.25.

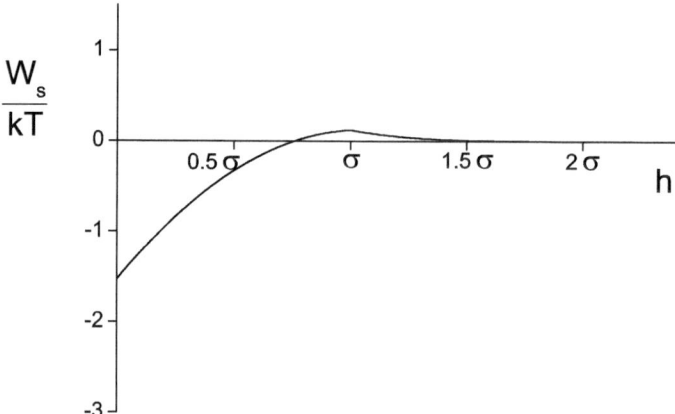

Fig. 2.24 Depletion potential between two hard spheres ($R = 5\sigma$) mediated by small hard spheres with $\phi = 0.1$

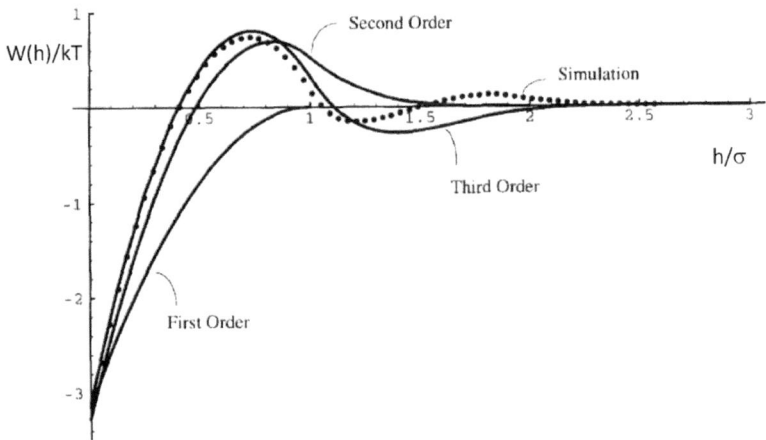

Fig. 2.25 A comparison of the calculations of Mao et al. [51] with the computer simulations of Biben et al. [59] for $\phi = \pi/15$ and $q = 0.1$. Reprinted with permission from Ref. [51]. Copyright 1995 Elsevier

---

Exercise 2.12. (a) Derive the interaction potential between spheres (Eq. (2.105)) from the extended Gibbs adsorption equation in the limit $R \gg \sigma$ (as is implicit when using the Derjaguin approximation).
(b) Show that Eq. (2.105) in the limit of first order in ϕ equals Eq. (2.22).

Mao [58] also considered the interaction between two spheres in a bath of polydisperse small hard spheres using the Derjaguin approximation. The accumulation effects due to nonadsorbing small hard spheres become much less pronounced with increasing polydispersity. Goulding and Hansen [60] used DFT theory to investigate the case of depletion due to small hard spheres. Their findings correlate with Mao's analytical results but also show that Mao's Derjaguin approximation already deviates from the DFT result for a size ratio of 5.

2.4   Depletion Interaction Due to Colloidal Hard Rods

Asakura and Oosawa [1,61] already recognised that rod-like macromolecules are very efficient depletion agents. In retrospect, the observations of Fåhraeus [62] (that the rod-like protein fibrinogen has, on a weight basis, the strongest effect on the aggregation of red blood cells) can be understood on the basis of its high efficiency as a depletion agent associated with its rod-like shape. Here, we consider the interaction caused by rod-like colloids as depletants and focus on a simple case: infinitely thin hard rods of length L. These rods have no excluded volume with respect to each other and hence, behave thermodynamically ideally.

2.4.1 Depletion Interaction Between Two Flat Plates

Interaction Potential Between Two Flat Plates Using the Force Method
As we are dealing with hard plates and a hard wall, we can again use Eq. (2.85) to calculate the force. The contact densities arise this time by considering the angles of the rods as a function of distance from the wall that leads to contact of an end point with the wall. First of all to make contact with the wall, the distance of the centre of the rod from the wall should be smaller than $L/2$. At a distance from the wall $x < L/2$, the angle that leads to contact is given by

$$\theta_x = \arccos \frac{x}{L/2} \tag{2.108}$$

(see Fig. 2.26). Outside the confining walls x runs from $L/2$ to 0. Hence, θ_x runs from 0 to $\pi/2$, so using spherical coordinates, we obtain

$$n_o = n_b \int_0^{\pi/2} \sin \theta \, d\theta = n_b \tag{2.109}$$

giving

$$P_o = n_b kT. \tag{2.110}$$

This result could have been written down at once as infinitely thin rods behave ideally.

Between two confining walls separated by a distance $h < L$, the second wall prevents contact configurations with the first wall for distances $x \geq h/2$. Hence,

$$n_i = n_b \int_{\theta_{h/2}}^{\pi/2} \sin \theta \, d\theta = \begin{cases} n_b \frac{h}{L} & 0 \leq h \leq L, \\ n_b & h > L. \end{cases} \tag{2.111}$$

From this result, and applying Eq. (2.86), it follows that

$$P_i = \begin{cases} n_b kT \frac{h}{L} & 0 \leq h \leq L, \\ n_b kT & h > L. \end{cases} \tag{2.112}$$

Fig. 2.26 Hard rod at a (hard) wall (*left*) and confined between two walls (*right*)

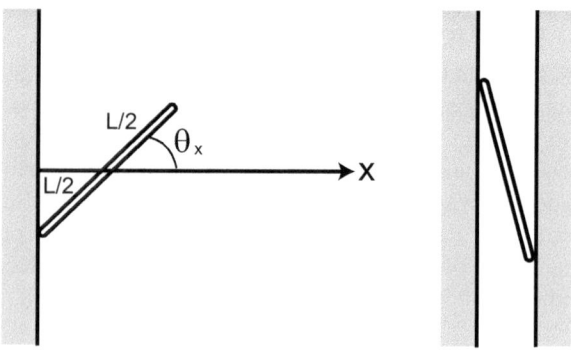

By combining Eqs. (2.110) and (2.112), we obtain the following for the force [1]:

$$K(h) = P_i - P_o \tag{2.113}$$

$$= \begin{cases} -n_b kT \left[1 - h/L\right] & 0 \le h \le L, \\ 0 & h > L. \end{cases} \tag{2.114}$$

Integration of the force in Eq. (2.113) yields the interaction potential per unit area $W(h)$ between the plates

$$W(h) = \begin{cases} -\frac{1}{2} n_b kT \frac{(L-h)^2}{L} & 0 \le h \le L, \\ 0 & h > L. \end{cases} \tag{2.115}$$

This result was first obtained by Asakura and Oosawa [61] (see also Auvray [63]). Mao et al. [64,65] considered the depletion interaction due to long thin rods with a finite diameter D. Then the system is no longer ideal and the interaction potential contains higher order terms in n_b. Like in the case of hard spheres, the interactions between the rods themselves and with the wall result in the accumulation of rods at the wall, which in turn leads to a repulsive contribution to the depletion interaction. Further details can be found in the papers by Mao et al. [64,65].

Interaction Potential Between Two Flat Plates From the Extended Gibbs Equation

The concentration profile of the rods near a wall also follows by considering the allowed angles. The angles ranging from θ_x (defined by Eq. (2.108)) to $\pi/2$ are allowed for a rod at a distance $x < L/2$ from a single wall (Fig. 2.26). Hence, the density profile is given by

$$n(x) = n_b \int_{\theta_x}^{\pi/2} \sin\theta \, d\theta = n_b \frac{x}{L/2}, \tag{2.116}$$

and is shown in Fig. 2.27. This provides an adsorbed amount at one wall

$$\Gamma_{\text{single wall}} = \int_0^{L/2} [n(x) - n_b] \, dx = -n_b L/4, \tag{2.117}$$

and thus

$$\Gamma(\infty) = 2\Gamma_{\text{single wall}} = -n_b L/2. \tag{2.118}$$

For two confining walls separated by a distance $h < L$, the density profile is given by

$$n(x) = \begin{cases} n_b \frac{x}{L/2} & 0 \le x \le h/2, \\ n_b \frac{h-x}{L/2} & h/2 \le x \le h. \end{cases} \tag{2.119}$$

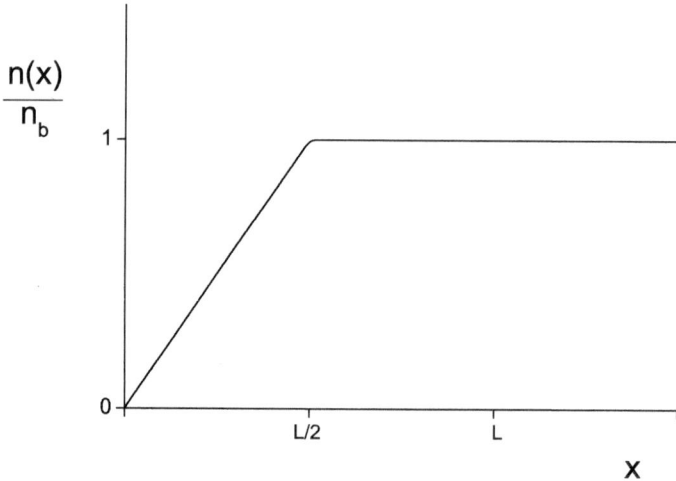

Fig. 2.27 Density profile of hard rods at a hard wall

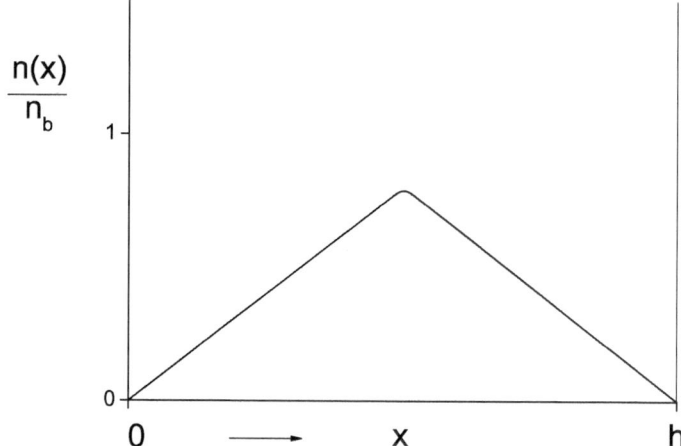

Fig. 2.28 Density of hard rods between two hard walls for $h = 4L/5$

This rod density profile between two walls is shown in Fig. 2.28 for $h = 0.8L$. Hence,

$$\Gamma(h) \int_0^h [n(x) - n_b]\,dx = \begin{cases} n_b \left[\frac{h^2}{2L} - h \right] & 0 \le h \le L, \\ -n_b \frac{L}{2} & h > L. \end{cases} \qquad (2.120)$$

Consequently,

$$\Gamma(h) - \Gamma(\infty) = \begin{cases} n_b \frac{[L-h]^2}{2L} & 0 \le h \le L, \\ 0 & h > L. \end{cases} \qquad (2.121)$$

Combining Eqs. (2.121) and (2.13) and carrying out the integration immediately results in the interaction potential of Eq. (2.115) when taking into account that $\mu = kT \ln n_b$ (ideal behaviour). For the calculation and simulation of concentration profiles at walls of rods of finite thickness and the evaluation of the resulting depletion interaction to higher orders of n_b, we refer to Mao et al. [66].

2.4.2 Interaction Between Two (Big) Colloidal Spheres Using the Derjaguin Approximation

Using the Derjaguin approximation (Eq. (2.31)), we obtain the interaction between two big spheres with radius $R \gg L$ by integration

$$W_s(h) = \pi R \int_h^L W(h')dh'. \tag{2.122}$$

Using Eq. (2.115) for the interaction potential per unit area in the above integration, we obtain for $0 \leq h \leq L$

$$W_s(h) = -n_b kT \pi R \frac{(L-h)^3}{6L}. \tag{2.123}$$

This expression for the interaction potential is also valid to order n_b for long thin rods with a finite diameter D, and we can then write Eq. (2.123) in the form

$$W_s(h) = -\frac{2}{3}kT\phi \frac{R}{D}\frac{L}{D}\left(1 - \frac{h}{L}\right)^3, \tag{2.124}$$

where

$$\phi = n_b \frac{\pi}{4}LD^2 \tag{2.125}$$

is the volume fraction of the rods. Comparing this expression for the depletion interaction between two big spheres with that for small spheres as depletant (Eq. (2.105)) for low ϕ reveals that the factor L/D, which is usually significantly larger than unity, is an important difference. Take as an example $R = 1 \, \mu m$, $L = 200$ nm and $D = 10$ nm. Then the factor

$$\frac{R}{D}\frac{L}{D} = 2000, \tag{2.126}$$

which implies that, for a volume fraction of rods as low as 0.1%, the depletion interaction will already be of order kT. For small colloidal spheres with $\sigma = 10$ nm, this would require a volume fraction of about 1%. The higher order terms calculated by Mao et al. [65] result (as in the case of small spheres as a depletion agent) in a repulsive barrier in the depletion interaction.

Above, we used the Derjaguin approximation. It is noted that Yaman et al. [67,68] directly computed the depletion interaction between two hard spheres mediated by an ideal gas of infinitely thin rods numerically. This method was extended by Lang [69] to calculate the depletion interaction mediated by polydisperse rods.

2.5   Depletion Interaction Due to Thin Colloidal Hard Discs

Thin colloidal hard discs provide another example of an anisometric colloidal particle that acts as an efficient depletion agent. The depletion interaction mediated by discs was first considered by Piech and Walz [70]. At the end of this section, where we compare spheres, rods and discs as depletion agents, we will see that the disc is intermediate in efficiency for inducing depletion attraction between spheres and rods. Here, we consider discs of diameter D and thickness L (Eq. (2.29)). Notice that, for the simplest case (i.e., infinitely thin hard discs), the excluded volume of the discs with respect to each other is non-zero, and only in the limit of the concentration going to zero will the discs behave thermodynamically ideally. We restrict ourselves to this limiting case.

2.5.1   Depletion Interaction Between Two Flat Plates

Interaction Potential Between Two Flat Plates From the Force Method
We again use Eq. (2.85) as the starting point for the calculation of the force between two parallel flat plates. To make contact with the wall, he distance of the centre of the discs from the wall should be smaller than $D/2$. At a distance from the wall $x < D/2$ (Fig. 2.29), the angle between the normal of the disc and the normal of the wall that leads to contact is now given by

Fig. 2.29 Discs at a hard wall. The grey area represents the thickness L

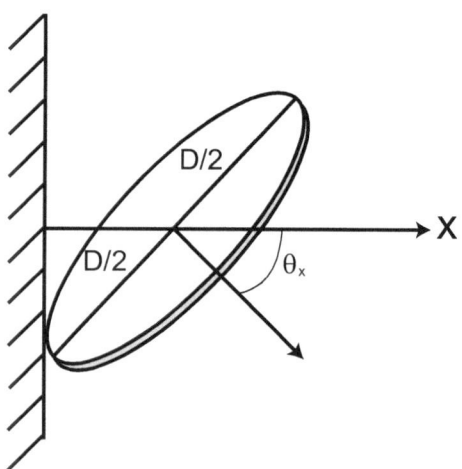

$$\theta_x = \arcsin\left(\frac{x}{D/2}\right).$$

(2.127)

Outside the confining walls, x runs from 0 to $D/2$. The contact density n_o follows as

$$n_o = n_b \int_0^{\pi/2} \sin\theta \, d\theta = n_b$$

(2.128)

and Eq. (2.86) is again found:

$$P_o = n_b kT.$$

Between two confining walls separated by a distance $h < D$, the second wall prevents contact configurations with the first wall for distances $x \geq h/2$. Hence,

$$n_i = n_b \int_0^{\theta_h/2} \sin\theta \, d\theta = \begin{cases} n_b\left[1 - \sqrt{1 - (h/D)^2}\right] & 0 \leq h \leq D, \\ n_b & h > D, \end{cases}$$

and hence,

$$P_i = \begin{cases} n_b kT\left[1 - \sqrt{1 - (h/D)^2}\right] & 0 \leq h \leq D, \\ n_b kT & h > D. \end{cases}$$

This leads to the following expression for the force between the plates:

$$K(h) = P_i - P_o$$

(2.129)

$$= \begin{cases} -n_b kT\sqrt{1 - (h/D)^2} & 0 \leq h \leq D \\ 0 & h > D. \end{cases}$$

(2.130)

Integration of the force Eq. (2.129) yields the interaction potential per unit area $W(h)$ between the plates

$$W(h) = -n_b kT \frac{D}{2}\left[\frac{\pi}{2} - \frac{h}{D}\sqrt{1 - \left(\frac{h}{D}\right)^2} - \arcsin\left(\frac{h}{D}\right)\right].$$

(2.131)

Exercise 2.13. Derive the interaction potential Eq. (2.131) from the force, Eq. (2.130).

Interaction Potential Between Two Flat Plates From the Extended Gibbs Equation

The concentration profile of the discs near a wall can also be found by considering the allowed angles [63]. For a disc at a distance $x < D/2$ from a single wall, the angles ranging from 0 to θ_x (defined by Eq. (2.127)) are allowed (Fig. 2.29). Hence,

$$n(x) = n_b \int_0^{\theta_x} \sin \theta \, d\theta = n_b \left[1 - \sqrt{1 - \left(\frac{x}{D/2} \right)^2} \right], \tag{2.132}$$

which is shown in Fig. 2.30. Hence,

$$\Gamma_{\text{single wall}} = \int_0^{D/2} [\hat{n}(x) - n_b] \, dx = -n_b D \frac{\pi}{8}$$

and thus,

$$\Gamma(\infty) = 2\Gamma_{\text{single wall}} = -n_b D \frac{\pi}{4}. \tag{2.133}$$

For two confining walls separated by a distance $h < D$

$$n(x) = \begin{cases} n_b \left[1 - \sqrt{1 - \left(\frac{x}{D/2} \right)^2} \right] & 0 \le x \le \frac{h}{2}, \\[2ex] n_b \left[1 - \sqrt{1 - \left(\frac{h-x}{D/2} \right)^2} \right] & \frac{h}{2} \le x \le h, \end{cases} \tag{2.134}$$

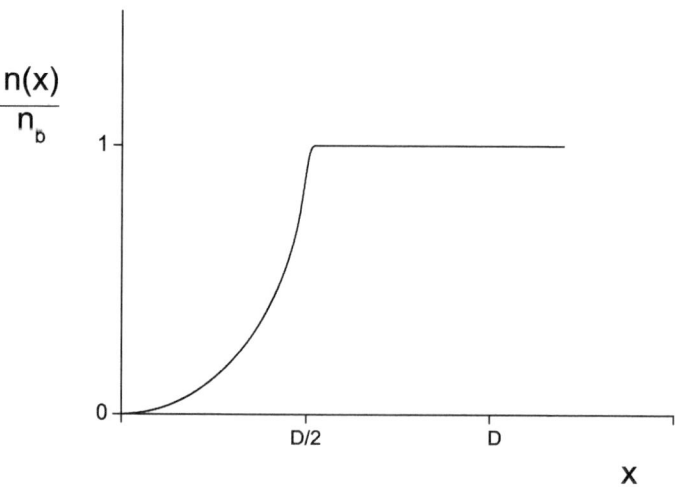

Fig. 2.30 Density profile of hard platelets at a hard wall from Eq. (2.132)

Fig. 2.31 Density profile of hard discs between two walls for $h = 4D/5$

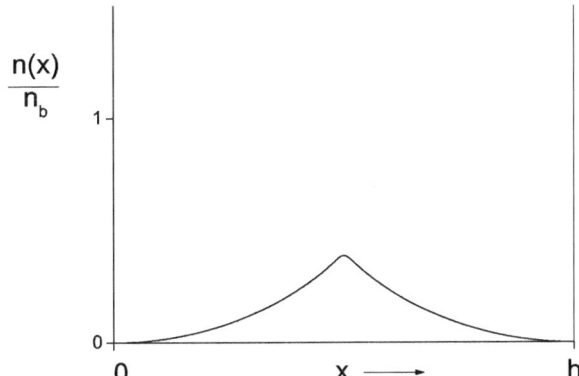

shown in Fig. 2.31. Hence,

$$\Gamma(h) = \int_0^h [n(x) - n_b]\, dx \tag{2.135}$$

$$= \begin{cases} -n_b \dfrac{D}{2} \left[\left(\dfrac{h}{D}\right) \sqrt{1 - \left(\dfrac{h}{D}\right)^2} + \dfrac{1}{2} \arcsin\left(\dfrac{h}{D}\right) \right] & 0 \le h \le D \\ n_b D \dfrac{\pi}{4} & h > D \end{cases} \tag{2.136}$$

By combining Eqs. (2.133) and (2.136), we obtain the following result for $0 \le h \le D$:

$$\Gamma(h) - \Gamma(\infty) = n_b \frac{D}{2} \left[\frac{\pi}{2} - \left(\frac{h}{D}\right) \sqrt{1 - \left(\frac{h}{D}\right)^2} - \arcsin\left(\frac{h}{D}\right) \right]. \tag{2.137}$$

Combining Eqs. (2.137) and (2.13) and carrying out the integration with $\mu = kT \ln n_b$ once again yields the interaction potential of Eq. (2.131).

2.5.2 Interaction Between Two (Big) Colloidal Spheres Using the Derjaguin Approximation

Combining the interaction potential per unit area (Eq. (2.131)) and the Derjaguin approximation (Eq. (2.31)) yields

$$W_s(h) = -n_b kT \frac{\pi}{3} R D^2$$

$$\times \left[-\frac{3}{4}\pi \frac{h}{D} + \frac{3}{2}\frac{h}{D} \arcsin\left(\frac{h}{D}\right) + \left(1 + \frac{1}{2}\left(\frac{h}{D}\right)^2\right) \sqrt{1 - \left(\frac{h}{D}\right)^2} \right]. \tag{2.138}$$

This result obtained for infinitely thin discs will presumably also be valid for discs with finite thickness L to the lowest order in n_b (although such a calculation has not been carried out), and we can then write Eq. (2.138) in the form

$$W_s(h) = -\frac{4}{3}kT\phi\frac{R}{L}$$

$$\times \left[-\frac{3}{4}\pi\frac{h}{D} + \frac{3}{2}\frac{h}{D}\arcsin\left(\frac{h}{D}\right) + \left(1 + \frac{1}{2}\left(\frac{h}{D}\right)^2\right)\sqrt{1 - \left(\frac{h}{D}\right)^2}\right],$$

(2.139)

where the disc volume fraction is given by

$$\phi = n_b\frac{\pi}{4}LD^2.$$

(2.140)

For thin discs, L is small so R/L is large. Assume $R = 1\,\mu m$, $D = 200$ nm and $L = 1$ nm then $R/L = 1000$ implies that, for volume fractions ϕ of the discs of 0.1%, the depletion interaction is already of the order kT.

Comparison of the depletion potentials due to spheres (Eq. (2.105)), rods (Eq. (2.124)), discs (Eq. (2.138)) and ideal polymers (Eq. (2.65)) reveals that to the lowest order in the depletant density they all can be written in the general form

$$W_s(h) = -n_b kTRC\ell^2 f\left(\frac{h}{\ell}\right),$$

(2.141)

where ℓ is the characteristic length scale of the depletion agent, the prefactor C determines the depth of the potential and the function f sets the distance dependence normalised such that $f(0) = 1$ and $f(1) = 0$. This is summarised in Table 2.1.

Because higher order h/R_g terms are not available for the ideal chain result, the $f(1) = 0$ limit can not be accessed. The functions f for ideal chains (small h), spheres, rods and plates are presented in Fig. 2.32. It is clear that the dependence on the interparticle separation $f(h/\ell)$ is similar for greatly different depletants. The results for the depletion interaction between big spheres discussed here are based on the Derjaguin approximation valid for $R \gg \ell$ ($\ell = \sigma$, L, D for spheres, rods and discs). An analysis of the accuracy of the Derjaguin approximation for depletion

Table 2.1 Characteristic parameters for C, ℓ and f in Eq. (2.141)

Depletion agent	C	ℓ	f
Sphere	$\pi/2$	σ	$(1 - h/\ell)^2$
Rod	$\pi/6$	L	$(1 - h/\ell)^3$
Disc	$\pi/3$	D	$\frac{3}{2}\left(\frac{h}{\ell}\right)\arcsin\left(\frac{h}{\ell}\right) - \frac{3}{4}\pi\left(\frac{h}{\ell}\right) + \left(1 + \frac{1}{2}\left(\frac{h}{\ell}\right)^2\right)\sqrt{1 - \left(\frac{h}{\ell}\right)^2}$
Ideal polymer	$\pi/2$	$R_g\sqrt{8\ln 2}$	$\left[1 - \sqrt{\frac{8}{\pi\ln 2}}(h/\ell) + (h/\ell)^2 + \cdots\right]$

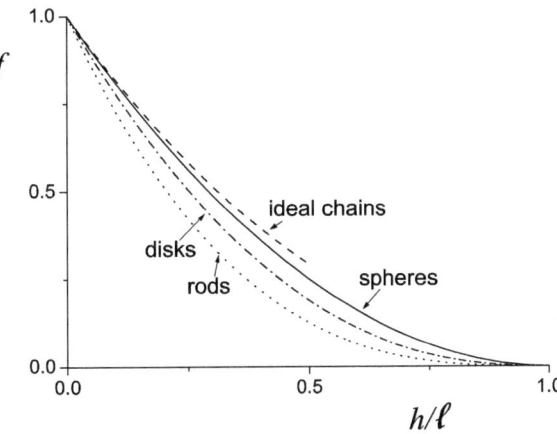

Fig. 2.32 Plot of the function f in Eq. (2.141) for rods (dotted), discs (dashed-dotted), spheres (solid) and ideal chains (dashed), see Table 2.1

potentials is presented in Ref. [71]. Using the Derjaguin approximation to analyse the depletion potential between large spheres leads to an underestimation of the interaction potential mediated by small spheres, a surprisingly accurate description for discs, and an overestimation for rod-like depletion agents. For an in-depth statistical mechanical analysis of the Derjaguin approximation applied to depletion interactions in colloidal fluids, we refer to the work of Henderson [53].

2.6   Measurements of Depletion Interactions

In this section, we summarise experimental methods that enable (depletion) interaction potentials between particles to be measured [72–77]. We distinguish pair interactions (Sects. 2.6.1–2.6.3) and many-body interactions (Sect. 2.6.4). The latter can be measured indirectly using scattering techniques or microscopy, whereas direct methods are available to measure the former. Common instruments for investigating such pair interactions are the surface force apparatus (SFA) [78], optical tweezers [79,80], atomic force microscopy (AFM) [81] and total internal reflection microscopy (TIRM) [82,83].

The SFA was the first method that enabled the forces between particles to be measured. It was developed by Tabor and Winterton [84] for two cylindrical surfaces in air or vacuum. An upgrade of the apparatus enabling measurements in liquids was constructed by Israelachvili and Adams [85,86]. An advantage of SFA is the high spatial resolution of 0.1 nm when using molecularly smooth mica sheets; SFA is mainly used for model surfaces. Unfortunately, the force resolution is low ($\mathcal{O}(10^{-8}$ N$)$) and the contact area between the surfaces needs to be very large ($\mathcal{O}(1$ mm$^2)$). Overall, it turned out SFA is less suitable for measuring depletion forces, and we therefore restrict ourselves to discussing AFM, TIRM and optical tweezers. Below we provide a brief introduction of these techniques. A few arbitrarily chosen experimental examples of potentials in the presence of depletants are given as illustrations.

The effective pair interactions measured with these techniques are the direct pair interactions between two colloidal particles plus the interactions mediated by the depletants. In practice, depletants are polydisperse, for which theoretical results are sometimes available. Expressions are available in literature for the interaction potential between hard spheres in the presence of polydisperse PHSs [60], polydisperse ideal chains [87], polydisperse hard spheres [58] and polydisperse thin rods [69].

2.6.1 Atomic Force Microscope

The atomic force microscope (AFM) was designed for high-resolution surface topography analysis. The basic measuring principle is sketched in Fig. 2.33. A sample is scanned by a sharp tip attached to a sensitive cantilever spring via a piezoelectric positioner. Forces on the tip lead to spring deflection, which is detected optically [88]. Topographic images of the sample are obtained by plotting the deflection of the cantilever as a function of the sample position. Alternatively, a feedback loop can be used to fix the spring deflection, and response of the piezoelectric positioner generates the image [81]. The pair interactions between a colloidal sphere and a surface by free depletants can be studied with a colloidal probe particle attached to the cantilever tip [89].

Interactions between a spherical colloid and a wall can be measured by bringing probe and substrate together and monitoring the cantilever deflection as a function of the interparticle distance. The photodetector voltage versus piezo position curve can be converted into a force–distance curve. The force acting on the cantilever is a result of the deflection of the cantilever and its known spring constant. The zero

Fig. 2.33 Representation of an atomic force microscope. The sample of interest is placed on the piezoelectric scanner and a laser is reflected off the upper side of the cantilever and guided to a split photo detector. In this way, vertical and horizontal deflection signals can be measured. A well-defined colloidal particle can be glued to the tip of the cantilever to measure the force between that particle and the surface

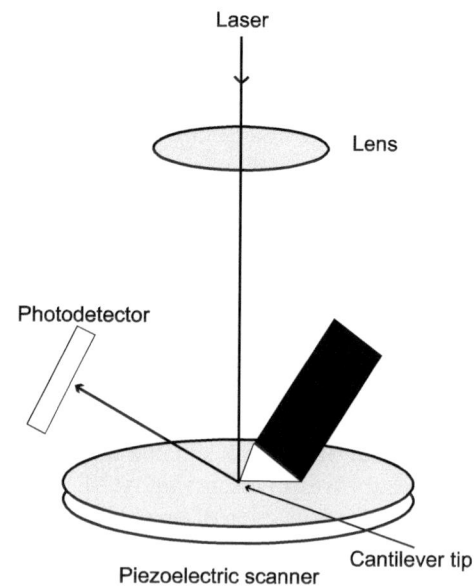

force is defined by the deflection of the cantilever as the colloidal probe is far from the surface of the substrate. To obtain the force–distance dependence on an absolute scale, the zero distance (i.e., where the colloid touches the wall) has to be determined. Commonly, the zero distance is obtained from the force curve itself and not through an independent method [81].

In practice, the position where the motion of the probe complies with the piezo movement defines the point of zero distance. Force–distance curves recorded with AFM depend on the specific geometry of the probe and the surface. Usually, the interaction is displayed as the force divided by the radius of the colloidal sphere, R, in units N/m. The Derjaguin approximation relates this quantity to the interaction potential per unit area between equivalent flat surfaces at a given separation distance (Eq. (2.33)).

Since AFM is widely used for imaging, the technology is well-developed. Due to its high lateral resolution of ~ 1 nm, nanoparticles can be used and material inhomogeneities can be mapped and imaged. The small contact areas (~ 10 nm^2) reduce the probability of experimental artifacts due to surface contamination and roughness [81]. The high spatial resolution capability makes AFM a complementary approach to the SFA, which has been used to measure interfacial forces between proximal surfaces over areas on the order of ~ 1 mm^2. Moreover, the force resolution of AFM is a couple of orders of magnitude better than that of the SFA.

The determination of the zero separation distance using AFM remains a complicated issue in some cases. This often makes it difficult to fully quantify depletion interactions. The achievable force sensitivity is limited when compared to TIRM. This makes AFM suitable only for measuring strong depletion forces.

In Fig. 2.34, we show the measured force oscillating between a silicon wafer and a silica sphere (radius $R = 2.2 \, \mu$m) that is attached to a cantilever spring in the presence of Ludox silica spheres with a radius of 11 nm [90]. The volume fraction of the Ludox spheres was 1.5%. The effective volume fraction is much larger due to repulsive double layer interactions.

2.6.2 Total Internal Reflection Microscopy

The interaction potentials between a single colloidal particle and a wall can be obtained using evanescent field scattering in total internal reflection microscopy (TIRM) [75,82,83]. The fluctuations of the separation distance resulting from thermal motion can be directly detected from the scattered intensity. In a typical TIRM set-up, a laser beam is directed via a prism to the glass/solution interface as shown in Fig. 2.35, with an incident angle that is chosen such that total reflection occurs.

The electric field of the laser beam penetrates the interface and causes an evanescent wave, the amplitude of which decays exponentially along the normal to the

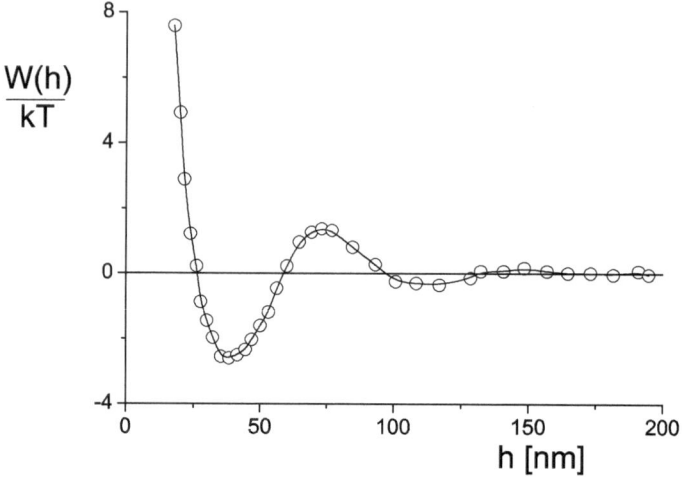

Fig. 2.34 Interaction potential derived from force measurements between a flat silica surface and a silica sphere ($R = 2.2\,\mu$m) in the presence of 1.5 vol% Ludox spheres (radius 11 nm) as depletants at pH 5.6. The ionic strength was 0.76 mM. Data replotted from Piech and Walz [90]. Curve guides the eye

Fig. 2.35 Sketch of a TIRM set-up. Whenever the incident angle is larger than the critical angle, the incident beam is totally reflected at the glass–fluid interface and an evanescent wave penetrates into the fluid. A colloidal particle located close to the surface will scatter light from the evanescent wave, which is collected by a photomultiplier and provides the probability density of separation distances between the particle and the wall. A CCD camera is used to image the field of view

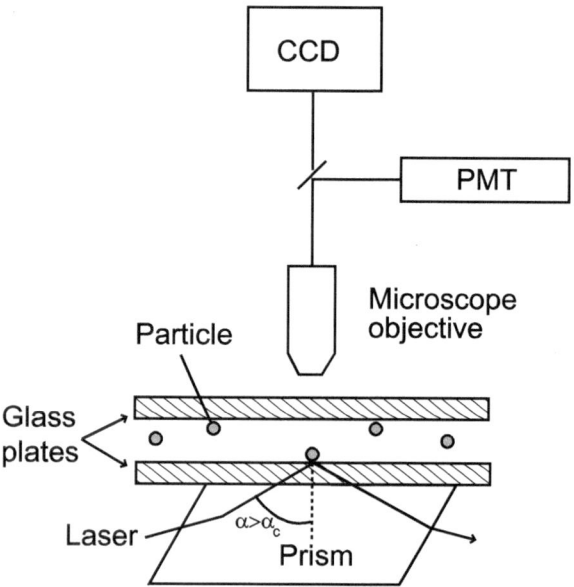

interface. A single colloidal sphere in the field of gravity that interacts with this evanescent wave will scatter light depending on its position h, as in [91]

$$I_s(h) = I(h = 0) \exp[-h/\varrho], \qquad (2.142)$$

where ϱ is the penetration depth of the evanescent wave. A photomultiplier is used to monitor the time dependence of the scattered intensity, with a resolution in the millisecond range. A sufficient number of data points allows a histogram of intensities to be converted to the probability density distribution010 of the intensity. Through Eq. (2.142), the intensity histogram can be converted to a probability density distribution (pdf) of separation distances. Using Boltzmann's law $\ln[\text{pdf}(h)] \sim -U(h)/kT$, this pdf provides the potential energy $U(h)$. Usually, a charged sphere is used with a size of the order of a µm. The solvent is often an aqueous salt solution. In this way, double layer repulsion between particle and a like-charged surface counterbalances gravity, enabling the particle to fluctuate near the wall. From $U(h)$ the bare pair depletion potential can be found by subtraction of double layer repulsion and gravity.

Exercise 2.14. Why does the scattered intensity caused by a colloid decrease with increasing distance from the surface?

An optical trap can be set up to prevent the colloidal particle from moving out of the microscope's observation area. For this purpose, a second laser beam has to be focused directly at the particle. It is recommended to use p-polarised light and a penetration depth below 150 nm.

The major advantages of using TIRM over AFM and SFA to study depletion potentials [92] are its outstanding force sensitivity and its non-invasive nature. With TIRM, it is possible to investigate the interactions of a single, freely moving, Brownian particle. This method enables measurements of forces as small as 10^{-14} N. The reason for this high sensitivity is the use of a molecular gauge for energy kT instead of a mechanical gauge for the force determined by a spring constant, as is used in AFM and SFA [83].

TIRM is less suited for measuring strong depletion potentials. When the repulsion between the particle and the wall is bigger than $5kT$ the pdf for finding the particle in this range becomes virtually zero. Therefore, the error in determining $\text{pdf}(h)$ becomes very large. If the attraction between the sphere and the wall becomes too strong, the intensity histogram becomes narrower than the range set by the electronic noise of the photomultiplier [83].

An example of a pair potential measured using TIRM is given in Fig. 2.36. The data measured are wall–sphere potentials between a flat silica surface and a polystyrene sphere ($R = 1.85\,\mu\text{m}$) in the presence of nonadsorbing polydisperse, charged boehmite rods (averaged length $L = 200$ nm) [93]. The range of the potential is obviously close to the length of the rods. The volume fraction of the rods is 0.09%.

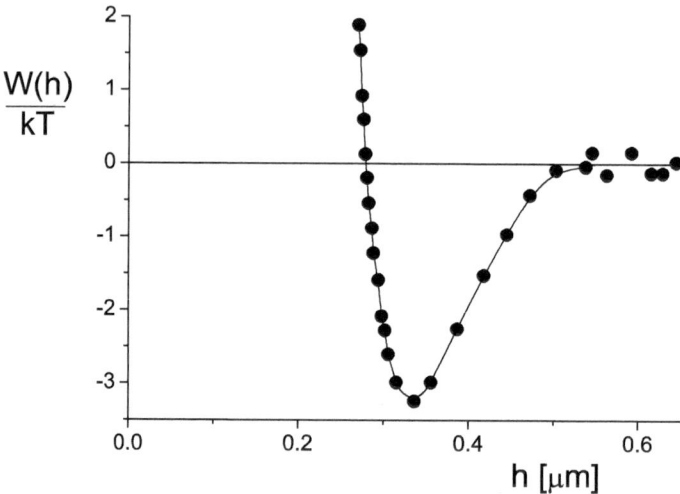

Fig. 2.36 Interaction potential between a flat silica surface and a polystyrene sphere ($R = 1.85\,\mu$m) mediated by polydisperse boehmite rods (0.09 vol%) with averaged length of 200 nm. Data replotted from Helden et al. [93]

2.6.3 Optical Tweezers

Around 1970, it was found that laser radiation forces can be used to trap and manipulate small dielectric particles [94]. A laser beam can push a particle towards the centre of the beam, provided the particle has a higher refractive index than the surrounding medium. Thus, optical tweezers can pick up and manipulate colloidal particles; experts nowadays can even spell your name with colloidal particles by using a single optical tweezer. This technique found a broad application in biology as well as in colloid science [95,96]. Figure 2.37 illustrates a typical optical tweezer arrangement. The laser beam is tightly focused by the microscope's objective lens, which also makes it possible to image trapped particles with a camera. Optical tweezers can be configured using multiple beams to trap many particles simultaneously. This can be implemented in the following manner: firstly, a single beam is used to rapidly scan two or more trap positions. Next, the beam is split at an early stage in the optical circuit to produce two separate light paths that are then recombined before entering the microscope. Finally, computer-generated holograms are used to generate multiple beams simultaneously.

Boltzmann's law is used to find the interaction potential between the trapped particles using the measured probability density as a function of separation distance. Position detection results either from particle tracking via video microscopy or back focal plane interferometry [80]. Accurate video microscopy requires the acquisition of bright field or fluorescence images from the microscope [97]. Particle centre separations can then be determined with a sub-pixel resolution through image-processing operations [97]. A spatial resolution of $\sim$10 nm can be achieved. Back focal plane interferometry enables the spatial resolution to be reduced to $\sim$1 nm.

Fig. 2.37 A simple diagram
of optical tweezers. The
microscope objective lens
enables the laser beam to be
tightly focused and trapped
particles to be imaged

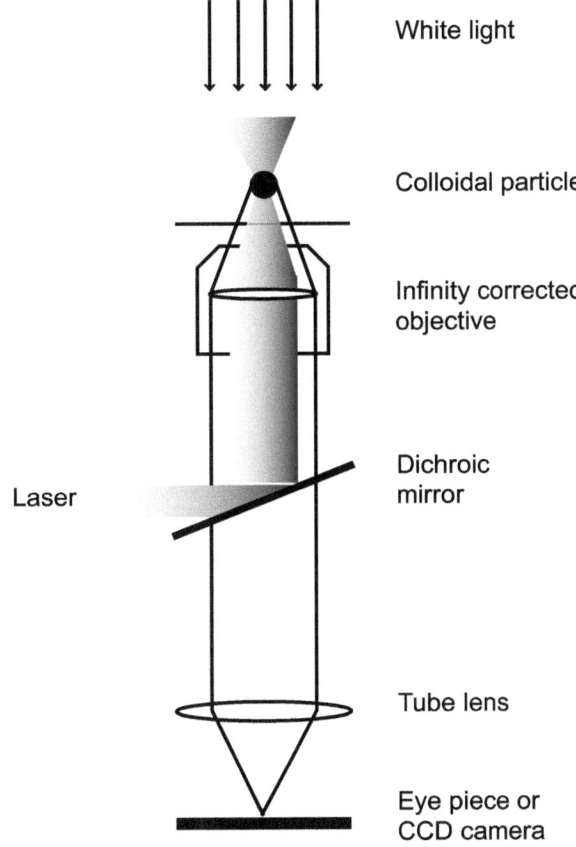

White light

Colloidal particle

Infinity corrected
objective

Laser

Dichroic
mirror

Tube lens

Eye piece or
CCD camera

A major advantage of optical tweezers is that the detected forces range between 10^{-13} and 10^{-10} N. Like TIRM, optical tweezers enable colloidal interactions to be studied in a non-invasive manner. When optical tweezers are used with TIRM the interaction potentials between two colloidal particles can be measured, whereas TIRM and AFM are restricted to wall–particle potentials. The main problems that can arise when taking measurements with optical tweezers is that the results are susceptible to misinterpretation due to image processing problems [98,99].

The pair interaction between two silica spheres in the presence of rather monodisperse, nonadsorbing DNA chains measured by using optical tweezers [100] is plotted in Fig. 2.38. Data are given for three DNA concentrations beyond the coil overlap concentration indicated in the plot. For a theoretical quantification of these data, see Ref. [43].

Fig. 2.38 The interaction
potential between two silica
spheres ($R = 0.63 \, \mu\text{m}$)
mediated by DNA chains
($R_g = 0.50 \, \mu\text{m}$). Data
replotted from Ref. [100].
The DNA concentrations are
indicated in the plot

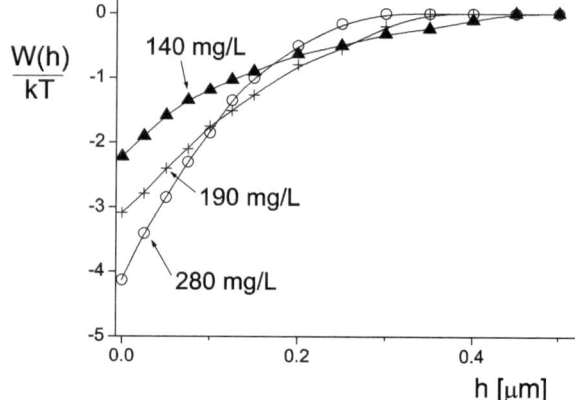

2.6.4 Scattering and Microscopy

One of the manifestations of depletion effects in a colloidal dispersion is that its
fluid structure is affected by the presence of nonadsorbing depletants (for instance,
polymer chains). This is reflected in the radial distribution function $g(r)$—the local
concentration of particle centres from a distance r to a fixed particle centre. Statistical
mechanics links $g(r)$ to the potential of mean force W_{mf} [101],

$$W_{\text{mf}}(r) = -kT \ln g(r). \tag{2.143}$$

For a dilute colloidal dispersion, $g(r) = \exp[-W(r)/kT]$, where $W(r)$ is the *pair*
interaction. The quantity $g(r)$ can be measured using confocal laser scanning
microscopy. This method performs quantitative three-dimensional real space measurements of the positions of the (fluorescently labelled) colloidal particles. Analysis
of the positions of the particles yields $g(r)$. This means that confocal microscopy
enables indirect measurement of both the potential of mean force and (using a dilute
dispersion) the pair interaction in a mixture of colloids and depletants. Royall et al.
[102] have performed such a study in a colloid–polymer mixture with free polymers
as depletants.

Scattering techniques allow the structure factor $S(Q)$ to be measured as a function
of the wave vector Q of colloidal dispersions, defined as

$$Q = \frac{4\pi}{\lambda_{\text{m}}} \sin\left(\frac{\theta_{\text{s}}}{2}\right). \tag{2.144}$$

Here, λ_{m} is the wavelength of radiation through the medium and θ_{s} the scattering
angle. Statistical mechanics relates the structure factor $S(Q)$ to the radial distribution
function $g(r)$ [103]:

$$S(Q) = 1 + \frac{\phi}{v_0} \int_0^\infty 4\pi r^2 [g(r) - 1] \frac{\sin Qr}{Qr} \, dr. \tag{2.145}$$

Hence, via Eq. (2.143), Eq. (2.145) reveals that $S(Q)$ contains the potential of mean force in the long wavelength limit ($Q \to 0$).

In the case of a colloid–depletant mixture, in which the depletant is made 'invisible' by contrast matching, the scattered intensity $I(Q)$ reads

$$I(Q) \sim \phi P(Q)S(Q), \tag{2.146}$$

where $P(Q)$ is the particle scattering form factor. The proportionality constant is the squared particle scattering amplitude. The structure factor then follows from Eq. (2.146) as

$$S(Q) = \frac{I(Q)}{I_0(Q)} \frac{\phi_0}{\phi}, \tag{2.147}$$

where ϕ_0 is the volume fraction of a very dilute dispersion and $I_0(Q)$ is its scattered intensity. Here, the fact was used that $S(Q)$ equals unity in a very dilute dispersion.

Following an early light scattering study of de Hek and Vrij [104], Ye et al. [105] made a small-angle neutron scattering (SANS) study on dispersions of $CaCO_3$ particles (effective hard sphere radius $R = 4.85$ nm) stabilised by alkylbenzene sulfonate to which polyethylene propylene (PEP) copolymers ($R_g = 8.3$ nm) were added. SANS allows contrast matching so that the structure factors of the free polymers and the colloids can be measured independently (see Refs. [106,107] for a theoretical analysis). Further, SANS is much less sensitive to the multiple scattering problems encountered in light scattering.

In Fig. 2.39, a few representative measured structure factors $S(Q)$ of colloidal spheres at a colloid volume fraction $\phi = 0.086$ are plotted at various PEP concentrations. Clearly, the measured structure factor increases upon adding more free polymer at $Q < 0.2$ nm^{-1}, corresponding to an increase of the attraction between the colloids. This increase of $S(Q)$ at small Q has also been found in a few other studies [108–110]. Mutch et al. [110] showed that it is possible to rescale structure factors at high q (relatively large polymers) to obtain a universal $S(Q)$ behaviour. PRISM [111,112] would be quite useful in quantifying these experimental data.

Fig. 2.39 Measured colloidal structure factor $S(Q)$ of colloid–PEP mixtures ($R = 4.85$ nm, $q = R/R_g = 1.7$) at 17.5 wt% of colloids (corresponding to $\phi = 0.086$), and at polymer concentrations of 3.8 (+), 23.3 (o) and 65.2 g/L (■). Data replotted from Ref. [105]. Curves are drawn to guide the eye

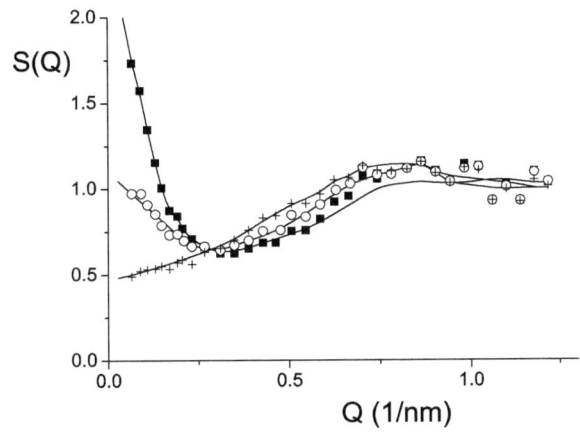

Static and dynamic light scattering can also be used on colloid–polymer mixtures to quantify the spinodal, defined at $1/S(Q \to 0) \equiv 0$ using extrapolation [108]. When the attraction becomes very strong, the structure factor diverges at small Q and the dispersion starts to decompose. This demixing will be considered in Sect. 4.4.

Exercise 2.15. Why does $1/S(Q \to 0) \equiv 0$ correspond to the spinodal? *Hint*: check Eq. (1.27).

References

1. Asakura, S., Oosawa, F.: J. Chem. Phys. **22**, 1255 (1954)
2. Vrij, A.: Pure Appl. Chem. **48**, 471 (1976)
3. Widom, B., Rowlinson, J.S.: J. Chem. Phys. **52**, 1670 (1970)
4. Onsager, L.: Chem. Rev. **13**, 73 (1933)
5. Onsager, L.: Ann. NY. Acad. Sci. **51**, 627 (1949)
6. D.G. Hall, J. Chem. Soc., Faraday Trans. 1 **68**, 2169 (1972)
7. S.G. Ash, D.H. Everett, C. Radke, J. Chem. Soc., Faraday Trans. 1 **69**, 1256 (1973)
8. Evans, R., Marconi, U.M.B.: J. Chem. Phys. **86**, 7138 (1987)
9. It is noted that n refers to the density profile when applying the extended Gibbs equation, while contact densities n are used when applying the force method
10. Odijk, T.: J. Chem. Phys. **106**, 3402 (1997)
11. Derjaguin, B.V.: Kolloid-Z. **69**, 155 (1934)
12. Flory, P.J.: Principles of Polymer Chemistry. Cornell University Press, New York (1953)
13. Yamakawa, H.: Modern Theory of Polymer Solutions. Harper and Row, New York (1971)
14. De Gennes, P.G.: Scaling Concepts in Polymer Physics. Cornell University Press, Ithaca (1979)
15. Doi, M., Edwards, S.F.: The Theory of Polymer Dynamics. Clarendon Press, Oxford (1986)
16. J. des Cloizeaux, G. Jannink, *Polymers in Solution* (Oxford University Press, Oxford, 1990)
17. Fujita, H.: Polymer Solutions. Elsevier, New York (1990)
18. Grosberg, A.Y., Khoklov, A.R.: Statistical Physics of Macromolecules. AIP, New York (1994)
19. Doi, M.: Introduction to Polymer Physics. Oxford University Press, Oxford (1995)
20. L. Schäfer, *Excluded Volume Effects in Polymer Solutions* (Springer, Heidelberg, 1999)
21. Rubinstein, M., Colby, R.: Polymer Physics. Oxford University Press, New York (2003)
22. Edwards, S.F.: Proc. Phys. Soc. **85**, 613 (1965)
23. Edwards, S.F., Freed, K.F.: J. Phys. A **2**, 145 (1969)
24. Casassa, E.F.: J. Polym. Sci. **5**, 773 (1967)
25. Casassa, E.F., Tagami, Y.: Macromolecules **2**, 14 (1969)
26. Richmond, P., Lal, M.: Chem. Phys. Lett. **24**, 594 (1974)
27. Tuinier, R., Fleer, G.J.: Macromolecules **37**, 8764 (2004)
28. Einstein, A.: Ann. Phys. (IV Folge) **22**, 569 (1907)
29. Einstein, A.: Ann. Phys. (IV Folge) **33**, 1275 (1910)
30. Tuinier, R., Vliegenthart, G.A., Lekkerkerker, H.N.W.: J. Chem. Phys. **113**, 10768 (2000)
31. Eisenriegler, E.: Phys. Rev. E **55**, 3116 (1997)
32. Eisenriegler, E.: J. Chem. Phys. **79**, 1052 (1983)
33. Marques, C.M., Joanny, J.F.: Macromolecules **23**, 268 (1990)
34. Fleer, G.J., Skvortsov, A.M., Tuinier, R.: Macromolecules **36**, 7857 (2003)

35. Taniguchi, T., Kawakatsu, T., Kawasaki, K.: Slow dynamics in condensed matter–proceedings of the 1st Tohwa University International Symposium. AIP Conference Proceedings, vol. 256, Kawasaki, K., Kawakatsu, T., Tokuyama, M. (eds.), , p. 503. AIP, New York (1992)
36. Eisenriegler, E., Hanke, A., Dietrich, S.: Phys. Rev. E **54**, 1134 (1996)
37. Odijk, T.: Macromolecules **29**, 1842 (1996)
38. Aarts, D.G.A.L., Tuinier, R., Lekkerkerker, H.N.W., Phys, J.: Condens. Matter **14**, 7551 (2002)
39. Louis, A.A., Bolhuis, P.G., Meijer, E.J., Hansen, J.P.: J. Chem. Phys. **116**, 10547 (2002)
40. Tuinier, R.: Eur. Phys. J. E **10**, 123 (2003)
41. Tuinier, R., Taniguchi, T., Wensink, H.H.: Eur. Phys. J. E. **23**, 355 (2007)
42. Louis, A.A., Bolhuis, P.G., Meijer, E.J., Hansen, J.P.: J. Chem. Phys. **117**, 1893 (2002)
43. Tuinier, R., Aarts, D.G.A.L., Wensink, H.H., Lekkerkerker, H.N.W.: Phys. Chem. Chem. Phys. **5**, 3707 (2003)
44. Tuinier, R., Fleer, G.J.: Macromolecules **37**, 8754 (2004)
45. Hill, T.L.: An Introduction to Statistical Thermodynamics. Addison-Wesley, New York (1962)
46. Widom, B.: Statistical Mechanics: A Concise Introduction for Chemists. Cambridge University Press, Cambridge (2002)
47. As dimensionless concentration variable ϕ is used throughout. In case of hard colloidal particles the quantity ϕ is the volume fraction. For polymers and penetrable hard spheres ϕ refers to the relative concentration with respect to overlap (see (1.21))
48. Fisher, I.Z.: Statistical Theory of Liquids. The University of Chicago Press, Chicago (1964)
49. Glandt, E.D.: J. Colloid Interface Sci. **77**, 512 (1980)
50. Antonchenko, V.Y., Ilyin, V.V., Makovsky, N.N., Pavlov, A.N., Sokhan, V.P.: Mol. Phys. **52**, 345 (1984)
51. Mao, Y., Cates, M.E., Lekkerkerker, H.N.W.: Physica A **222**, 10 (1995)
52. Walz, J.Y., Sharma, A.: J. Colloid Interface Sci. **168**, 485 (1994)
53. Henderson, J.R.: Mol. Phys. **59**, 89 (1986)
54. Holyst, R.: Mol. Phys. **68**, 391 (1989)
55. Louis, A.A., Bolhuis, P.G., Hansen, J.P., Meijer, E.J.: Phys. Rev. Lett. **85**, 2522 (2000)
56. Fleer, G.J., Cohen Stuart, M.A., Scheutjens, J.M.H.M., Cosgrove, T., Vincent, B.: Polymers at Interfaces. Chapman and Hall, New York (1993)
57. van der Gucht, J., Besseling, N.A.M., van Male, J., Cohen Stuart, M.A.: J. Chem. Phys. **112**, 2886 (2000)
58. Mao, Y.: J. Phys. II France **5**, 1761 (1995)
59. Biben, T., Bladon, P., Frenkel, D.: J. Phys.: Condens Matter **8**, 10799 (1996)
60. Goulding, D., Hansen, J.P.: Mol. Phys. **99**, 865 (2001)
61. Asakura, S., Oosawa, F.: J. Pol. Sci. **33**, 183 (1958)
62. Fåhraeus, R.: Physiol. Rev. **9**, 241 (1929)
63. Auvray, L.: J. Phys. France **42**, 79 (1981)
64. Mao, Y., Cates, M.E., Lekkerkerker, H.N.W.: Phys. Rev. Lett. **75**, 4548 (1995)
65. Mao, Y., Cates, M.E., Lekkerkerker, H.N.W.: J. Chem. Phys. **106**, 3721 (1997)
66. Mao, Y., Bladon, P., Lekkerkerker, H.N.W., Cates, M.E.: Mol. Phys. **92**, 151 (1997)
67. Yaman, K., Jeng, M., Pincus, P., Jeppesen, C., Marques, C.M.: Phys. A **247**, 159 (1997)
68. Yaman, K., Jeppesen, C., Marques, C.M.: Europhys. Lett. **42**, 221 (1998)
69. Lang, P.R.: J. Chem. Phys. **127**, 124906 (2007)
70. Piech, M., Walz, J.Y.: J. Colloid Interface Sci. **232**, 86 (2000)
71. Oversteegen, S.M., Lekkerkerker, H.N.W.: Physica A **341**, 23 (2004)
72. Kleshchanok, D., Tuinier, R., Lang, P.R.: J. Phys.: Condens. Matter **20**, 073101 (2008)
73. Ji, S., Walz, J.Y.: Curr. Opin. Colloid Interface Sci. **20**, 39 (2015)
74. Briscoe, W.H.: Curr. Opin. Colloid Interface Sci. **20**, 46 (2015)
75. Xing, X., Hua, L., Ngai, T.: Curr. Opin. Colloid Interface Sci. **20**, 54 (2015)
76. Ludwig, M., von Klitzing, R.: Current Opin. Colloid. Interface Sci. **47**, 137 (2020)
77. Scarratt, L.R., Trefalt, G., Borkovec, M.: Curr. Opin. Colloid Interface Sci. **55**, 101482 (2021)
78. Israelachvili, J.N.: Intermolecular and Surface Forces, 3rd edn. Academic, Amsterdam (2011)
79. Grier, D.G.: Curr. Opin. Colloid Interface Sci. **2**, 264 (1997)

80. Furst, E.M.: Soft Mater. **1**, 167 (2003)
81. Butt, H.J., Cappella, B., Kappl, M.: Surf. Sci. Rep. **59**, 1 (2005)
82. Walz, J.Y.: Curr. Opin. Colloid Interface Sci. **2**, 600 (1997)
83. Prieve, D.C.: Adv. Colloid Interface Sci. **82**, 93 (1999)
84. Tabor, D., Winterton, R.H.S.: Nature **219**, 1120 (1968)
85. Israelachvili, J.N., Adams, G.E.: Nature **262**, 774 (1976)
86. Israelachvili, J.N., Adams, G.E.: J. Chem. Soc. Faraday Trans. **74**, 975 (1978)
87. Tuinier, R., Petukhov, A.V.: Macromol. Theory Simul. **11**, 975 (2002)
88. Ralston, J., Larson, I., Rutland, M.W., Feiler, A.A., Kleijn, M.: Pure Appl. Chem. **77**, 2149 (2005)
89. Ducker, W.A., Senden, T.J., Pashley, R.M.: Nature **353**, 239 (1991)
90. Piech, M., Walz, J.Y.: J. Phys. Chem. B **108**, 9177 (2004)
91. Prieve, D.C., Walz, J.Y.: Appl. Opt. **32**, 1629 (1993)
92. Edwards, T.D., Bevan, M.A.: Macromolecules **45**, 585 (2012)
93. Helden, L., Koenderink, G.H., Leiderer, P., Bechinger, C.: Langmuir **20**, 5662 (2004)
94. Askin, A.: Phys. Rev. Lett. **24**, 156 (1970)
95. Svoboda, K., Block, S.M.: Annu. Rev. Biophys. Biomol. Struct. **23**, 247 (1994)
96. Schmidt, C.F.: In: Poon, W.C.K., Andelman, D. (eds.) Soft Condensed Matter Physics in Molecular and Cell Biology, p. 279. Taylor and Francis, New York (2006)
97. Crocker, J.C., Grier, D.G.: J. Colloid Interface Sci. **179**, 298 (1996)
98. Han, Y.L., Grier, D.G.: Phys. Rev. Lett. **91**, 038302 (2003)
99. Baumgärtl, J., Bechinger, C.: Europhys. Lett. **71**, 487 (2005)
100. Verma, R., Crocker, J.C., Lubensky, T.C., Yodh, A.G.: Macromolecules **33**, 177 (2000)
101. McQuarrie, D.A.: Statistical Mechanics. University Science Books, Sausalito (2000)
102. Royall, C.P., Louis, A.A., Tanaka, H.: J. Chem. Phys. **127**, 044507 (2007)
103. Kirkwood, J.G.: J. Chem. Phys. **7**, 919 (1939)
104. De Hek, H., Vrij, A.: J. Colloid Interface Sci. **88**, 258 (1982)
105. Ye, X., Narayanan, T., Tong, P., Huang, J.S., Lin, M.Y., Carvalho, B.L., Fetters, L.J.: Phys. Rev. E **54**, 6500 (1996)
106. Vrij, A.: J. Chem. Phys. **112**, 9489 (2000)
107. Vrij, A.: Colloids Surf. A **213**, 117 (2003)
108. Bodnár, I., Dhont, J.K.G., Lekkerkerker, H.N.W.: J. Phys. Chem. **100**, 19614 (1994)
109. Tuinier, R., Dhont, J.K.G., de Kruif, C.G.: Langmuir **16**, 1497 (2000)
110. Mutch, K.J., van Duijneveldt, J.S., Eastoe, J., Grillo, I., Heenan, R.K.: Langmuir **26**, 1630 (2010)
111. Fuchs, M., Schweizer, K.S.: Phys. Rev. E **64**, 021514 (2001)
112. Fuchs, M., Schweizer, K.S., Phys, J.: Condens. Matter **14**, R239 (2002)

Phase Transitions of Hard Sphere–Depletant Mixtures—The Basics

3

Phase transitions are the result of the physical properties of a collection of particles and depend on their interactions. In Chap. 2, we focused on two-body interactions. As we shall see, depletion interactions are usually not pair-wise additive. Therefore, the prediction of phase transitions of particles with depletion interaction is not straightforward. A description of the thermodynamic properties of the pure colloidal dispersion is required as a starting point. Here, the colloid–atom analogy, recognised by Einstein and exploited by Perrin in his classical experiments, is very useful. Subsequently, we explain the basics of the free volume theory for the phase behaviour of colloid–depletant systems. In this chapter, we only treat the simplest type of depletant—the penetrable hard sphere (PHS).

3.1   Introduction: The Colloid–Atom Analogy

In his seminal 1905 paper on Brownian motion, Einstein [1] recognised and used the fact that colloidal particles in a suspension obey the same statistical thermodynamics as atoms in an assembly of atoms. A well-known example of this colloid–atom analogy is the striking similarity between the ideal gas law for the pressure of a dilute gas and the van 't Hoff law for the osmotic pressure of a dilute suspension. The colloid–atom analogy was exploited by Perrin [2] with simple, yet brilliant, experiments. Using an ordinary light microscope, Perrin verified that the equilibrium concentration of colloidal particles in a dilute suspension in the gravitational field varies exponentially with height. By applying Boltzmann's law to this height distribution, he was able to determine the Boltzmann constant k and Avogadro's number N_{Av}. For this work Perrin received the 1926 Nobel Prize for Physics.

The colloid–atom analogy can also be applied to interacting systems. The direct interaction potentials between atoms then have to be replaced by the *potential of mean*

© The Author(s) 2024
H. N. W. Lekkerkerker, R. Tuinier, M. Vis, *Colloids and the Depletion Interaction*,
Lecture Notes in Physics 1026, https://doi.org/10.1007/978-3-031-52131-7_3

force between the dispersed colloidal particles. In the calculation of the potential of mean force, one takes a statistical average over all possible configurations of the solvent components. In the previous chapter, we treated the calculation of the potential of mean force due to dissolved nonadsorbing polymers and (small) colloidal particles.

The concept 'potential of mean force' was used by Onsager [3] in his theory for the isotropic–nematic phase transition in suspensions of rod-like particles. Since the 1980s, the field of phase transitions in colloidal suspensions has developed tremendously. The fact that the potential of mean force can be varied both in range and depth has given rise to new and fascinating phase behaviour in colloidal suspensions [4]. In particular, sterically stabilised colloidal spheres with interactions close to those between hard spheres [5] have received much attention.

The phase behaviour of such colloidal suspensions should be nearly the same as those of the hypothetical hard-sphere atomic system. Kirkwood [6] stated that when a hard sphere system is gradually compressed, the system will show a transition towards a state of long-range order long before close packing is reached. In 1957, Wood and Jacobson [7] and Alder and Wainwright [8] showed by computer simulations that systems of purely repulsive hard spheres indeed exhibit a well-defined fluid–crystal transition. It took some time before the fluid–crystal transition of hard spheres became widely accepted. There is no exact proof that the transition occurs. Its existence has been inferred from numerical simulations or from approximate theories as treated in this chapter. However, this transition has been observed in hard-sphere-like colloidal suspensions [9].

The hard sphere fluid–crystal transition plays an important role as a reference point in the development of theories for the liquid and solid states and their phase behaviour [10]. We consider it in some detail in the next section. For hard spheres, the phase behaviour is relatively simple as there is no gas–liquid (GL) coexistence. After that we discuss the phase behaviour under the influence of the attraction caused by the depletion interaction; then a GL transition can occur. We illustrate the enrichment of the phase behaviour in the somewhat hypothetical system consisting of hard spheres and PHSs.

3.2   The Hard-Sphere Fluid–Crystal Transition

Following the work of Wood and Jacobson [7] and Alder and Wainwright [8], the location of the hard sphere fluid–crystal transition was determined from computer simulations by Hoover and Ree [11]. They found that the volume fractions of the coexisting fluid (f) and face-centred cubic crystal (s) are given by $\phi_f = v_0 n = 0.494$ and $\phi_s = v_0 n = 0.545$ [12] at a coexistence pressure $P v_0/kT = 6.12$. The quantity v_0 is the volume of a colloidal particle, so here $v_0 = (4\pi/3) R^3$, with R the radius of the hard sphere, is the hard sphere volume. As in Chap. 2, $n = N/V$ refers to the number density of N particles in a volume V.

Fig. 3.1 The pressure of hard spheres. Shown are (solid curves) the Carnahan–Starling expression (Eq. (3.1)) for a fluid ($\phi \leq 0.494$), the cell model result (Eq. (3.13)) for an FCC crystal ($\phi \geq 0.545$) and (•) Monte Carlo computer simulation results [14]. Fluid–solid coexistence is also indicated from (dotted line) theory (see Sect. 3.2.3) and (○) simulations [11]

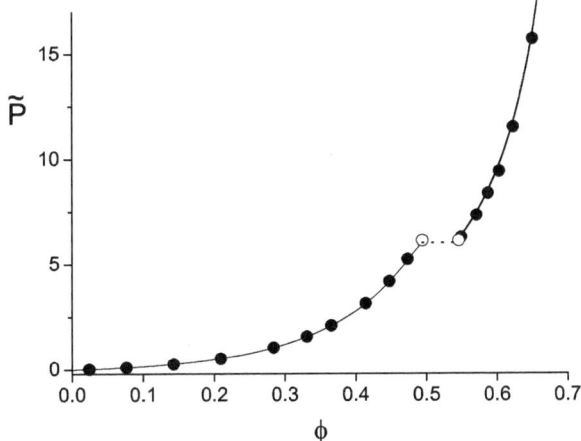

We present a simple theoretical treatment of the hard-sphere fluid–crystal transition that will also serve as a reference framework for our treatment of phase transitions in a system of colloids with depletion attraction.

3.2.1 Hard-Sphere Fluid

The Carnahan–Starling equation of state [13] is a useful expression to describe a fluid of hard spheres. It can be written in terms of the dimensionless pressure $\tilde{P}_{\mathrm{f}}$ as

$$\tilde{P}_{\mathrm{f}} = \frac{P v_0}{kT} = \frac{\phi + \phi^2 + \phi^3 - \phi^4}{(1 - \phi)^3}. \tag{3.1}$$

In Fig. 3.1, we compare the pressure given by the Carnahan–Starling equation of state (Eq. (3.1)) up to $\phi \lesssim 0.5$ with computer simulations. We see that Eq. (3.1) is very accurate.

A 'simple' way to derive this equation of state is to start from the virial expansion of the pressure [15],

$$\frac{P}{nkT} = 1 + \sum_{i=2} B_i n^{i-1}, \tag{3.2}$$

and use the fact that, to a good approximation, the virial coefficients can be written as [13]

$$\frac{B_i}{v_0^{i-1}} = (i - 1)(i + 2). \tag{3.3}$$

With Eq. (3.2), this yields Eq. (3.1). For hard spheres, it is possible to calculate exact values of B_2–B_4 and to perform numerical calculations for B_5 and beyond using statistical mechanics [16]. In Table 3.1, we compare exact virial coefficients

Table 3.1 Comparison between the state-of-the-art values [17] and the Carnahan–Starling equation of state [13] for the virial coefficients of hard spheres. The virial coefficients B_2 to B_{10} are normalised by the particle volume as B_i/v_0^{i-1}

i	B_i/v_0^{i-1}	
	Actual	Eq. (3.3)
2	4	4
3	10	10
4	18.36	18
5	28.22	28
6	39.82	40
7	53.34	54
8	68.53	70
9	85.81	88
10	105.8	108

Exercise 3.1. Show that the summation on the right-hand side of Eq. (3.2) with Eq. (3.3) for the virial coefficients indeed leads to the equation of state of Eq. (3.1)

(B_2, B_3, B_4) and those of numerically high accuracy [17] $(B_5, \ldots, B_{10})$ with the approximation given by Eq. (3.3).

From the Gibbs–Duhem relation $S dT - V dP + N d\mu = 0$, we can calculate the chemical potential from the pressure (Eq. (A.12)). For constant temperature T, this relation may be written as

$$dP = n d\mu = \frac{\phi}{v_0} d\mu \tag{3.4}$$

so that μ follows as:

$$\mu = kT \ln\left(\frac{\Lambda^3}{v_0}\right) + v_0 \int_0^\phi \frac{1}{\phi'} \frac{dP}{d\phi'} d\phi', \tag{3.5}$$

where $dP/d\phi$ can be calculated from Eq. (3.1) for a fluid of hard spheres. The first (constant) term follows from the ideal gas reference state [16]; Λ is the de Broglie wavelength $\Lambda = h/\sqrt{2\pi m_c kT}$, with the colloid mass m_c and Planck's constant h. The result for the chemical potential of a hard sphere in a fluid with volume fraction of hard spheres ϕ is

$$\frac{\mu_f}{kT} = \ln\left(\frac{\Lambda^3}{v_0}\right) + \ln\phi + \frac{(8 - 9\phi + 3\phi^2)\phi}{(1 - \phi)^3}. \tag{3.6}$$

After simplification and defining the dimensionless chemical potential $\widetilde{\mu} = \mu/kT$ the simpler form

$$\widetilde{\mu}_{\mathrm{f}} = \ln\left(\frac{\Lambda^3}{v_0}\right) + \ln\phi + \frac{3-\phi}{(1-\phi)^3} - 3 \tag{3.7}$$

is obtained. Finally, using the standard thermodynamic result $\widetilde{P} = \phi\widetilde{\mu} - \widetilde{F}$ (Appendix A), the resulting canonical free energy of the pure hard-sphere dispersion of a fluid is

$$\widetilde{F} = \phi\ln\left(\frac{\phi\Lambda^3}{v_0}\right) - \phi + \frac{4\phi^2 - 3\phi^3}{(1-\phi)^2}. \tag{3.8}$$

Here, we have introduced the normalised Helmholtz energy $\widetilde{F} = Fv_0/kTV$. The first two terms on the right-hand side of Eq. (3.8) are the ideal contribution, while the last hard-sphere interaction term originates from the Carnahan–Starling equation of state [13].

3.2.2   Hard-Sphere Crystal

To obtain the thermodynamic functions of the hard-sphere crystal we use the cell model of Lennard-Jones and Devonshire [18]. The idea of the cell model is that a given particle moves in a free volume v^* set by its neighbours which are located on their lattice positions (see Fig. 3.2). Then the partition function Q takes the form

$$Q = \frac{(v^*)^N}{\Lambda^{3N}}. \tag{3.9}$$

The 'exact' free volumes have a complicated geometry [19], but here we will use the simple approximation of the inscribed sphere. This yields

$$v^* = \frac{4\pi}{3}(r - 2R)^3, \tag{3.10}$$

Fig. 3.2 The free volume of a hard sphere (hatched area) in the cage of its nearest neighbours in the approximation of the inscribed sphere. The hatched area identifies the available volume for the centre of the central sphere and has a radius $r - 2R$

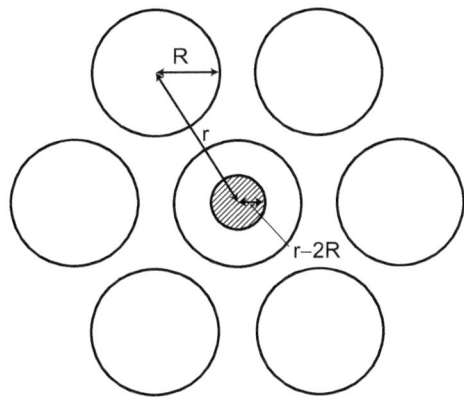

where r is the centre-to-centre distance between the nearest neighbours. Using the relations

$$n\frac{\pi}{6}(2R)^3 = \phi$$

and

$$n\frac{\pi}{6}r^3 = \phi_{cp},$$

where $\phi_{cp} = \pi/3\sqrt{2} \simeq 0.74$ is the volume fraction at close packing, the free volume can be written as

$$v^* = 8v_0\left[\left(\frac{\phi_{cp}}{\phi}\right)^{1/3} - 1\right]^3.$$

We now obtain for the free energy

$$F = -kT \ln Q \tag{3.11}$$

$$= NkT\left\{\ln\left(\frac{27\Lambda^3}{8v_0}\right) - 3\ln\left[\left(\frac{\phi_{cp}}{\phi}\right) - 1\right]\right\}. \tag{3.12}$$

In writing down Eq. (3.12), we used the approximation

$$\left(\frac{\phi_{cp}}{\phi}\right)^{1/3} - 1 \simeq \frac{1}{3}\left(\frac{\phi_{cp}}{\phi} - 1\right).$$

Using the standard thermodynamic relations

$$P = -\left(\frac{\partial F}{\partial V}\right)_{N,T},$$

$$\mu = \left(\frac{\partial F}{\partial N}\right)_{V,T},$$

we obtain

$$\widetilde{P}_s = \frac{3\phi}{1 - \phi/\phi_{cp}}, \tag{3.13}$$

$$\widetilde{\mu}_s = \ln\frac{\Lambda^3}{v_0} + \ln\left[\frac{27}{8(\phi_{cp})^3}\right] + 3\ln\left[\frac{\phi}{1 - (\phi/\phi_{cp})}\right] + \frac{3}{1 - (\phi/\phi_{cp})}. \tag{3.14}$$

The pressure given by Eq. (3.13) can be compared to computer simulation data (e.g., [20]) and turns out to be highly accurate, as can be seen in Fig. 3.1 (for $\phi \gtrsim 0.55$). The constant on the right-hand side

$$\ln\left[\frac{27}{8(\phi_{cp})^3}\right] = 2.1178$$

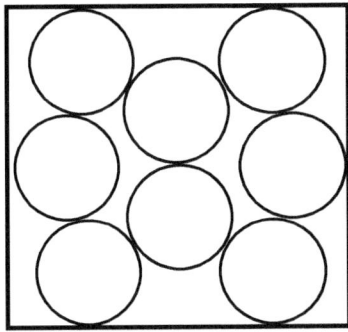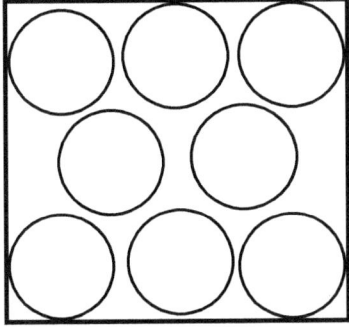

Fig. 3.3 A hard-sphere fluid (*left*) and hard spheres with 'crystalline' order (*right*); free volume entropy drives freezing

is quite close to 2.1306, which can be abstracted from the computer simulation results of Frenkel and Ladd [21]. The full free energy expression for the hard-sphere solid phase is

$$\widetilde{F} = \phi \ln \left(\frac{\Lambda^3}{v_0} \right) + 2.1178\phi + 3\phi \ln \left(\frac{\phi}{1 - \phi/\phi_{cp}} \right). \tag{3.15}$$

3.2.3 Fluid–Crystal Coexistence

Solving the coexistence conditions (Appendix A)

$$\widetilde{P}_f(\phi_f) = \widetilde{P}_s(\phi_s) \tag{3.16a}$$

$$\widetilde{\mu}_f(\phi_f) = \widetilde{\mu}_s(\phi_s) \tag{3.16b}$$

yields coexisting volume fractions $\phi_f = 0.491$, $\phi_s = 0.541$ and a coexistence pressure $\widetilde{P} = 6.01$. These values are indeed very close to the computer simulation results (see the comparison in Fig. 3.1).

The equilibrium configuration of hard spheres is the one that maximises the entropy of the system. At low densities, the configurations of maximum entropy correspond to disordered arrangements. As the density increases, the number of disordered arrangements is severely reduced due to the inefficiency of 'packing' them into the fixed volume. Then, crystalline arrangements lead to a more efficient packing and make more arrangements possible. This is schematically depicted in Fig. 3.3.

Hence, the thermodynamic stability of the hard sphere crystal can be 'explained' on a purely entropic basis. Since the 1950s, the fluid–crystal transition has been observed in suspensions of monodisperse repulsive colloidal particles [22–24]. Particularly, the work on sterically stabilised silica particles [25] and sterically stabilised PMMA particles [9] has served as a reference point. Figure 3.4a, b illustrate the phase behaviour of dispersed PMMA colloids as studied by Pusey and van Megen [9]. Above a volume fraction $\phi = 0.58$, these authors observed an amorphous glassy

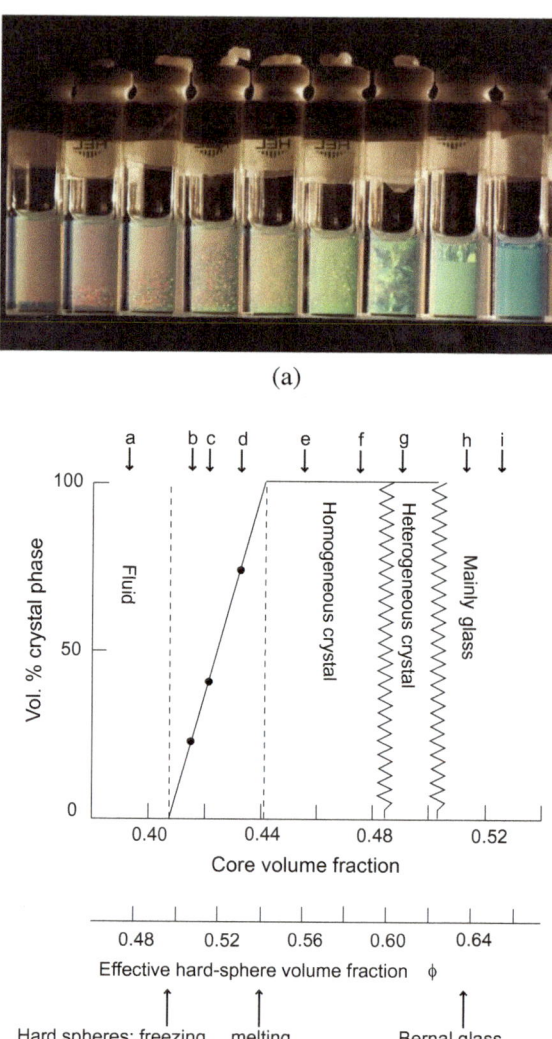

Fig. 3.4 **a** Dispersions with hard-sphere-like PMMA spheres at volume fractions around the fluid–solid phase transition [9]. Reprinted with permission from Ref. [9]. Copyright 1986 Nature. High-quality image kindly provided by P.N. Pusey. **b** The states of these dispersions; the labels a–i refer to the samples from left to right in (**a**). The abscissa indicates the measured volume fraction of PMMA cores, which is smaller than the effective volume fraction of hard spheres that includes the short stabilising brushes

phase that did not crystallise over several months as well as the fluid–crystal transition. The explanation for this phenomenon is that, for these high-volume fractions, the particles become so tightly trapped or caged by their neighbours that they are unable to move far enough to nucleate crystallisation. Instead, long-lived metastable states called colloidal glasses are obtained. We return to glasses in Sect. 4.4.2.

In practice, colloids are polydisperse. Computer simulations show that crystallisation of hard spheres does not occur above a polydispersity of 11.8% in diameter [26]. Pusey [27] suggested that the maximum polydispersity, in terms of the relative standard deviation σ_{max}, depends on the close packing and melting volume fractions ϕ_{cp} and ϕ_m, respectively,

$$\sigma_{max} = \left(\frac{\phi_{cp}}{\phi_m}\right)^{1/3} - 1. \tag{3.17}$$

For hard spheres with $\phi_{cp} = 0.74$ and $\phi_m = 0.545$, Eq. (3.17) provides $\sigma_{max} = 0.11$, so 11%. To keep the description simple, we further focus on monodisperse hard spheres.

Exercise 3.2. Rationalise Eq. (3.17).

3.3 Free Volume Theory of Hard Spheres and Depletants

3.3.1 System

Several theories have been developed that enable calculations of phase transitions in systems with depletion interactions. An important successful treatment accounting for depletion interactions in a many-body system [28,29] is thermodynamic perturbation theory ([16], or see Chap. 6 in Ref. [15]). In this classical approach, depletion effects can be treated as a perturbation to the hard-sphere free energy, as was done by Gast, Hall and Russel [28]. Their important work predicted that, for a sufficient depletant concentration, the depletion interaction leads to a phase diagram with stable colloidal gas, liquid and solid phases for $\delta/R \geq 0.3$. For small depletants with $\delta/R \leq 0.3$ only colloidal fluid and solid phases are thermodynamically stable, and the gas–liquid transition is meta-stable. Although implementation of this theory is straightforward, it has the drawback that it does not account for depletant partitioning over the coexisting phases.

A theory that accounts for depletant partitioning over the coexisting phases was developed in the early 1990s [30,31], which nowadays is commonly referred to as free volume theory (FVT) [32]. This theory is based on the osmotic equilibrium between a (hypothetical) depletant and the colloid–depletant system. The depletants were simplified as PHSs. A pictorial representation is given in Fig. 3.5.

This theory has the advantage that the depletant concentrations in the coexisting phases follow directly from the (semi-)grand potential that describes the colloid–depletant system. As illustrated in Fig. 3.6, the system tries to arrange itself such as to provide a large free volume for the depletant. This (entropic) physical origin of the phase transitions induced by depletion interactions is incorporated into the theory in a natural way.

Fig. 3.5 A system (*right*) that contains colloids and penetrable hard spheres (PHSs) in osmotic equilibrium with a reservoir (*left*) only consisting of PHSs. A hypothetical membrane that allows permeation of solvent and PHSs but not of colloids is indicated by the dashed line. Solvent is considered as 'background'

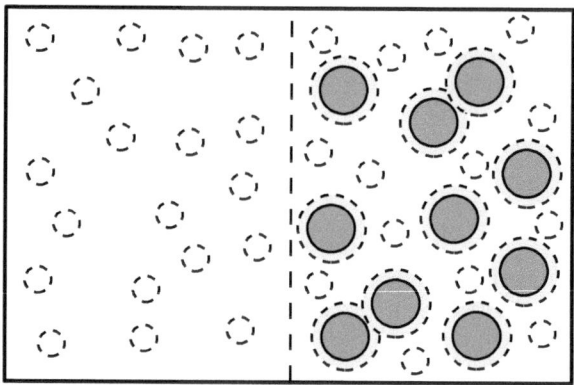

Fig. 3.6 Illustration of the free volume V_{free}: it is the unshaded volume not occupied by the colloids and (partially overlapping) depletion layers

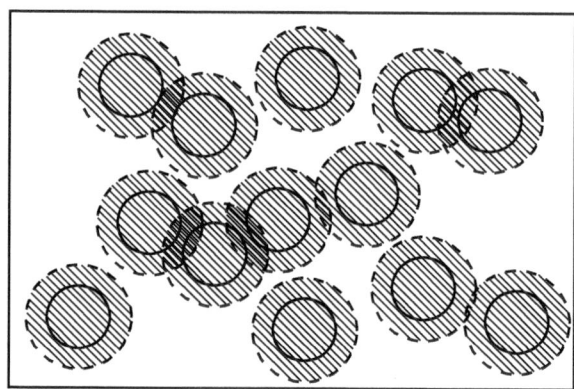

In FVT, the multiple overlap of depletion zones with thickness δ is taken into account (see Fig. 3.7). Multiple overlap occurs for

$$\frac{\delta}{R} > \frac{2}{\sqrt{3}} - 1 \simeq 0.15,$$

where three depletion zones start to overlap (Fig. 3.7). Only for $\delta/R < 0.15$ is a colloid–depletant mixture pair-wise additive. This has a considerable influence on the topology of the phase diagram [33]. Multiple overlap of depletion layers widens the liquid window, which is the parameter range with phase transitions that include a stable liquid, in comparison with a pair-wise additive system [32].

Exercise 3.3. Show that multiple overlap only occurs for $\delta/R > 2/\sqrt{3} - 1$.

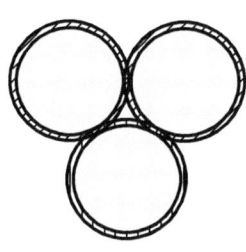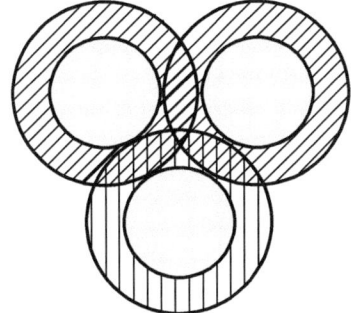

Fig. 3.7 Three hard spheres surrounded by depletion layers (hatched areas). When the depletion layers are thin (*left*) there is no multiple overlap of depletion layers; the system is pair-wise additive. For thicker depletion layers (*right*), multiple overlap of depletion layers occurs and depends on more than two-body contributions. The lowest value for δ/R, where multiple overlap occurs, follows from considering the triangle formed by the three particle centres; its edge is $2R + h$ at particle separation h. Multiple overlap starts when the centre of the triangle is a distance $R + \delta$ from the corners

3.3.2 Thermodynamics

The starting point of FVT is the calculation of the semi-grand potential describing the system of N_c colloidal spheres mixed with N_d depletants as depicted in Fig. 3.5.

$$\Omega(N_c, V, T, \mu_d) = F(N_c, N_d, V, T) - \mu_d N_d. \tag{3.18}$$

Using the thermodynamic relationship

$$\left(\frac{\partial \Omega}{\partial \mu_d}\right)_{N_c, V, T} = -N_d, \tag{3.19}$$

we can write

$$\Omega(N_c, V, T, \mu_d) = F_0(N_c, V, T) - \int_{-\infty}^{\mu_d} N_d(\mu_d')\mathrm{d}\mu_d'. \tag{3.20}$$

Here, $F_0(N_c, V, T)$ is the free energy of the colloidal particle system without added depletant as given by Eq. (3.8) (fluid) or Eq. (3.15) (solid).

The key step now is the calculation of the number of depletants in the colloid–depletant system as a function of the chemical potential μ_d imposed by the depletants in the reservoir. In the calculations presented below, we model the colloidal particles as hard spheres with diameter $2R$, and the depletants by PHSs with diameter σ.

For the calculation of N_d, we make use of the Widom insertion theorem [34], according to which the chemical potential of the depletants in the hard sphere–depletant system can be written as

$$\mu_d = \mu_d^0 + kT \ln \frac{N_d}{\langle V_{\text{free}} \rangle}. \tag{3.21}$$

Here, μ_d^0 is the reference chemical potential of the depletants and $\langle V_{\text{free}} \rangle$ is the ensemble-averaged free volume for the depletants in the system of hard spheres, illustrated in Fig. 3.6.

The chemical potential of the ideal depletants in the reservoir is simply

$$\mu_d = \mu_d^0 + kT \ln n_d^R, \tag{3.22}$$

where n_d^R is the number density of the depletants in the reservoir. By equating the depletant chemical potentials Eqs. (3.21) and (3.22), we obtain

$$N_d = n_d^R \langle V_{\text{free}} \rangle. \tag{3.23}$$

The average free volume obviously depends not only on the volume fraction of the hard spheres in the system but also on the chemical potential of the depletants. The activity of the depletants affects the average configuration of the hard spheres. We now make the key approximation to replace $\langle V_{\text{free}} \rangle$ by the free volume in the pure hard sphere dispersion $\langle V_{\text{free}} \rangle_0$:

$$N_d \approx n_d^R \langle V_{\text{free}} \rangle_0. \tag{3.24}$$

This expression is correct in the limit of low depletant activity but is only an approximation for higher depletant concentrations. Substituting the approximation Eq. (3.24) in Eq. (3.20) and using the Gibbs–Duhem relation (Eq. (A.12)),

$$n_d^R d\mu_d = dP^R, \tag{3.25}$$

gives

$$\Omega(N_c, V, T, \mu_d) = F_0(N_c, V, T) - P^R \langle V_{\text{free}} \rangle_0, \tag{3.26}$$

where $P^R = n_d^R kT$ is the (osmotic) pressure of the depletants in the reservoir. It is noted that this expression was formally derived by Dijkstra, Brader and Evans [35].

As we have expressions for the free energy of the hard-sphere system (both in the fluid and solid state, see Sect. 3.2 and for the pressure of the reservoir, the only remaining quantity to calculate is $\langle V_{\text{free}} \rangle_0$. If we also replace $\langle V_{\text{free}} \rangle$ in Eq. (3.21) with the free volume in the pure hard sphere dispersion $\langle V_{\text{free}} \rangle_0$ we obtain:

$$\mu_d = \mu_d^0 + kT \ln \frac{N_d}{\langle V_{\text{free}} \rangle_0}. \tag{3.27}$$

But we can also write the chemical potential μ_d as

$$\mu_d = \mu_d^0 + kT \ln \frac{N_d}{V} + W, \tag{3.28}$$

where W is the reversible work required for inserting the depletant in the hard sphere dispersion. Combining Eqs. (3.27) and (3.28), we find for the free volume fraction α:

$$\alpha = \frac{\langle V_{\text{free}} \rangle_0}{V} = e^{-W/kT}. \tag{3.29}$$

3.3.3 Scaled Particle Theory

An expression for the work of insertion W can be obtained from scaled particle theory (SPT) [36]. SPT was developed to derive expressions for the chemical potential and pressure of hard sphere fluids by relating them to the reversible work needed to insert an additional particle in the system. This work W is calculated by scaling the size of the sphere to be inserted: the size of the scaled particle is $\lambda\sigma$, with λ being the scaling parameter.

In the limit $\lambda \to 0$, the inserted sphere approaches a point particle. In this limiting case, it is very unlikely that the depletion layers overlap. The free volume fraction in this limit can therefore be written as

$$\alpha = \frac{V - N_c\frac{\pi}{6}(2R + \lambda\sigma)^3}{V}$$

$$= 1 - n_c\frac{\pi}{6}(2R + \lambda\sigma)^3.$$

It then follows from Eq. (3.29) that

$$W = -kT \ln\left[1 - n_c\frac{\pi}{6}(2R + \lambda\sigma)^3\right] \qquad \text{for } \lambda \ll 1. \tag{3.30}$$

In the opposite limit $\lambda \gg 1$, when the size of the inserted scaled particle is very large, W (to a good approximation) is equal to the volume work needed to create a cavity $\frac{\pi}{6}(\lambda\sigma)^3$ and is given by

$$W = \frac{\pi}{6}(\lambda\sigma)^3 P \qquad \text{for } \lambda \gg 1, \tag{3.31}$$

where P is the (osmotic) pressure of the hard-sphere dispersion.

In SPT, the above two limiting cases are connected by expanding W as a series in λ:

$$W(\lambda) = W(0) + \left(\frac{\partial W}{\partial \lambda}\right)_{\lambda=0}\lambda + \frac{1}{2}\left(\frac{\partial^2 W}{\partial \lambda^2}\right)_{\lambda=0}\lambda^2 + \frac{\pi}{6}(\lambda\sigma)^3 P. \tag{3.32}$$

This yields

$$\frac{W(\lambda = 1)}{kT} = -\ln[1 - \phi] + \frac{3q\phi}{1 - \phi} + \frac{1}{2}\left[\frac{6q^2\phi}{1 - \phi} + \frac{9q^2\phi^2}{(1 - \phi)^2}\right] + \frac{\frac{\pi}{6}q^3(2R)^3 P}{kT}, \tag{3.33}$$

where q is the size ratio between the depletant with diameter σ and the hard sphere with diameter $2R$

$$q = \frac{\sigma}{2R}. \tag{3.34}$$

As was the original objective of SPT [36], the pressure P of the pure hard sphere system can be obtained from the reversible work of inserting an identical sphere $(q = 1)$

$$\frac{W}{kT} = -\ln[1 - \phi] + \frac{6\phi}{1 - \phi} + \frac{9\phi^2}{2(1 - \phi)^2} + \frac{\pi(2R)^3 P}{6kT}, \qquad (3.35)$$

to obtain the chemical potential of the hard spheres

$$\mu_c = \mu_c^0 + kT \ln \frac{N_c}{V} + W. \qquad (3.36)$$

Applying the Gibbs–Duhem relation (Eq. (A.12))

$$\frac{\partial P}{\partial n_c} = n_c \frac{\partial \mu_c}{\partial n_c},$$

one obtains

$$\frac{P v_0}{kT} = \frac{\phi + \phi^2 + \phi^3}{(1 - \phi)^3}, \qquad (3.37)$$

which is the famous SPT expression for the pressure of a hard sphere fluid [36]. This preceded the slightly more accurate Carnahan–Starling equation Eq. (3.1), which contains an additional term ϕ^4.

Inserting Eq. (3.37) into Eq. (3.33) and using Eq. (3.29) yields

$$\alpha = (1 - \phi) \exp[-Q], \qquad (3.38)$$

where

$$Q = ay + by^2 + cy^3, \qquad (3.39a)$$

$$a = 3q + 3q^2 + q^3 = (1 + q)^3 - 1, \qquad (3.39b)$$

$$b = \frac{9}{2}q^2 + 3q^3, \qquad (3.39c)$$

$$c = 3q^3, \qquad (3.39d)$$

$$y = \frac{\phi}{1 - \phi}. \qquad (3.39e)$$

In Fig. 3.8, we present a comparison of the free volume fraction predicted by SPT (Eq. (3.38)) and computer simulations [37] on hard sphere–PHS mixtures for $q = 0.5$ as a function of ϕ. As can be seen, the agreement is very good. We now have all the ingredients to compile the semi-grand potential Ω given by Eq. (3.26).

Fig. 3.8 Free volume fraction for penetrable hard spheres in a hard sphere dispersion for $q = \sigma/2R = 0.5$ as a function of the hard-sphere concentration. Data points are redrawn from Meijer [37]. Curve is the SPT prediction, Eq. (3.38)

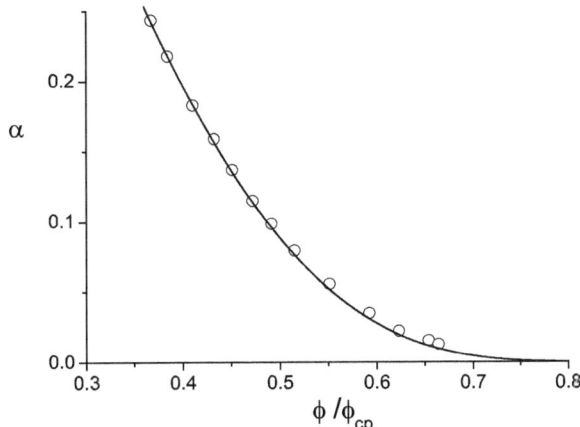

From Ω (given by Eq. (3.26)), the pressure and chemical potential of the hard spheres in the hard sphere–depletant system at given μ_d are obtained

$$P = -\left(\frac{\partial\Omega}{\partial V}\right)_{N_c, T, \mu_d} = P^0 + P^R\left(\alpha - n_c\frac{\partial\alpha}{\partial n_c}\right), \qquad (3.40)$$

$$\mu_c = \left(\frac{\partial\Omega}{\partial N_c}\right)_{V, T, \mu_d} = \mu_c^0 - P^R\frac{\partial\alpha}{\partial n_c}. \qquad (3.41)$$

For non-interacting depletants, P^R is simply given by van 't Hoff's law $P^R = n_d^R kT$ or

$$\widetilde{P}^R = \frac{P^R v_0}{kT} = n_d^R v_d q^{-3} = \phi_d^R q^{-3}, \qquad (3.42)$$

with $\phi_d^R = n_d^R v_d$ the relative reservoir depletant concentration, where $v_d = \frac{\pi}{6}\sigma^3$ is the volume of a depletant sphere [12]. As PHSs can, by definition, freely interpenetrate each other, it is common to define the overlap condition via $n^* v_d = 1$, i.e., at $n^* = 1/v_d$ the spheres fill the available space. Hence, ϕ_d also denotes the concentration of PHSs relative to the overlap or, more briefly, their volume fraction.

3.3.4 Phase Diagrams

We can now calculate the phase behaviour of a system of hard spheres and depletants (Appendix A) by solving the coexistence equations for a phase I in equilibrium with a phase II

$$\mu_c^I(n_c^I, \mu_d) = \mu_c^{II}(n_c^{II}, \mu_d), \qquad (3.43a)$$

$$P^I(n_c^I, \mu_d) = P^{II}(n_c^{II}, \mu_d). \qquad (3.43b)$$

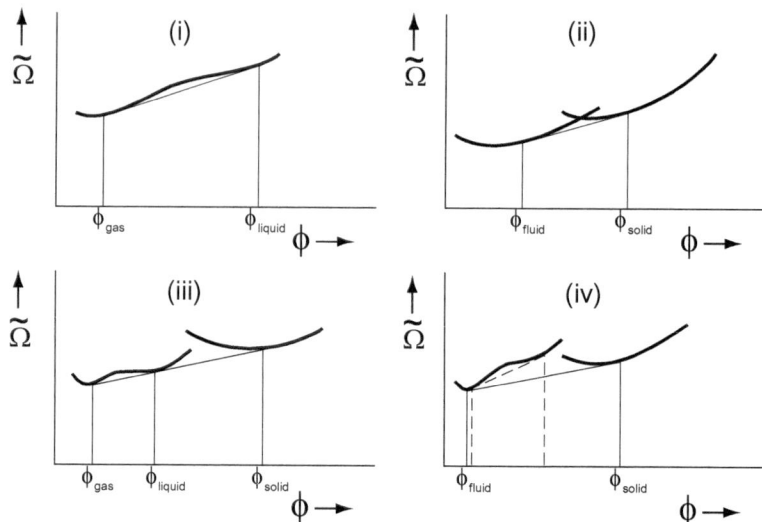

Fig. 3.9 The dimensionless semi-grand potential $\widetilde{\Omega}$ as a function of volume fraction ϕ. Schematic view of the common tangent construction (straight lines) to determine the phase coexistence in mixtures of colloidal hard spheres and PHSs. (i): gas–liquid coexistence, (ii): fluid–solid coexistence, (iii): gas–liquid–solid triple coexistence, and (iv): fluid–solid coexistence near a metastable gas–liquid coexistence (dashed lines represent the common tangent construction for this case)

For numerical computations of phase coexistence, it is convenient to work with dimensionless quantities. The dimensionless version of Eq. (3.26), the free volume expression for the grand potential, is

$$\widetilde{\Omega} = \widetilde{F}_0 - \alpha \widetilde{P}^{\mathrm{R}}, \tag{3.44}$$

where $\widetilde{\Omega} = \Omega v_0 / kTV$.

In Fig. 3.9, the semi-grand potential is presented as a function of the colloid volume fraction for a given depletant reservoir concentration and size ratio q. In this figure, thin straight lines are drawn, which denote a so-called common tangent construction. This allows (graphical) determination of conditions where phases coexist. The first criterion for two coexisting binodal compositions is equality of the slope because it corresponds to the chemical potential (Appendix A):

$$\widetilde{\mu}_{\mathrm{c}} = \left(\frac{\partial \widetilde{\Omega}}{\partial \phi} \right)_{T,V,\widetilde{\mu}_{\mathrm{d}}}. \tag{3.45}$$

Equality of the chemical potential of the depletants is ensured by the reservoir, which has a fixed depletant concentration. Therefore, when two compositions can be connected through a common tangent, the binodal points are found. As the pressure is given by

$$\widetilde{P} = \phi \widetilde{\mu}_{\mathrm{c}} - \widetilde{\Omega}, \tag{3.46}$$

the extrapolation of the common tangent to $\phi = 0$ corresponds to the total pressure $-\widetilde{P}$ of the system.

Four possible scenarios are considered in Fig. 3.9. Scenario (i) corresponds to the possibility of gas–liquid coexistence, which follows from $\widetilde{\Omega}(\phi)$ with $\widetilde{F}_0(\phi)$ for the colloidal fluid state. In situation (ii) $\widetilde{\Omega}(\phi)$ is given for both the fluid state and for the solid state and the common tangent shows the compositions where fluid and solid coexist. A combination of (i) and (ii) is possible under conditions where the curve for the fluid state shows an instability itself and the gas and liquid compositions coexist with a solid phase as exemplified by situation (iii). Finally, situation (iv) refers to an instability of the fluid state within the concentration region where fluid and solid coexist. Here, the values for the chemical potential of the colloidal particles are larger for gas–liquid coexistence than for fluid–solid coexistence, as follows from the slopes; as a consequence, the gas–liquid coexistence is metastable. The binodal compositions for each polymer concentration can be found in this manner, and full-phase diagrams can be constructed.

For non-interacting depletants such as PHSs μ's and P's in Eqs. (3.43a) and (3.43b) can be written such that binodal colloid concentrations follow from solving one equation in a single unknown [32]. We rewrite Eqs. (3.40) and (3.41) as

$$\widetilde{\mu} = \widetilde{\mu}^0 + \widetilde{P}^R g(\phi), \tag{3.47}$$

$$\widetilde{P} = \widetilde{P}^0 + \widetilde{P}^R h(\phi), \tag{3.48}$$

where $g = -\partial\alpha/\partial\phi$ and $h = \alpha + g\phi$. The functions g and h may be written as

$$g(\phi) = e^{-Q(\phi)}[1 + (1 + y)(a + 2by + 3cy^2)] \tag{3.49}$$

and

$$h(\phi) = e^{-Q(\phi)}(1 + ay + 2by^2 + 3cy^3). \tag{3.50}$$

The gas–liquid binodal can be solved from the second and third parts of

$$\widetilde{P}^R = \frac{\widetilde{\mu}_f^0(\phi_l) - \widetilde{\mu}_f^0(\phi_g)}{g(\phi_g) - g(\phi_l)} = \frac{\widetilde{P}_f^0(\phi_l) - \widetilde{P}_f^0(\phi_g)}{h(\phi_g) - h(\phi_l)}, \tag{3.51}$$

where $\widetilde{\mu}_f^0$ and $\widetilde{P}_f^0$ are only a function of ϕ (see Eqs. (3.1) and (3.7)). Hence, Eq. (3.51) gives a unique correlation for $\phi_l(\phi_g)$ at given q: for some value of ϕ_g, within the region of ϕ_g values where a colloidal gas coexists with a colloidal liquid, the corresponding value of ϕ_l follows from the second equality of Eq. (3.51). The corresponding binodal depletant reservoir pressure $\widetilde{P}^R$ then follows from the first equality.

Exercise 3.4. Derive Eqs. (3.49) to (3.51).

Similarly, the fluid–solid binodal can be obtained from

$$\widetilde{P}^R = \frac{\widetilde{\mu}_s^0(\phi_s) - \widetilde{\mu}_f^0(\phi_f)}{g(\phi_f) - g\phi_s)} = \frac{\widetilde{P}_s^0(\phi_s) - \widetilde{P}_f^0(\phi_f)}{h(\phi_f) - h(\phi_s)}, \qquad (3.52)$$

where again $\widetilde{\mu}_f^0$ is given by Eq. (3.7) and $\widetilde{P}_f^0$ by Eq. (3.1); these are the fluid contributions. For the solid phase $\widetilde{P}_s^0(\phi)$ and $\widetilde{\mu}_s^0(\phi)$ are given by Eqs. (3.13) and (3.14).

Triple points have equal pressures and chemical potentials simultaneously for colloidal gas, liquid *and* solid phases. At the triple point, Eqs. (3.51) and (3.52) are connected through equal values for $\widetilde{P}^R$ and, in principle, form a set of four equations from which the four coordinates of the triple point (ϕ_g, ϕ_l, ϕ_s, $\widetilde{P}^R$) follow. However, for the present PHS system the problem can be reduced to solving one equation with one unknown [32].

For large q ($q \geq 0.6$), the triple point can be approximated easily from Eqs. (3.47) and (3.48). It can be observed that the fluid–solid coexistence of the triple point occurs at very similar colloid concentrations as the pure hard sphere phase transition. For large q values, Eqs. (3.47) and (3.48) can be written as $\widetilde{\mu}_f = \widetilde{\mu}_f^0 = \widetilde{\mu}_s^0$ and $\widetilde{P}_f = \widetilde{P}_f^0 = \widetilde{P}_s^0$ because $g(\phi)$ and $h(\phi)$ vanish for large q. In the coexisting colloidal gas phase the colloid concentration is then extremely small, such that $\widetilde{P}_g = \widetilde{P}^R$ since $h(\phi) \to 1$. This implies that $\widetilde{P}^R = \widetilde{P}_f^0 = \widetilde{P}_s^0 = 6.01$ at the triple point. Hence, for large q, the fluid–solid coexistence of the triple point occurs at nearly the same colloid concentrations as for the pure hard-sphere phase transition. The relative depletant concentration at the triple point now follows as $\phi_d^R \simeq \widetilde{P}^R q^3 = 6.01 q^3$. As can be seen in Figs. 3.10 and 3.11 ($q = 1.0$ and 0.6, respectively) this is rather accurate.

The critical point can also be found as one equation with one unknown. For details, we refer the reader to Ref. [32]. The same applies to the *critical endpoint* (CEP), which corresponds to the q value where CP and TP coincide; it is the lowest q where a stable liquid is possible. See the extended discussions on liquid windows with relation to the CEP in Refs. [32,33].

In Fig. 3.10, we present phase diagrams for $q = 0.1, q = 0.4$ and $q = 1.0$. As was already found by Gast, Hall and Russel [28], for $q = 0.1$, there is only a fluid–crystal transition. For $\phi_d = 0$, the demixing gap is $0.491 < \phi < 0.541$ (see Sect. 3.2.3). With increasing depletant concentration, this gap widens. For the phase diagram with $q = 0.4$, both a critical point (CP) and triple point (TP) are present, analogous to those found in simple atomic systems. At high depletant concentrations in the reservoir (above TP), a very dilute fluid (colloidal gas) coexists with a highly concentrated colloidal solid. A colloidal gas (dilute fluid) coexists with a colloidal liquid (more concentrated fluid) between TP and CP. At high packing fractions below the triple line, a colloidal liquid coexists with a colloidal solid phase. Increasing the depletant activity now plays a role similar to lowering the temperature in atomic systems. For larger q (see $q = 1.0$) the qualitative picture remains the same, while the liquid window expands. As expected, in the absence of depletant only the fluid–solid phase transition of a pure hard sphere dispersion remains.

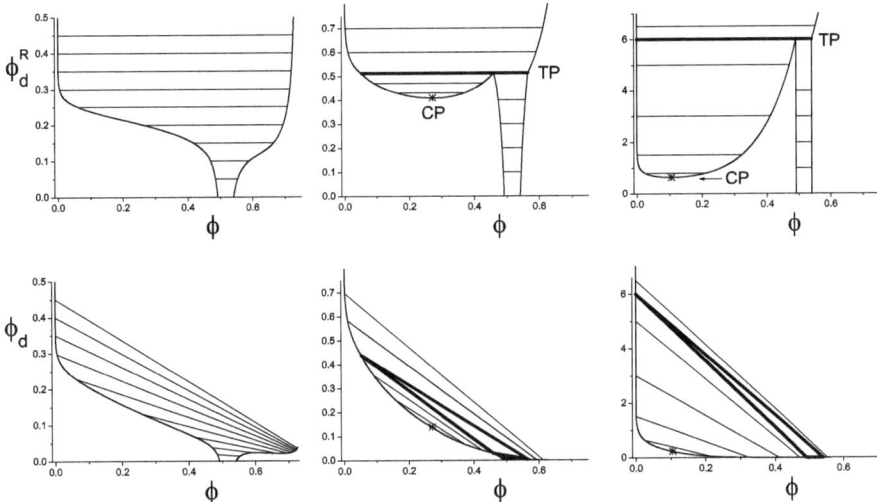

Fig. 3.10 Free volume theory predictions for the phase diagrams for hard spheres as depletants following Lekkerkerker et al. [31]. The diagrams are for $q = 0.1$ (*left*), $q = 0.4$ (*middle*) and $q = 1.0$ (*right*). The top row of diagrams have depletant reservoir concentrations ϕ_d^R as ordinates, and the bottom row of diagrams are in system depletant concentrations. Triple lines and triangles are indicated as thick lines. Triple point is indicated by "TP"; critical point is indicated by "CP" and an asterisk. A few representative tie-lines are plotted as thin lines

In the top diagrams of Fig. 3.10, the ordinate axis is the depletant concentration in the reservoir. The depletant concentrations in the system of coexisting phases can be obtained by using the relation

$$\frac{N_d}{V} = -\frac{1}{V}\left(\frac{\partial \Omega}{\partial \mu_d}\right)_{N_c, V, T} = \alpha n_d^R$$

or

$$\phi_d = \alpha \phi_d^R.$$

Coexisting phases of course have the same μ_d and hence the same n_d^R. However, since the volume fractions of hard spheres and, subsequently, the free volume fractions α can be substantially different in the coexisting phases, the depletant concentration n_d in the two (or three) phases is not the same, so the tie-lines are no longer horizontal. This is illustrated in the bottom diagrams of Fig. 3.10: now the ordinate axis gives the relative 'internal' or system concentrations ϕ_d. A few selected tie-lines are drawn to give an impression of depletant partitioning over the phases. Interestingly, the horizontal triple line in the presentation of the phase diagram at constant chemical potential μ_d (field-density representation) is now converted into a three-phase triangle system representation. It should be noted that within this triangle, the composition of the three coexisting phases is constant; merely their relative volumes change.

As discussed in Sect. 3.2.3, the free volume theory is approximate in the sense that $\langle V_{\text{free}} \rangle$ is replaced by $\langle V_{\text{free}} \rangle_0$. To get an idea of the accuracy of the phase diagrams

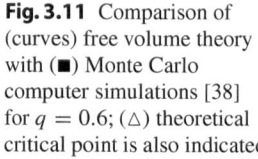

Fig. 3.11 Comparison of (curves) free volume theory with (■) Monte Carlo computer simulations [38] for $q = 0.6$; ($\triangle$) theoretical critical point is also indicated

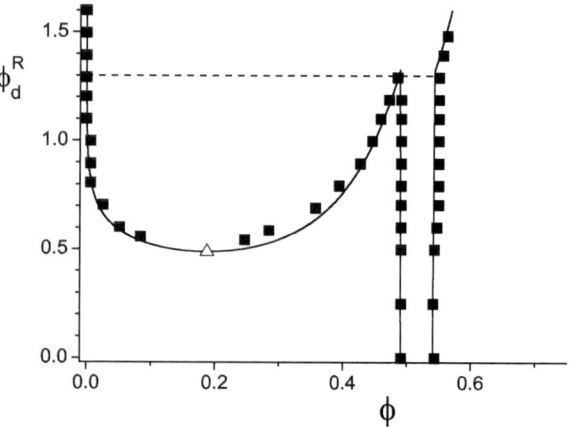

calculated with free volume theory, we compare the results for $q = 0.6$ in Fig. 3.11 with computer simulations [38]. The agreement is very good, given the fact that the free volume theory is approximate. Also for $q = 0.1$–1.0 [38] and large q values [39] the agreement with simulations is striking.

In this chapter, we have presented the free volume theory for hard sphere–depletant systems, and focused on the simplest possible case of hard sphere–PHS mixtures. In the next chapters, we will extend the free volume theory to more realistic situations (Chapter 4, hard spheres and polymers; Chap. 6, large and small hard spheres; Chap. 7, hard spheres and hard rod-like colloids; Chap. 8, hard rods and depletants; Chap. 9, hard platelets and depletants; Chap. 10, hard superballs (cubes) and depletants). We will also compare the results with experiments and computer simulations.

References

1. Einstein, A.: Ann. Phys. **17**, 549 (1905)
2. Perrin, J.: Ann. de Chem. Et De Phys. **18**, 5 (1909)
3. Onsager, L.: Ann. NY. Acad. Sci. **51**, 627 (1949)
4. Baus, M., Rull, L.F., Ryckaert, J.P. (eds.): Observation and Simulation of Phase Transitions in Complex Fluids. Kluwer Academic Publishers, Dordrecht (1995)
5. Arora, A.K., Tata, B.V.R. (eds.): Phase Transitions in Charge Stabilized Colloids. New York VCH Publishers, New York (1996)
6. Kirkwood, J.G.: J. Chem. Phys. **7**, 919 (1939)
7. Wood, W.W., Jacobson, J.D.: J. Chem. Phys. **27**, 1207 (1957)
8. Alder, B.J., Wainwright, T.E.: J. Chem. Phys. **27**, 1208 (1957)
9. Pusey, P.N., Van Megen, W.: Nature **320**, 340 (1986)
10. Russel, W.B., Saville, D.A., Schowalter, W.R.: Colloidal Dispersions. Cambridge University Press (1989)
11. Hoover, W.G., Ree, F.H.: J. Chem. Phys. **49**, 3609 (1968)
12. As dimensionless concentration variable ϕ is used throughout. In case of hard colloidal particles the quantity ϕ is the volume fraction. For polymers and penetrable hard spheres ϕ refers to the relative concentration with respect to overlap (see (1.21))
13. Carnahan, N.F., Starling, K.E.: J. Chem. Phys. **51**, 635 (1969)
14. Fortini, A., Dijkstra, M., Tuinier, R.: J. Phys.: Condens. Matter **17**, 7783 (2005)

15. Hansen, J.P., McDonald, I.R.: Theory of Simple Liquids, 2nd edn. Academic Press, San Diego, CA, USA (1986)
16. McQuarrie, D.A.: Statistical Mechanics. University Science Books, Sausalito, CA, USA (2000)
17. Malijevský, A., Kolafa, J.: In: Mulero, A. (ed.) Theory and Simulation of Hard-Sphere Fluids and Related Systems. Lecture Notes in Physics, vol. 753, p. 546. Springer, Berlin, Heidelberg (2008)
18. Lennard-Jones, J.E., Devonshire, A.F.: Proc. Roy. Soc. **163A**, 53 (1937)
19. Buehler, R.J., Wentorf, R.H., Hirschfelder, J.O., Curtis, C.F.: J. Chem. Phys. **19**, 61 (1951)
20. Alder, B.J., Hoover, W.G., Young, D.A.: J. Chem. Phys. **49**, 3688 (1968)
21. Frenkel, D., Ladd, A.J.C.: J. Chem. Phys. **81**, 3188 (1984)
22. Alfrey, T., Bradford, E.B., Vanderhof, J.F., Oster, G.: J. Opt. Soc. Am. **44**, 603 (1954)
23. Fischer, E.W.: Kolloid Z. **160**, 120 (1958)
24. Luck, W., Klier, M., Weslau, H.: Ber. Buns. Phys. Chem. **67**, 75 (1963)
25. de Kruif, C.G., Rouw, P.W., Jansen, J.W., Vrij, A.: J. Phys. Colloques **46**, C3 (1985)
26. Bolhuis, P.G., Kofke, D.A.: Phys. Rev. E **54**, 634 (1996)
27. Pusey, P.N.: J. Phys. France **48**, 709 (1987)
28. Gast, A.P., Hall, C.K., Russel, W.B.: J. Colloid Interface Sci. **96**, 251 (1983)
29. Vincent, B., Edwards, J., Emmett, S., Croot, R.: Colloids Surf. **31**, 267 (1988)
30. Lekkerkerker, H.N.W.: Colloids Surf. **51**, 419 (1990)
31. Lekkerkerker, H.N.W., Poon, W.C.K., Pusey, P.N., Stroobants, A., Warren, P.B.: Europhys. Lett. **20**, 559 (1992)
32. Fleer, G.J., Tuinier, R.: Adv. Colloid Interface Sci. **143**, 1 (2008)
33. Fleer, G.J., Tuinier, R.: Physica A **379**, 52 (2007)
34. Widom, B.: J. Chem. Phys. **39**, 2808 (1963)
35. Dijkstra, M., Brader, J.M., Evans, R., Phys, J.: Condens. Matter **11**, 10079 (1999)
36. Reiss, H., Frisch, H.L., Lebowitz, J.L.: J. Chem. Phys. **31**, 369 (1959)
37. Meijer, E.J.: Computer simulation of molecular solids and colloidal dispersions. Ph.D. thesis, Utrecht University (1993)
38. Dijkstra, M., van Roij, R., Roth, R., Fortini, A.: Phys. Rev. E **73**, 041409 (2006)
39. Moncho-Jordá, A., Louis, A.A., Bolhuis, P.G., Roth, R.: J. Phys.: Condens. Matter **15**, S3429 (2003)

Phase Separation and Long-Lived Metastable States in Colloid–Polymer Mixtures

4

When a dispersion containing spherical colloids is mixed with a polymer solution two kinds of instabilities can occur, as depicted in Fig. 4.1: (1) bridging flocculation caused by adsorbing polymer chains or (2) unmixing driven by the depletion force. The type of instability encountered depends on whether or not the polymers adsorb onto the colloidal surfaces. When nonadsorbing polymers are added, depletion forces lead to demixing (situation '2'), which is the main topic of this book. Polymer adsorption, however, occurs when the contact between the colloid surface and the polymer segments is energetically favourable to such a degree that the loss of configurational entropy is compensated [1]. When the amount of adsorbing polymer in the system is insufficient to fully cover all of the available surface area on the colloidal particles, so-called 'bridging flocculation' occurs [2]. Some polymers then attach to more than one particle, leading to aggregates or complexes (Fig. 4.1) that tend to sediment when they are large (situation '1'). Characteristic of this type of flocculation is that both colloids and polymers are concentrated in one part of a container. When all particle surfaces are saturated with adsorbed polymers in a good solvent (Fig. 4.2), the particle interactions are effectively repulsive because dense polymer layers overlap upon close approach giving rise to steric repulsion. This kinetically stabilises the dispersion (see Chap. 10 in Ref. [1]).

In Chap. 1, we saw that many systematic depletion studies were performed on mixtures of spherical colloids and nonadsorbing or free polymers. The reason is obvious: spherical colloids can be prepared in a relatively controlled way (rather monodisperse, hard-sphere like) and so are of industrial and fundamental relevance, and polymers are ubiquitous and efficient depletants. As a result, these compounds are often mixed in product formulations.

We stress that any study on colloid–polymer mixtures should be preceded by an analysis of whether the polymers adsorb or not. For instance, an analysis of the composition of the two phases can be used to verify whether depletion interaction is

© The Author(s) 2024
H. N. W. Lekkerkerker, R. Tuinier, M. Vis, *Colloids and the Depletion Interaction*,
Lecture Notes in Physics 1026, https://doi.org/10.1007/978-3-031-52131-7_4

Fig. 4.1 Types of instability
that occur after mixing a
colloidal dispersion with a
polymer solution. When
adsorption occurs and the
polymer concentration is
low, bridging between
different colloidal particles
can induce flocculation (1).
When the polymer chains do
not adsorb, depletion leads to
the partitioning of colloids
and polymers over different
phases (2)

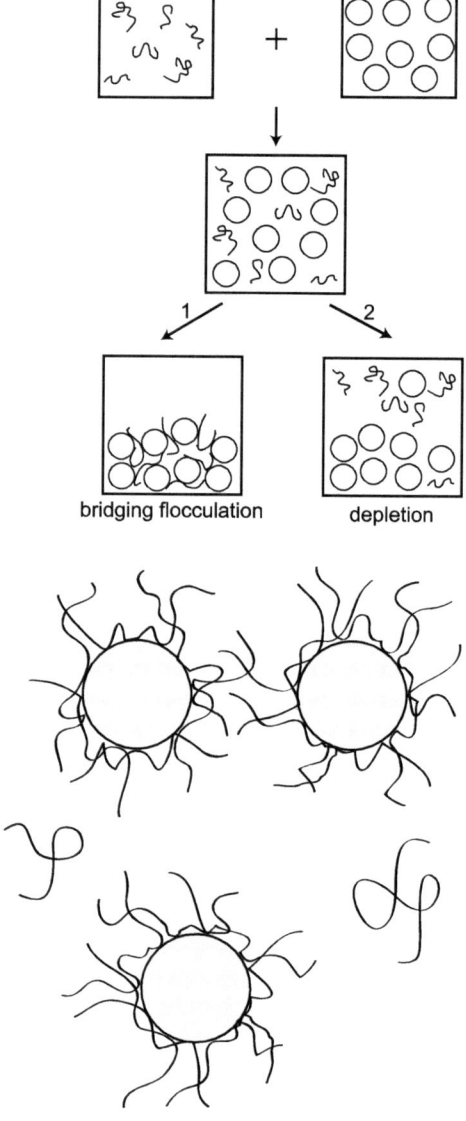

bridging flocculation   depletion

Fig. 4.2 Colloidal particles
that are sterically stabilised
through polymer adsorption

responsible for demixing. Polymer adsorption also increases the friction coefficient
of colloidal particles. Therefore, using the Stokes friction coefficient $f = 6\pi\eta^{\text{eff}} R_{\text{h}}$
[3] leads to a larger observed hydrodynamic radius R_{h}. This can be measured, for
instance, by sedimentation or dynamic light scattering [4,5]. The typical adsorbed
amount at saturation Γ is ~ 1 mg·m^{-2} [1].

Exercise 4.1. Verify that the polymer concentration c_p (in mass per volume) required to fully cover all spheres with radii R in a dispersion with volume fraction ϕ can be expressed as

$$c_p = \frac{3\phi\Gamma}{R}.$$

Hint: assume that all added polymers adsorb.

When the colloidal particles are completely covered with adsorbing polymer, adding more polymer gives rise to excess polymer in the bulk solution, which is thus not adsorbed. This nonadsorbing polymer may lead to depletion interaction as well [6,7]. Here, depletion effects are weaker for two reasons. Firstly, more polymer is required before depletion-induced instability of the dispersion occurs because polymer is first consumed in order to cover the particles [1]. Secondly, the depletion interaction is weak due to the soft repulsion between the adsorbed polymer layers. It is known that depletion effects between such soft surfaces are rather small [1,8].

When polymer depletion occurs near hard surfaces, exceeding a certain polymer concentration may lead to phase separation: a polymer-enriched phase coexists with a colloidal particle-enriched phase (Fig. 4.1). In Chap. 3, we introduced the phase behaviour of hard spheres mixed with penetrable hard spheres (PHSs). This provides a starting point for describing the phase behaviour of colloid–polymer mixtures. In Sect. 4.1, we show that the PHS description is adequate for mixtures in the *colloid-limit* of small size ratio $q = R_g/R$, with R_g denoting the polymer's radius of gyration. In this limit, the polymer chains are smaller than the particle radius of the colloidal spheres. In Sect. 4.2, we treat the modifications for the case that the polymers are treated as ideal chains. More advanced treatments that account for non-ideal behaviour of depletion thickness and osmotic pressure for interacting polymer chains also enable intermediate and large q situations to be described. This is the topic of Sect. 4.3. We also pay attention to nonequilibrium phenomena, ranging from polymer depletion-induced phase separation kinetics to colloidal (transient) gel and glass formation. Such effects are of significant practical relevance and are discussed in Sect. 4.4.

4.1 Experimental State Diagrams of Model Colloid–Polymer Mixtures

Well-defined colloids with a lyophilic surface coating and a steep repulsive interaction have been developed in several laboratories. Dispersions of such colloidal particles in appropriate solvents can be approximated as hard-sphere fluids (or solids above some concentration). Spherical silica particles in cyclohexane [9–11], in which the particles were made lyophilic by covering the surface with a layer of terminally anchored octadecyl chains, are a primary example of such a model dispersion. In this system, the refractive indices of silica (1.45) and cyclohexane (1.42) are close, so light scattering effects are minimised and the van der Waals attraction between the particles is small. Cyclohexane was chosen since it is a good solvent for octadecyl chains. The surface layers of two encountering particles then repel each other sterically (see Sect. 1.2.4). This results in a fairly steep pseudo-hard-sphere interaction that can be described by

$$W(r) = \begin{cases} \infty & r \leq 2R_{\text{eff}} \\ 0 & r > 2R_{\text{eff}}, \end{cases} \tag{4.1}$$

where r is the centre-to-centre distance between the spheres ($= 2R_{\text{eff}} + h$) (see also the discussion near Sect. 1.16). Here, R_{eff} is the effective hard-sphere radius: the sphere radius plus the thickness of the terminally anchored chains.

Another interesting model system is a dispersion of polymethylmethacrylate (PMMA) particles that are sterically stabilised with poly-12-hydroxy stearic acid in solvents such as decalin, sometimes mixed with tetralin in order to match solvent and particle refractive indices. Early synthesis of and studies with these particles were performed in Bristol [12, 13]. These systems exhibit fluid to solid phase transitions when the particle volume fraction exceeds about 0.5 [14].

Well-defined dispersions of hard-sphere-like PMMA colloids and nonadsorbing polymers were extensively studied in Edinburgh [15, 16]. These PMMA particles can be synthesised with a size dispersity below 5%, and behave almost like perfect hard spheres [17, 18] see Sect. 3.2.

Polystyrene (PS) is one of the well-studied random coil polymers used in combination with PMMA spheres. PS can be synthesised with polydispersities as small as $M_w/M_n \approx 1.02$, with M_n and M_w as the number- and weight-averaged molar masses, respectively. The physical properties of PS in solution have been characterised in a wide range of solvents [19]. Optical tweezer experiments [20] on a pair of PMMA spheres in a PS solution were consistent with the presence of depletion layers of PS surrounding the spheres. Also, DLS measurements showed that adsorption does not occur [4]. Hence, the model system of PMMA with PS offers an excellent tool for studying the phase behaviour of hard spheres mixed with free polymers [16].

In Fig. 4.3 state diagrams are plotted that were measured by Poon et al. [15, 21] for three size ratios $q = R_g/R = 0.08, 0.57$ and 1. Here, ϕ_p is the polymer concentration relative to overlap (Eq. 1.21). At $\phi < 0.49$ and low polymer concentrations the mixtures appear as single-state fluid phases. At zero polymer content, the hard-sphere fluid–crystal phase transition is found when the colloids occupy about half of the volume. Upon addition of polymer the fluid–crystal coexistence region expands for $q = 0.08$; then a colloidal fluid at smaller volume fraction coexists with a denser colloidal crystal. Slanted tie-lines were observed that indicate polymer partitioning over the two phases [15]. These findings are consistent with predictions in Chap. 3 for hard spheres mixed with PHSs [22] for small q. For larger q values ($q \gtrsim 0.3$), a critical point appears in the phase diagram, identifying the onset of the gas–liquid coexistence region. This is observed in the phase diagrams for $q = 0.57$ and $q = 1$ in

Fig. 4.3 State diagrams of colloid–polymer mixtures for $q = 0.08$ (*top*), $q = 0.57$ (*middle*) and $q = 1.0$ (*bottom*). Experimental data: PMMA spheres mixed with polystyrene polymers in *cis*-decalin [15,21]. Curves depict free volume theory calculations [22], with $\delta = R_g$. For high q a triple triangle (hatched) is predicted by the theory

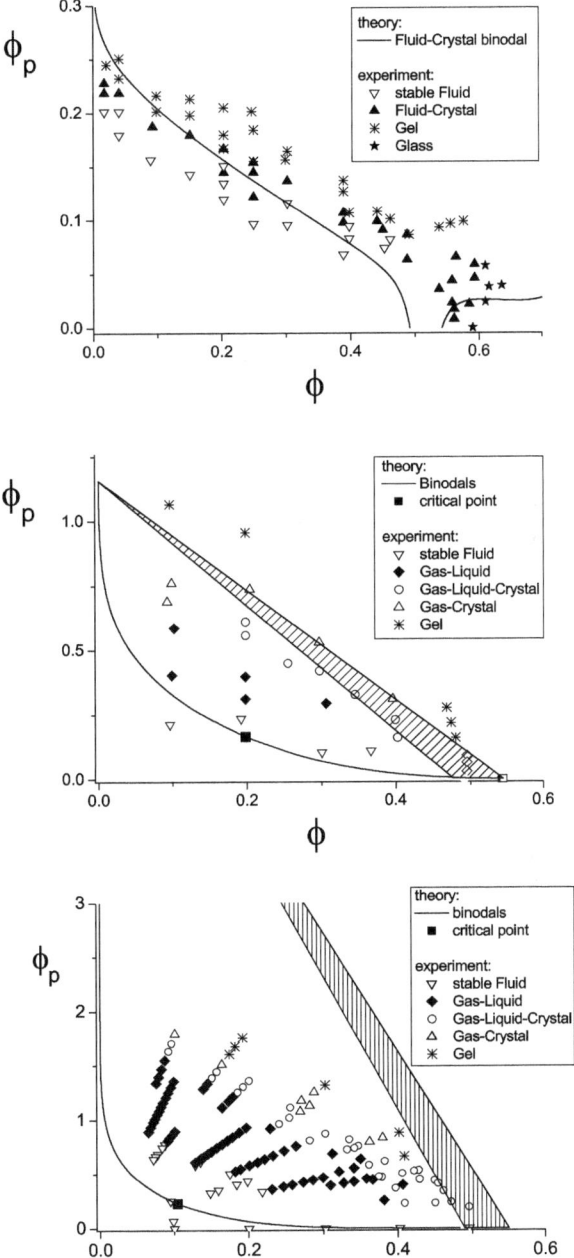

Fig. 4.3 and is also found for larger q values [23–27]. Large q implies a long-ranged attraction. For colloidal gas–liquid phase separation, the degree of partitioning over the two phases depends on how far the system is from the critical point.

The absence of a liquid state in phase diagrams for a collection of particles with short-ranged attractions is a general finding [28,29], for which Pusey and Poon gave the following simple physical argument [16,30]: consider a close-packed crystal $\phi \approx 0.74$ of adhesive hard spheres that have a mutual attraction (Fig. 4.4). Upon adding solvent, the crystalline structure expands in volume. Below $\phi = 0.545$ (the volume of the close-packed crystal has expanded by about 1/3), the point of loss of rigidity is attained and a fluid state becomes possible. Liquid configurations require that the particles attract one another with sufficient strength as they are moving. For weak attraction (weaker than the attraction at the critical point), thermal energy overcomes this attraction and a liquid state is impossible. For stronger attractions (exceeding the critical value), the state depends on the range of attraction.

For short-ranged attractions, the particles are directly out of their range of attraction, and so a low-density gas is the most stable situation upon dilution. Gas–liquid equilibria are then metastable (Fig. 3.9(iv), see also the discussion of Fig. 11.4). It has been shown that such 'hidden' gas–liquid coexistence regions have an impact on dynamics and phase separation kinetics, and play a role in crystallisation phenomena [31–35]. Fortini et al. [36] studied the relationship between equilibrium and nonequilibrium phase diagrams of a system of hard spheres with a short-ranged attraction using Monte Carlo and Brownian Dynamics simulations. They found that crystallisation is enhanced for attractions that are sufficiently strong to enter the metastable gas–liquid binodal. Then formation of a dense liquid is observed, followed by nucle-

Fig. 4.4 The expansion of a close-packed colloidal crystal phase (1) towards (2), at the melting volume fraction (~ 0.55). Where there is a long-range attraction with appropriate strength, a colloidal liquid (3) is the stable state after melting (i). For a short-ranged attraction after melting (ii) a colloidal gas (4) is more favourable upon further expansion

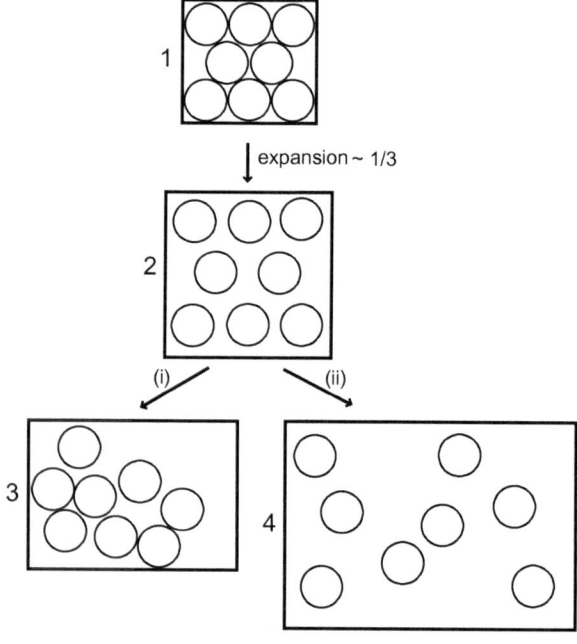

ation of the crystallites within the dense fluid. Only at larger colloid concentrations is a percolating network structure due to an arrested gas–liquid phase separation found.

For sufficiently long-ranged attraction, the particles still attract one another at the loss of rigidity point, so a liquid state is then possible. In colloid–polymer mixtures, there is no direct attraction between the colloidal particles but the attraction enters through repulsion. Attraction is caused by the overlap of depletion layers rather than through direct pair interactions. For $q > 0.15$, multiple overlap of depletion layers occurs (Fig. 3.7), which is expected to promote the occurrence of a colloidal liquid. The critical point at which the range of attraction is just sufficient enough for a stable liquid state is termed the critical end point (CEP) [29]. For a shorter range of attraction, no critical point borders a stable fluid phase. Theory and computer simulations point out that the CEP generally corresponds to a range of attraction close to $1/3$ of the particle diameter [29].

Exercise 4.2. What does it take to make a stable liquid?

We now return to Fig. 4.3 and focus on the state diagrams for $q = 0.57$ and $q = 1$. Adding polymer leads to gas–liquid coexistence as discussed, followed by a region where a gas–liquid–crystal equilibrium (○) is found. For $q = 0.57$, this three-phase coexistence region corresponds roughly to the theoretical prediction of the free volume theory (FVT) as outlined in Chap. 3, with PHSs playing the role of the polymer chains. Above the three-phase coexistence region, a gas–crystal binodal is found, which is also predicted by FVT.

At even higher polymer concentrations crystallisation does not occur anymore, while dense solid sediments of particles can be seen [15]. This nonequilibrium behaviour is also found for $q = 0.08$ at high polymer concentrations where (metastable) gel or glassy states are observed. A colloidal glass refers to a state where the particles are topologically trapped ('caged') by their neighbours. The term gel is identified as a disordered arrested state which does not flow but exhibits solid-like rheological properties such as a significant value for the elastic shear modulus [37]. We return to these nonequilibrium states of colloid–polymer mixtures in Sect. 4.4.

In Fig. 4.3, we also plot the (equilibrium) binodals by using the FVT outlined in Chap. 3 for hard spheres mixed with PHSs with diameters of $2R_g$. Qualitatively, the phase diagram topology is quite well predicted. For $q = 0.08$, only equilibrium fluid, crystal and fluid–crystal regions are found and predicted. For both $q = 0.57$ and $q = 1$, the phase diagram contains fluid, gas, liquid, and crystalline (equilibrium) phases. In the different unmixing regions, one now finds gas–liquid coexistence with a critical point, a three-phase gas–liquid–crystal, and gas–crystal coexistences. The observed q-dependence of the phase diagram topology outlined above is not limited to the PMMA model system. Similar findings were reported for the phase behaviour of polystyrene latex spheres mixed with hydroxy-ethyl cellulose in water [38,39] (see also Refs. [40,41] for other examples).

Next, we make a more quantitative comparison between theory and experiment. We observe in Fig. 4.3 that the quantitative agreement between FVT and the experimental data becomes less upon increasing q. For $q = 0.08$, the fluid–crystal binodal is in nearly quantitative agreement with the experimental results with a slight, nearly imperceptible overestimation of the binodal.

For $q = 0.57$, the data are in fair agreement with the FVT predictions (middle panel of Fig. 4.3). The triple region lies somewhat above the experimental data, especially at low ϕ, and the FVT gas–liquid binodal curve lies slightly below the experimental binodal. Hence, the width of the gas–liquid coexistence region is overestimated. The reason for this lies in the fact that, whereas, for $q = 0.08$, the relevant polymer concentrations are $\phi_p < 0.3$, for $q = 0.57$, the polymer concentrations are significantly larger and are much closer to the overlap concentration. This means that one has to take into account the excluded volume interactions between the polymer chains: regarding them as PHSs only suffices for small q where ϕ_p is small too.

For $q = 1.0$ (lower panel of Fig. 4.3) classical FVT also fails in a quantitative sense. The gas–liquid binodal predicted by FVT now lies far below the experimental phase boundary. Also, the predicted triple region (with ϕ_p reaching ≈ 6, see also Fig. 3.10) largely exceeds the experimental one. In Sect. 4.3, we generalise FVT by incorporating polymeric interactions between the polymer chains and compare these results to the experimental equilibrium phase diagrams for $q = 0.57$ and 1.0.

To summarise, theory and experiment clearly demonstrate that the types of phase equilibria encountered in unmixed colloid–polymer mixtures are rather sensitive to the size ratio q. For sufficiently large q ($\gtrsim 0.3$), a colloidal gas–liquid phase separation is encountered. For $q \gtrsim 0.4$, the simple model of hard sphere–PHS mixtures fails to accurately describe the phase behaviour of well-defined hard-sphere colloid–polymer mixtures. For large q-values ($q \gtrsim 0.6$), it is essential to improve the simple description of polymer chains as PHSs.

4.2   Phase Behaviour of Colloid–Ideal Polymer Mixtures

The first step in taking into account more appropriate polymer physics (as opposed to the simple description of PHSs) is to consider the polymers as ideal chains. This can be evaluated by incorporating the depletion thickness of nonadsorbing ideal chains near a colloidal hard sphere into free volume theory.

In Chap. 2, an analytical expression (Eq. (2.71)) was derived for the depletion thickness around a sphere that is due to ideal polymer chains:

$$\frac{\delta_s}{R} = \left(1 + \frac{6q}{\sqrt{\pi}} + 3q^2\right)^{1/3} - 1. \tag{4.2}$$

It was plotted in Fig. 2.15 with a slightly different representation. Figure 2.15 showed that the depletion thickness, normalised as δ_s/R_g, drops with increasing q. For $q < 1$, the depletion thickness is close to or larger than R_g, but not much: the maximum is $2R_g/\sqrt{\pi} \approx 1.13R_g$ in the limit $q \to 0$. For $q \gtrsim 1$, the depletion thickness gets significantly smaller than R_g. In Fig. 4.5, δ_s is normalised with the sphere radius R and plotted as a function of q.

Fig. 4.5 The depletion thickness around a sphere as a function of the polymer-to-sphere size ratio q. *Solid curve*: ideal polymer chains, Eq. (4.2); *dashed line*: penetrable hard sphere approach, $\delta_s = R_g$; *dotted curve*: the approximation $\delta_s/R = 0.938q^{0.9}$, see Eq. 4.25

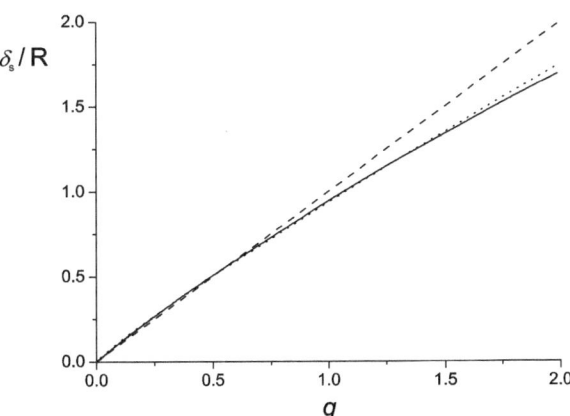

Fig. 4.6 Osmotic equilibrium between a reservoir containing polymer chains and a colloid–polymer system where, as an example, a colloidal gas is in equilibrium with a colloidal liquid

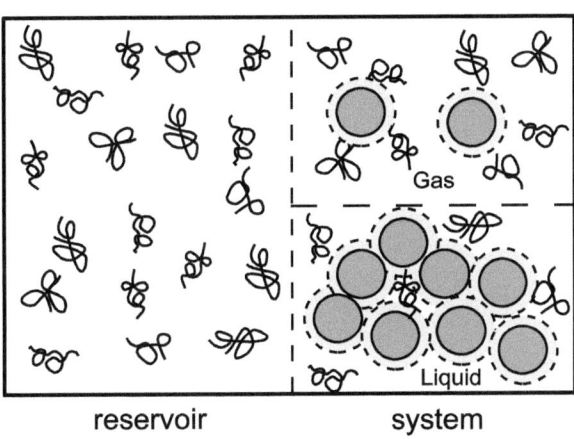

reservoir system

We now incorporate the correct depletion thickness into the free volume theory that was presented in Sect. 3.3. We consider the osmotic equilibrium between a polymer solution (reservoir) and the colloid–polymer mixture (system) of interest (Fig. 4.6). The general expression for the semi-grand potential for N_c hard spheres mixed with interacting polymers as depletants (see Eq. (3.20)), is

$$\Omega(N_c, V, T, \mu_p) = F_0(N_c, V, T) + \int_{-\infty}^{\mu_p^R} \frac{\partial \Omega}{\partial \mu_p^{R'}} \, d\mu_p^{R'}, \qquad (4.3)$$

with F_0 the free energy of the pure hard-sphere dispersion and μ_p^R the chemical potential of the polymer chains. Just as in Chap. 3 (see also Appendix A), we define the dimensionless free energies, chemical potential and pressure:

$$\widetilde{F}_0 = \frac{F_0 v_0}{VkT}, \qquad \widetilde{\Omega} = \frac{\Omega v_0}{VkT}, \qquad \widetilde{\mu} = \frac{\mu}{kT}, \qquad \widetilde{P} = \frac{P v_0}{kT}.$$

We drop the explicit dependencies (N_c, V, T, μ_p) in Eq. (4.3) and arrive at

$$\widetilde{\Omega} = \widetilde{F}_0 - \int_0^{\phi_p^R} \alpha \left(\frac{\partial \widetilde{P}^R}{\partial \phi_p^{R'}} \right) d\phi_p^{R'}, \tag{4.4}$$

using the free volume theory approximations discussed in Chap. 3. For ideal polymers $\widetilde{P}^R = \phi_p^R q^{-3}$ and α is independent of ϕ_p we arrive at

$$\widetilde{\Omega} = \widetilde{F}_0 - \alpha \phi_p^R q^{-3}, \tag{4.5}$$

an expression similar to Eq. (3.44).

Exercise 4.3. Starting from Eq. (3.18), derive Eq. (4.4) by using $\partial \Omega / \partial \mu_p = -N_p$, the Gibbs–Duhem relation $d\mu_p^R = (v_p/\phi_p^R)dP^R$, and $\widetilde{P}^R = P^R v_0/kT$.

For the free volume fraction α, we recall Eq. (3.38),

$$\alpha = (1 - \phi) \exp[-ay - by^2 - cy^3], \tag{4.6}$$

with

$$y = \frac{\phi}{1 - \phi}$$

and revised definitions of a, b and c:

$$a = 3\frac{\delta_s}{R} + 3\left(\frac{\delta_s}{R} \right)^2 + \left(\frac{\delta_s}{R} \right)^3, \tag{4.7a}$$

$$b = \frac{9}{2}\left(\frac{\delta_s}{R} \right)^2 + 3\left(\frac{\delta_s}{R} \right)^3, \tag{4.7b}$$

$$c = 3\left(\frac{\delta_s}{R} \right)^3. \tag{4.7c}$$

Inserting Eq. (4.2) into Eq. (4.6) gives the corrected free volume fraction in a mixture of ideal chains and colloidal spheres. Inspection of the gas–liquid coexistence binodals for $q = 1$ and smaller q reveals that replacing PHSs with ideal chains does not give significant differences. It follows from Fig. 4.5 and the detailed comparison in [42] that this corrected description for $q > 1$ does affect the phase diagram. For such high q values, however, it becomes essential to account for interactions between the polymer segments, whereby δ becomes a function of the polymer concentration and the osmotic pressure is no longer ideal.

Exercise 4.4. At low colloid volume fractions ϕ, why does replacing PHSs with ideal chains result in a polymer concentration shift upwards of the gas–liquid binodals for large q?

4.3 Mixtures of Spheres and Interacting Polymer Mixtures

In this section we consider the phase behaviour of dispersions containing spherical colloidal spheres and *interacting* polymer chains in a common solvent. For small polymer-to-colloid size ratios ($q \lesssim 0.4$) the relevant part of the phase diagram lies below the polymer overlap concentration ($\phi_p^R < 1$). There is then no need to account for interactions between the polymers in order to provide a proper description of the phase diagram, and it is still sufficient [42] to approximate the polymer-induced osmotic pressure by the ideal gas law as assumed within free volume theory [22]. However, for $q \gtrsim 0.4$ the polymer concentrations at which phase transitions occur are of the order of and above the polymer overlap concentration. In that case, interactions between the polymer segments should be accounted for.

Here we will extend FVT to incorporate the correct expressions for the (polymer concentration-dependent) depletion thickness and osmotic pressure, resulting in so-called generalised free volume theory (GFVT). Equation (4.4) for the semi-grand potential is still valid in GFVT: it does not, as yet, contain any assumption about the physical properties of the depletants or the colloids. But now we need to specify the quantities α and $\widetilde{P}^R$ for interacting polymers, so we need the osmotic pressure and depletion thickness (that determines α, see Eqs. (4.6) and (4.7)). These will be considered in Sects. 4.3.2 and 4.3.3. First, we start in Sect. 4.3.1 with some basics on the physics of polymer solutions in the dilute and semidilute concentration regimes.

4.3.1 Characteristic Length Scales in Polymer Solutions

In Sect. 2.2 we considered the concept of ghost or ideal chains. The segments of such chains do not feel each other. Here, we consider excluded volume interactions between the segments. Throughout Sect. 4.3 we consider two limiting cases of interacting polymer chains: the excluded volume limit (good solvent) and the Θ-solvent situation. The good solvent condition refers to the situation where the segments of the polymer chains effectively repel other segments so that chains in a good solvent will swell due to excluded volume interactions.

When the solvent-mediated attraction between the segments *exactly* compensates for the (hard-core) excluded volume effect the polymer chains behave quasi-ideally for dilute polymer solutions. This situation is commonly referred to as the Θ-solvent condition.

The physical properties of the polymer solution depend on concentration under both good and Θ-solvent conditions. The characteristic length scale is the correla-

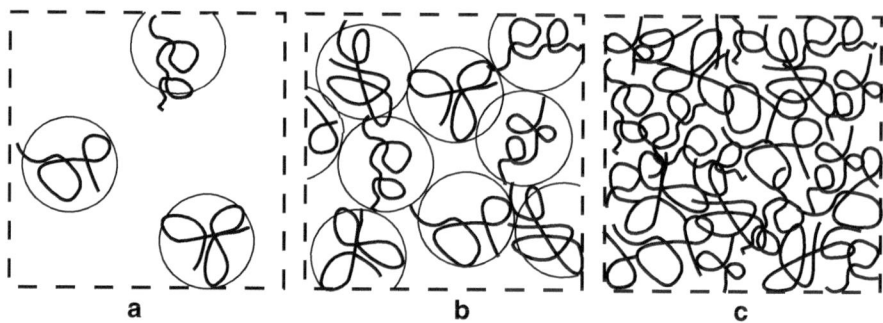

Fig. 4.7 Depiction of the various concentration regimes in polymer solutions, **a** dilute ($\phi_p < 1$), **b** near overlap ($\phi_p \approx 1$), **c** semidilute ($\phi_p > 1$)

tion length ξ, the length scale over which the polymer segments show a correlated behaviour. In the dilute concentration regime this is the radius of gyration of the coils, which depends on chain length M but not on concentration. Beyond the overlap concentration of polymer coils the correlation length decreases with increasing polymer concentration and is independent of M. Figure 4.7 shows the various polymer concentration regimes.

4.3.1.1 Dilute Polymer Concentration

For long chains, the following scaling relationship holds [43]:

$$R_g \sim M^\nu, \tag{4.8}$$

with

$$\nu = \begin{cases} \frac{1}{2} & \Theta\text{-solvent,} \\ 0.588 & \text{good solvent,} \end{cases} \tag{4.9}$$

where the scaling exponent ν is known as the Flory exponent. In a Θ-solvent the chains are ideal, so R_g is proportional to $\sqrt{M}$. Then we have,

$$R_g = b\sqrt{\frac{M}{6}}, \tag{4.10}$$

where b is the segment size. For a good solvent, the exponent ν follows from Renormalisation Group Theory (RGT) [44].

For shorter chains an approximate expression for any solvency may be derived using the Flory excluded volume parameter **v**, which is unity when the segments experience hard-core repulsion, and vanishes in the case of a Θ-solvent. Based on Flory's result [43] for the expansion coefficient, the coil size can be written as [41],

$$R_g = 0.31bM^{1/2}\left[1 + \sqrt{1 + 6.5\mathbf{v}M^{1/2}}\right]^{0.352}, \tag{4.11}$$

where we adjusted the Flory scaling exponent $2/5$ in the last factor to 0.352 to achieve the correct scaling behaviour [44] of Eq. 4.9.

Exercise 4.5. Show that Eq. 4.11 reduces to the scaling limits of Eqs. (4.8) and (4.9) for $M \gg 1$.

4.3.1.2 Semidilute Polymer Solutions

Above the polymer overlap concentration we enter the semidilute regime (Fig. 4.7c). In the dilute regime (Fig. 4.7a), each polymer coil occupies a volume $v_p = (4\pi/3) R_g^3$. When $v_p n_b$ becomes unity the solution is completely filled with polymer coils (Fig. 4.7b). For $v_p n_b > 1$ the chains overlap. Therefore, overlap in terms of the number density is defined as $n_b^* = 1/v_p$. It is convenient to define a relative polymer concentration:

$$\phi_p = v_p n_b = \frac{\varphi}{\varphi^*}, \tag{4.12}$$

which is unity at the overlap concentration (see also Eq. 1.21). Here, φ is the polymer segment volume fraction, often used in polymer physics. Using this polymer segment volume fraction $\varphi = n_b M v_s$ the overlap volume fraction φ^* follows as

$$\varphi^* = M \frac{v_s}{v_p}, \tag{4.13}$$

with v_s denoting the volume of a segment. Next, the relationship between correlation length and polymer concentration is considered. Below overlap ($\varphi < \varphi^*$) the correlation length ξ is the coil size R_g, which depends only on M (and solvency). Above overlap ($\varphi > \varphi^*$) we have the famous De Gennes scaling law [45]:

$$\xi \sim \varphi^{-\gamma}, \tag{4.14}$$

which does not depend on chain length. The De Gennnes scaling exponent γ is given by

$$\gamma = \begin{cases} 1 & \Theta\text{-solvent,} \\ 0.77 & \text{good solvent.} \end{cases} \tag{4.15}$$

Near the overlap concentration we have $\xi \approx R_g$ and $\varphi \approx \varphi^*$, so $R_g \sim (\varphi^*)^{-\gamma}$. Consequently,

$$\frac{\xi}{R_g} \sim \phi_p^{-\gamma}. \tag{4.16}$$

Since $\varphi^* \sim M/v_p \sim R_g^{1/\nu - 3}$ (see Eqs. (4.8) and (4.13)) and $\varphi^* \sim R_g^{-1/\gamma}$, we have the following general relationship between the Flory and De Gennes exponents:

$$\frac{1}{\gamma} + \frac{1}{\nu} = 3. \tag{4.17}$$

To incorporate the crossover from dilute to the semidilute polymer solutions, Fleer et al. [46] derived approximate but accurate expressions for the polymer concentration-dependent depletion thickness δ and osmotic pressure P. They did this by interpolating between the exactly known dilute limit and scaling relations at semidilute polymer solutions. These interpolations are discussed in Sect. 4.3.2 (δ) and Sect. 4.3.3 (P).

4.3.2   Depletion Thickness

4.3.2.1 Concentration Profile at a Hard Wall

In the semidilute limit, De Gennes [45] made a mean-field analysis of the polymer concentration at a nonadsorbing hard wall. He used the Ground State Approximation (GSA) to approximate the Edwards equation [47–51] for polymer trajectories in an external field. The GSA basically simplifies chains in the sense that the spatial distribution of the segments is assumed to be independent of the ranking number of the segments: there is no difference between, for instance, end segments and middle segments of the chains. The GSA is especially powerful in the semidilute concentration regime [52]. The GSA concentration profile for the volume fraction of segments $\varphi(x)$ is simple:

$$f(x) = \frac{\varphi(x)}{\varphi_b} = \tanh^2 \left(\frac{x}{\xi} \right), \tag{4.18}$$

where $\varphi(x)$ is the local polymer segment concentration, φ_b its value in the bulk and x the distance from the colloidal surface.

Exercise 4.6. Rationalise the shape of the profile $f(x) = \tanh^2(x/\xi)$ from the van der Waals density profile at a gas–liquid interface. *Hint*: use the relationship between the order parameter and the density profile (see also Chap. 5).

Applying Eq. (2.60),

$$\delta = \int_0^\infty dx \, [1 - f(x)],$$

results in $\delta = \xi$. It follows that the correlation length sets the length scale over which polymer segments are depleted from the wall. Beyond a distance of the correlation length from a colloidal surface the segments are 'unaware' of a nonadsorbing surface.

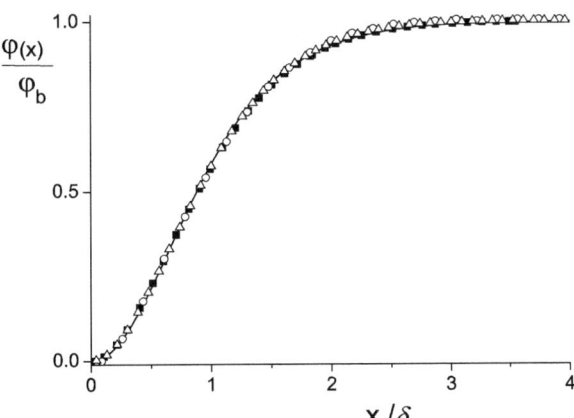

Fig. 4.8 Segment density profiles $\varphi(x)$ as a function of the distance from a flat wall x for various descriptions (data points) of the nonadsorbing polymer chains compared to the $\tanh^2$ profile (solid curve) of Eq. 4.19: ideal chains (■, [53]), mean-field chains (○, [50]) and excluded volume chains (△, [54])

The expressions for the density profiles may be substantially more involved than a simple $\tanh^2$ profile for other cases (e.g., depleted dilute ideal or excluded volume chains, semidilute chains with fluctuation effects). Surprisingly,

$$f(x) = \tanh^2\left(\frac{x}{\delta}\right) \qquad (4.19)$$

turns out to be very accurate, in general [50,53], as long as the correct depletion thickness δ is inserted (Fig. 4.8).

4.3.2.2 Depletion Thickness at a Hard Wall

For ideal chains (or dilute chains in a Θ-solvent, see Eqs. (2.57) and (2.60)), the depletion thickness at a hard wall equals $\delta_0 = 2R_g\sqrt{\pi} \approx 1.13R_g$ [53,55], where the subscript 0 now refers to the dilute (ideal) limit.

A general expression for dilute polymer solutions is

$$\delta_0 = pR_g \quad \text{with} \quad p \approx \begin{cases} 1.13 & \Theta\text{-solvent,} \\ 1.07 & \text{good solvent.} \end{cases} \qquad (4.20)$$

The good solvent result was derived by Hanke, Dietrich and Eisenriegler [56] using RGT.

As we have seen, the De Gennes result for the semidilute limit provides $\delta = \xi$. A GSA analysis of mean-field polymer chains in a slit [50] enables the dilute and semidilute limits to be combined, and provides a very simple and accurate relationship:

$$\delta^{-2} = \delta_0^{-2} + \xi^{-2}. \qquad (4.21)$$

This result was derived from a mean-field treatment, where the semidilute scaling behaviour is $\xi \sim \phi_p^{-1}$ (Θ-solvent) or $\xi \sim \phi_p^{-1/2}$ (good solvent). The scaling exponent -1 is valid for chains in a Θ-solvent but $-1/2$ is incorrect for good solvent conditions. Expression (4.21) can, however, be generalised to include the correct scaling exponents by inserting the correct scaling from Eq. 4.16 with Eq. 4.15 with the appropriate numerical prefactor into Eq. 4.21. The result is [46]:

$$\left(\frac{\delta_0}{\delta}\right)^2 = 1 + \beta\phi_{\mathrm{p}}^{2\gamma} \;, \tag{4.22}$$

with

$$\beta = \begin{cases} 6.02 & \Theta\text{-solvent} \\ 3.95 & \text{good solvent} \end{cases}.$$

For a Θ-solvent with $\gamma = 1$ Eq. 4.22 is in quantitative agreement with numerical self-consistent field results. For a good solvent with $\gamma = 0.77$ Eq. 4.22 compares favourably with computer simulation results [46].

4.3.2.3 Depletion Thickness Around a Hard Sphere

Converting the depletion thickness δ at a hard wall to its value δ_s around a hard sphere is a geometrical issue. In Sect. 2.2.2 we saw that the concentration profile of ideal polymer chains around a sphere gives Eq. (2.71), which can be rewritten in the form:

$$\frac{\delta_s}{R} = \left[1 + 3\frac{\delta}{R} + \frac{3\pi}{4}\left(\frac{\delta}{R}\right)^2\right]^{1/3} - 1. \tag{4.23}$$

For dilute chains in the excluded volume limit the following expansion has been found:

$$\frac{\delta_s}{R} = \left[1 + C_1\frac{\delta}{R} + C_2\left(\frac{\delta}{R}\right)^2 + C_3\left(\frac{\delta}{R}\right)^3 + \ldots\right]^{1/3} - 1, \tag{4.24}$$

with the flat wall result $C_1 = 3$. The curvature terms $C_2 \approx 2.273$ (which is close to $3\pi/4 \approx 2.356$) and $C_3 \approx -0.0975$ were computed using RGT [56]. Although higher order C_i terms are yet unknown, it is clear that the curvature effects for excluded volume and ideal chains in Eqs. (4.23) and (4.24) are rather similar.

These expressions for δ_s/R can be easily approximated as the simple accurate power-laws [41]:

$$\frac{\delta_{0,s}}{R} = \begin{cases} 0.938q^{0.9} & \Theta\text{-solvent}, \\ 0.865q^{0.88} & \text{good solvent}. \end{cases} \tag{4.25}$$

These power-laws hold for a wide range of q-values [41] (see, for instance, Fig. 4.5). In combination with the concentration dependence Eq. 4.22, this approximation leads to

$$\frac{\delta_s}{R} = \begin{cases} 0.938\left(q^{-2} + 6.02q^{-2}\phi_{\mathrm{p}}^2\right)^{-0.45} & \Theta\text{-solvent}, \\ 0.865\left(q^{-2} + 3.95q^{-2}\phi_{\mathrm{p}}^{1.54}\right)^{-0.44} & \text{good solvent}, \end{cases} \tag{4.26}$$

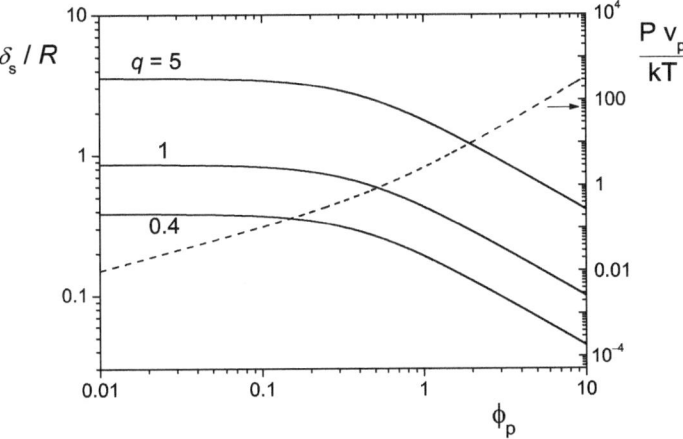

Fig. 4.9 Depletion thickness (solid curves) for three size ratios as indicated, and osmotic pressure (dashed curve) of polymer chains in the excluded volume limit as a function of the relative polymer concentration

expressing the concentration- and curvature-dependence of the depletion thickness around a sphere in a solution with interacting polymers under Θ-solvent and good solvent conditions. For the good solvent situation, the depletion thickness is plotted in Fig. 4.9 for three q-values.

4.3.3 Osmotic Pressure of Polymer Solutions

In the limit of dilute polymer solutions, the osmotic pressure is given by the ideal Van 't Hoff law $P_{id} = n_b kT$ (see Sect. 2.2). For the osmotic pressure of non-ideal polymers in solution one can write down a general virial series:

$$\frac{P}{n_b kT} = 1 + A_2 n_b + A_3 n_b^2 + \cdots , \qquad (4.27)$$

where A_2 and A_3 are the second and third osmotic virial coefficients. Note that we use B_i for colloids (see, for instance, Eq. (1.26), Sects. 2.3.1 and 3.2.1) and A_i for polymers. The second virial coefficient is proportional to the effective excluded volume per polymer segment: $A_2 \sim \mathbf{v}$. In the good solvent limit (where $\mathbf{v} = 1$), this excluded volume equals the physical volume of a segment, so A_2 attains a finite positive value. In a poor solvent $A_2 < 0$. For polymer chains in a Θ-solvent, the excluded volume of a segment is exactly compensated by the attractions between the segments and $A_2 \equiv 0$; in a Θ-solvent, van 't Hoff's law $P = n_b kT$ holds up to quadratic order in n_b. Higher-order virial coefficients A_3 and beyond are non-zero; so, for higher polymer concentrations, deviations from ideal behaviour are also found in a Θ-solvent.

Perturbation expansions in terms of the excluded volume in principle yield the second and higher order osmotic virial coefficients [44,57]. This procedure becomes rather cumbersome for A_4 and higher-order coefficients and established scaling exponents [45] for the semidilute polymer concentration regime can not be reproduced in a virial expansion, as could have been expected as this is also true for simple fluids.

In fact, in the semidilute case, the picture is simple; the chains overlap to such a degree that the characteristic length scale is determined by the correlation length ξ rather than the coil size set by the chain length M. The corresponding volume ξ^3 is denoted as a blob. The osmotic pressure can then be viewed upon as an ideal gas of blobs, so $P_{sd} \sim \xi^{-3}$, with the number of blobs $\sim \xi^{-3}$. Therefore, the scaling result becomes $P_{sd}/kT \sim \phi_p^{3\gamma}$.

A convenient expression that enables description of both the dilute and semidilute polymer concentration regimes follows from a simple additivity rule: $P = P_{id} + P_{sd}$. This additivity follows from the Flory–Huggins theory [43] for a Θ-solvent but appears to be an excellent approximation for good solvents as well [46]. This leads to the following expression for the ratio P/P_{id}:

$$\frac{P}{P_{id}} = \frac{Pv_p}{\phi_p kT} = 1 + \zeta \phi_p^{3\gamma - 1}, \tag{4.28}$$

with

$$\zeta = \begin{cases} 4.10 & \Theta\text{-solvent,} \\ 1.62 & \text{good solvent.} \end{cases} \tag{4.29}$$

Under Θ-solvent conditions Flory–Huggins theory reproduces Eq. 4.28. The numerical coefficient ζ follows from Flory–Huggins theory for a Θ-solvent and from RGT for a good solvent. Equation 4.29 turns out to be extremely accurate when compared with experimental and computer simulation data [46]. The result for a good solvent is plotted in Fig. 4.9.

The osmotic compressibility that we need in Eq. (4.4) now follows straightforwardly:

$$\frac{\partial (Pv_p/kT)}{\partial \phi_p} = 1 + 3\gamma \zeta \phi_p^{3\gamma - 1} \tag{4.30}$$

$$= \begin{cases} 1 + 12.3 \phi_p^2 & \Theta\text{-solvent,} \\ 1 + 3.73 \phi_p^{1.31} & \text{good solvent.} \end{cases} \tag{4.31}$$

Exercise 4.7. Derive Eq. 4.30 from Eqs. (4.28), (4.29) and (4.15).

4.3.4 Generalised Free Volume Theory Phase Behaviour

Below, we summarise the results for the osmotic pressure and depletion thickness, which are inserted into the Generalised Free Volume Theory (GFVT) expressions. Subsequently, we consider the implications for the phase behaviour.

4.3.4.1 GFVT Ingredients; Θ-solvent

For polymer chains in a Θ-solvent the scaling exponent γ takes the mean-field value $\gamma = 1$. The polymer concentration derivative of the reduced osmotic pressure follows from Eq. 4.30 as

$$q^3 \frac{\partial \widetilde{P}}{\partial \phi_p} = 1 + 12.3\phi_p^2, \tag{4.32}$$

and the ratio between the depletion thickness and the colloid radius is

$$\frac{\delta_s}{R} = 0.938 \left(\frac{q}{\sqrt{1 + 6.02\phi_p^2}} \right)^{0.9}. \tag{4.33}$$

Equation (4.33) follows directly from Eq. 4.26.

4.3.4.2 GFVT Ingredients for Polymers in a Good Solvent

The De Gennes scaling exponent γ equals 0.77 under good solvent conditions. Therefore, from Eq. 4.30 we have

$$q^3 \frac{\partial \widetilde{P}}{\partial \phi_p} = 1 + 3.73\phi_p^{1.31}, \tag{4.34}$$

and from Eq. 4.26 we have

$$\frac{\delta_s}{R} = 0.865 \left(\frac{q}{\sqrt{1 + 3.95\phi_p^{1.54}}} \right)^{0.88}. \tag{4.35}$$

Note that, in contrast with classical FVT [22], $\partial \widetilde{P}/\partial \phi_p$ and δ_s/R in Eqs. (4.32) and (4.35) now depend on the polymer concentration ϕ_p.

In Fig. 4.10 we compare the free volume fraction α calculated from Eq. (4.6), in the good solvent limit with δ_s/R in Eq. (4.7) from Eq. (4.35), with Monte Carlo simulation results of Fortini et al. [58] for $q = 1.05$ along the binodal gas liquid curve. Except for some deviation at large colloid volume fractions the agreement is excellent.

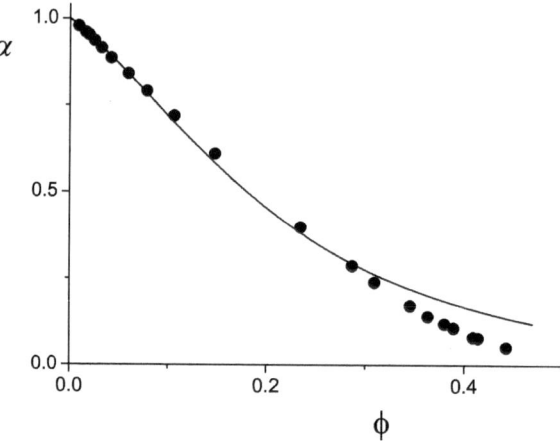

Fig. 4.10 Free volume fraction α as a function of colloid volume fraction for $q = 1.05$. GFVT (curve) compared to Monte Carlo computer simulations (data points) [58]

4.3.4.3 GFVT Phase Diagrams

We can now compute the phase diagrams for hard spheres with interacting polymers using the general expression Eq. (4.4) and its ingredients by computing the chemical potential $\widetilde{\mu} = (\partial \widetilde{\Omega}/\partial \phi)$ and total pressure $\widetilde{P} = -\partial(\widetilde{\Omega}/\phi)/\partial(1/\phi) = \phi\widetilde{\mu} - \widetilde{\Omega}$,

$$\widetilde{\mu} = \widetilde{\mu}^0 + \int_0^{\phi_p^R} g\left(\frac{\partial \widetilde{P}^R}{\partial \phi_p^{R'}}\right) d\phi_p^{R'}, \tag{4.36}$$

$$\widetilde{P} = \widetilde{P}^0 + \int_0^{\phi_p^R} h\left(\frac{\partial \widetilde{P}^R}{\partial \phi_p^{R'}}\right) d\phi_p^{R'}. \tag{4.37}$$

Here, g and h are given by Eqs. (3.49) and (3.50), with a, b and c defined in Eq. (4.7). Coexistence curves then follow from Eqs. (3.43a) and (3.43b).

In Fig. 4.11, we compare gas–liquid coexistence curves from GFVT in the good solvent limit with Monte Carlo simulation results of Bolhuis et al. [59] for $q = 0.67$ and 1.05. These simulations were performed using hard spheres mixed with polymer chains with hard core excluded volume between the segments. It is clear GFVT is capable of predicting the location of the phase boundaries reasonably well.

The critical end points for Θ-solvent and good solvent conditions are rather close to the one for PHSs [41,60]:

$$q_{\text{cep}} = \begin{cases} 0.328 & \text{penetrable hard spheres} \\ 0.337 & \Theta\text{-solvent} \\ 0.388 & \text{good solvent} \end{cases} \tag{4.38}$$

We turn back to Fig. 4.3 and make a comparison of the experimental phase diagrams with GFVT under good solvent conditions. We show again the experimental data

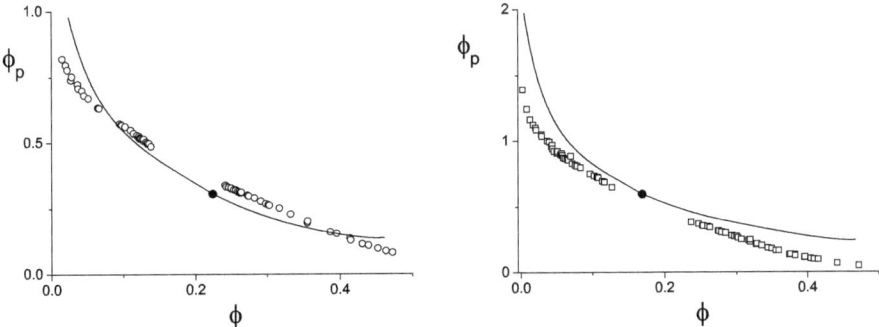

Fig. 4.11 Phase diagram (colloidal gas–liquid equilibria) for hard spheres (in terms of their volume fraction ϕ) mixed with interacting polymers (in terms of their relative concentration ϕ_p) in a good solvent. Monte Carlo simulations (○) [59] for $q = 0.67$ (*left*) and $q = 1.05$ (*right*) are compared to GFVT predictions (curves; critical point: •)

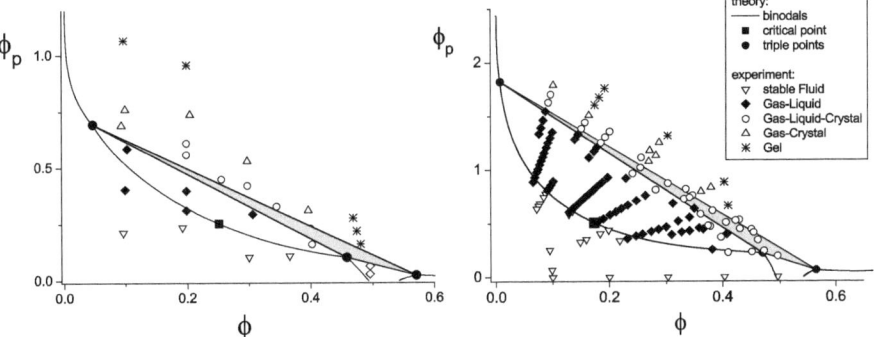

Fig. 4.12 State diagrams of colloid–polymer mixtures for $q = 0.57$ (*left*) and $q = 1.0$ (*right*) as in Fig. 4.3, but now compared to theoretical GFVT predictions (curves) from Fleer and Tuinier [41]. The experimental data (see key) is from a system of PMMA spheres mixed with polystyrene polymers in *cis*-decalin [15,21]

in Fig. 4.12 and inserted GFVT predictions (good solvent) for the binodals as the curves and the triple triangle as filled region.

GFVT is capable of accurately describing the experimental equilibrium phase diagrams for $q = 0.57$ and $q = 1$, and constitutes a major improvement with respect to FVT for $q > 0.5$. GFVT proves to be very useful, especially for $q = 1$, since FVT completely fails to quantitatively describe the phase diagram here.

For $q = 0.57$, the triple point composition of the colloidal liquid that coexists with a colloidal gas and crystal was determined by Moussaïd et al. [30]. In Table 4.1, we compare these data with FVT and GFVT predictions. The experimental colloid volume fraction and polymer concentration clearly deviate significantly from FVT. In particular, the polymer concentration of the coexisting colloidal liquid phase is about 30 times larger than the prediction using classical FVT. GFVT gives a much better prediction of the composition, especially if the polymers are assumed to be in a good solvent.

Table 4.1 Liquid composition at the triple point for $q = 0.57$

	ϕ	ϕ_p
FVT	0.489	0.0037
GFVT Θ	0.470	0.048
GFVT good	0.452	0.108
experiment	0.444	0.1

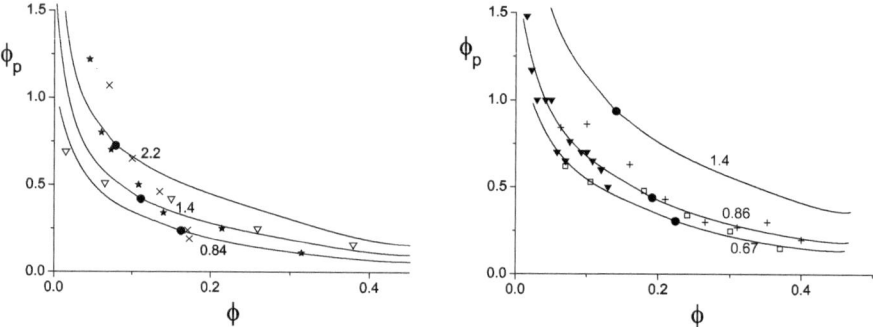

Fig. 4.13 Comparison of experimental gas–liquid coexistence binodals (data) compared to GFVT (curves; critical point: •). *Left*: spherical colloids mixed with polymer chains in a Θ-solvent for $q = 0.84$ ($\triangledown$, [24]), 1.4 ($\star$, [25]) and 2.2 ($\times$, [25]). *Right*: colloidal spheres mixed with polymers in a good solvent for $q = 0.67$ ($\square$, [24]), 0.86 ($\blacktriangledown$, [62]) and 1.4 ($+$, [24])

One might think GFVT is only useful for describing the phase diagrams of well-defined experimental hard-sphere–random coil systems (such as PMMA with PS) or phase equilibria from computer simulations. GFVT, however, also helps to give reasonable predictions for many other colloid–nonadsorbing polymer mixtures [41]. In Fig. 4.13 we compare GFVT binodals for gas–liquid coexistence with experimental data on colloid–polymer mixtures under Θ- (left panel) and good solvent (right panel) conditions. The order of magnitude of the predicted binodals is accurate in all cases. Sometimes, the agreement is nearly quantitative. Aspects that give deviations are related to polymer chain length dispersity and the non-hard-sphere character of the colloidal particles. The colloids may, for instance, be somewhat sticky, or they may repel one another to some extent due to anchored brushes that are not very short, or due to repulsive double layer interactions. In principle, it is possible to include these effects into GFVT [61] but we shall not consider this here.

Exercise 4.8. In what directions will gas–liquid binodals and fluid–solid binodals at low ϕ shift when there is an additional weak short-ranged double layer repulsion between the spheres?

4.3.4.4 Scaling Behaviour in the Semidilute Regime

We now consider the large q-limit of mixtures containing long polymer chains and relatively small colloidal spheres. This is the regime where ϕ_p^R along the binodals exceeds unity; so at the binodal, we have semidilute polymer solutions. The characteristic length scale in semidilute polymer solutions is ξ, which scales as in Eq. 4.16, $\xi/R_g \sim \phi_p^{-\gamma}$. For colloid–polymer mixtures this expression can be rewritten in terms of $\xi/R = q\xi/R_g$ to become

$$\phi_p q^{-1/\gamma} \sim \left(\frac{\xi}{R}\right)^{-1/\gamma}. \tag{4.39}$$

It is important to note that ξ is independent of R_g. This implies that the right-hand side of Eq. 4.39 is independent of q. Therefore, $\phi_p q^{-1/\gamma}$ is also independent of q, yielding

$$\phi_p \sim q^{1/\gamma}. \tag{4.40}$$

For large q values, it is therefore efficient to introduce a parameter Y as a rescaled polymer concentration:

$$Y = \phi_p q^{-1/\gamma}. \tag{4.41}$$

In the large q-limit Y is a constant (independent of q). It follows that $\phi_p = Yq^{1/\gamma}$ diverges as $q^{1/\gamma} = q^{1.3}$ for large q under good solvent conditions. This predicted $q^{1.3}$ scaling [60] of large q binodals is corroborated by simulation [63] and experiment [64,65], as demonstrated in Fig. 4.14. In the plot on the left, we show rescaled computer simulation data for the gas–liquid binodal [63] for hard spheres mixed with polymer chains in the good solvent limit (long chains consisting of hard spherical segments). The data are binodal points from Fig. 1.23 for $q > 3$. In the rescaled form they collapse onto a universal curve. Clearly, in the colloid limit, where Y depends on q, this scaling does not apply.

In the right-hand plot in Fig. 4.14, we show experimental data [64,65] for the gas–liquid coexistence for two large q values. Also, these data collapse onto a single curve after rescaling according to Eq. (4.41). Hence, this predicted $q^{1.3}$ scaling is corroborated by both simulations and experiments.

The parameter Y is a convenient normalised polymer concentration, which has the important property of becoming independent of the size ratio q in the high q limit [41,60], where the polymer concentrations along the binodals are in the semidilute regime. The Y values along the binodals always remain of order unity. Then $\delta = \xi \sim \varphi^{-\gamma}$ [45], which does not depend on R_g. Hence, δ/R does not depend on $q = R_g/R$, and δ/R reaches a constant, q-independent level.

Analytical approximations for the phase behaviour of colloid–polymer mixtures can be found in [41], providing simple, approximate yet reasonably accurate descriptions of equilibrium phase diagrams. Using Y instead of ϕ_p turns all phase diagrams to more universal ones with a polymer concentration variable that is always of order unity for the relevant characteristic parts of the phase behaviour.

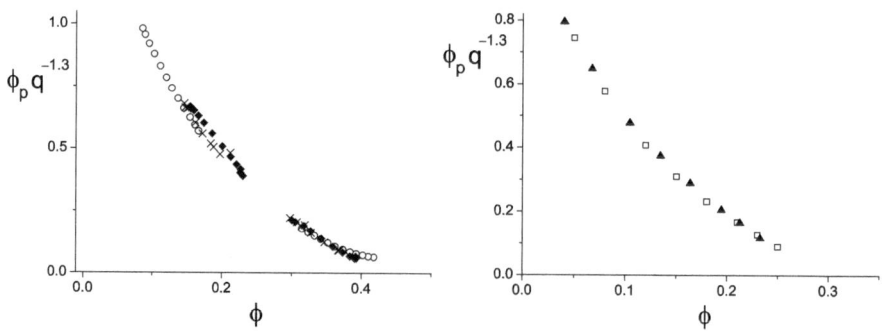

Fig. 4.14 Scaling of (*left*) Monte Carlo computer simulation results (see Fig. 1.23) for $q = 3.86$ (○), 5.58 (×) and 7.78 (♦) by Bolhuis [63], and experimental results (*right*) on micro-emulsion droplets mixed with free polyisoprene polymer chains ($q = 10$ (□) and $q = 16$ (▲)) by Mutch et al. [64, 65] for the gas–liquid coexistence in the protein limit regime

4.4   Phase Separation Kinetics and Long-Lived Metastable States in Colloid–Polymer Mixtures

So far, we have considered the equilibrium phase behaviour in colloidal suspensions that results from depletion interactions. Fascinating phase behaviour can be observed, such as a metastable colloidal gas–liquid phase separation and a three-phase gas–liquid–crystal region. This is because the range and strength of the depletion interactions in colloidal suspensions can be readily tuned, as opposed to those in atomic and molecular systems. The predictions of phase diagrams are, however, not always realised.

Systems often become trapped in metastable nonequilibrium gel and glass states. In several cases, experiments reveal that the states of the mixtures strongly depend on the starting position in the phase diagram; and discrepancies between predictions and actual observations are due to the intricacies of the dynamics of phase transitions. In this section, we briefly consider the phase separation process and the nonequilibrium states in colloid–polymer mixtures. Taking advantage of the (large) length scales and (long) time scales involved allows us to reveal some of the secrets of the complex pathways involved in the formation of gels and glasses. The discussion below is organised into two parts. First, Sect. 4.4.1 covers systems with a size ratio of the polymer-to-colloid size q larger than 0.3; and secondly, Sect. 4.4.2 focuses on situations where q is smaller than 0.3 (relatively large colloidal spheres). For a comprehensive overview of the topics treated in this section, see the review by Poon [66].

4.4.1 $q > 0.3$

Quantitative predictions of demixing kinetics in colloid–polymer mixtures are fairly complicated; but frequently, insight on demixing mechanisms [67] can be obtained by careful inspection of the equilibrium phase diagrams of colloid–polymer mixtures. The equilibrium phase diagram for $q > 0.3$ (as follows from (G)FVT) is summarised in Fig. 4.15. In the absence of polymer, only a fluid–crystal phase coexistence is found; for $\phi < 0.49$ colloids are in a fluid phase. As ϕ_p is increased, this fluid becomes unstable above a certain concentration. Then, phase separation occurs towards colloidal gas–liquid coexistence. At high ϕ_p, a colloidal gas coexists with a colloidal crystal. Following Gibbs' phase rule, there must be a three-phase gas–liquid–crystal coexistence region in between, which is indeed predicted and observed experimentally.

4.4.1.1 Gas–Liquid Demixing

We first focus on the phase separation in the (colloidal) gas–(colloidal) liquid region. Above the spinodal curve, long wavelength fluctuations in colloid or polymer concentration lower the Helmholtz energy. After a quench in this unstable two-phase region, spontaneous long wavelength density fluctuations are no longer stable with respect to a homogeneous distribution. Concentration fluctuations with large wavelengths have a stronger thermodynamic driving force, whereas the diffusion process is faster for short wavelengths. This competition, which results in a 'fastest growing mode', is characteristic of spinodal decomposition, and can be described using Cahn–Hilliard theory [68–71]. For a didactic account of the Cahn–Hilliard theory of spinodal decomposition in colloidal dispersions, see Chap. 9 of Ref. [72] or see Ref. [73].

The time and length scales involved in colloidal systems allow the relevant phenomena to be probed by small-angle light scattering and optical microscopy [23,74]. In Fig. 4.16 (left panel), we plot the measured scattered intensity $I(Q)$ during the

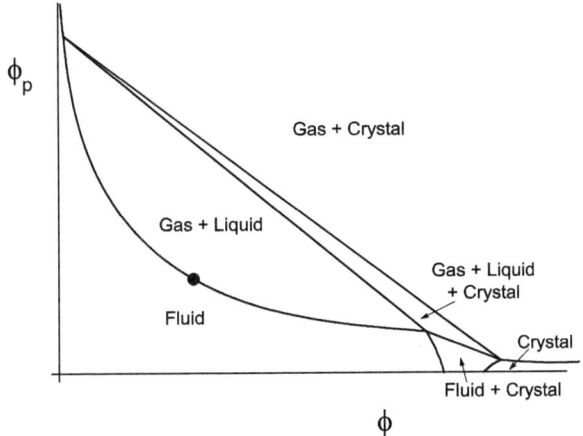

Fig. 4.15 Overview of the phase behaviour of a colloid–polymer mixture for large q. Full curves are binodals. The different phase states are indicated. The critical point is indicated by the filled circle

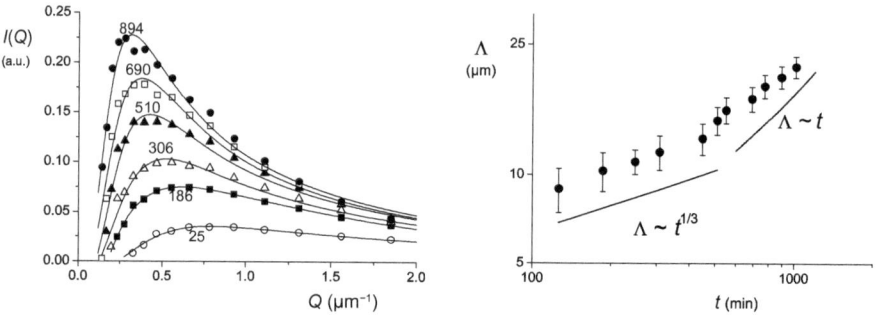

Fig. 4.16 *Left*: scattered intensity $I(Q)$ as a function of the scattering wave vector Q of an unmixing dispersion of whey protein colloids ($R \approx 27$ nm) mixed with exocellular polysaccharides ($R_g = 86$ nm; $q = 3.2$). The time (in minutes) after mixing is indicated. *Right*: Characteristic length scale $\Lambda = 2\pi/Q_m$ obtained from the $I(Q)$ curves as a function of time. The diffusive growth Eq. 4.42 and viscous hydrodynamic growth Eq. 4.43 scaling regimes are indicated as straight lines. Data replotted from Ref. [23]

unmixing of a dispersion of whey protein colloids ($R = 27$ nm) mixed with polysaccharides ($R_g = 86$ nm), so $q \approx 3$. The overall values of the scattered intensity increase with time, which implies that larger structures are being formed. At each time frame the scattered intensity passes through a maximum as a function of Q. The quantity Q_m denotes the scattering wave factor at this maximum intensity. The value of Q_m decreases with time, corroborating an increase of the characteristic length scale $\Lambda = 2\pi/Q_m$. In Fig. 4.16 (right panel) we present this length scale Λ as a function of time. Two regimes can be distinguished in this figure:

$$\Lambda \sim t^{1/3} \tag{4.42}$$

and

$$\Lambda \sim t. \tag{4.43}$$

These two time scales are characteristic of the diffusive growth (Eq. 4.42) and viscous hydrodynamic growth (Eq. 4.43) regimes in spinodal decomposition [75].

As indicated above, the phase separation process can also be studied using optical microscopy as has been done by Verhaegh et al. [74]. They observed spinodal decomposition in a well-defined model colloid–polymer mixture of silica spheres in cyclohexane with dissolved polydimethyl siloxane chains of $q \approx 1$, and found similar results to those plotted in Fig. 4.16. Their findings agree with the scenario described above and the characteristic length scale Λ follows the regimes of Eqs. (4.42) and (4.43).

The spinodal decomposition can be studied more directly using confocal scanning laser microscopy (CSLM). Aarts et al. [76] studied the phase separation kinetics of a PMMA colloid–PS mixture in decalin with $q = 0.56$. Typical spinodal structures that coarsen in time are shown in Fig. 4.17.

Fig. 4.17 CSLM images (each side is 1400 μm) of a phase-separating mixture of polystyrene polymers and fluorescently labelled PMMA spheres, exhibiting the typical spinodal structure. The images correspond to $t = 3$ s (*left*), 11 s (*middle*) and 22 s (*right*) after homogenisation. Images kindly provided by D.G.A.L. Aarts. See also the movies in the SI of Ref. [76]

4.4.1.2 Demixing in the Three-Phase Region

We now consider the phase separation process in the three-phase colloidal gas–liquid–crystal region. This case has been analysed experimentally and theoretically in great detail by Poon et al. [77–79]. By consideration of the free energy landscape, they were able to distinguish several pathways for this phase separation process. The pathways were shown to depend on the location of the starting position in the three-phase region. In the central section of the three-phase triangle in Fig. 4.15, Poon et al. predicted and observed the scenario presented in Fig. 4.18.

Initially, the sample is a colloidal fluid (i) that phase separates into a polymer-rich colloidal gas and a colloidal liquid that is dilute in polymer (ii). This gas–liquid coexistence, however, is metastable. Soon after the formation of a (sharp) gas–liquid interface, 'flashes' of light appear from the liquid (lower) phase (iii). These flashes are caused by homogeneously nucleating and growing crystallites. Subsequently, the crystallites sink to the bottom, giving rise to the final gas–liquid–crystal coexistence (iv). This is a classic example of a multi-step kinetic pathway.

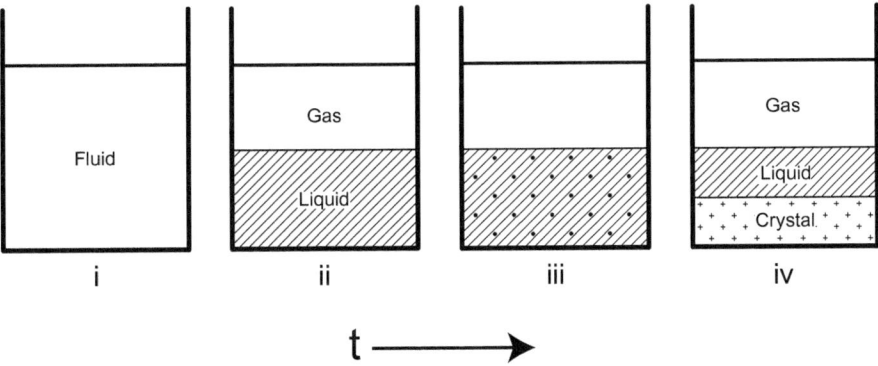

Fig. 4.18 Time evolution of phase separation kinetics in the three-phase region as observed by Poon et al. [77]

4.4.1.3 Gas–crystal Phase Transition and Nonequilibrium States

Finally, we consider the large q situation at high polymer concentrations. Gas–crystal phase equilibria were observed (Fig. 4.12) above the three-phase region over a narrow polymer concentration regime [15,21]. A kinetically arrested (gel-like) nonequilibrium state with no visible crystallites appeared for higher polymer concentrations. The appearance of a particle gel can be explained as follows. When the depletion attraction becomes sufficiently strong the colloidal particles stick irreversibly and form a space-spanning network. In such cases, the phase separation process will not reach its equilibrium state but rather a nonequilibrium gel-like state is encountered. This has been well-studied for small q (see Sect. 4.4.2) but its appearance for $q > 0.3$ has not received much attention yet.

In summary, the observed state diagram for large q is sketched in Fig. 4.19. The only difference with the equilibrium phase diagram presented in Fig. 4.15 is that at high polymer concentrations ϕ_p (or large depletion attraction) a nonequilibrium gel-like state is observed [23,80]. At $\phi \approx 0.58$ there is a glassy state. The influence of a depletion attraction on colloidal glasses has not yet been studied systematically for high q.

4.4.2 $q < 0.3$

At first sight, the equilibrium phase diagram at low q (depicted in Fig. 4.20) appears to be dull when compared to the phase diagram for $q > 0.3$ in Fig. 4.15. However, while for large q the predictions of the equilibrium phase diagram are generally realised, for small q nonequilibrium and metastable states dominate in large regions of the state diagram. A premonition that the pathways involved in the phase separation for small q can be intricate is provided by the presence of a metastable gas–liquid phase separation in the fluid–crystal domain of the phase diagram.

Fig. 4.19 State diagram of a colloid–polymer mixture for large q. The different observed equilibrium and long-lived nonequilibrium phase states are indicated. The critical point is indicated by the filled circle

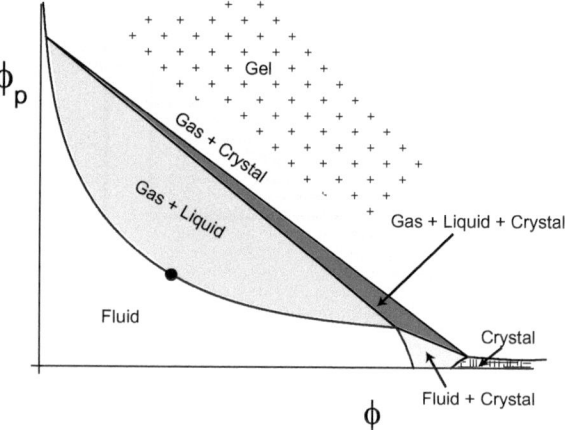

Fig. 4.20 Equilibrium phase
diagram of a
colloid–polymer mixture for
small q. The solid curve is
the fluid–solid coexistence
curve; the dashed curve is
the metastable gas–liquid
coexistence region

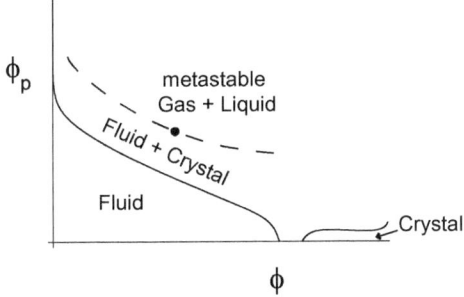

In order to sample these intricacies, we consider the phase separation pathways upon increasing the polymer concentration at three colloid volume fractions: low colloid volume fractions ($\phi \lesssim 0.05$), intermediate volume fractions ($0.05 \lesssim \phi \lesssim 0.4$), and high volume fractions ($\phi \gtrsim 0.4$).

4.4.2.1 Low Colloid Volume Fractions

De Hoog et al. [81] studied the phase behaviour of a mixture containing sterically stabilised fluorescent PMMA spheres ($R \approx 600$ nm) and PS polymer chains ($M_p \approx 2000$ kg/mol) in a mixed solvent consisting of tetralin, *cis*-decalin and carbon tetrachloride. In this solvent mixture, the PMMA spheres are nearly refractive index matched, which enables fluorescence confocal scanning light microscopy (CSLM) measurements to be taken deep into the sample. This solvent mixture has the additional advantage that the density difference with the PMMA spheres is small to such a degree that significant sedimentation of PMMA particles only becomes apparent after months. The radius of gyration of the polymers was determined to be 46 nm. Hence, $q = 0.08$ and the polymer overlap concentration of the PS chains was estimated to be 8 g/L. The colloid–polymer mixture was studied at a fixed colloid volume fraction of $\phi \approx 0.02$ and PS concentrations up to 10 g/L, i.e., just above the overlap concentration. From Eq. (2.65) the contact potential follows as

$$W_{\text{dep}} = W(r = 2R) = -3 \ln(2) \frac{\phi_p}{q} kT. \qquad (4.44)$$

Hence, the investigated polymer concentrations imply strengths of the attraction up to $25kT$. De Hoog et al. observed four characteristic scenarios, and the polymer concentration determines which scenario is observed. The corresponding concentration regimes (A)–(D) are identified in Fig. 4.21.

Regime (A) is the one-phase fluid region, whereas regimes (B–D) are in the two-phase region. Representative micrographs of the structures found in these regions are given in Fig. 4.22. These were all taken at $t \simeq 400 \, \tau_B$, where $\tau_B = R^2/D_s$ is the Brownian time scale, D_s is the self-diffusion coefficient, and R is the sphere radius. In the narrow region (B), the formation of nucleation clusters can be observed (Fig. 4.22). These clusters eventually sediment and form a colloidal crystal. In this regime, just across the fluid–crystal binodal, the contact potentials are $W_{\text{dep}} \approx -5kT$.

Fig. 4.21 State diagram for the $q = 0.08$ colloid–polymer mixture investigated by De Hoog et al. [81]. The different regions A–D are indicated in the plot. The PS concentration is plotted on the ordinate versus the colloid concentration on the abscissa. Reprinted with permission from Ref. [81]. Copyright 2001 American Physical Society (APS)

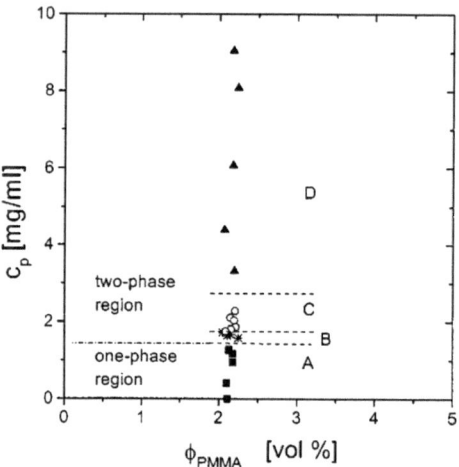

In region (C), centred at about 2 g/L (corresponding to $W_{dep} \approx -6.5kT$), aggregates are formed from single particles, followed by growth of the clusters via subsequent aggregation. These clusters are dense but not crystalline. The sediment formed is dense but no crystallinity is observed. Above 3.3 g/L in region (D), which corresponds to $W_{dep} < -10kT$, aggregation is also observed; but here, the clusters have a more ramified or elongated string-like shape. The sediment formed is dilute and gets denser within a few days. Figure 4.23 shows xy-scans after one or several days of the different regions (A)–(D) indicated in Fig. 4.21.

4.4.2.2 Intermediate Colloid Volume Fractions

For intermediate colloid volume fractions ($0.05 \lesssim \phi \lesssim 0.4$), gel formation is observed at sufficiently high polymer concentrations. Here, we consider the sedimentation of these gels under gravity. In [82–85] it was demonstrated that the settling behaviours at low intermediate colloid volume fractions ($0.05 \lesssim \phi \lesssim 0.2$) and at high intermediate colloid volume fractions ($0.2 \lesssim \phi \lesssim 0.4$) are distinctly different. The drastic change in settling kinetics takes place over a fairly narrow concentration range.

Low Intermediate Volloid Volume Fractions

As an example of gel collapse at low intermediate concentrations we consider the work of Poon, Pirie and Pusey [86]. They studied a similar system to De Hoog et al. [81]—this time with PMMA particles ($R = 238$ nm) and PS polymers ($M_p = 370$ kg/mol). We focus on their experiments carried out at a colloid volume fraction $\phi \approx 0.1$. Just like De Hoog et al., they observed four regimes. A colloidal fluid was observed at low concentration. Across the phase boundary, a narrow concentration regime of equilibrium fluid–crystal phase behaviour was found.

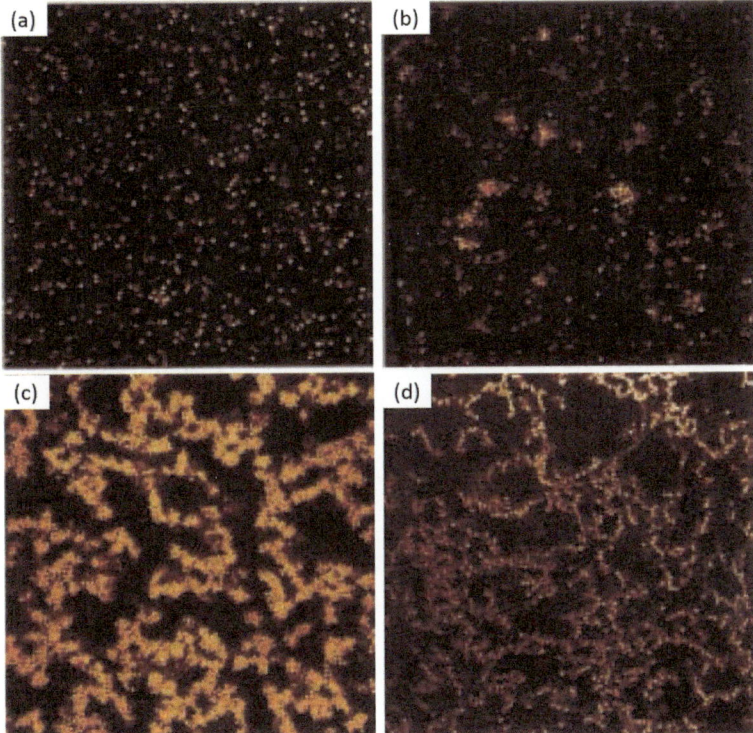

Fig. 4.22 CSLM images taken after mixing colloids and polymer at four concentrations [81] at a colloid volume fraction of 2 vol %. Polymer concentrations were (**a**): 1.2 g/L (region A), (**b**): 1.7 g/L (region B), (**c**): 2.1 g/L (region C) and (**d**): 8.1 g/L (region D). The image size represents 100 μm by 100 μm. These images were taken at $t \approx 400\tau_B$. Reprinted with permission from Ref. [81]. Copyright 2001 APS

At higher concentrations, they observed a spinodal-like small-angle light scattering pattern in the region where De Hoog et al. observed aggregation. While at first surprising, Rouw et al. [87] had already noted in the late 1980s that the computer simulations of colloidal aggregation phenomena by Ziff [88] appear to show a long-wavelength spinodal-like modulation of the aggregate density. This perception was turned into a quantitative framework by Carpineti and Giglio [89], who proved experimentally that colloidal aggregation exhibits the same features as spinodal decomposition, be it that the scaled structure factor $\widetilde{S}(Q/Q_m, t)$ is now described by

$$\widetilde{S}(Q/Q_m, t) = Q_m(t)^{-d} \widetilde{f}(Q/Q_m), \tag{4.45}$$

where $\widetilde{f}(Q/Q_m)$ is a time-independent scaling function. For ordinary spinodal decomposition $d = 3$; while for aggregating systems, Eq. (4.45) holds if we take $d = d_f$, where d_f is the fractal dimension of the aggregates. Poon et al. [86] showed that the small-angle light scattering data of their depletion-induced aggregating system are described by Eq. 4.45 with a value of d that depends on the polymer concentration, i.e., on the strength of the depletion interaction. With increasing strength

Fig. 4.23 CSLM images of PMMA colloid–PS polymer mixtures taken at different periods of time, as indicated below each image. The images show xy scans at $\phi \approx 0.02$. From top to bottom: regime D, polymer concentration $c_p = 8.1$ g/L; regime C, $c_p = 2.1$ g/L; regime B, $c_p = 1.7$ g/L; regime A, $c_p = 1.2$ g/L; regime A, $c_p = 0$ g/L. The particles have a radius of approximately 600 nm. Reprinted with permission from Ref. [81]. Copyright 2001 APS

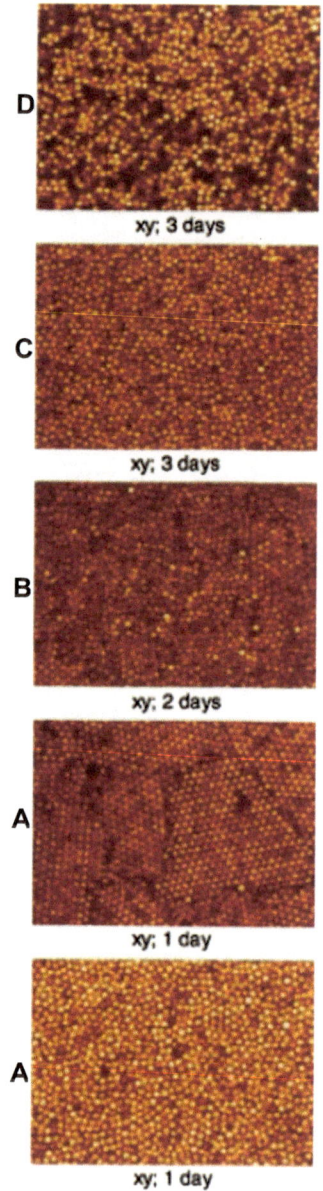

of the depletion interaction, the fractal dimension decreases from $d_f = 3$ (dense clusters) to $d_f = 1.7$ (ramified clusters) (Fig. 4.24), which is in agreement with the visual observations of De Hoog et al. [81].

For still higher polymer concentrations Poon et al. [86] observed the formation of a transient gel. Such a particle gel is characterised by a rapid collapse of its structure followed by a delay period where no significant sedimentation occurs. This delay time can range from seconds to many months, depending upon the strength of the gel.

Fig. 4.24 Fractal dimension d_f inside a gel of aggregated PMMA spheres and PS polymer chains ($q = 0.08$) as a function of the estimated depletion attraction at contact. Data replotted from Ref. [86]

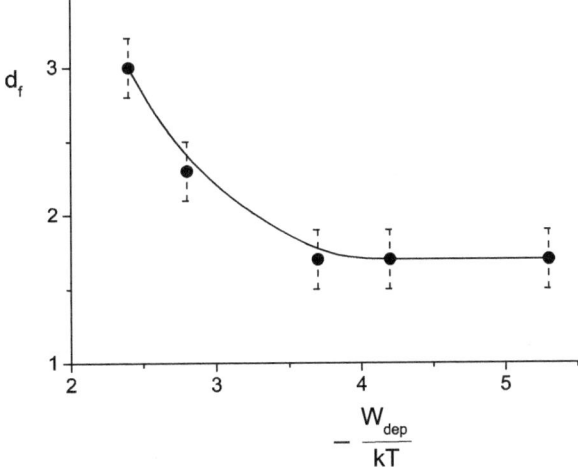

Fig. 4.25 Sketch of a gelation formed by arrested spinodal phase separation. A space-spanning network of colloidal spheres is aggregated through depletion of nonadsorbing polymer chains. Drawing inspired by Verhaegh et al. [90]

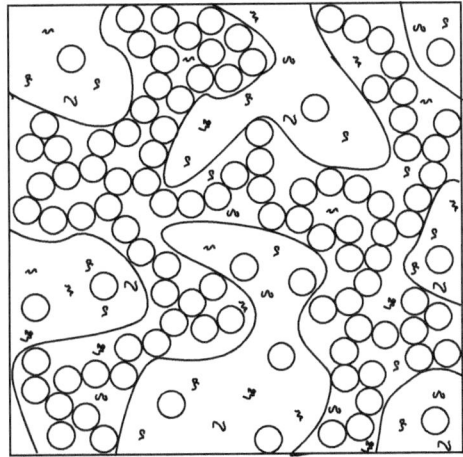

Verhaegh et al. [90, 91] studied transient gelation with a combination of small-angle light scattering, light microscopy and confocal scanning laser microscopy in a system of sterically stabilised silica spheres mixed with PDMS polymer chains in cyclohexane at colloid volume fractions of $\phi \approx 0.1$. Early time small-angle light scattering curves show a peak at Q_m which shifts in time to smaller values and increases slightly in intensity. This indicates the presence of a coarsening bicontinuous structure (Fig. 4.25). Alternating dark and bright domains observed in light microscopy confirm the existence of this bicontinuous network of colloid-rich and colloid-poor domains. A slight coarsening of the domains was detected, along with an increased contrast. After this initial stage, which only lasts a few seconds, the shift in the light scattering peak is arrested. Also, the speckle fluctuations are arrested in time, implying that the system now has a gel character. The above observations suggest that gelation results from a spinodal gas–liquid (also termed fluid–fluid) phase separation, which is arrested at some intermediate stage and leaves the system in

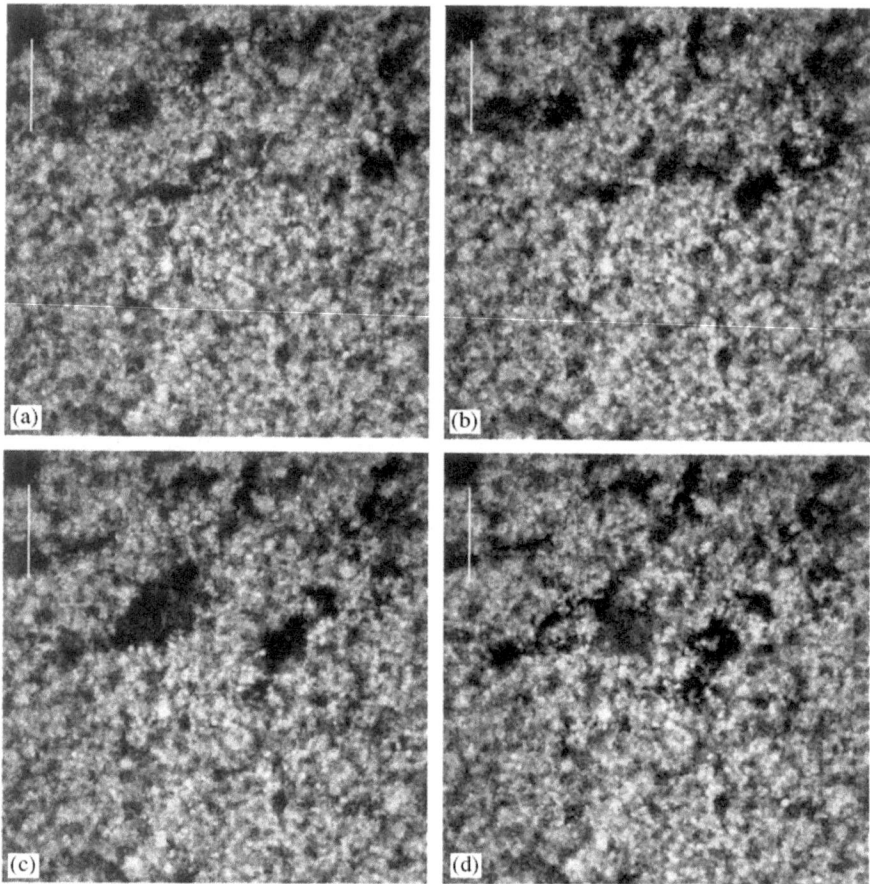

Fig. 4.26 CSLM images of a mixture of fluorescent silica spheres ($R = 115$ nm) and PDMS polymers ($R_g = 23$ nm) in cyclohexane. The images taken are mixtures with $\phi = 0.125$, $\phi_p = 1.23$ (85 g/L PDMS) during gel lifetimes $t = 230$ s (**a**), 240 s (**b**), 250 s (**c**) and 260 s (**d**). The vertical bar corresponds to 50 μm. Reprinted with permission from Ref. [91]. Copyright 1999 Elsevier

a 'frozen' state of microphase separation [90,92,93]. Polymer depletion-mediated gelation of conducting colloidal particles can significantly enhance the conductivity due to interparticle electron tunneling [94,95].

From CSLM analysis [91], it appears that the internal structure of the particle network becomes disrupted by the formation of fractures (Fig. 4.26). The number of fractures increases with time, as is in agreement with an increase of scattered light in the forward direction. This increase of the number of fractures weakens the gel strength until the elastic modulus becomes so small that, in the end, the gel collapses under gravity.

Using dark-field microscopy, Starrs et al. [82] studied this delayed sedimentation in a colloid–polymer mixture of PMMA spheres and PS polymers in tetralin and *cis*-decalin) with $R = 186$ nm and $R_g = 17$ nm, respectively ($q \approx 0.1$). The time evolution of the sedimentation height is shown in Fig. 4.27 for a colloid volume

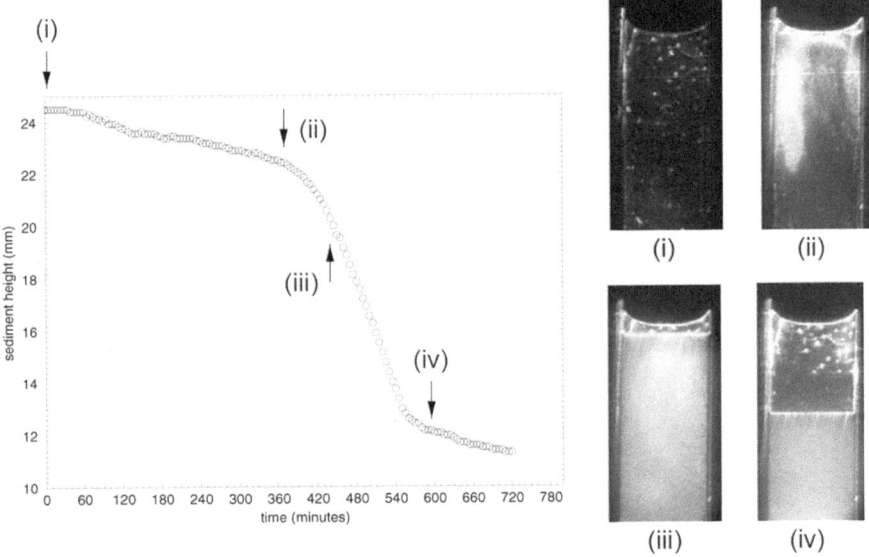

Fig. 4.27 *Left*: Sedimentation profile of PMMA spheres ($R = 186$ nm; $\phi = 0.2$) mixed with 5 g/L PS polymers ($R_g = 17$ nm) in a tetralin/*cis*-decalin solvent mixture. Initial sample height is 24.5 mm. *Right*: Corresponding dark-field images of the structures observed during time evolution of the transient gel. Reprinted with permission from Ref. [82]. Copyright 2002 IOP Publishing, Ltd

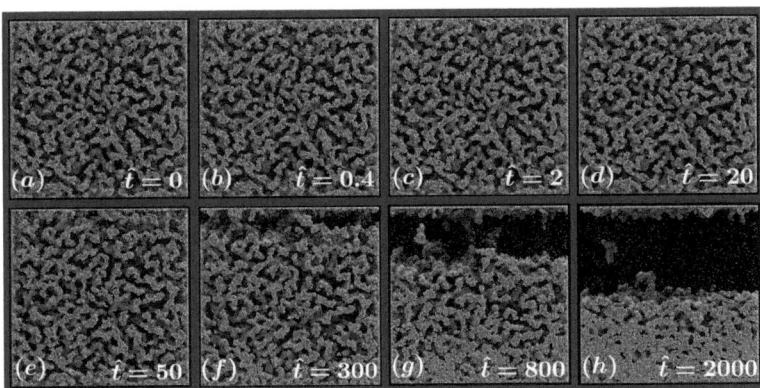

Fig. 4.28 Computer simulation snapshots (**a–h**) of the gel during sedimentation as a function of diffusion time $\hat{t} = R^2/D_s$ for a gel with particle depletion attraction at contact of $5kT$ and gravitational Peclet number $Pe = (4\pi/3)R^4\Delta\rho g/kT = 0.1$ [96]. Image kindly provided by R. Zia

fraction of 0.20 and PS concentration of 5 g/L. Note the brightening of the sample during the delay period from (i) to (ii). After a delay time of about 460 min the gel collapses.

Padmanabhan and Zia [96] studied the three stages of gel collapse by computer simulation and obtained the results plotted in Fig. 4.28 for a particle interaction of 5

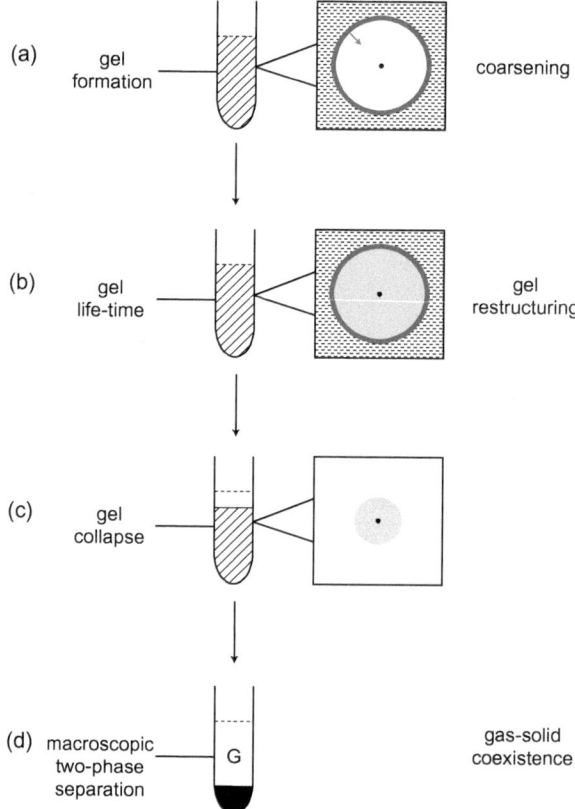

Fig. 4.29 The demixing process observed in a sample containing a colloid–polymer mixture at high polymer concentrations for $q < 0.3$. The corresponding light scattering patterns are indicated as well. **a** gel formation 'birth', **b** gel life-time 'life', **c** gel collapse 'death', **d** macroscopic phase separation. Drawing inspired by Verhaegh et al. [90]

kT and gravitational Peclet number of 0.1. Here again, we clearly see three stages: an induction period, fast collapse, and slow compaction.

The experimental results of Verhaegh et al. [91] suggest that it is possible to distinguish four stages in the evolution of a transient gel: birth, life (during which the gel ages), collapse, and finally macroscopic two-phase separation (Fig. 4.29).

The lifetime of the transient gel is determined by the strength of the depletion interaction and the colloid concentration, and plays a role in many practical systems. For example, in salad dressing, which is an oil-in-water emulsion, the depletion flocculation of the oil droplets induced by the addition of a polysaccharide such as xanthan leads to the formation of a particle network [97,98]. The yield stress of this network (in the sense of food science) 'stabilises' the dressing, i.e., prevents creaming. Buscall et al. [99] proposed a simple theory to rationalise collapse times for the delayed sedimentation of weakly aggregated colloidal gels.

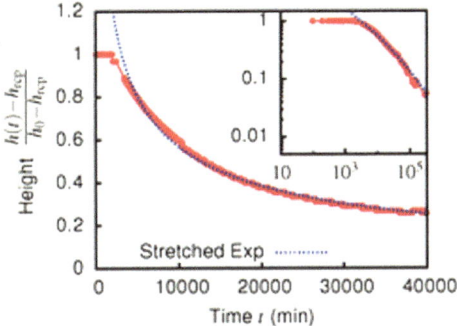

Fig. 4.30 Sedimentation profile of PMMA spheres ($R = 186$ nm; $\phi = 0.35$) mixed with 7 g/L PS polymers ($M_p = 6.0 \cdot 10^5$ g/mol and $R_g = 21$ nm) dispersed in cis-decalin. Blue dotted curves: stretched exponential fits to the slow sedimentation $h(t) = [\exp(t/\tau)]^\beta$ with exponent $\beta = 0.12$. The quantity τ is characteristic time scale of the sedimentation process. The inset shows the double logarithmic form of the main plot. Reprinted from Ref. [85] under the terms of CC-BY-3.0

High Intermediate Colloid Volume Fractions

Starrs et al. [82], Secchi et al. [84], and Harich et al. [85] all observed that at high intermediate colloid volume fractions ($0.2 \lesssim \phi \lesssim 0.4$) the settling behaviour is distinctly different from that at low intermediate colloid volume fractions ($0.05 \lesssim \phi \lesssim 0.2$).

In this case, the gels, after a quiescent period, compress smoothly without any sign of the "catastrophic" collapse observed for the gels at lower colloid concentration, which exhibit characteristic compression kinetics. The time evolution of the gel height h follows a stretched exponential $h(t) = \exp(t/\tau)^\beta$, as is illustrated in Fig. 4.30 from Harich et al. [85].

The gels at low intermediate concentrations are referred to as 'collapsing gels' based on their settling behaviour; and the gels at high intermediate concentrations are referred to as 'creeping gels' [84] when in a system with a very small density difference ($\Delta\rho = 0.063$ g/cm^3) between the colloidal particles (casein micelles) and the solvent (the aqueous solution).

For aqueous mixtures of casein micelles and poly(ethylene oxide) (PEO), Mahmoudi and Stradner [100] presented a state diagram that includes various equilibrium and nonequilibrium phase states that were discussed in Fig. 4.31. The phase states they observed are homogeneous fluid, fluid–fluid phase separation, rapidly collapsing gels, delayed collapsing gels, and stable gels.

4.4.2.3 High Colloid Volume Fractions

We now focus on colloid volume fractions above the fluid–crystal phase transition. As discussed in Sect. 3.2.3, Pusey and van Megen [14] observed that, above a volume fraction of about 0.58, suspensions of hard-sphere like PMMA particles do not crystallise over several months. The explanation for this phenomenon is that particles become increasingly tightly caged by their neighbours as the volume fraction

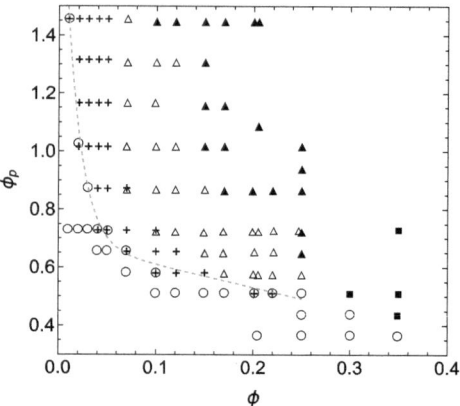

Fig. 4.31 Redrawn state diagram of mixtures of casein micelles (in volume fractions along the abscissa) and poly(ethylene oxide) (normalised by the overlap concentration along the ordinate) in aqueous 0.1 M NaCl solutions. The size ratio $q \approx 0.31$. Symbols: (○) homogeneous fluid, (+) fluid–fluid phase separation, (△) rapidly collapsing gels, (▲) delayed collapsing gels, and (■) stable gels; dotted curve: experimental phase boundary. Data replotted from Ref. [100]

increases. For sufficiently high volume fractions, this caging reaches such a degree that the particles are unable to move far enough to nucleate crystallisation and the system is termed glassy. Mode coupling theory (MCT) [101] supports the existence of such a glass transition. Also further experimental results on hard-sphere colloidal glasses [102] have been successfully interpreted with MCT [103].

Ilett et al. [15] observed that adding 1–2 g/L PS ($M_p = 390$ kg/mol, $R_g = 18$ nm) to a concentrated ($\phi \approx 0.6$) suspension of sterically stabilised PMMA colloids ($R = 217$ nm and hence, $q = 0.08$) leads to crystallisation of an initially glassy suspension. At higher polymer concentrations (above 3 g/L) the system again becomes kinetically arrested. Systems composed of hard spheres with short-range attraction display two glass states: one referred to as 'repulsive glass' (no or very weak attraction), and one referred to as 'attractive glass' (strong attraction), with a metastable fluid (weak attraction) in between. These two types of glasses were subsequently predicted by MCT [104–106] and confirmed experimentally by Pham et al. [107] and Eckert and Bartsch [108].

A simple physical picture of the repulsive and attractive glass and the metastable fluid was given by Poon [109] and by Pusey [110], and is depicted in Fig. 4.32. The first interpretation of the attractive glass given in Fig. 4.32 has been refined by Zaccarelli and Poon [111] and Royall, Williams and Tanaka [112]. Zaccarelli and Poon [111] studied the interplay between bonding and caging by examining the long-time dependence of the single particle mean-squared displacement using molecular dynamics simulations. From this, they found that each particle is indeed trapped by bonds with a particular set of neighbours. However, these bonds break after a certain time and bonds reform with a different set of neighbours. Royall, Williams

Fig. 4.32 The influence of (short-ranged) attraction on the glassy state. **a** *Repulsive glass.* This corresponds to the situation where attractions are absent or very weak. There is significant free volume in the cage, but the particle cannot escape this cage. **b** *Metastable fluid that may crystallise.* Adding a weak attraction leads to particles clustering in the cage. Now, holes open up and particles can escape. **c** *Attractive glass.* Upon further increasing the attraction an attractive glass is formed. The attraction is now so strong that particles are tightly bound, so a cage is again formed

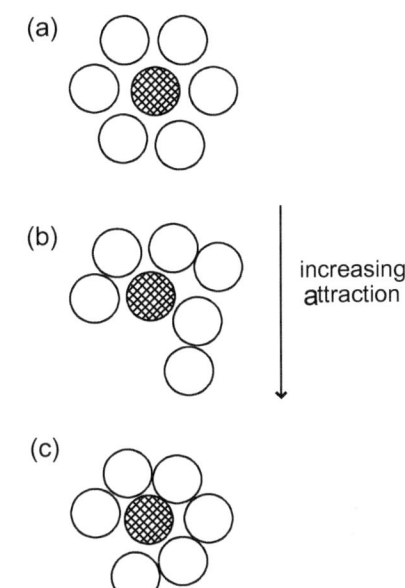

(a)

(b)

increasing attraction

(c)

and Tanaka [112] see hints from both experiments and simulations of the attractive glass transition, but it is ultimately superseded by a dense gel.

The observed overall state diagram is shown in Fig. 4.33. It is intriguing and challenging that the depletion force, which allows independent control of the range and strength of the attraction, opens up new ways of structuring soft matter. It can lead to phase transitions as well as to long-lived metastable states.

Fig. 4.33 Observed state diagram of a colloid–polymer mixture for small q

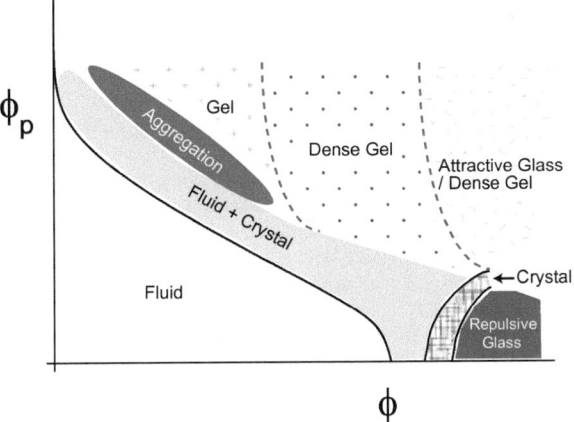

References

1. Fleer, G.J., Cohen Stuart, M.A., Scheutjens, J.M.H.M., Cosgrove, T., Vincent, B.: Polymers at Interfaces. Chapman and Hall, New York (1993)
2. Pelssers, E.G.M., Cohen Stuart, M.A., Fleer, G.J.: J. Chem. Soc. Faraday Trans. **86**, 1355 (1990)
3. Hoogendam, C.W., Peters, J.C.W., Tuinier, R., de Keizer, A., Cohen Stuart, M.A., Bijsterbosch, B.H.: J. Colloid Interface Sci. **207**, 309 (1998)
4. Golz, P.M.: Dynamics of colloids in polymer solutions. Ph.D. thesis, University of Edinburgh (1999)
5. Gögelein, C., Nägele, G., Buitenhuis, J., Tuinier, R., Dhont, J.K.G.: J. Chem. Phys. **130**, 204905 (2009)
6. McFarlane, N.L., Wagner, N.J., Kaler, E.W., Lynch, M.L.: Langmuir **26**, 13823 (2010)
7. McFarlane, N.L., Wagner, N.J., Kaler, E.W., Lynch, M.L.: Langmuir **26**, 6262 (2010)
8. Wijmans, C.M., Zhulina, E.B., Fleer, G.J.: Macromolecules **27**, 3238 (1994)
9. Van Helden, A.K., Jansen, J.W., Vrij, A.: J. Colloid Interface Sci. **81**, 354 (1981)
10. de Kruif, C.G., Rouw, P.W., Jansen, J.W., Vrij, A.: J. Phys. Colloques **46**, C3 (1985)
11. de Kruif, C.G., Briels, R.P., May, A.: Langmuir **4**, 668 (1988)
12. Antl, L., Goodwin, J.W., Hill, R.D., Ottewill, R.H., Owens, S.M., Papworth, S., Waters, J.A.: Colloids Surf. **17**, 67 (1986)
13. Bartlett, P., Ottewill, R.H., Pusey, P.N.: J. Chem. Phys. **93**, 1299 (1990)
14. Pusey, P.N., Van Megen, W.: Nature **320**, 340 (1986)
15. Ilett, S.M., Orrock, A., Poon, W.C.K., Pusey, P.N.: Phys. Rev. E **51**, 1344 (1995)
16. Poon, W.C.K., Phys, J.: Condens. Matter **14**, R859 (2002)
17. Castello, B.A.L., Luckham, P.F., Tadros, T.F.: Langmuir **8**, 464 (1992)
18. Royall, C.P., Charbonneau, P., Dijkstra, M., Russo, J., Smallenburg, F., Speck, T., Valeriani, C.: (2023). https://doi.org/10.48550/arXiv.2305.02452
19. Berry, C.G.: J. Phys. Chem. **44**, 1550 (1966)
20. Starrs, L., Bartlett, P.: Faraday Disc. **51**, 123 (2003)
21. Tuinier, R., Smith, P.A., Poon, W.C.K., Egelhaaf, S.U., Aarts, D.G.A.L., Lekkerkerker, H.N.W., Fleer, G.J.: Europhys. Lett. **82**, 68002 (2008)
22. Lekkerkerker, H.N.W., Poon, W.C.K., Pusey, P.N., Stroobants, A., Warren, P.B.: Europhys. Lett. **20**, 559 (1992)
23. Tuinier, R., Dhont, J.K.G., de Kruif, C.G.: Langmuir **16**, 1497 (2000)
24. Ramakrishnan, S., Fuchs, M., Schweizer, K.S., Zukoski, C.F.: J. Chem. Phys. **116**, 2201 (2002)
25. Lynch, I., Cornen, S., Piculell, L.: J. Phys. Chem. B **108**, 5443 (2004)
26. Hennequin, Y., Evens, M., Quezada Angulo, C.M., Van Duijneveldt, J.S.: J. Chem. Phys. **123**, 054906 (2005)
27. Zhang, Z.X., van Duijneveldt, J.S.: Langmuir **22**, 63 (2006)
28. Tejero, C.F., Daanoun, A., Lekkerkerker, H.N.W., Baus, M.: Phys. Rev. Lett. **73**, 752 (1994)
29. Fleer, G.J., Tuinier, R.: Physica A **379**, 52 (2007)
30. Moussaïd, A., Poon, W.C.K., Pusey, P.N., Soliva, M.F.: Phys. Rev. Lett. **82**, 225 (1999)
31. Thomson, J.A., Schurtenberger, P., Thurston, G.M., Benedek, G.B.: Proc. Natl. Acad. Sci. **84**, 7079 (1987)
32. Ten Wolde, P.R., Frenkel, D.: Science **277**, 1975 (1997)
33. Haas, C., Drenth, J.: J. Cryst. Growth **196**, 388 (1999)
34. Auer, S., Frenkel, D.: Nature **409**, 1020 (2001)
35. Sear, R.: J. Chem. Phys. **114**, 3170 (2001)
36. Fortini, A., Sanz, E., Dijkstra, M.: Phys. Rev. E **78**, 041402 (2008)
37. Zaccarelli, E.: J. Phys.: Condens. Matter **19**, 323101 (2007)
38. Leal-Calderon, F., Bibette, J., Biais, J.: Europhys. Lett. **23**, 653 (1993)
39. Faers, M.A., Luckham, P.F.: Langmuir **13**, 2922 (1997)
40. Tuinier, R., Rieger, J., de Kruif, C.G.: Adv. Colloid Interface Sci. **103**, 1 (2003)

41. Fleer, G.J., Tuinier, R.: Adv. Colloid Interface Sci. **143**, 1 (2008)
42. Aarts, D.G.A.L., Tuinier, R., Lekkerkerker, H.N.W., Phys, J.: Condens. Matter **14**, 7551 (2002)
43. Flory, P.J.: Principles of Polymer Chemistry. Cornell University Press, New York (1953)
44. Schäfer, L.: Excluded Volume Effects in Polymer Solutions (Springer, Heidelberg, 1999)
45. De Gennes, P.G.: Scaling Concepts in Polymer Physics. Cornell University Press, Ithaca (1979)
46. Fleer, G.J., Skvortsov, A.M., Tuinier, R.: Macromol. Theory Sim. **16**, 531 (2007)
47. Edwards, S.F.: Proc. Phys. Soc. **85**, 613 (1965)
48. Edwards, S.F., Freed, K.F.: J. Phys. A **2**, 145 (1969)
49. Doi, M., Edwards, S.F.: The Theory of Polymer Dynamics. Clarendon Press, Oxford (1986)
50. Fleer, G.J., Skvortsov, A.M., Tuinier, R.: Macromolecules **36**, 7857 (2003)
51. Surve, M., Pryamitsyn, V., Ganesan, V.: J. Chem. Phys. **122**, 154901 (2005)
52. Joanny, J.F., Leibler, L., De Gennes, P.G.: J. Polymer Sci.: Polym. Phys. **17**, 1073 (1979)
53. Tuinier, R., Vliegenthart, G.A., Lekkerkerker, H.N.W.: J. Chem. Phys. **113**, 10768 (2000)
54. Bolhuis, P.G., Louis, A.A., Hansen, J.P., Meijer, E.J.: J. Chem. Phys. **114**, 4296 (2001)
55. Eisenriegler, E.: J. Chem. Phys. **79**, 1052 (1983)
56. Hanke, A., Eisenriegler, E., Dietrich, S.: Phys. Rev. E **59**, 6853 (1999)
57. Yamakawa, H.: Modern Theory of Polymer Solutions. Harper and Row, New York (1971)
58. Fortini, A., Bolhuis, P.G., Dijkstra, M.: J. Chem. Phys. **128**, 024904 (2008)
59. Bolhuis, P.G., Louis, A.A., Hansen, J.P.: Phys. Rev. Lett. **89**, 128302 (2002)
60. Fleer, G.J., Tuinier, R.: Phys. Rev. E **76**, 041802 (2007)
61. Gögelein, C., Tuinier, R.: Eur. Phys. J. E. **27**, 171 (2008)
62. Tuinier, R., de Kruif, C.G.: J. Chem. Phys. **110**, 9296 (1999)
63. Bolhuis, P.G., Meijer, E.J., Louis, A.A.: Phys. Rev. Lett. **90**, 068304 (2003)
64. Mutch, K.J., van Duijneveldt, J.S., Eastoe, J., Grillo, I., Heenan, R.K.: Langmuir **25**, 3944 (2009)
65. Mutch, K.J., van Duijneveldt, J.S., Eastoe, J., Grillo, I., Heenan, R.K.: Langmuir **26**, 1630 (2010)
66. Poon, W.C.K.: Curr. Opin. Colloid Interface Sci. **3**, 593 (1998)
67. Anderson, V.J., Lekkerkerker, H.N.W.: Nature **416**, 811 (2002)
68. Cahn, J.W., Hillard, J.E.: J. Chem. Phys. **28**, 258 (1958)
69. Cahn, J.W., Hillard, J.E.: J. Chem. Phys. **31**, 688 (1959)
70. Cahn, J.W.: Acta Metall. **9**, 795 (1961)
71. Cahn, J.W.: J. Chem. Phys. **42**, 93 (1965)
72. Dhont, J.K.G.: An Introduction to Dynamics of Colloids. Elsevier, Amsterdam (1996)
73. Verhaegh, N.A.M., Lekkerkerker, H.N.W.: The physics of complex systems. In: Mallamace, F., Stanley, H. (eds.), Proceedings of the International School of Physics "Enrico Fermi" Course CXXXIV. IOS Press, Amsterdam (1997)
74. Verhaegh, N.A.M., van Duijneveldt, J.S., Dhont, J.K.G., Lekkerkerker, H.N.W.: Physica A **230**, 409 (1996)
75. Siggia, E.D.: Phys. Rev. A **20**, 595 (1979)
76. Aarts, D.G.A.L., Dullens, R.P.A., Lekkerkerker, H.N.W.: New J. Phys. **7**, 40 (2005)
77. Poon, W.C.K., Renth, F., Evans, R.M.L., Fairhurst, D.J., Cates, M.E., Pusey, P.N.: Phys. Rev. Lett. **83**, 1239 (1999)
78. Renth, F., Poon, W.C.K., Evans, R.M.L.: Phys. Rev. E **64**, 031402 (2001)
79. Evans, R.M.L., Poon, W.C.K., Renth, F.: Phys. Rev. E **64**, 031403 (2001)
80. Smith, P.A., Egelhaaf, S.U., Poon, W.C.K.: Personal Communication. Springer (2008)
81. de Hoog, E.H.A., Kegel, W.K., van Blaaderen, A., Lekkerkerker, H.N.W.: Phys. Rev. E **64**, 021497 (2001)
82. Starrs, L., Poon, W.C.K., Hibberd, D.J., Robins, M.M.: J. Phys.: Condens. Matter **14**, 2485 (2002)
83. Huh, J.Y., Lynch, M.L., Furst, E.M.: Phys. Rev. E **76**, 051409 (2007)
84. Secchi, E., Buzzaccaro, S., Piazza, R.: Soft Matter **10**, 5296 (2014)
85. Harich, R., Blythe, T.W., Hermes, M., Zaccarelli, E., Sederman, A.J., Gladden, L.F., Poon, W.C.K.: Soft Matter **12**, 4300 (2016)

86. Poon, W.C.K., Pirie, A.D., Pusey, P.N.: Faraday Disc. **101**, 65 (1995)
87. Rouw, P.W., Woutersen, A.T.J.M., Ackerson, B.J., de Kruif, C.G.: Physica A **56**, 876 (1989)
88. Ziff, R.M.: In: Family, F., Landau, D.P. (eds.) Kinetics of Aggregation and Gelation, p. 191. Elsevier, Amsterdam (1984)
89. Carpineti, M., Giglio, M.: Phys. Rev. Lett. **68**, 3327 (1992)
90. Verhaegh, N.A.M., Asnaghi, D., Lekkerkerker, H.N.W., Giglio, M., Cipelletti, L.: Physica A **242**, 104 (1997)
91. Verhaegh, N.A.M., Asnaghi, D., Lekkerkerker, H.N.W.: Physica A **264**, 64 (1999)
92. Miller, C.A., Miller, D.D.: Colloids Surf. **16**, 219 (1985)
93. Lu, P.J., Zaccarelli, E., Ciulla, F., Schofield, A., Sciortino, F., Weitz, D.A.: Nature **453**, 499 (2008)
94. Nigro, B., Grimaldi, C., Ryser, P., Varrato, F., Foffi, G., Lu, P.J.: Phys. Rev. E **87**, 062313 (2013)
95. Guillemeney, L., Lermusiaux, L., Landaburu, G., Wagnon, B., Abécassis, B.: Commun. Chem. **5**, 7 (2022)
96. Padmanabhan, P., Zia, R.: Soft Matter **14**, 3265 (2018)
97. Parker, A., Gunning, P.A., Ng, K., Robbins, M.M.: Food Hydrocolloids **9**, 333 (1995)
98. Tuinier, R., de Kruif, C.G.: J. Colloid Interface Sci. **218**, 201 (1999)
99. Buscall, R., Choudhury, T.H., Faers, M.A., Goodwin, J.W., Luckham, P.A., Partridge, S.J.: Soft Matter **5**, 1345 (2009)
100. Mahmoudi, N., Stradner, A.: J. Phys. Chem. B **119**, 15522 (2015)
101. Bengtzelius, U., Götze, W., Sjölander, A.: J. Phys. C **17**, 5915 (1984)
102. Pusey, P.N., van Megen, W.: Phys. Rev. Lett. **59**, 2083 (1987)
103. Götze, W.: J. Phys.: Condens. Matter **11**, A1 (1999)
104. Bergenholtz, J., Fuchs, M.: Phys. Rev. E **59**, 5706 (1999)
105. Fabbian, L., Götze, W., Sciortino, F., Tartaglia, P., Thiery, F.: Phys. Rev. E **59**, R1347 (1999)
106. Fabbian, L., Latz, A., Schilling, R., Sciortino, F., Tartaglia, P., Theis, C.: Phys. Rev. E **60**, 5768 (1999)
107. Pham, K.N., Puertas, A.M., Bergenholtz, J., Egelhaaf, S.U., Moussaïd, A., Pusey, P.N., Schofield, A.B., Cates, M.E., Fuchs, M., Poon, W.C.K.: Science **296**, 104 (2002)
108. Eckert, T., Bartsch, E.: Phys. Rev. Lett. **89**, 125701 (2002)
109. Poon, W.C.K.: Mater. Res. Bull. 96 (2004)
110. Pusey, P.N.: Lecture at the 7th Liquid Matter Conference (Lund, Sweden, 27 June–1 July 2008)
111. Zaccarelli, E., Poon, W.C.K.: Proc. Natl. Acad. Sci. **106**, 15203 (2009)
112. Royall, C.P., Williams, S.R., Tanaka, H.: J. Chem. Phys. **148**, 044501 (2018)

The Interface in Demixed Colloid–Polymer Dispersions

5

In Chaps. 3 and 4, the focus was on theory and experiments related to the phase behaviour of mixtures containing colloidal spheres and nonadsorbing polymers. As we have seen, when the polymer coils are sufficiently large relative to the colloidal spheres, a colloidal gas–liquid (fluid–fluid) phase separation may occur. The two phases that appear differ in composition. One phase is a dilute colloidal fluid (a colloidal 'gas') dispersed in a concentrated polymer solution. This phase coexists with a concentrated colloidal fluid (a colloidal 'liquid') dispersed in a dilute polymer solution. Besides the phase behaviour, the properties of the interface between such coexisting phases have gained interest [1–10]. The interface can be characterised by a number of quantities, such as the interfacial tension and the interfacial thickness. Perrin's atom–colloid analogy suggests similarities with the molecular gas–liquid interface. However, as we will see, there are also differences driven by the vastly different length scales of the systems.

As discussed in Sect. 1.3.2, we expect the interfacial tension in demixed colloid–polymer systems to be ultra low. Its magnitude can be estimated from [11]

$$\gamma \approx \frac{kT}{d^2}, \tag{5.1}$$

where d is the particle diameter. For simple molecular systems, where d is less than roughly a nanometer, this yields values for the surface tension γ of about 10–100 mN/m [12], which agrees well with experimental results. For the colloidal domain, interfacial tensions are predicted to be orders of magnitude smaller; for instance, for particles of $d = 100$ nm an interfacial tension of about 0.4 µN/m is predicted. As we shall see, this is indeed the order of magnitude of the tension of the colloidal gas–liquid interface that is found in experiments and theory.

In Sect. 5.1, we focus on experiments that have been conducted to measure the interfacial tension. Subsequently, a theoretical approach is presented in Sect. 5.2 that

© The Author(s) 2024
H. N. W. Lekkerkerker, R. Tuinier, M. Vis, *Colloids and the Depletion Interaction*,
Lecture Notes in Physics 1026, https://doi.org/10.1007/978-3-031-52131-7_5

enables one to predict the interfacial tension. This is accompanied by a quantification of the thickness of the interface, which will be compared with experiments. In Sect. 5.3, we show that the ultra-low tension gives rise to an interfacial roughness due to thermal fluctuations, which can be observed visually. The implications of these aspects for the hydrodynamics of droplet coalescence are also discussed.

5.1   Interfacial Tension Measurements

Although various approaches to determine interfacial tensions exist, such as the Wilhelmy plate, the Du Noüy ring or the pendant drop techniques [13], these common methods generally do not allow (accurate) measurement of the ultra-low tensions occurring in demixed colloid–polymer mixtures. Here, we highlight two methods that have been used successfully in the past to measure these interfacial tensions: the spinning drop and the interfacial profile methods.

5.1.1   The Spinning Drop Method

In the spinning drop method [14, 15], a droplet of the phase with the lowest density (usually the colloidal gas phase) is suspended in the phase with the highest density (usually the colloidal liquid phase) in a tube (Fig. 5.1). When spinning this tube around its axis, the elongation of this droplet induced by centrifugal forces is balanced by interfacial forces. As shown below, it is possible to quantify the interfacial tension using this force balance from an analysis of the droplet deformation.

When the length L of the droplet is significantly larger than its diameter D, the droplet geometry approaches the so-called Vonnegut limit [14]. In this case, the balance between centrifugal and interfacial forces can be quantified as follows: consider two points (C) and (D) located at a distance $D/2$ from the axis of rotation, as

Fig. 5.1 Measurement of the tension of the colloidal gas–liquid interface using spinning drop measurements. A droplet of the colloidal gas phase with the indicated dimensions is dispersed in the colloidal liquid phase. The system is composed of stearyl silica spheres in cyclohexane ($d = 26$ nm) with poly(dimethylsiloxane) ($q \approx 1.1$). Parts reproduced with permission from Ref. [16]. Copyright 2008 Springer

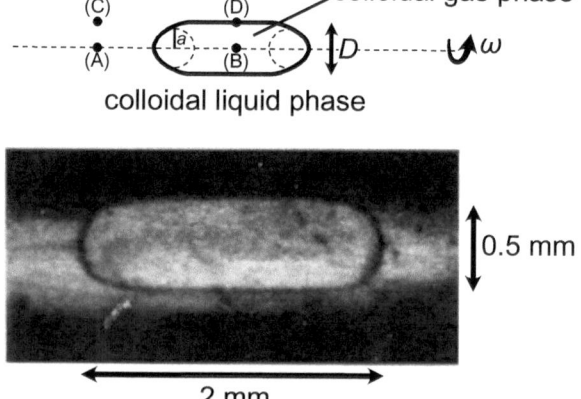

depicted in Fig. 5.1. For mechanical equilibrium, the pressure at these two points must be equal: $P_{(C)} = P_{(D)}$. These pressures $P_{(C)}$ and $P_{(D)}$ are related to the pressures at points (A) and (B) that lie on the axis of rotation. The pressure at (C) is the pressure at (A) plus a contribution by the centrifugal pressure, $P_{(C)} = P_{(A)} + \frac{1}{8}\rho_L \omega^2 D^2$, where ω is the angular velocity and ρ_L is the mass density of the liquid phase. Similarly, the pressure at point (D) is the pressure at (B) plus the centrifugal pressure *minus* the Laplace pressure due to the cylindrically shaped interface at this position; as such, $P_{(D)} = P_{(B)} + \frac{1}{8}\rho_G \omega^2 D^2 - \frac{2\gamma}{D}$, where ρ_G is the mass density of the colloidal gas phase. In turn, the pressure at point (B) is the pressure at point (A) plus the Laplace pressure due to the curvature of the spherically shaped end-cap of the droplet, which has a radius a, so that $P_{(B)} = P_{(A)} + \frac{2\gamma}{a}$. It can be shown that, in the Vonnegut limit, $a = \frac{1}{3}D$ (see, for instance, Chap. 1 of Ref. [13]). Combining these ingredients, we find [14, 15]

$$\gamma = \frac{\Delta\rho\omega^2 D^3}{32},\tag{5.2}$$

where the mass density difference $\Delta\rho \equiv \rho_L - \rho_G$. This Vonnegut limit is accurate when $L \gtrsim 4D$ [14]; a more general expression by Princen et al. [15] is available that also works for less elongated droplets, for which a is a function of the droplet aspect ratio L/D.

Exercise 5.1. Derive Eq. (5.2) based on the conditions for mechanical equilibrium, centrifugal pressure and Laplace pressure, as outlined in the text.

The spinning drop method was applied to measure the tension of colloidal gas–liquid interfaces by Vliegenthart and Lekkerkerker [17] in 1997. Two years later, more systematic experiments were carried out by de Hoog and Lekkerkerker [18] and Chen et al. [3]. De Hoog and Lekkerkerker studied a system composed of sterically stabilised silica spheres (with diameter $d = 26$ nm $\pm$ 19%) dispersed in cyclohexane and mixed with poly(dimethylsiloxane) (PDMS, $R_g = 14$ nm), and found values between approximately 3.0 and 4.5 µN/m depending on the colloid and polymer volume fractions (Table 5.1). Chen et al. studied a similar system with slightly smaller silica spheres ($d = 20.2$ nm $\pm$ 8%) mixed with PDMS in cyclohexane, especially focused on a wide range of polymer concentrations. They found values ranging from about 0.6 to 17 µN/m, close to and far from the critical point,, respectively. These values are remarkably close to those predicted by the crude estimate of Eq. (5.1) (6 and 10 µN/m, respectively), provided the system is sufficiently far from the critical point.

Table 5.1 Measurements of the colloidal gas–liquid interfacial tension by de Hoog and Lekkerkerker [18] for a system comprising sterically stabilised silica spheres with $d = 26$ nm $\pm$ 19% dispersed in cyclohexane with poly(dimethylsiloxane) ($R_g = 14$ nm; $q = 1.1$). Note that ϕ and ϕ_p indicate the overall colloid volume fraction and relative polymer concentration c_p/c_p^*, respectively

sample	ϕ	ϕ_p	$\Delta\rho$ (kg/m^3)	γ (μN/m)
A	0.246	1.016	175	3.0 ± 0.7
B	0.264	1.132	240	3.8 ± 0.6
C	0.268	1.189	256	3.2 ± 0.2
D	0.283	1.133	268	4.2 ± 0.4
E	0.301	1.143	291	4.5 ± 0.5

Exercise 5.2. Based on Eq. (5.2), show that interfacial tensions of the order of micronewtons per metre are indeed experimentally measurable for droplets with a diameter of the order of hundreds of micrometres and ω of the order of 100 rad/s.

5.1.2   The Meniscus Method

Another method used to measure ultra-low interfacial tensions is analysing the shape of the interface, known as a meniscus, near a vertical surface. Far from this surface, the interface is macroscopically horizontal as gravity dominates. Near the wall, however, interfacial effects start to dominate as the interface must meet the wall under a certain contact angle, set by the interfacial tension between the two bulk phases and between each of the bulk phases and the surface. As a result, the interface deforms over a certain length scale (Fig. 5.2). This length scale is known as the capillary length and quantifies the balance of gravitational and interfacial forces; it can be found by analysing the shape of the meniscus. Aarts et al. [19] presented the first such measurements for a colloidal gas–liquid interface in the early 2000s. The capillary length is defined as

$$\ell_c \equiv \sqrt{\frac{\gamma}{\Delta\rho g}}, \tag{5.3}$$

where $\Delta\rho$ again is the mass density difference between the two phases and g is the gravitational acceleration.

The balance between the gravitational pressure on one hand and the Laplace pressure due to the deformation of the interface on the other hand can be mathematically expressed as

$$-\Delta\rho g z(y) = \frac{\gamma}{R_c(y)}, \tag{5.4}$$

where $z(y)$ is the height of the interface at a vertical distance y from the wall and $R_c(y)$ is the radius of curvature at that position. Here, we define $z(y)$ such that it is zero far from the wall ($y \to \infty$), where the interface is flat. The minus sign on the left-hand side signifies that the gravitational contribution to pressure decreases with increasing height.

From geometrical arguments, it can be shown that the radius of curvature can be expressed as

$$\frac{1}{R_c(y)} = \frac{-z''(y)}{[1 + z'(y)^2]^{3/2}}, \tag{5.5}$$

where $z'(y)$ and $z''(y)$ represent the first and second derivatives of $z(y)$ with respect to y. The minus sign on the right-hand side is due to the convention that an interface that is curved towards the dense bottom phase has a positive radius of curvature. Combining with Eqs. (5.3) and (5.4) gives

$$z(y) = \ell_c^2 \frac{z''(y)}{[1 + z'(y)^2]^{3/2}}, \tag{5.6}$$

which can be integrated once to yield

$$\frac{1}{2} z^2 = -\ell_c^2 \left(\frac{1}{\sqrt{1 + z'^2}} - 1 \right), \tag{5.7}$$

where the term -1 between the parentheses is an integration constant that follows from the boundary condition that $z'(y \to \infty) = 0$. In turn, this can be rearranged to give

$$\frac{dz}{dy} = -\sqrt{\left(\frac{1}{1 - \frac{z^2}{2\ell_c^2}} \right)^2 - 1}. \tag{5.8}$$

Unfortunately, this differential equation needs to be solved numerically for $z(y)$. However, its inverse,

$$\frac{dy}{dz} = -\left[\left(\frac{1}{1 - \frac{z^2}{2\ell_c^2}} \right)^2 - 1 \right]^{-1/2}, \tag{5.9}$$

can be solved analytically. The result is [20]:

$$\frac{y(z)}{\ell_c} = \mathrm{acosh}\left(\frac{2\ell_c}{z} \right) - \mathrm{acosh}\left(\frac{2\ell_c}{h} \right) - \sqrt{4 - \frac{z^2}{\ell_c^2}} + \sqrt{4 - \frac{h^2}{\ell_c^2}}. \tag{5.10}$$

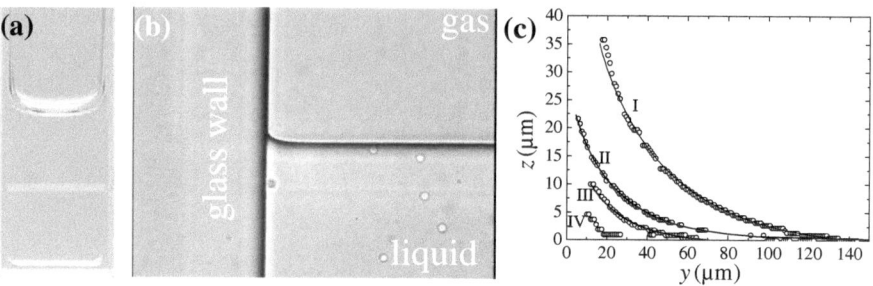

Fig. 5.2 Measurement of the tension of a colloidal gas–liquid interface through analysis of the meniscus shape near a vertical wall. **a** Macroscopic observation of a sample with a width of 1 cm; **b** micrograph of an area of 0.706 mm × 0.528 mm; **c** interfacial profiles fitted to Eq. (5.10). Reproduced with permission from Ref. [19]. Copyright 2003 IOP Publishing, Ltd. [19]

Here, $h = z(y = 0)$ is the contact height (i.e., the elevation of the interface at the wall), which is given by $h^2 = 2\ell_c^2(1 - \sin\theta)$, where θ is the contact angle.

The physical interpretation of the capillary length ℓ_c becomes clearer if one considers the shape of the profile somewhat away from the wall, where the slope $|z'(y)| \ll 1$. Then Eq. (5.5) may be approximated as $1/R_c(y) = -z''(y)$, and Eq. (5.6) takes the form $z''(y) = z/\ell_c^2$. As a result, $z(y)$ can now explicitly be approximated as

$$z(y) = z_0 \exp(-y/\ell_c), \tag{5.11}$$

where z_0 is a numerical pre-factor. Thus, it follows that the capillary length can be seen as a transverse (lateral) correlation length.

Exercise 5.3. Make a plot of Eqs. (5.10) and (5.11) for a contact angle of $\theta = 0°$. What is the y/ℓ_c-range of validity of Eq. (5.11)? Estimate the value of z_0. How is the validity of the approximation affected if the contact angle is increased towards 90°?

As we have seen, γ is quite small for colloidal systems; as such, ℓ_c is typically a few orders of magnitude smaller than for molecular systems. The consequence is that perturbations of the interface decay over a short distance, which makes the interface seem flat even close to a wall, as shown in Fig. 5.2a.

However, on a scale of tens of micrometres the meniscus is still present (Fig. 5.2b). Using microscopic observations on an appropriate model system, Aarts et al. [19] were able to apply this approach for the first time on a demixed colloid–polymer system. The system was the same as that described for the spinning drop method, composed of sterically stabilised silica spheres ($d = 26$ nm $\pm$ 19%) dispersed in cyclohexane, to which PDMS with $R_g = 14$ nm was added. Their fits of the experimentally observed menisci are shown in Fig. 5.2c and the results are summarised in

Table 5.2 Measurements of the colloidal gas–liquid interfacial tension by Aarts et al. [19] for a system comprising sterically stabilised silica spheres with $d = 26$ nm $\pm 19\%$ dispersed in cyclohexane with poly(dimethylsiloxane) ($R_g = 14$ nm; $q = 1.1$). The state points I–IV correspond to those in Fig. 5.2

Sample	ϕ	ϕ_p	$\Delta\rho$ (kg/m^3)	ℓ_c (µm)	γ (µN/m)
I	0.254	1.94	320	32.9	3.38
II	0.217	1.66	256	26.5	1.76
III	0.208	1.59	233	16.0	0.58
IV	0.197	1.50	197	–	–

Table 5.2, showing tensions that are approximately the same as those obtained via the spinning drop method.

5.2 Prediction of Interfacial Properties Using Free Volume Theory

In this section, we will focus on predicting properties of colloidal gas–liquid interfaces based on (G)FVT; in particular, we will treat the interfacial tension and interfacial width. It should be stressed that other useful approaches, such as density functional theory (DFT) [21–23], provide additional detailed microscopic information. Given the scope of the previous chapters, here we focus mainly on FVT and its extensions. We first discuss the interfacial tension and, subsequently, the interfacial width. Predictions of both these quantities are compared to experiments.

5.2.1 Interfacial Tension

It is possible to extend free volume theory, which usually focuses on the bulk phase behaviour, in such a way that it predicts some properties of interfaces too. This can be done by supplementing the bulk free energy by an additional interfacial term that is proportional to *gradients* in concentration. This approach was pioneered by van der Waals for molecular systems as early as 1893 [24]. In 2004, Aarts et al. [25] applied the van der Waals approach to FVT for describing the colloidal gas–liquid interface. Here, we will not go into significant theoretical detail; instead, the focus will be on the general principle. For a translation of van der Waals' original work, see Ref. [26]; for a modern account of van der Waals' theory, see Chap. 3 of the book of Rowlinson and Widom [11].

The van der Waals approach starts by defining a function $w(\phi)$, which expresses the variation of the free energy per unit volume in the direction perpendicular to the interface, as a function of the local colloid volume fraction $\phi(z)$. It is convenient to work with a dimensionless version of this function, $\widetilde{w}(\phi) = w v_0/(kT)$, which is defined as [11,24]

$$\widetilde{w}(\phi) = \widetilde{\Omega}(\phi) - \phi\widetilde{\mu}_{\mathrm{coex}} + \widetilde{\Pi}_{\mathrm{coex}}, \tag{5.12}$$

with $\widetilde{\Omega} = \Omega v_0/(kTV)$ a dimensionless version of the semi-grand potential common in (G)FVT. Additionally, $\widetilde{\mu}_{\mathrm{coex}} = \mu/kT$ is the colloid chemical potential and $\widetilde{\Pi}_{\mathrm{coex}} = \Pi v_0/(kT)$ is the osmotic pressure under the conditions of the two-phase coexistence; these should be computed beforehand using the phase diagram. This definition ensures that $\widetilde{w}(\phi) = 0$ if ϕ equals either of the two coexisting colloid volume fractions, and that it is larger than zero otherwise.

Van der Waals showed that, following this approach, the interfacial tension may be calculated immediately: prior knowledge of the volume fraction profiles $\phi(z)$ is not needed. The interfacial tension can be calculated from $\widetilde{w}(\phi)$ according to

$$\gamma = 2\left(\frac{6}{\pi}\right)^{3/2} \frac{kT}{d^2} \int_{\phi^{\mathrm{g}}}^{\phi^{\mathrm{l}}} \sqrt{\widetilde{m}\,\widetilde{w}(\phi)}\,\mathrm{d}\phi, \tag{5.13}$$

where ϕ^{g} and ϕ^{l} are the coexisting colloid volume fractions of the colloidal gas and liquid phases, respectively. Notice that the prefactor contains the term kT/d^2, which was the simple estimation for the magnitude of the interfacial tension introduced in Eq. (5.1). The integral contains a coefficient $\widetilde{m}$, which dictates how much the system (dis)likes deviations from the bulk composition and which is related to the interactions between the colloids. (For simplicity, we use the dimensionless $\widetilde{m}$, defined as $\widetilde{m} = m/(kTd^5)$ where m is the dimensionful quantity.) Effectively, $\widetilde{m}$ quantifies how particles are affected by being inside a concentration gradient at an interface, having fewer interactions on one side than on the other. Mathematically, $\widetilde{m}$ is related to the direct correlation function $c(\widetilde{r})$, which quantifies the direct interactions between two colloidal spheres in the Ornstein–Zernike equation (i.e., excluding the influence of a third colloid), and is related to the probability of finding two particles at a certain distance. The parameter $\widetilde{m}$ is the second moment of this direct correlation function:

$$\widetilde{m} = \frac{\pi}{3} \int_0^\infty c(\widetilde{r})\widetilde{r}^4\mathrm{d}\widetilde{r}. \tag{5.14}$$

Here, $\widetilde{r} \equiv r/d$ is the normalised centre-to-centre distance between two colloidal spheres.

In the case of long-ranged interactions, which are typical for colloidal gas–liquid coexistence, the mean-spherical approximation can be accurate [25,27,28], which entails $c(\widetilde{r}) = -W(\widetilde{r})/kT$ for $\widetilde{r} \geq 1$, where the depletion pair potential $W(\widetilde{r})$ reads

$$\frac{W(\widetilde{r})}{kT} = -\int_0^{\phi_{\mathrm{d}}^{\mathrm{R}}} \mathrm{d}\phi_{\mathrm{d}}^{\mathrm{R}'} \left(\frac{\partial\widetilde{\Pi}_{\mathrm{d}}^{\mathrm{R}}}{\partial\phi_{\mathrm{d}}^{\mathrm{R}'}}\right) \frac{V_{\mathrm{overlap}}}{v_{\mathrm{d}}}, \tag{5.15}$$

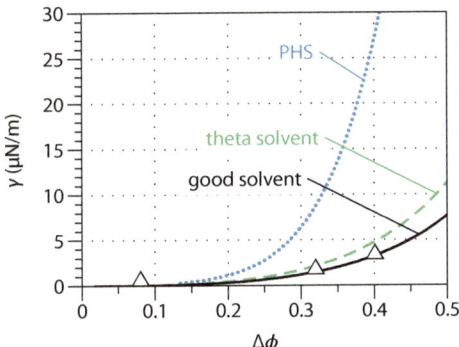

Fig. 5.3 Tension of the colloidal gas–liquid interface as calculated using (G)FVT for penetrable hard spheres, polymers in Θ-solvent and polymers in good solvent ($q = 1.1$), compared to measurements on a system of stearyl silica spheres in cyclohexane ($d = 26$ nm) mixed with poly(dimethylsiloxane) (points) by Aarts et al. [19]

where $V_{overlap}/v_d$ is the overlap volume of depletion zones on two spheres normalised by the volume of the depletants. It is given by

$$\frac{V_{overlap}}{v_d} = q^{-3}(1 + q^* - \tilde{r})^2 \left(1 + q^* + \frac{\tilde{r}}{2}\right), \tag{5.16}$$

for $1 \leq \tilde{r} \leq 1 + q^*$ (it is zero otherwise). The parameter $q^* = 2\delta_s/d$ denotes the actual size of the depletion zones under the conditions at hand, whereas $q = 2R_g/d$ has its usual meaning.

With these ingredients, the interfacial tension may be computed for the various (G)FVT flavours that have been discussed in the preceding chapters, i.e., the penetrable hard sphere (PHS) model and the Θ- and good-solvent conditions for the interacting polymers model. Figure 5.3 shows the interfacial tension as a function of the colloidal liquid–gas volume fraction difference $\Delta\phi = \phi^l - \phi^g$. The quantity $\Delta\phi$ is a useful measure to quantify how strongly a system is phase separated, i.e., how far it is from the critical point, because it often can be measured quite accurately experimentally and is by definition zero at the critical point.

The calculations are compared to results by Aarts et al. [19], again for a system comprising sterically stabilised silica and PDMS in cyclohexane (a good solvent for PDMS). The results show that the precise polymer model being used strongly affects the magnitude of the calculated interfacial tension; when an appropriate model is used, a good description of experimental results appears to be possible. Using both a van der Waals approach and a DFT approach, Moncho-Jordá et al. [5,29] have also noted a similar decrease of the interfacial tension when taking into account polymer interactions in their theoretical predictions.

The (too) high interfacial tensions predicted by the PHS model can be understood from the phase diagrams, which are shown in Fig. 5.4. These reveal that, compared to the other two models, the PHS description has a triple point that is located at substantially higher polymer reservoir volume fractions. In the interacting polymer

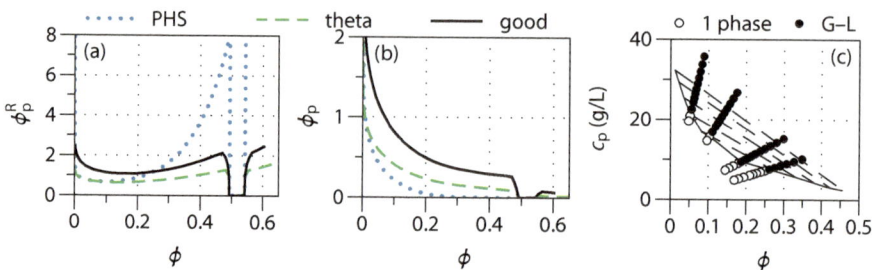

Fig. 5.4 Phase diagrams (**a**) and (**b**) are calculated using (G)FVT for penetrable hard spheres, polymers in a Θ solvent and polymers in a good solvent ($q = 1.1$). **c** Phase diagram of sterically stabilised silica spheres in cyclohexane ($d = 29.4 \pm 2.2$ nm) mixed with poly(dimethylsiloxane) ($R_g \approx 16.4$ nm, 117 kDa), shown for qualitative comparison only. Some data was replotted from Ref. [30]

description, the depletion thickness is reduced upon increasing polymer concentration (see Sect. 4.3), which disfavours a stable colloidal liquid phase at elevated polymer concentrations. This effect is not part of the PHS description, thus yielding a much higher triple point, especially at relatively large q.

The location of the triple point for PHSs at high ϕ_p^R implies that large values of $\Delta\phi$ also occur at high polymer reservoir concentrations. In turn, this means that strong attractions are operational between the colloidal spheres and that the factor $\widetilde{m}$ is also large. Therefore, the PHS description predicts significantly larger interfacial tensions than the interacting polymer models. This once again stresses the importance of being careful in selecting an appropriate polymer model to describe experimental results on colloid–polymer mixtures.

5.2.2 Interfacial Density Profiles

Through the van der Waals approach interfacial density profiles (or volume fraction profiles) may also be computed. This enables a subsequent quantification of the width of such interfaces. In the spirit of the van der Waals approach, the colloid density profile $\phi(z)$ may be found by (numerically) solving the following differential equation:

$$- \widetilde{w}[\phi(\widetilde{z})] + \frac{6}{\pi}\widetilde{m}\left(\frac{d\phi(\widetilde{z})}{d\widetilde{z}}\right)^2 = 0, \tag{5.17}$$

or, in dimensionful quantities,

$$- w[n(z)] + m\left(\frac{dn(z)}{dz}\right)^2 = 0. \tag{5.18}$$

Here, $\widetilde{z} = z/d$ denotes the distance perpendicular to the interface. Note the appearance of the factor $\widetilde{m}$, which links the strength of the depletion interaction to the broadness of the interface. It follows that small values of $\widetilde{m}$ favour sharper interfaces, and larger values favour broader interfaces. It should be stressed that it is the

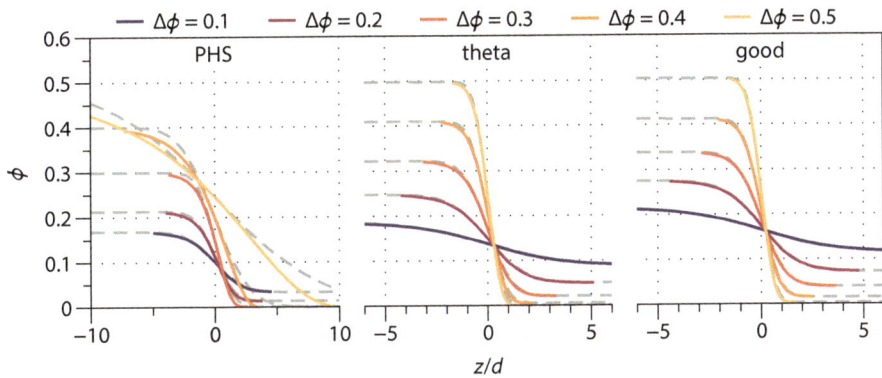

Fig. 5.5 Colloid density profiles of the gas–liquid interface for $q = 1.1$ for penetrable hard spheres (*left*), polymers in Θ-solvent, (*middle*) and polymers in good solvent (*right*). The dashed curves are fits to the error function Eq. (5.20) to extract a measure for the interfacial width

combination of $\widetilde{w}$ and $\widetilde{m}$ that determines the actual width of the interface: without the contribution of $\widetilde{m}$, profiles would be infinitely sharp; without the contribution of $\widetilde{w}$, profiles would be infinitely wide.

In Fig. 5.5, we show the colloid density profiles (solid curves) as calculated for $q = 1.1$, where $z < 0$ denotes the colloidal liquid phase and $z > 0$ denotes the colloidal gas phase. Note that the definition of $z = 0$ is arbitrary; here, we have opted to define $z = 0$ such that $\phi(z = 0)$ is exactly the average of ϕ^g and ϕ^l. These density profiles have been computed for the PHS model (left panel), and the Θ (middle) and good (right) solvent scenarios for the interacting polymers description. In each case, the polymer reservoir volume fraction was chosen to obtain (approximately) the indicated colloid density differences $\Delta\phi$. Qualitatively, it is already evident that, far from the critical point, the density profiles have a width comparable to the colloid diameter except for the PHS model: the large value of $\widetilde{m}$ starts to dominate over $\widetilde{w}$ far from the critical point and disfavours strong gradients. Closer to the critical point the width significantly increases. Also, it appears that for the same $\Delta\phi$ value there are differences between the ensuing density profiles for the various depletant models.

Before turning towards a more quantitative interpretation of the density profiles, it is worth mentioning that, in principle, one could use the colloid density profiles to also compute the polymer density profiles by employing an appropriate expression for the free volume fraction α. It is to be expected that the profiles are qualitatively similar. We are, however, not aware of any experimental data on such polymer density profiles and, therefore, we do not go into further detail on this aspect.

Direct measurements of the full colloid density profiles $\phi(z)$ have also proven challenging. Experiments have been conducted that do not directly assess the full shape of $\phi(z)$ but still shed light on the broadness [30,31]. The question is, however, how to assign a (unique) thickness to profiles such as those in Fig. 5.5. The answer is, in fact, that there are various ways to do this and, indeed, different experimental approaches may probe different measures for the interfacial width. Therefore, let us first focus on a number of ways in which the width of $\phi(z)$ can be quantified mathematically.

Table 5.3 Parameters obtained by fitting Eq. (5.20) to the profiles in Fig. 5.5

$\Delta\phi$	h_{erf}/d		
	PHS	Θ	good
0.1	1.83	3.01	3.02
0.2	1.26	1.49	1.49
0.3	1.11	0.95	0.98
0.4	2.23	0.68	0.70
0.5	6.90	0.51	0.50

A measure for the interfacial broadness is the so-called 10–90% width, here denoted as h_{10-90}, which can be found by finding the positions z_{10} and z_{90} in the profiles for which the condition holds that $\phi(z_{10}) = \phi^{\mathrm{g}} + 0.1\Delta\phi$ and $\phi(z_{90}) = \phi^{\mathrm{g}} + 0.9\Delta\phi$, where $h_{10-90} = |z_{10} - z_{90}|$. The advantage of this approach is that it does not make assumptions about the shape of $\phi(z)$ [32].

One may also fit the profile $\phi(z)$ to a given function and quantify the width through the resulting fitting parameters. For instance, close to the critical point in mean-field theories, $\phi(z)$ is exactly described by a tanh profile [33], which may be expressed as

$$\phi(z) = \phi^{\mathrm{g}} + \frac{1}{2}\Delta\phi\left[1 + \tanh\left(\frac{z}{h_{\mathrm{tanh}}}\right)\right], \tag{5.19}$$

which has a width given by h_{tanh} [34]. It can be shown that the 10–90% width of such a profile is about $h_{10-90} \approx 2.2h_{\mathrm{tanh}}$.

A similar fit function that can be convenient is the cumulative normal distribution, given by

$$\phi(z) = \phi^{\mathrm{g}} + \frac{1}{2}\Delta\phi\left[1 + \mathrm{erf}\left(\frac{1}{\sqrt{2}}\frac{z}{h_{\mathrm{erf}}}\right)\right], \tag{5.20}$$

where the width h_{erf} is in fact the standard deviation of the distribution. The broadness of this profile can also be expressed in terms of a 10–90% width, $h_{10-90} \approx 2.6h_{\mathrm{erf}}$. The tanh and erf profiles are qualitatively very similar in shape, and their widths are approximately related as $h_{\mathrm{tanh}} \approx 1.2h_{\mathrm{erf}}$. As such, one can see that although the interfacial width can be quantified in various ways, the resulting measures are, in fact, closely related. In Fig. 5.5, the density profiles are fitted to Eq. (5.20) (grey dashed curves); the fit results are being shown in Table 5.3.

We now turn our attention to the few experiments that have been devoted to this topic. De Hoog et al. [31] carried out a pioneering ellipsometric study on a colloidal gas–liquid interface for sterically stabilised silica with $d = 26$ nm $\pm$ 19% in cyclohexane, mixed with PDMS ($q \approx 1.1$). For a specific sample, which according to the phase diagram in their work has $\Delta\phi \approx 0.3$, they could obtain a value for the interfacial width. In their analysis, a tanh profile was assumed, which yielded $h_{\mathrm{tanh}} = 5.2$ nm. Such a value for the interfacial width implies $h_{\mathrm{erf}}/d \approx 0.17$ (see also

Table 5.4 Overview of measurements by de Hoog et al. [31] and Vis et al. [30] of the width of colloidal gas–liquid interfaces

	ϕ	c_p (g/L)	$\Delta\phi$	h_{erf}/d
De Hoog et al. [31]	0.21	28.3	~ 0.3	0.17
Vis et al. [30]	~ 0.06	19.9	0.20	0.96 ± 0.02
	~ 0.06	20.1	0.23	0.76 ± 0.04
	~ 0.06	22.9	0.31	0.66 ± 0.03
	~ 0.06	25.8	0.31	0.66 ± 0.01
	~ 0.06	28.7	0.50	0.42 ± 0.02
	~ 0.06	31.5	0.52	0.36 ± 0.01

Fig. 5.6 Comparison of the interfacial width normalised to a particle diameter as predicted by (G)FVT (curves, [30]) and measured using X-ray reflectometry (○, [30]) and ellipsometry (■, [31]) for systems comprising sterically stabilised silica spheres mixed with poly(dimethlysiloxane) in cyclohexane ($q = 1.1$)

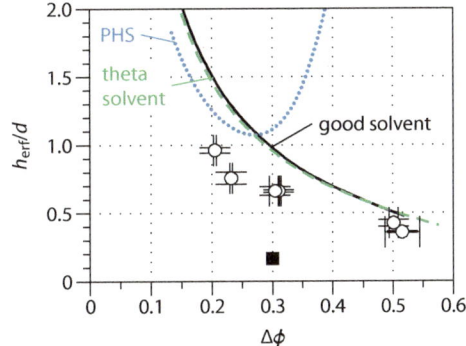

Table 5.4), and that the interface is quite a bit sharper than expected, when compared to the 'good solvent' scenario in Table 5.3.

More recently, Vis et al. [30] characterised the interfacial structure of a similar system of sterically stabilised silica ($d = 29.4 \pm 2.2$ nm) mixed with PDMS ($R_g \approx 16.4$ nm) in cyclohexane ($q = 1.1$, good solvent conditions) using X-ray reflectometry. The measured phase diagram for that system is shown in Fig. 5.4 (right panel). In the analysis, it was assumed that the interface followed a cumulative normal distribution, i.e., Eq. (5.20). The results are summarised in Table 5.4.

A graphical overview of the (G)FVT predictions and the experiments by de Hoog et al. [31] and Vis et al. [30] is given in Fig. 5.6. Overall, the agreement between the latter experiments and the predictions by GFVT (the interacting polymer description) is quite good; the polymer solvency conditions have minimal influence on the theoretically predicted width. In comparison, the experiments by de Hoog et al. [31] seem to find a relatively narrow interface.

In terms of the phase behaviour, we have seen that the PHS model mostly resembles the Θ-solvent model (Fig. 5.4) in the vicinity of the critical point. For the interfacial width, also the good solvent model gives quite similar results to the other two descriptions. However, at $\Delta\phi \gtrsim 0.3$, the PHS model starts to display unphysical behaviour in the form of a strongly *increasing* width. This again has to do with the high polymer concentration of the triple point: the resulting large values of $\tilde{m}$ yield broader interfaces through Eq. (5.17). Such behaviour is not expected for actual colloid–polymer mixtures, showing that accounting for the influence of polymer physics on (interfacial) properties is essential.

Finally, it should be remarked that several authors [22,23,29,35] have reported on DFT calculations that show the presence of oscillations in the colloid density profiles on the colloidal liquid side, especially near the triple point. The simple square gradient van der Waals approach discussed here cannot reproduce such features. However, the work of Moncho-Jordá et al. [29] suggests that these oscillations are only present in case polymers are described as PHSs, and disappear due to a lowering of the triple point when taking polymer interactions into account. On the other hand, DFT computations of Bryk [23] do predict oscillatory density profiles when the polymers are described as freely jointed tangentially bonded hard-sphere chains. In practice, the presence of capillary waves, which will be discussed in the next section, may further hinder the observation of such oscillations. Therefore, whether these oscillations can be observed in an experimental colloid–polymer mixture remains an open question.

5.3   Some Dynamic Properties of the Colloidal Gas–Liquid Interface

5.3.1   Thermal Capillary Waves

In the previous section it was shown that gas–liquid interfaces in general, and those of colloid–polymer mixtures in particular, are not infinitely sharp but have a finite width, even though, macroscopically, they appear to be sharp. Macroscopically, these interfaces also appear to be *flat* far from any (solid) surfaces. However, the thermal energy unavoidably distorts the local interface position on a more microscopic level, leading to a corrugated or rough interface on the colloidal length scale.

These fluctuations in the position of the interface are known as thermally excited capillary waves, or (thermal) capillary waves in short. They have been predicted for molecular systems by von Smoluchowski in 1908 [36], and were theoretically quantified by Mandelstam a few years later [37]. Since then, their existence has been confirmed for molecular liquids with light, neutrons and X-rays [38,39].

The mean-squared interfacial roughness due to thermal fluctuations is given approximately by

$$\langle h^2 \rangle \sim \frac{kT}{\gamma} = L_{\mathrm{T}}^2. \tag{5.21}$$

The thermal length $L_{\mathrm{T}} \equiv \sqrt{kT/\gamma}$ for a molecular gas–liquid interface with an interfacial tension of 50 mN/m is of the order of 0.3 nm and cannot be visually observed directly. However, in the case of coexisting colloid–polymer mixtures the interfacial tension can easily reach values as low as 50 nN/m. For these values, the thermal length becomes about 0.3 μm and can be quantified using optical techniques.

Table 5.5 Characteristic magnitudes for molecular and colloidal systems (assuming colloids of about $d = 0.3\,\mu m$)

	Molecular	Colloidal
Interfacial tension γ	50 mN/m	50 nN/m
Density difference $\Delta\rho$	1000 kg/m^3	100 kg/m^3
Viscosity η	1 mPa s	10 mPa s
Capillary length $\ell_c = \sqrt{\gamma/(g\Delta\rho)}$	2 mm	7 μm
Thermal length $L_T = \sqrt{kT/\gamma}$	0.3 nm	0.3 μm
Capillary velocity $v_c = \gamma/\eta$	50 m/s	5 μm/s
Correlation time $\tau = \ell_c/v_c$	50 μs	1.5 s

Another important interfacial aspect that dictates whether these fluctuations may be observed visually is their dynamics. In the viscous hydrodynamic regime, this is governed by the balance between interfacial tension and viscous forces. The typical velocity can be estimated from the quasi-static Stokes equation:

$$\nabla P = \eta \nabla^2 v, \tag{5.22}$$

where P is the pressure, η is the viscosity and v is the velocity. The capillary pressure P is proportional to γ/L, with γ the interfacial tension and L a typical length scale. Additionally, the gradient terms (∇) scale as $1/L$. This leads to the definition of the so-called capillary velocity,

$$v_c = \frac{\gamma}{\eta}. \tag{5.23}$$

A correlation time τ can be defined as the time it takes for capillary waves travelling at a velocity v_c to travel the capillary length ℓ_c:

$$\tau = \frac{\ell_c}{v_c}. \tag{5.24}$$

In Table 5.5, we have collected the characteristic magnitudes of various relevant interfacial quantities for a molecular and a colloidal system. From this table, it becomes clear that, for colloidal systems, not only the magnitude of the thermal capillary waves can be brought into reach of optical techniques but also their velocity and correlation time become accessible through relatively standard microscopy approaches.

A study was performed by Aarts et al. [40], who made a direct visual observation of thermal capillary waves. They used fluorescent PMMA colloidal spheres with $d = 142$ nm dispersed in decalin, to which polystyrene chains with $R_g = 44$ nm were added ($q = 0.6$). The measured phase diagram is shown in Fig. 5.7. The system was further studied using confocal laser scanning microscopy on the state points labelled I–V.

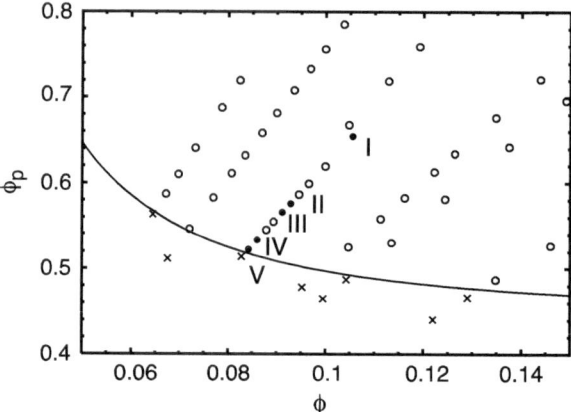

Fig. 5.7 Phase diagram of the system studied by Aarts et al. [40] composed of PMMA colloidal spheres ($d = 142$ nm) dispersed in decalin, mixed with polystyrene polymers with $R_g = 44$ nm ($q = 0.6$). Samples in a single-phase fluid state ($\times$) and in two-phase gas–liquid coexistence ($\circ$) are indicated, together with the state points ($\bullet$) of Figs. 5.8 and Table 5.6. Reprinted with permission from Ref. [40]. Copyright 2004 American Association for the Advancement of Science (AAAS)

Confocal micrographs of the interface at various state points are shown in Fig. 5.8. It is clear that the height of the fluctuations increases closer to the critical point, where the interfacial tension is lower. From the confocal data, the static and dynamic height correlation functions can be determined, which in turn can be used to obtain the interfacial tension. In this way, Aarts et al. found that the interfacial tension of these state points decreases from about 100 nN/m (state point I) to about 1 nN/m (state point V) (see Table 5.6). For a more detailed discussion of these correlation functions, the reader is referred to Ref. [40].

5.3.2 Droplet Coalescence

The process of droplet coalescence is frequently observed in everyday life. Whenever two liquid drops, or a liquid drop and bulk liquid, come into contact, coalescence may occur. Coalescence is favourable since it reduces the total interfacial area and is driven by interfacial tension. The phenomenon has been studied at least since the nineteenth century [41]. The breakup of free-surface flows under the influence of surface tension (e.g., the breakup of a liquid jet) has witnessed renewed interest [42]. Notably, significant progress has been made in the study of the hydrodynamic singularities that occur in these problems [43]. In the case of droplet coalescence, three stages may be identified, as illustrated in Fig. 5.9: first the liquid film between two interfaces drains. Subsequently, this film spontaneously ruptures in a single spot, and finally, this spot ('neck') grows.

As we have seen in Table 5.5, the capillary velocity for colloidal systems is of the order of micrometres per second, whereas in molecular systems it is tens of metres per second. This has implications for the coalescence of droplets and can be seen

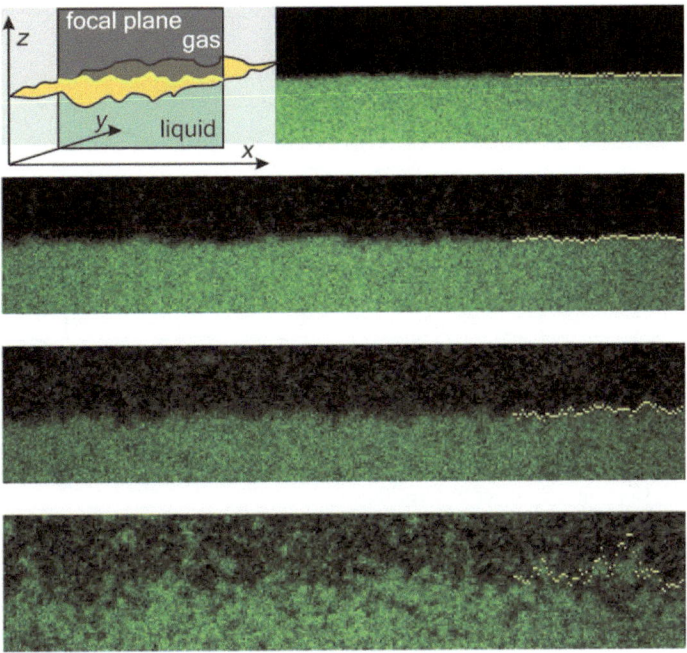

Fig. 5.8 Observation of thermal capillary waves at the colloidal gas–liquid interface at state points I, III, IV and V (top to bottom), as defined in Fig. 5.7. The size of the area displayed in each image is 17.5 μm × 85 μm. The yellow points indicate the position of the interface. Reprinted with permission from Ref. [40]. Copyright 2004 AAAS

Table 5.6 Interfacial tension for the state points denoted in the phase diagram of Fig. 5.7. Data taken from Ref. [40]

State point	ϕ	ϕ_p	γ (nN/m)
I	0.105	0.654	100.8 ± 1.9
II	0.093	0.576	34 ± 5
III	0.091	0.566	18.2 ± 0.3
IV	0.086	0.534	3.6 ± 0.4
V	0.084	0.522	$\lesssim 1$

through the Reynolds number

$$Re = \rho v L / \eta, \qquad (5.25)$$

where v is the characteristic velocity and L the characteristic length. For Re larger than ~ 1 inertial effects start to dominate. We can assume that $v \sim v_c = \gamma/\eta$; hence, $Re \sim \rho \gamma L / \eta^2$. Thus, due to the ultra-low interfacial tensions of colloidal systems, it is evident that the viscous hydrodynamic regime $Re < 1$ is significantly expanded compared to molecular systems when droplet coalescence is concerned.

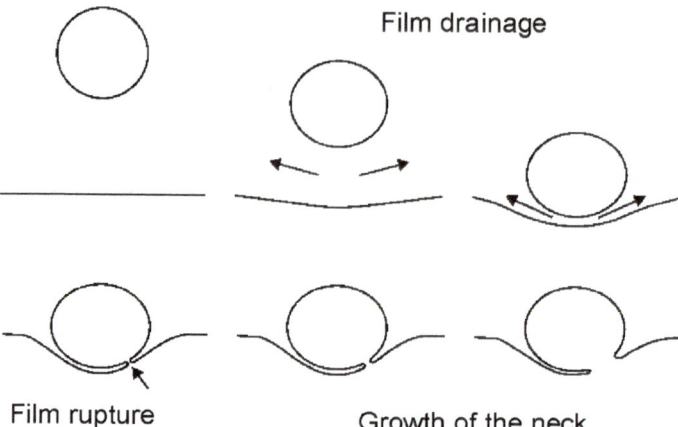

Film drainage

Film rupture   Growth of the neck

Fig. 5.9 Representation of the various stages of droplet coalescence. Reprinted with permission from Ref. [16]. Copyright Springer Nature 2008

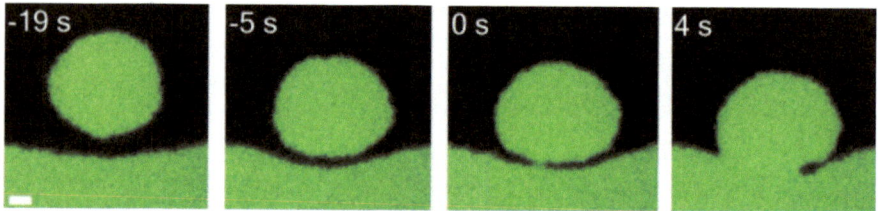

Fig. 5.10 Observation of the various stages of droplet coalescence. $t = -19$ s and -5 s: film drainage; 0 s: film rupture; 4 s: growth of the neck. The scale bar denotes 5 μm. Reprinted with permission from Ref. [40]. Copyright 2004 AAAS

Due to the slow-down of interfacial dynamics in phase-separated colloid–polymer mixtures, it is possible to observe the process of droplet coalescence visually in great detail, as is shown in Fig. 5.10 [40]. This allows for a more complete understanding of the hydrodynamics of droplet coalescence [44]. For instance, the Brownian interfacial fluctuations can be analysed microscopically to quantify the stochastic nature of the film rupture [45].

Exercise 5.4. Inertia becomes important if the Reynolds number (Eq. (5.25)) becomes larger than ~ 1. For $v = v_c$, we can estimate this happens for length scales larger than L_0 and time scales larger than t_0:

$$L > L_0 = \frac{\eta^2}{\rho\gamma} \quad \text{and} \quad t > t_0 = \frac{\eta^3}{\rho\gamma^2}. \tag{5.26}$$

(A) Calculate L_0 and t_0 for molecular and colloidal systems using the values in Table 5.5. Is it realistic to observe viscous coalescence in ordinary molecular liquids?
(B) Aarts et al. [44] observed viscous coalescence in silicon oil with $\eta = 1$ Pa s and $\gamma = 20$ mN/m. What are L_0 and t_0 for this system? What aspect makes the viscous regime observable in this system, and how is that different from colloidal systems?

To conclude, the ultra-low tensions of the interfaces in demixed colloid–polymer systems predicted by Eq. (5.1) have indeed been experimentally observed. We have seen in this chapter that this leads to new and important findings, such as the direct visual observation of capillary waves and the low Reynolds regime in droplet coalescence.

References

1. Vrij, A.: Phys. A **235**, 120 (1997)
2. van der Schoot, P.: J. Phys. Chem. B **103**, 8804 (1999)
3. Chen, B.H., Payandeh, B., Robert, M.: Phys. Rev. E **62**, 2369 (2000)
4. Wijting, W.K., Besseling, N.A.M., Cohen Stuart, M.A.: Phys. Rev. Lett. **90**, 196101 (2003)
5. Moncho-Jordá, A., Louis, A.A., Bolhuis, P.G., Roth, R.: J. Phys.: Condens. Matter **15**, S3429 (2003)
6. Vink, R.L.C., Horbach, J., Binder, K.: J. Chem. Phys. **122**, 134905 (2005)
7. Royall, C.P., Aarts, D.G.A.L., Tanaka, H.: Nature Phys. **3**, 636 (2007)
8. Blokhuis, E.M., Kuipers, J., Vink, R.L.C.: Phys. Rev. Lett. **101**, 086101 (2008)
9. Binder, K., Virnau, P., Statt, A.: J. Chem. Phys. **141**, 559 (2014)
10. Koß, P., Statt, A., Virnau, P., Binder, K.: Mol. Phys. **116**, 2977 (2018)
11. Rowlinson, J.S., Widom, B.: Molecular Theory of Capillarity. Clarendon Press, Oxford (1982)
12. Lide, D.R. (ed.): CRC Handbook of Chemistry and Physics, 90th edn. CRC Press, Boca Raton, FL (2009)
13. Lyklema, J.: Fundamentals in Colloid and Interface Science, vol. 3. Elsevier, Amsterdam (2000)
14. Vonnegut, B.: Rev. Sci. Instrum. **13**, 6 (1942)
15. Princen, H.M., Zia, I.Y.Z., Mason, S.G.: J. Colloid Interface Sci. **23**, 99 (1967)
16. Lekkerkerker, H.N.W., de Villeneuve, V.W.A., de Folter, J.W.J., Schmidt, M., Hennequin, Y., Bonn, D., Indekeu, J.O., Aarts, D.G.A.L.: Eur. Phys. J. B **64**, 341 (2008)
17. Vliegenthart, G.A., Lekkerkerker, H.N.W.: Prog. Colloid Polym. Sci. **105**, 27 (1997)
18. de Hoog, E.H.A., Lekkerkerker, H.N.W.: J. Phys. Chem. B **103**, 5274 (1999)
19. Aarts, D.G.A.L., van der Wiel, J.H., Lekkerkerker, H.N.W., Phys, J.: Condens. Matter **15**, S245 (2003)

20. Batchelor, G.K.: An Introduction to Fluid Dynamics. Cambridge University Press, Cambridge, U.K. (2002)
21. Brader, J.M., Evans, R.: Europhys. Lett. **49**, 678 (2000)
22. Brader, J.M., Evans, R., Schmidt, M., Löwen, H.: J. Phys.: Condens. Matter **14**, L1 (2002)
23. Bryk, P.: J. Chem. Phys. **122**, 064902 (2005)
24. van der Waals, J.D.: Verhandel. Konink. Akad. Weten. Amsterdam (Sect. I) **1** (1893)
25. Aarts, D.G.A.L., Dullens, R.P.A., Lekkerkerker, H.N.W., Bonn, D., van Roij, R.: J. Chem. Phys. **120**, 1973 (2004)
26. Rowlinson, J.S.: J. Stat. Phys. **20**, 197 (1979)
27. Madden, W.G., Rice, S.A.: J. Chem. Phys. **72**, 4208 (1980)
28. Svensson, B., Jönsson, B.: Mol. Phys. **50**, 489 (1983)
29. Moncho-Jordá, A., Dzubiella, J., Hansen, J.P., Louis, A.A.: J. Phys. Chem. B **109**, 6640 (2005)
30. Vis, M., Brouwer, K.J.H., González García, Á., Petukhov, A.V., Konovalov, O., Tuinier, R.: J. Phys. Chem. Lett. **11**, 8372 (2020)
31. de Hoog, E.H.A., Lekkerkerker, H.N.W., Schulz, J., Findenegg, G.H.: J. Phys. Chem. B **103**, 10657 (1999)
32. It should be noted that, if $\phi(z)$ is non-monotonic, h_{10-90} may be not uniquely defined
33. Cahn, J.W., Hillard, J.E.: J. Chem. Phys. **28**, 258 (1958)
34. It should be noted that, for various reasons, in literature sometimes a numerical prefactor is added to the denominator; in this case the width extracted through fitting obviously decreases by the same factor
35. Brader, J.M., Evans, R., Schmidt, M.: Mol. Phys. **101**, 3349 (2003)
36. von Smoluchowski, M.: Ann. Phys. **330**, 205 (1908)
37. Mandelstam, L.: Ann. Phys. **346**, 609 (1913)
38. Vrij, A.: Adv. Colloid Interface Sci. **2**, 39 (1968)
39. Daillant, J., Gibaud, A. (eds.): X-Ray and Neutron Reflectivity: Principles and Applications. Lecture Notes in Physics, vol. 770. Springer (2009)
40. Aarts, D.G.A.L., Schmidt, M., Lekkerkerker, H.N.W.: Science **304**, 847 (2004)
41. Thomson, J., Newall, H.: Proc. Roy. Soc. London **39**, 417 (1885)
42. Eggers, J.: Rev. Mod. Phys. **69**, 865 (1997)
43. Eggers, J., Lister, J.R., Stone, H.A.: J. Fluid Mech. **401**, 293 (1999)
44. Aarts, D.G.A.L., Lekkerkerker, H.N.W., Guo, H., Wegdam, G.H., Bonn, D.: Phys. Rev. Lett. **95**, 164503 (2005)
45. de Villeneuve, V.W.A., van Leeuwen, J.M.J., de Folter, J.W.J., Aarts, D.G.A.L., van Saarloos, W., Lekkerkerker, H.N.W.: Europhys. Lett. **81**, 60004 (2008)

Phase Behaviour of Colloidal Binary Hard Sphere Mixtures

6

6.1   Introduction to Binary Mixtures of Hard Spheres

In the previous chapters we considered the effect of added nonadsorbing polymers on the phase behaviour (Chap. 4) and interface (Chap. 5) appearing in suspensions of spherical colloids. The depletion effect is also operational in other types of mixtures, such as binary mixtures composed of large and small (hard) spheres where two big spheres in a sea of small spheres are brought together (Fig. 6.1). As the big spheres get close, the smaller spheres can no longer enter the gap between the big ones. The small particles then push the big spheres together.

The addition of nonadsorbing small hard spheres to a dispersion of big hard spheres can be treated within free volume theory (FVT) [1–3]. Original FVT treatments [1,2] were limited to a specific range of sufficiently asymmetric hard sphere mixtures, say, $0.05 \lesssim q \lesssim 0.2$, with $q = d_2/d_1$. For larger q values, binary colloidal crystals AB_2 and AB_{13} (consisting of large colloids A with diameter d_1 and small colloids B with diameter d_2) have been observed in the size range $0.425 \leq d_2/d_1 \leq 0.60$ (AB_2) and $0.485 \leq d_2/d_1 \leq 0.62$ (AB_{13}) [4–6], and larger size ratios [7–9]. The situation gets even more complex in the case of binary mixtures of charged spheres [10,11]. Such structures cannot easily be treated within FVT.

The added small colloids may be of a similar colloid shape (i.e. spheres) or a different shape such as rod-like colloids, as will be discussed in Chap. 7. The focus in this chapter is on rather asymmetric binary hard sphere mixtures, i.e. $q \lesssim 0.2$, although extensions towards larger q values are possible [3]. In Sect. 6.2 two free volume theory approaches are outlined and compared to computer simulation results, followed by comparisons with experiments in Sect. 6.3.

© The Author(s) 2024
H. N. W. Lekkerkerker, R. Tuinier, M. Vis, *Colloids and the Depletion Interaction*,
Lecture Notes in Physics 1026, https://doi.org/10.1007/978-3-031-52131-7_6

  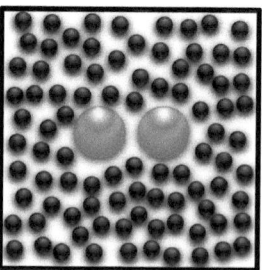

Fig. 6.1 Illustration of the depletion effect in a mixture of two big hard spheres and small hard spheres in 2D. As the big spheres approach each other the small spheres are no longer able to enter the gap between them. As a consequence, the small spheres impose an effective attractive force between the big spheres

6.2   Free Volume Theory for Binary Hard Sphere Mixtures

In 1964 Lebowitz and Rowlinson [12] showed that within the Percus–Yevick treatment of hard sphere fluids [13], binary hard sphere mixtures are completely miscible for all concentrations and size ratios. This proof was later extended by Vrij [14] to hard sphere mixtures with an arbitrary number of components. Until 1990, it was generally accepted that hard sphere mixtures do not phase separate into two fluid phases. In 1991 Biben and Hansen [15] showed, on the basis of a thermodynamically self-consistent theory, that dense binary mixtures of hard spheres with diameters d_1 and d_2 with a size ratio $q \lesssim 0.2$ show a spinodal instability, while phase separation into two fluid phases was not predicted.

Two years later Lekkerkerker and Stroobants [1], guided by their work on colloid–polymer mixtures [16], conjectured that the addition of small hard colloidal spheres will lead to a fluid–solid phase separation, preempting a metastable gas–liquid phase separation that includes a spinodal instability. Such metastable gas–liquid phase transitions for polymer–colloid mixtures have already been discussed in Chap. 4. In 1999 this conjecture of a metastable gas–liquid phase coexistence was confirmed by computer simulations of Dijkstra, van Roij and Evans [17].

Exercise 6.1. The effective depletion interaction mediated by hard spheres compared to penetrable hard spheres (PHSs) was discussed in Sects. 2.1 and 2.3. Based upon the difference between these interactions, why can one expect that the physical properties of a binary hard sphere mixture are more complex than those of hard sphere–PHS mixtures?

The physical origin of phase separation in highly asymmetric hard sphere mixtures is the depletion interaction, similar to what we encountered in Chaps. 3 and 4. Throughout this chapter we refer to small hard spheres in the reservoir (R) as the depletants. The free volume treatment given in Chap. 3 for mixtures of hard spheres and PHSs can be extended to the case of (highly) asymmetric hard sphere mixtures [1,2]. Here, we first present the most straightforward extension possible [1], followed by the more rigorous approach [18] of Opdam et al. [3].

6.2.1 Simple FVT Extension for a Binary Hard Sphere Mixture

The osmotic equilibrium system considered is depicted in Fig. 6.2. We assume the depletion layers are equal to the radii of the small hard spheres. As discussed in Chap. 3, the semi-grand potential of a system with volume V and temperature T of a mixture of N_1 colloidal particles and N_2 depletants with chemical potential μ_2 can be obtained by applying the exact expression Eq. (3.20) to this case:

$$\Omega(N_1, V, T, \mu_2) = F_0(N_1, V, T) - \int_{-\infty}^{\mu_2} N_2(\mu_2')\mathrm{d}\mu_2'. \qquad (6.1)$$

Here, $F_0(N_1, V, T)$ is the Helmholtz energy of the pure system of hard colloidal particles 1, while $\Omega(N_1, V, T, \mu_2)$ is the grand potential of a mixture of N_1 hard spheres 1 and N_2 hard spheres 2 in a volume V at given chemical potential of the depletant hard spheres 2.

Fig. 6.2 Osmotic equilibrium system for a dispersion of big and small hard spheres in the system in equilibrium with a reservoir that consists of a small hard sphere dispersion. The semi-permeable membrane (dashes) allows permeation of small hard spheres but is impermeable to the big hard spheres. The shells indicate the excluded volume surrounding the particles for the centres of the small hard spheres. Reprinted with permission from Ref. [3]. Copyright AIP Publishing 2021

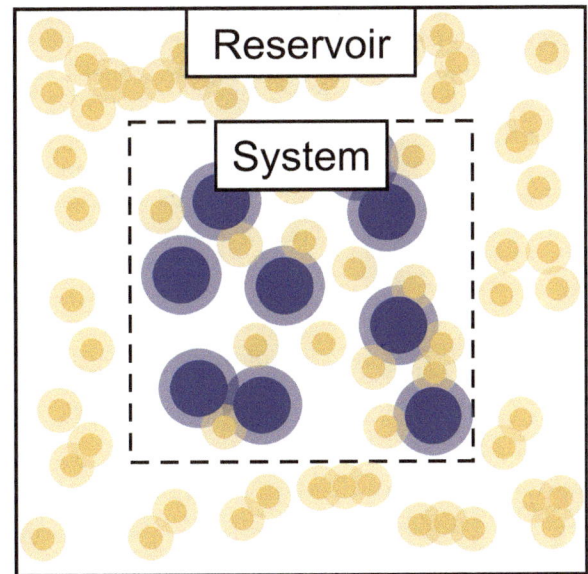

Using the Widom insertion theorem [19] (see Sect. 3.3.2), the chemical potential of small hard spheres 2 in the system can be written as

$$\mu_2 = \mu_2^0 + kT \ln \frac{N_2}{\langle V_{\text{free}}(N_1, N_2) \rangle}, \tag{6.2}$$

where $\langle V_{\text{free}}(N_1, N_2) \rangle$ is the ensemble-averaged free volume for the small hard spheres 2 in the system, containing hard spheres 1 *and* hard spheres 2.

For the reservoir,

$$\mu_2 = \mu_2^0 + kT \ln \frac{N_2^R}{\langle V_{\text{free}}(N_2^R) \rangle}, \tag{6.3}$$

with $\langle V_{\text{free}}(N_2^R) \rangle$ the ensemble-averaged free volume for the small hard spheres 2 in the reservoir of hard spheres 2.

By equating the chemical potentials of component 2 in the system (6.2) and in the reservoir (6.3), we obtain

$$N_2 = N_2^R \frac{\langle V_{\text{free}}(N_1, N_2) \rangle}{\langle V_{\text{free}}(N_2^R) \rangle}. \tag{6.4}$$

We now make the following approximations:

$$\frac{\langle V_{\text{free}}(N_1, N_2) \rangle}{\langle V_{\text{free}}(N_2^R) \rangle} \approx \frac{\langle V_{\text{free}}(N_1) \rangle}{V^R} \approx \frac{\langle V_{\text{free}} \rangle_0}{V^R}, \tag{6.5}$$

where $\langle V_{\text{free}} \rangle_0$ is the undistorted free volume of an added small hard sphere in the system of N_1 large hard spheres in a volume V. The first approximation implies that the free volume available for hard spheres 2 in the reservoir, $\langle V_{\text{free}}(N_2^R) \rangle$, equals the total reservoir volume V^R. Secondly, the free volume available for hard spheres 2 in the system, $\langle V_{\text{free}}(N_1, N_2) \rangle$, is assumed to only depend on the number of hard spheres 1, so it equals $\langle V_{\text{free}}(N_1) \rangle$. The third approximation says that the configurations of the hard spheres 1 are not affected by the hard spheres 2.

Combination of Eqs. (6.4) and (6.5) results in

$$N_2 = n_2^R \langle V_{\text{free}} \rangle_0, \tag{6.6}$$

with $n_2^R = N_2^R/V^R$. Insertion of the Gibbs–Duhem equation (see Eq. (A.12)),

$$n_2^R d\mu_2 = dP^R, \tag{6.7}$$

into Eq. (6.1) now leads to the following simple expression for the semi-grand potential of the asymmetric hard sphere mixture:

$$\Omega(N_1, V, T, \mu_2) = F_0(N_1, V, T) - P^R \langle V_{\text{free}} \rangle_0, \tag{6.8}$$

with P^R as the pressure of the small hard spheres in the reservoir. The quantity $\langle V_{\text{free}} \rangle_0$ can now be approximated by the same expression as the free volume of an added PHS (Eq. (3.38)),

$$\langle V_{\text{free}} \rangle_0 = \alpha V. \tag{6.9}$$

For the free volume fraction α, one can use expression Eq. (3.38):

$$\alpha = (1 - \phi) \exp \left[(-ay - by^2 - cy^3) \right], \tag{6.10}$$

with

$$\begin{aligned} a &= 3q + 3q^2 + q^3 \\ b &= \tfrac{9}{2}q^2 + 3q^3 \\ c &= 3q^3, \end{aligned} \tag{6.11}$$

and

$$y = \frac{\phi_1}{1 - \phi_1}, \tag{6.12}$$

with $\phi_1 = n_1 v_1 = n_1 \pi d_1^3 / 6$ denoting the volume fraction of the large spheres. The volume of a hard sphere 1 is defined as v_1.

In dimensionless form, Eq. (6.8) can be written as

$$\widetilde{\Omega} = \widetilde{F}_0 - \frac{\alpha}{q^3} \widetilde{P}^R, \tag{6.13}$$

with $\widetilde{\Omega} = \Omega v_1 / (kTV)$, $\widetilde{F}_0 = F_0 v_1 / (kTV)$ and $kT\widetilde{P}^R = P^R v_2$, with v_2 the volume of hard sphere 2. Basically, we account for the hard interactions between the small spheres via P^R. For the pressure P^R in the reservoir (which for the case of PHSs is given by the ideal gas law) we now use the SPT expression Eq. (3.37),

$$\frac{P^R}{n_2^R kT} = \frac{1 + \phi_2^R + (\phi_2^R)^2}{(1 - \phi_2^R)^3}. \tag{6.14}$$

Here, n^R is the number density of small hard spheres in the reservoir and $\phi^R = n^R v_2 = n^R \pi d_2^3 / 6$ the volume fraction of the small spheres in the reservoir. Hence we can rewrite Eq. (6.14) as

$$\widetilde{P}^R = \frac{\phi_2^R + (\phi_2^R)^2 + (\phi_2^R)^3}{(1 - \phi_2^R)^3}. \tag{6.15}$$

We now have all the ingredients that make up the semi-grand potential (Eq. (6.8)) of the asymmetric hard sphere mixture. From it we obtain the pressure of the system P

and the chemical potential μ_1 of the large hard spheres using standard thermodynamic relations:

$$P = -\left(\frac{\partial \Omega}{\partial V}\right)_{N_1,T,\mu_2} = P^0 + P^R\left(\alpha - n_1\frac{\partial \alpha}{\partial n_1}\right), \tag{6.16}$$

and

$$\mu_1 = \left(\frac{\partial \Omega}{\partial N_1}\right)_{V,T,\mu_2} = \mu_1^0 - P^R\frac{\partial \alpha}{\partial n_1}, \tag{6.17}$$

where P^0 and μ_1^0 are the pressure and chemical potential of the pure (big) hard sphere system (for which we use the expressions derived in Chap. 3). The dimensionless forms of Eqs. (6.16) and (6.17) are given in Eqs. (3.45) and (3.46). We can now calculate the phase behaviour of the asymmetric hard sphere mixture from the coexistence equations

$$\mu_1^{I}(n_1^{I}, \mu_2) = \mu_1^{II}(n_1^{II}, \mu_2) \tag{6.18}$$

and

$$P^{I}(n_1^{I}, \mu_2) = P^{II}(n_1^{II}, \mu_2). \tag{6.19}$$

Analogously to Eqs. (3.47) and (3.48), the expressions for μ and P can be simplified to

$$\tilde{\mu} = \tilde{\mu}^0 + \tilde{P}^R\, g(\phi_1) \tag{6.20}$$

and

$$\tilde{P} = \tilde{P}^0 + \tilde{P}^R\, h(\phi_1). \tag{6.21}$$

The fluid–solid binodal can be obtained from

$$\tilde{P}^R = \frac{\tilde{\mu}_s(\phi_{1,s}) - \tilde{\mu}_f(\phi_{1,f})}{g(\phi_{1,f}) - g(\phi_{1,s})} = \frac{\tilde{P}_s(\phi_{1,s}) - \tilde{P}_f(\phi_{1,f})}{h(\phi_{1,f}) - h(\phi_{1,s})}. \tag{6.22}$$

From an experimental point of view, we are interested in phase diagrams in the (ϕ_1, ϕ_2) representation. By using the relation

$$n_2 = -\frac{1}{V}\left(\frac{\partial \Omega}{\partial \mu_2}\right)_{N_1,V,T} = \alpha n_2^R,$$

or

$$\phi_2 = \alpha\phi_2^R,$$

we can directly convert the (ϕ_1, ϕ_2^R) phase diagram to the (ϕ_1, ϕ_2) representation. In Fig. 6.3 we give the results for $q = 0.05, 0.1$ and 0.2. The Monte Carlo computer simulation results of Dijkstra et al. [17] have been added to the figure for comparison. The agreement is reasonable, although not as good as the agreement between FVT and computer simulations for the hard sphere–PHS system. For low q, the FVT predictions actually start to deviate quite significantly, as we shall see in Fig. 6.8. A

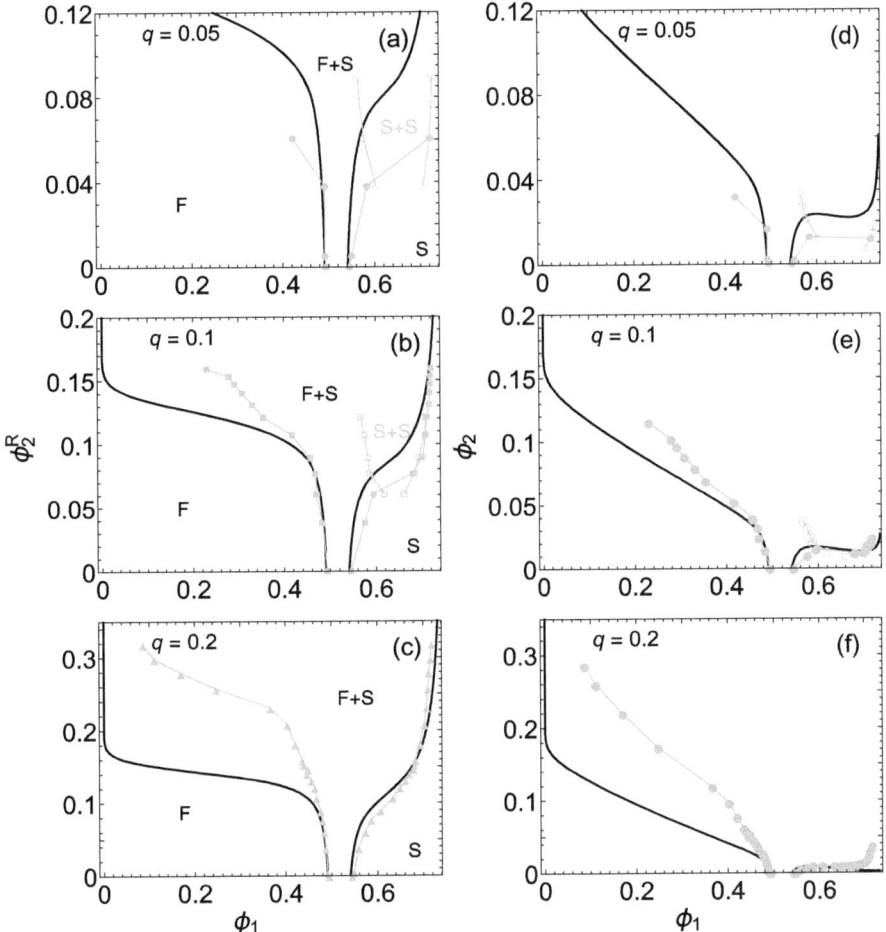

Fig. 6.3 Phase diagrams of big hard sphere–small hard sphere mixtures. Data points are redrawn Monte Carlo simulation results [17] guided by grey curves. Phase diagrams are given for $q = 0.05$ (**a, d**), $q = 0.1$ (**b, e**) and $q = 0.2$ (**c, f**) in the reservoir representation (**a–c**) and the system representation (**d–f**). The black curves show the phase coexistence concentrations predicted by FVT using the approach of [1]. Phase regions are indicated in (**a–c**): the fluid phase (F), the solid phase (S), the fluid–solid coexistence region (F+S) and the isostructural solid–solid coexistence region (S+S)

qualitative difference also is that computer simulations reveal solid–solid equilibria at high ϕ_1 for small q, which are not predicted by FVT.

Note that for the small size ratios, for which the FVT of asymmetric hard sphere mixtures is applicable, gas–liquid demixing (also predicted by FVT, not shown) is metastable with respect to the fluid–solid transition. The presence of this metastable phase does, however, affect the physical properties of the mixtures. Similar to mixtures of colloidal hard spheres and nonadsorbing polymers [20–22], asymmetric hard sphere mixtures display interesting gel and glass states that are supposed to be connected with the metastable gas–liquid phase transition [23,24].

It follows that the simple predictions of this free volume theory approach are reasonable, but become especially inaccurate at small q. This is mainly due to the fact that the expression for the free volume fraction (Eq. (6.10)) already deviates somewhat from computer simulations. This is shown in Fig. 6.4a, in which the predictions of Eq. (6.10) (dashed) are compared to computer simulation results (data points) by Dijkstra et al. [17].

The description of hard spheres as depletants has been accounted for in a limited manner for a number of aspects. Ideally, one would like to:

- Account for the fact that the configurations of the big hard spheres are distorted by the small hard spheres,
- Describe the free volume fraction in the solid since it fundamentally differs from that in a fluid phase, and
- Incorporate excluded volume interactions between the hard depletants in both reservoir and system.

The last aspect also accounts for accumulation effects, leading to repulsive contributions to the depletion interaction as we saw in Sect. 2.3, which are not incorporated in the theory described above. Parts of these improvements were incorporated by Opdam et al. [3] and are discussed next.

6.2.2 Rigorous FVT Approach for a Binary Hard Sphere Mixture

Consider again the osmotic equilibrium system depicted in Fig. 6.2. Careful inspection of this sketch and comparison with Fig. 3.5 shows that the small hard sphere depletants in Fig. 6.2 now also have a hard-core excluded volume interaction with each other. This implies that the FVT treatment should be adapted not only in the sense that the depletant osmotic pressure is larger than that given by Van't Hoff's law, but also one would like to account for the excluded volume interactions *between* the hard sphere depletants. These hard-core interactions influence the thermodynamic properties already in the reservoir. This affects the free volume fraction of small hard spheres in *both* reservoir and system.

We therefore return to the general expression for the semi-grand potential (Eq. (6.1)) and focus on Eq. (6.4). Using the definitions $\phi_2 = N_2 v_2 / V$, $\phi_2^R = N_2^R v_2 / V^R$, $\alpha(\phi_1, \phi_2) = \langle V_{\text{free}}(N_1, N_2) \rangle / V$ and $\alpha^R(\phi_2^R) = \langle V_{\text{free}}(N_2^R) \rangle / V^R$, the following implicit expressions [3] are obtained for the fluid phase:

$$\phi_2 = \phi_2^R \frac{\alpha_f(\phi_1, \phi_2)}{\alpha^R(\phi_2^R)}, \tag{6.23}$$

and for the solid phase:

$$\phi_2 = \phi_2^R \frac{\alpha_s(\phi_1, \phi_2)}{\alpha^R(\phi_2^R)}. \tag{6.24}$$

The free volume fractions α^R in Eqs. (6.23) and (6.24) are no longer unity and α_f and α_s depend on the volume fractions of both the small hard sphere depletants ϕ_2 and of large hard sphere ϕ_1. The volume fraction of depletants in the system ϕ_2, in coexistence with the reservoir with a certain depletant volume fraction ϕ_2^R, can be found by solving Eq. (6.23) and/or (depending on the phase states involved) Eq. (6.24). Substituting Eqs. (6.23) and (6.24) into the definition of the semi-grand potential given by Eq. (6.1) and applying the Gibbs–Duhem relation (see Appendix A.2) finally yields expressions for the semi-grand potential of a binary hard sphere mixture. For the fluid phase it yields

$$\widetilde{\Omega}_f = \widetilde{F}_{0,f} - \int_0^{\phi_2^R} \frac{\alpha_f}{\alpha^R} \left(\frac{\partial \widetilde{P}^R}{\partial \phi_2^{R'}} \right) d\phi_2^{R'}, \tag{6.25}$$

and for the solid phase it gives

$$\widetilde{\Omega}_s = \widetilde{F}_{0,s} - \int_0^{\phi_2^R} \frac{\alpha_s}{\alpha^R} \left(\frac{\partial \widetilde{P}^R}{\partial \phi_2^{R'}} \right) d\phi_2^{R'}, \tag{6.26}$$

where the dimensionless quantities from Appendix A are applied and the integration variable $d\mu_d'$ in Eq. (6.1) is changed to the volume fraction of depletants in the reservoir $d\phi_2^{R'}$ using the Gibbs–Duhem relation.

The free volume fraction for depletants in the reservoir α^R can be evaluated using the steps taken in Sect. 3.3.3. First, we apply Eq. (3.29) to relate α^R to the work of inserting a hard sphere into a hard sphere dispersion in the reservoir W^R:

$$\alpha^R = e^{-W^R/kT}. \tag{6.27}$$

For W^R Eq. (3.35) is used:

$$\frac{W^R}{kT} = -\ln[1 - \phi_2] + \frac{6\phi_2}{1 - \phi_2} + \frac{9\phi_2^2}{2(1 - \phi_2)^2} + \frac{\pi d_2^3 P}{6kT}. \tag{6.28}$$

One could take Eq. (3.37) for P, but we follow Opdam et al. [3] and use the more accurate Carnahan–Starling equation (Eq. (3.1)).

Next, expressions for the free volume fraction of hard sphere depletants in the binary system for both the fluid and solid phases are presented. The free volume fraction in the fluid phase of a binary hard sphere mixture can be determined using the work for depletant insertion in a binary mixture given by SPT [25] (see also Chaps. 3 and 4), resulting in

$$
\alpha_f = \exp -\left[\ln\left(\frac{1}{1-\phi_1-\phi_2}\right) + \frac{3q\phi_1+3\phi_2}{1-\phi_1-\phi_2}\right.
$$
$$
\left. + \frac{1}{2}\left\{\frac{6q^2\phi_1+6\phi_2}{1-\phi_1-\phi_2} + \left(\frac{3q\phi_1+3\phi_2}{1-\phi_1-\phi_2}\right)^2\right\} + q^3\widetilde{P}\right], \quad (6.29)
$$

where $\widetilde{P}$ is the osmotic pressure of the binary mixture of hard spheres. It is possible to use an SPT result for $\widetilde{P}$ [25]; however, we again follow [3] and use an expression for $\widetilde{P}$ given by the Boublík–Mansoori–Carnahan–Starling–Leland equation of state for binary hard sphere mixtures [26,27]:

$$
\widetilde{P} = \frac{\phi_1+\phi_2/q^3}{1-\phi_1-\phi_2} + 3\frac{\phi_1^2+\phi_1\phi_2/q+\phi_1\phi_2/q^2+\phi_2^2/q^3}{(1-\phi_1-\phi_2)^2}
$$
$$
+ (3-\phi_1-\phi_2)\left(\frac{\phi_1^3+3\phi_1^2\phi_2/q+3\phi_1\phi_2^2/q^2+\phi_2^3/q^3}{(1-\phi_1-\phi_2)^3}\right). \quad (6.30)
$$

Exercise 6.2. Think of arguments to explain why inclusion of excluded volume interaction in the reservoir *increases* ϕ_2 in the system at fixed ϕ_2^R.

The above approach cannot be followed for the solid phase since the osmotic pressure of a hard sphere solid containing smaller hard spheres is not known. The free volume fraction in the solid phase is approximated here by considering an FCC crystal of the larger spheres and assuming that the small spheres behave as a fluid in the free space left by the large spheres, which is valid for highly asymmetric binary sphere mixtures [17,28] with $q \lesssim 0.2$. With this assumption, the free volume fraction α_s can be approximated by a product of the free volume fraction of the hard sphere solid and the free volume fraction in the small sphere fluid that surrounds the larger spheres. This yields

$$
\alpha_s(q,\phi_1,\phi_2) = \alpha_s(q,\phi_1)\,\alpha_f(q=1,\phi_2^\dagger), \quad (6.31)
$$

where $\phi_2^\dagger = \phi_2/(1-\phi_1)$ is the effective volume fraction of the small spheres in the space that is not occupied by large spheres, and $\alpha_s(q,\phi_1)$ is given by the free volume fraction in the solid phase, which can be determined using geometrical arguments [29]. It reads

$$
\alpha_s = \begin{cases} 1-\phi_1\widetilde{v}_{exc}^0 & \text{for } \phi_1 < \phi_1^* \\ 1-\phi_1\widetilde{v}_{exc}^* & \text{for } \phi_1^* \le \phi_1 < 2^{3/2}\,\phi_1^* \\ 0 & \text{otherwise.} \end{cases} \quad (6.32)
$$

It is assumed that the centres of the spherical colloids are perfectly located on the FCC lattice points, where $\phi_1^* = \phi_1^{cp} / \tilde{v}_{exc}^0$ denotes the volume fraction of large spheres above which the depletion zones overlap, with $\tilde{v}_{exc}^0 = (1+q)^3$. Furthermore, the normalised excluded volume $\tilde{v}_{exc}^*$ in Eq. (6.32) is given by

$$\tilde{v}_{exc}^* = \tilde{v}_{exc}^0 - 6\left[1 + q - \left(\frac{\phi_1^{cp}}{\phi_1}\right)^{\frac{1}{3}}\right]^2 \left[1 + q + \frac{1}{2}\left(\frac{\phi_1^{cp}}{\phi_1}\right)^{\frac{1}{3}}\right]. \qquad (6.33)$$

Equation (6.32) formally only holds for small q, since it assumes there is no multiple overlap of depletion zones.

Furthermore, this expression for the free volume fraction in the solid does not accurately account for the overlap between the depletion zones of large and small spheres. To take this overlap into account one can assume that

$$\alpha_f(q, \phi_1, \phi_2) = \alpha_f(q, \phi_1)\,\alpha_f(q = 1, \phi_2^\dagger). \qquad (6.34)$$

Combining Eqs. (6.34) and (6.31) gives

$$\alpha_s(q, \phi_1, \phi_2) = \alpha_s(q, \phi_1)\frac{\alpha_f(q, \phi_1, \phi_2)}{\alpha_f(q, \phi_1)}. \qquad (6.35)$$

In Fig. 6.4a the predicted volume fraction of hard sphere depletants in the system ϕ_2 is plotted as a function of the reservoir volume fraction ϕ_2^R for the fluid phase, given by α/α^R (solid curves). The dashed curves are predictions using Eq. (6.10), derived for hard spheres mixed with PHS depletants (Eq. (3.38)). Also shown in Fig. 6.4 are computer simulation data (symbols) by Dijkstra et al. [17]. It is clear that inclusion of excluded volume interactions between the hard sphere depletants in the reservoir gives a more accurate description of the computer simulation data.

In Fig. 6.4b the hard sphere depletant volume fraction in the system ϕ_2 is plotted as a function of the reservoir volume fraction ϕ_2^R for the solid phase. As mentioned above, this relation is given by the ratio α/α^R. Also shown in Fig. 6.4b is the prediction from the original simple FVT extension (dashed) and the computer simulation data (symbols) by Dijkstra et al. [17]. The results from the rigorous FVT approach follow the simulation data remarkably well, which confirms the validity of the equations obtained for the free volume fractions in the fluid phase and the solid phase given by Eqs. (6.29) and (6.35).

All elements required to calculate the semi-grand potentials given in Eqs. (6.25 and (6.26) are now available. This enables one to compute the phase behaviour for binary mixtures of hard spheres by using the standard thermodynamics relations given in Appendix A and by solving Eqs. (6.18) and (6.19).

Predicted phase diagrams are shown in Fig. 6.5 for binary hard sphere mixtures with size ratios $q = 0.05, 0.1$ and 0.2 using the theory outlined in this subsection. The free volume fraction in the solid phase of the binary mixture is described using Eq. (6.35). For $q = 0.05$ and $q = 0.1$, the free volume fraction of the one-component

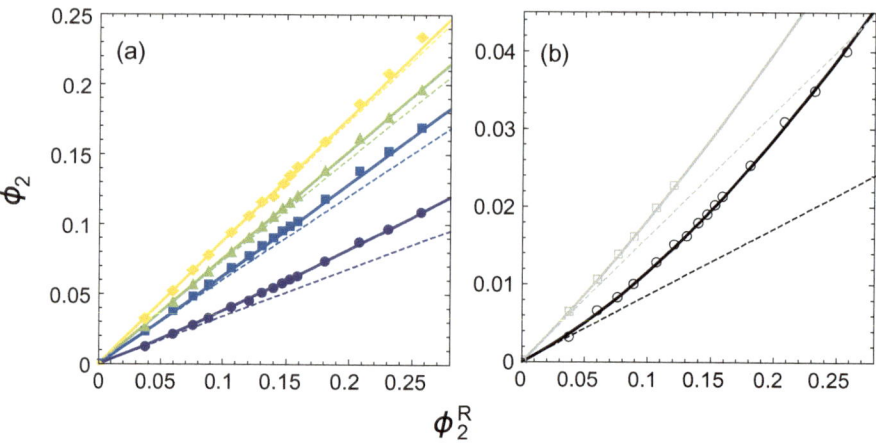

Fig. 6.4 Volume fraction of small spheres in the system ϕ_2 versus their reservoir volume fraction ϕ_2^R. **a** Fluid phase for $q = 0.1$. Solid curves are the result of Eqs. (6.23) and (6.29); symbols denote data from Monte Carlo simulations by Dijkstra et al. [17] for $\phi_1 = 0.1$ (◆), 0.2 (▲), 0.3 (■) and 0.5 (●). **b** Solid phase for $\phi_1 = 0.74$. Solid curves are from Eqs. (6.24) and (6.35); symbols are from Monte Carlo simulations by Dijkstra et al. [17] for $q = 0.05$ (□) and 0.1 (○). In both panels, the dashed curves are for penetrable hard spheres (Eq. (6.10) for fluid and Eq. (6.32) for solid)

solid α_s given by Eq. (6.32) is used. Multiple overlap of depletion zones is possible for $q = 0.2$ at high densities (for details see Ref. [3]). It is noted that the phase diagram for $q = 0.2$ determined with Eq. (6.32) showed no significant difference from the phase diagram calculated with the numerically computed α, which is most likely due to the fact that the deviations between both methods are quite small for this size ratio (see Ref. [3]).

A comparison of the theoretical phase diagrams and phase coexistence data obtained from direct coexistence simulations from Dijkstra, Van Roij and Evans [17] for both the reservoir and the system depletant representation is shown in Fig. 6.5.

Exercise 6.3. Compared to using PHSs as depletants one could expect the phase-transition concentrations could shift to either higher or to smaller depletant concentrations. Give an argument for both based upon the depletion potentials provided in Sects. 2.1 and 2.3.

The theoretical binodals calculated using rigorous FVT are in semi-quantitative agreement with the computer simulation results. The binodals shift to lower depletant concentrations when the size ratio q becomes smaller and an isostructural solid–solid coexistence region appears at high hard sphere densities and at low values of q. The solid–solid coexistence region in the theoretical phase diagram is metastable for $q = 0.1$, whereas a small stable isostructural solid–solid coexistence region was found in simulations. For $q = 0.2$ there is a slight underestimation of the fluid branch

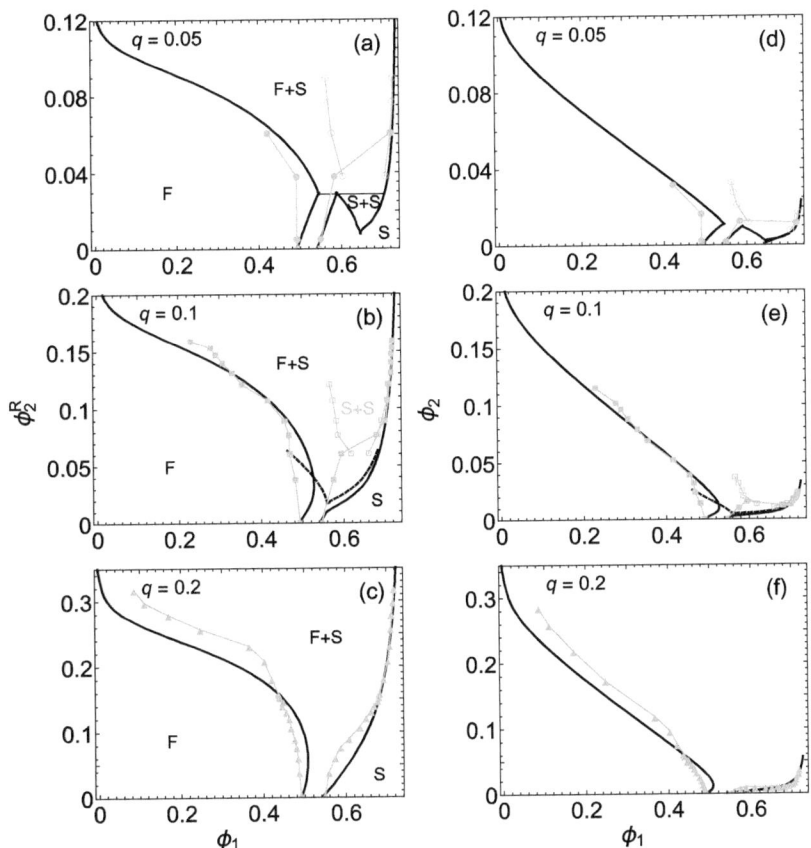

Fig. 6.5 Phase diagrams of binary hard sphere mixtures as in Fig. 6.3 for q's as indicated in the reservoir representation (**a–c**) and the system representation (**d–f**). The black curves show the binodals determined with rigorous FVT, dashed curves indicate metastable phase coexistence, and the grey data points are the results of direct coexistence simulations from Dijkstra et al. [17]. The grey lines guide the eye. For $q = 0.05$ and $q = 0.1$, Eq. (6.35) was used for α_s, and a numerical procedure was used for $q = 0.2$ [3]. Phase regions are indicated in (**a–c**): the fluid phase (F), the solid phase (S), the fluid–solid coexistence region (F+S) and the isostructural solid–solid coexistence region (S+S)

of the binodal compared to the simulation data. The agreement of the phase diagrams obtained with the FVT presented here with simulations [17] and previous perturbation and DFT studies [30,31] indicates that the excluded volume of the depletants is now more accurately taken into account and FVT can be accurately applied to hard depletants.

6.3   Phase Behaviour of Mixed Suspensions of Large and Small Spherical Colloids

6.3.1   Phase Separation in Binary Mixtures Differing Only in Diameter

Sanyal et al. [32] and Van Duijneveldt [33] were the first to present experimental evidence for phase separation in binodal suspensions of colloidal spheres with a large size difference. Since then, several studies [11,34–38] have appeared that present experimental phase diagrams for mixed suspensions of large and small colloids. It should be noted, however, that experimental model systems of mixtures in which both types of spherical colloidal particles are hard sphere-like do not (yet) exist, as far as we are aware.

In Fig. 6.6 we give the experimental phase diagram for $q \simeq 0.1$ by Imhof and Dhont [36], which is compared to free volume theory and Monte Carlo computer simulations. Rigorous FVT predictions (curves) and computer simulations (open symbols) overestimate the depletion activity of the small spheres at the binodal as compared to the experiments (closed symbols). The difference might be caused by charges on the colloidal particles in the experimental system not accounted for theoretically. Additional double layer repulsion does shift theoretical FVT binodals for fluid–solid coexistence at small q upwards [39].

Kaplan et al. [34] and Dinsmore et al. [35] observed crystallites at the sample walls at volume fractions of the small spheres significantly below the value required for the fluid–solid transition in the bulk (Fig. 6.7). This is a manifestation of the stronger depletion interaction between a colloidal sphere and a wall than the depletion interaction between two spheres (as was discussed in Chap. 2). This effect was also demonstrated using micrometre-sized silica spheres dispersed in cyclohexane in contact with hydrophobised silica substrates under the influence of nonadsorbing polymers by Ouhajji et al. [40] using confocal microscopy. A theoretical treatment for the wall phase behaviour based on the semi-grand potential of an adsorbed layer

Fig. 6.6 Fluid–solid coexistence curves established from experiments (■, guided by dotted lines) with sterically stabilised silica spheres of $q \simeq 0.1$ dispersed in DMF with 10^{-2} M LiCl [36]. Monte Carlo simulations [17] (□) and rigorous FVT [3] (solid curves) for $q = 0.1$ predict phase transitions at lower ϕ_2

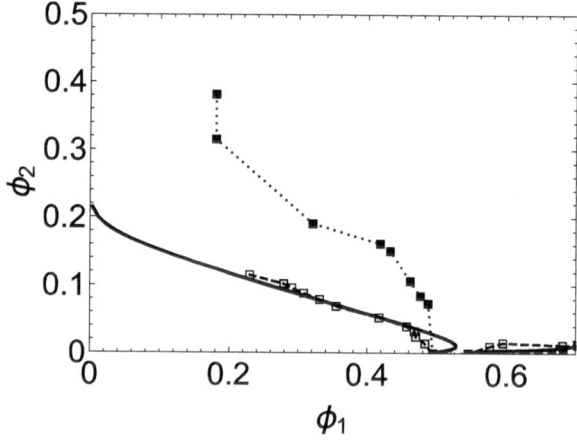

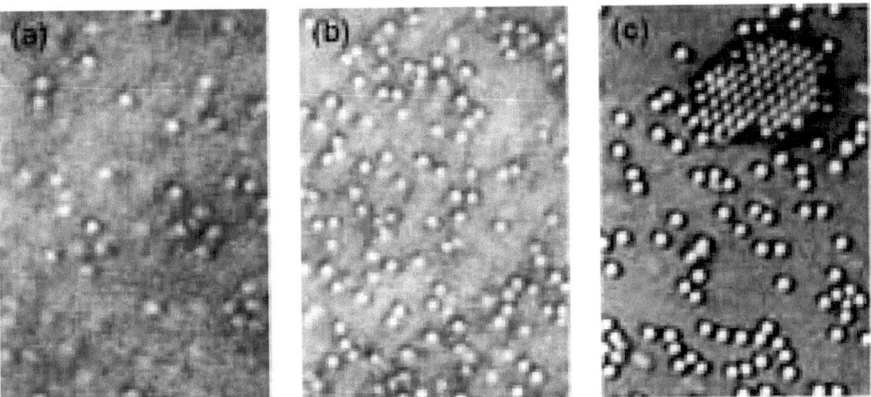

Fig. 6.7 Optical micrographs of polystyrene spheres ($d_1 = 0.8\ \mu$m) at a glass wall **a** without small spheres; **b** and **c** have small spheres of $d_2 = 70$ nm added of $\phi_2 = 0.08$ and $\phi_2 = 0.16$, respectively. The volume fraction of big spheres $\phi_1 = 0.02$. Reprinted with permission from Ref. [41]. Copyright 1997 IOP Publishing, Ltd

of colloids has been given by Poon and Warren [2]. Comparison with experiment [41] shows that this treatment also overestimates the depletion effect of the small spheres.

6.3.2 Mixtures of Latex Particles and Micelles

In 1980, Yoshimura, Takano and Hachisu [42] reported a fluid–solid phase separation in a dispersion of polystyrene latex ($d_1 = 510$ nm) spheres mixed non-ionic surfactant polyoxyethylene alkyl phenylether at KCl concentrations above 0.05 mol/l. Under these conditions the surfactants form spherical micelles. At a surfactant concentration of 2 wt% an iridescent bottom phase appeared, which increased in amount upon further increase of the surfactant concentration. At the same time, the latex concentration in the top phase decreased. The formation of colloidal crystals in the bottom phase, which causes the iridescence, could be confirmed by direct visual observation in the microscope.

A few years later, Ma [43] recognised that the origin of the phase separation is the depletion interaction between the latex particles caused by the micelles. Piazza and Di Pietro [44] have done quantitative measurements on the depletion-induced phase separation in mixtures of latex particles and micelles. In Fig. 6.8 we give their results for a mixture of colloidal polytetrafluoro-ethylene spheres with diameter $d_1 = 220$ nm and the non-ionic surfactant Triton X100 which forms globular micelles with diameters $d_2 = 6$–8 nm. In Fig. 6.8 we compare these experimental results (closed symbols) with the fluid binodal branch predicted from FVT. The simple FVT extension described in Sect. 6.2.1 is given as the dotted curve, while the rigorous FVT approach that explicitly includes excluded volume interactions in the reservoir presented in Sect. 6.2.2 is given as the solid curve. Computer simulations

Fig. 6.8 Phase diagram for asymmetric colloidal sphere mixtures for $q = 0.033$. Data points are experimental results from Piazza and Di Pietro [44], dotted curve is the simple FVT approach [1], solid curve is the rigorous FVT prediction [3] and the dashed curve represents Monte Carlo computer simulation results by Dijkstra et al. [17]

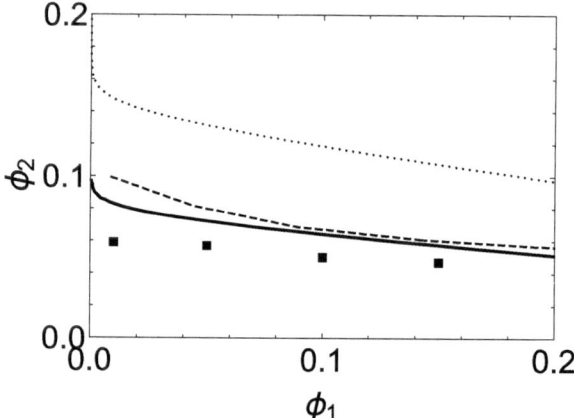

[17] are presented as the dashed curve. Clearly, rigorous FVT is close to both computer simulations and experiments for $q = 0.033$. Piazza et al. [45] have also shown that the fluid–solid phase transition induced by micellar depletants can be exploited to perform an efficient size fractionation of latex particles.

As early as 1952, Cockbain [46] observed the reversible aggregation and creaming of soap stabilised oil-in-water emulsion droplets at soap concentrations greater than the critical micelle concentration. Fairhurst et al. [47] suggested that this reversible aggregation and creaming arises from the depletion interaction between the oil droplets caused by the soap micelles. Quantitative measurements on depletion-induced phase separation by micelles were performed by Bibette and co-workers on silicone oil-in-water emulsions stabilised by sodium dodecylsulfate (SDS) [48]. Since the depletants now are charged it is more complicated to formulate simple models to quantify the effects, but similar phase diagrams to that in Fig. 6.8 have been measured. For small size ratios $q < 0.03$ it is clear from experiments that the phase-transition points shift to lower depletant volume fractions. This is also predicted by rigorous FVT but not by classical FVT. In conclusion, rigorous FVT [3] is in good agreement with computer simulations and is more accurate than simple FVT.

References

1. Lekkerkerker, H.N.W., Stroobants, A.: Physica A **195**, 387 (1993)
2. Poon, W.C.K., Warren, P.B.: Europhys. Lett. **28**, 513 (1994)
3. Opdam, J., Schelling, M.P.M., Tuinier, R.: J. Chem. Phys. **154**, 074902 (2021)
4. Sanders, J.V., Murray, M.J.: Nature **275**, 201 (1978)
5. Hachisu, S., Yoshimura, S.: Nature **283**, 188 (1980)
6. Bartlett, P., Ottewill, R.H., Pusey, P.N.: Phys. Rev. Lett. **68**, 3801 (1992)
7. Dijkstra, M.: In: Rice, S.A., Dinner, A.R. (eds.) Advances in Chemical Physics, vol. 156. Wiley, Chap. 2 (2014)
8. Schaertl, N., Botin, D., Palberg, T., Bartsch, E.: Soft Matter **14**, 5130 (2018)
9. Bommineni, P.K., Klement, M., Engel, M.: Phys. Rev. Lett. **124**, 218003 (2020)

10. Vermolen, E.C.M., Kuijk, A., Filion, L.C., Hermes, M., Thijssen, J.H.J., Dijkstra, M., van Blaaderen, A.: Proc. Natl. Acad. Sci. **106**, 16063 (2009)
11. Toyotama, A., Okuzono, T., Yamanaka, J.: Sci. Rep. **6**, 23292 (2016)
12. Lebowitz, J.L., Rowlinson, J.S.: J. Chem. Phys. **41**, 133 (1964)
13. Hansen, J.P., McDonald, I.R.: Theory of Simple Liquids, 2nd edn. Academic Press, San Diego, CA, USA (1986)
14. Vrij, A.: J. Chem. Phys. **69**, 1742 (1978)
15. Biben, T., Hansen, J.P.: Phys. Rev. Lett. **66**, 2215 (1991)
16. Lekkerkerker, H.N.W., Poon, W.C.K., Pusey, P.N., Stroobants, A., Warren, P.B.: Europhys. Lett. **20**, 559 (1992)
17. Dijkstra, M., Van Roij, R., Evans, R.: Phys. Rev. E **59**, 5744 (1999)
18. We use here the term rigorous since the excluded volume between depletants and between depletants and big hard spheres in the system are described much more accurately. It is however realised that the FVT approximation $\langle V_{\text{free}} \rangle \approx \langle V_{\text{free}} \rangle_0$ of Eq. (3.24) is still used
19. Widom, B.: J. Chem. Phys. **39**, 2808 (1963)
20. Zaccarelli, E.: J. Phys.: Condens. Matter **19**, 323101 (2007)
21. Anderson, V.J., Lekkerkerker, H.N.W.: Nature **416**, 811 (2002)
22. Lu, P.J., Zaccarelli, E., Ciulla, F., Schofield, A., Sciortino, F., Weitz, D.A.: Nature **453**, 499 (2008)
23. Hobbie, E.K., Holter, M.J.: J. Chem. Phys. **108**, 2618 (1998)
24. Hobbie, E.K.: Phys. Rev. Lett. **81**, 3996 (1998)
25. Lebowitz, J.L., Helfand, E., Praestgaard, E.: J. Chem. Phys. **43**, 774 (1965)
26. Boublík, T.: J. Chem. Phys. **53**, 471 (1970)
27. Mansoori, G.A., Carnahan, N.F., Starling, K.E., Leland, T.W.: J. Chem. Phys. **54**, 1523 (1971)
28. Filion, L.: Self-assembly in colloidal hard-sphere systems. Ph.D. thesis, Utrecht University (2011)
29. González García, Á., Opdam, J., Tuinier, R., Vis, M.: Chem. Phys. Lett. **709**, 16 (2018)
30. Velasco, E., Navascués, G., Mederos, L.: Phys. Rev. E **60**, 3158 (1999)
31. Roth, R., Evans, R.: Europhys. Lett. **53**, 271 (2001)
32. Sanyal, S., Easwar, N., Ramaswamy, S., Sood, A.K.: Europhys. Lett. **18**, 107 (1992)
33. Van Duijneveldt, J.S., Heinen, A.W., Lekkerkerker, H.N.W.: Europhys. Lett. **21**, 369 (1993)
34. Kaplan, P.D., Rouke, J.L., Yodh, A.G., Pine, D.J.: Phys. Rev. Lett. **72**, 582 (1994)
35. Dinsmore, A.D., Yodh, A.G., Pine, D.J.: Phys. Rev. E **52**, 4045 (1995)
36. Imhof, A., Dhont, J.K.G.: Phys. Rev. Lett. **75**, 1662 (1995)
37. Steiner, U., Meller, A., Stavans, J.: Phys. Rev. Lett. **74**, 4750 (1995)
38. Hennequin, Y., Pollard, M., Van Duijneveldt, J.S.: J. Chem. Phys. **120**, 1097 (2004)
39. Fortini, A., Dijkstra, M., Tuinier, R.: J. Phys.: Condens. Matter **17**, 7783 (2005)
40. Ouhajji, S., Nylander, T., Piculell, L., Tuinier, R., Linse, P., Philipse, A.P.: Soft Matter **12**, 3963 (2016)
41. Dinsmore, A.D., Warren, P.B., Poon, W.C.K., Yodh, A.G.: Europhys. Lett. **40**, 337 (1997)
42. Yoshimura, S., Takano, K., Hachisu, S.: In: Polymer Colloids, I.I. (ed.) R.M, pp. 139–151. Fitch (Plenum Publishing Company, New York (1980)
43. Ma, C.: Colloids Surf. **28**, 1 (1987)
44. Piazza, R., Di Pietro, G.: Europhys. Lett. **445**, 28 (1994)
45. Piazza, R., Iacopini, S., Pierno, M., Vignati, E.: J. Phys.: Condens Matter **14**, 7563 (2002)
46. Cockbain, E.G.: Trans. Faraday Soc. **48**, 185 (1952)
47. Fairhurst, D., Aronson, M., Ohm, M.L., Goddard, E.D.: Colloids Surf. **7**, 153 (1983)
48. Bibette, J., Roux, D., Nallet, F.: Phys. Rev. Lett. **65**, 2470 (1990)

Phase Behaviour of Colloidal Hard Spheres Mixed with Hard Rod-Like Colloids

7.1 Introduction

In Sect. 2.4, it was shown that, when compared to other types of depletants, rod-like colloids give rise to a strong depletion interaction at low concentration (Eq. 2.124). As a result, it is also expected that even adding a small amount of rods to a dispersion of colloidal spheres has a significant effect on the phase behaviour.

The high efficiency of using rods as depletants was addressed at an early stage by Asakura and Oosawa [1] in a theoretical study, and experimental studies have indeed demonstrated this. Vliegenthart et al. [2], Koenderink et al. [3] and Oversteegen et al. [4] studied mixtures of boehmite rods and silica spheres and observed phase separation in a colloidal fluid coexisting with a sphere-rich crystal-like solid phase at low rod concentrations. Yasarawan and Van Duijneveldt [5] found that mixtures of clay-rods and silica spheres tend to phase separate at low rod concentrations, but appear in an arrested state instead.

A colloidal rod-like model system that has been investigated in detail is fd-virus [6–8]. These are semi-flexible, have a length-over-diameter aspect ratio ≈ 130 and are charge stabilised at values of pH > 4.2 due to the negative surface charge of the coat proteins [9]. Filamentous fd-virus has been used to mediate an attraction between single spherical particles and a fixed wall, and to determine the corresponding depletion potential [10–12]. The equilibrium phase behaviour of mixtures of fd-virus and polystyrene spheres has been investigated by Adams et al. [13] for various sphere–rod size ratios, obtained by using spheres with different radii.

Guu et al. [14] studied the effect of the rod thickness on the phase behaviour of fd-virus and polystyrene spheres by varying the ionic strength, which affects the effective rod thickness. Guu et al. observed a transition from a single isotropic phase to an isostructural fluid/fluid coexistence upon increasing the fd-virus concentration. More details of this virus and other rod-like viruses will be discussed in Chap. 8.

© The Author(s) 2024

H. N. W. Lekkerkerker, R. Tuinier, M. Vis, *Colloids and the Depletion Interaction*,
Lecture Notes in Physics 1026, https://doi.org/10.1007/978-3-031-52131-7_7

In this chapter, we present an FVT approach for mixtures of hard spheres and hard rods. In these hard sphere–hard rod mixtures, the depletion interaction leads to interesting phase transitions, similar to what was discussed in Chaps. 3, 4 and 6. It will be demonstrated that FVT (correctly) captures the above-mentioned pronounced depletion effect caused by rod-like particles. It is noted that there are also useful alternative theoretical approaches [15–20], but we restrict ourselves to the simple, yet insightful FVT treatments. As in the previous chapter, we first focus on a simple FVT extension (Sect. 7.2) and compare these with experiments (Sect. 7.3). Finally, we discuss a more rigorous FVT approach (Sect. 7.4), in which excluded volume interactions between hard rod depletants are explicitly considered.

7.2   Free Volume Theory for Sphere–Rod Mixtures: Simple Extension

Again we start from an osmotic equilibrium, where the reservoir now contains colloidal rods and the system contains colloidal spheres and rods. The osmotic equilibrium system considered is depicted in a schematic way in Fig. 7.1. The system contains N_1 number of hard spheres, each having a volume $v_1 = \pi d^3/6$. F_0 is the free energy of the hard-sphere system without added rods, and $\langle V_{\text{free}} \rangle_0$ is the undistorted free volume available for an added rod in the system of N_1 hard spheres in a volume V. For the semi-grand potential for the system represented in Fig. 7.1, we obtain

$$\Omega(N_1, V, T, \mu_2) = F_0(N_1, V, T) - P^{\text{R}} \langle V_{\text{free}} \rangle_0 \qquad (7.1)$$

by following the same steps as in Sect. 6.2.1 (see the derivation of (Eq. 6.8), the only difference being that component 2 now refers to the rod-like depletants and μ_2 is the chemical potential of the N_2 hard rods imposed by the (hypothetical) reservoir). In (Eq. 7.1), P^{R} is the pressure of the hard rods in the reservoir. The rods are modelled

Fig. 7.1 Osmotic equilibrium system as in Fig. 6.2 but with rods replacing the small hard spheres

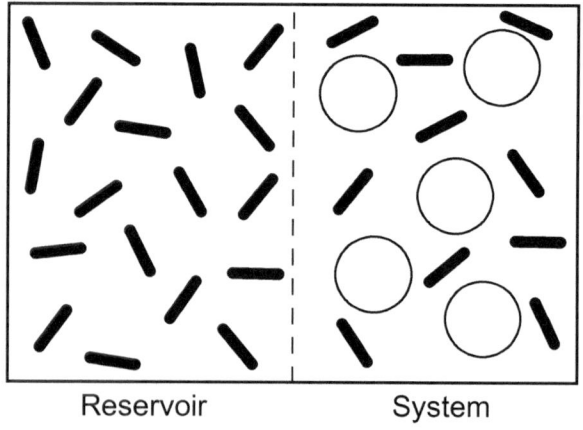

Reservoir   System

as hard spherocylinders (consisting of cylinders of diameter D and length L, capped with two hemispheres) with volume v_2 given by

$$v_2 = \frac{\pi}{6}D^3 + \frac{\pi}{4}LD^2. \tag{7.2}$$

Equation (7.1) is a reasonable application for thin rods, for which phase transitions occur at very low rod volume fractions. At low rod concentrations, the excluded volume interactions between the rods hardly affect the free volume fraction.

Since we are now dealing with hard rods as the depletion agent, both the pressure in the reservoir and the free volume differ from the case of (penetrable or hard) spheres as depletion agent. Both quantities can be calculated conveniently using SPT [21].

7.2.1 Free Volume Fraction

For the free volume, we again start from expression (Eq. 3.28):

$$\alpha = \frac{\langle V_{\text{free}}\rangle_0}{V} = e^{-W/kT}, \tag{7.3}$$

where W is now the reversible work to insert a rod in the hard-sphere system. As explained in Sect. 3.3, this work can be calculated by expanding the particle to be inserted from zero to its final size. As mentioned, the rods are described as hard spherocylinders. In the case of a spherocyclinder, the expansion can be described in terms of a scaling parameter λ for the length and v for the diameter, so the scaled particle has a length λL and diameter vD. In the limit $\lambda, v \to 0$, the inserted spherocyclinder approaches a point particle. In this limiting case, it is very unlikely that excluded volumes of a sphere and scaled spherocyclinder overlap. So,

$$\frac{W(\lambda, v)}{kT} = -\ln[1 - n_1 v_{\text{excl}}(\lambda, v)] \quad (\lambda, v \to 0), \tag{7.4}$$

where $v_{\text{excl}}(\lambda, v)$ is the excluded volume (Fig. 7.2) of the added scaled hard spherocylinder and a hard sphere with diameter $d = 2R$:

$$v_{\text{excl}}(\lambda, v) = \frac{\pi}{6}(d + vD)^3 + \frac{\pi}{4}\lambda L(d + vD)^2, \tag{7.5}$$

For large values of the scaling parameters λ and v, the work required to insert an additional spherocylinder is just the work to create the volume of the scaled particle against the pressure P of the hard-sphere fluid:

$$W(\lambda, v) = \left(\frac{\pi}{6}v^3 D^3 + \frac{\pi}{4}v^2 D^2 \lambda L\right)P \quad (\lambda, v \gg 1). \tag{7.6}$$

Fig. 7.2 Schematic of the excluded volume between a sphere with diameter d and a scaled spherocylinder with length λL and diameter νD

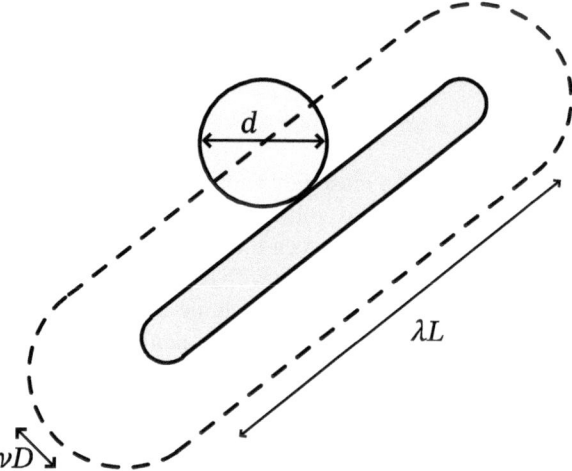

For intermediate values of the scaling parameters it is assumed that the work $W(\lambda, \nu)$ can be found from a Taylor expansion around $\lambda = \nu = 0$ up to the higher order terms that are being replaced by Eq. (7.6), resulting in

$$
W(\lambda, \nu) = \sum_{m=0}^{1} \sum_{n=0}^{1} \frac{1}{m!n!} \frac{\partial^{m+n} W}{\partial \lambda^m \partial \nu^n} \lambda^m \nu^n
$$
$$
+ \frac{1}{2} \frac{\partial^2 W}{\partial \nu^2} \nu^2 + \frac{1}{6} \pi \nu^3 D^3 \left(1 + \frac{3\lambda L}{2\nu D}\right) P. \tag{7.7}
$$

The expression for the work to insert a spherocylinder with length L and diameter D is obtained by setting $\lambda = \nu = 1$. By using the SPT expression (Eq. 3.37) for hard spheres for P

$$
\frac{P}{n_1 kT} = \frac{1 + \phi_1 + \phi_1^2}{(1 - \phi_1)^3}, \tag{7.8}
$$

we obtain

$$
\frac{W}{kT} = -\ln(1 - \phi_1) + a\left(\frac{\phi_1}{1 - \phi_1}\right) + b\left(\frac{\phi_1}{1 - \phi_1}\right)^2 + c\left(\frac{\phi_1}{1 - \phi_1}\right)^3. \tag{7.9}
$$

Here,

$$
a = 3q + \frac{3}{2}\mathcal{L} + 3q^2 + 3q\mathcal{L} + q^3 + \frac{3}{2}q^2\mathcal{L},
$$
$$
b = \frac{9}{2}q^2 + \frac{9}{2}q\mathcal{L} + 3q^3 + \frac{9}{2}q^2\mathcal{L},
$$
$$
c = 3q^3 + \frac{9}{2}q^2\mathcal{L},
$$

Fig. 7.3 Free volume fraction α for needles in a dispersion of hard spheres with volume fraction ϕ_1 for $\mathcal{L} = 2$. Data points are Monte Carlo computer simulation results at $n_2^R L^3 = 21.6$ redrawn from Bolhuis and Frenkel [22]. Solid curve is the FVT result (Eq. 7.11)

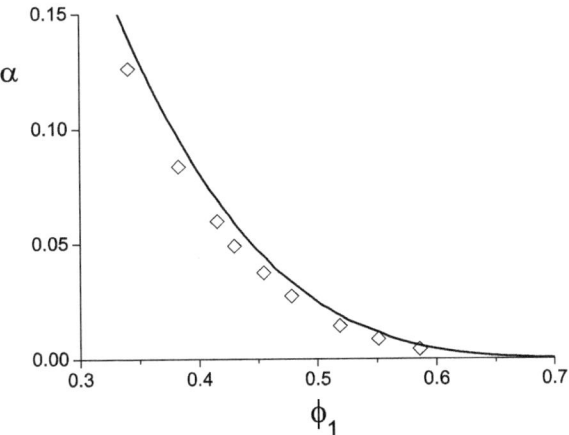

where the size ratios that characterise this mixture are defined as

$$q = \frac{D}{d} \quad \text{and} \quad \mathcal{L} = \frac{L}{d},$$

so $L/D = \mathcal{L}/q$.

Exercise 7.1. Verify that the above expressions for a, b and c match with those below Eq. (3.38) in the limit that a spherocylinder equals a sphere.

The free volume fraction now follows from Eq. (7.3) as

$$\alpha = (1 - \phi_1) \exp\left[- \left(ay_1 + by_1^2 + cy_1^3\right)\right], \qquad (7.10)$$

where

$$y_1 = \frac{\phi_1}{1 - \phi_1}.$$

In the limit $D = 0$ (and hence $q = 0$), where the spherocylinder reduces to an infinitely thin rod (also called needle), the free volume fraction takes the simple form

$$\alpha = (1 - \phi_1) \exp\left[-\frac{3}{2}\mathcal{L}\left(\frac{\phi_1}{1 - \phi_1}\right)\right]. \qquad (7.11)$$

This expression is compared to Monte Carlo computer simulations by Bolhuis and Frenkel [22] for the case $\mathcal{L} = 2$ in Fig. 7.3, and obviously agrees reasonably well.

Fig. 7.4 Sketch of the excluded volume between a spherocylinder and a scaled spherocylinder

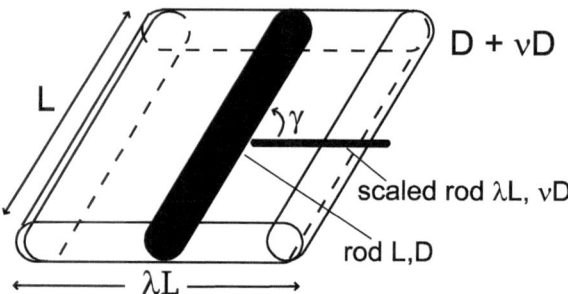

7.2.2   Osmotic Pressure of a Dispersion of Rods

We still have to find an expression for the pressure P^R of the rods in the reservoir in order to use Eq. (7.1). Since low concentrations of rods already induce phase transitions in dispersions of hard spheres, we only consider dispersions of isotropic rods here. We focus on the work of Cotter [23] (for a review on rod-like dispersions, see Vroege and Lekkerkerker [24]), who presented a thermodynamically consistent scaled particle treatment to derive an expression for the pressure of a system of hard spherocylinders. The starting point is again the calculation of the work W to insert an additional spherocylinder in the system of spherocylinders to obtain the excess part of the chemical potential. The pressure can then be obtained by using the Gibbs–Duhem equation (see Appendix A.2.).

Again, using the scaling parameter λ for the length and ν for the diameter we obtain in the limit $\lambda, \nu \to 0$,

$$W(\mathbf{\Omega}, \lambda, \nu) = -kT \ln \left[1 - n_2^R \int f(\mathbf{\Omega}') v_{\text{excl}}(\mathbf{\Omega}, \mathbf{\Omega}', \lambda, \nu) d\mathbf{\Omega}' \right]. \qquad (7.12)$$

The solid angle $\mathbf{\Omega}$ can be decomposed into a polar angle $\theta \in [0, \pi]$ and an azimuthal angle $\phi \in [0, 2\pi]$. In Eq. (7.12), $v_{\text{excl}}(\mathbf{\Omega}, \mathbf{\Omega}', \lambda, \nu)$ is the excluded volume of the added scaled spherocylinder with orientation $\mathbf{\Omega}$ and a spherocylinder of the fluid with orientation characterised by the solid angle $\mathbf{\Omega}'$,

$$
\begin{aligned}
v_{\text{excl}}(\mathbf{\Omega}, \mathbf{\Omega}', \lambda, \nu) = {} & \frac{\pi}{6}(D + \nu D)^3 + \frac{\pi}{4}(D + \nu D)^2(L + \lambda L) \\
& + (D + \nu D)\lambda L^2 \left| \sin \gamma(\mathbf{\Omega}, \mathbf{\Omega}') \right|,
\end{aligned}
\qquad (7.13)
$$

where $\gamma(\mathbf{\Omega}, \mathbf{\Omega}')$ is the angle between the axes of the two spherocylinders (Fig. 7.4).

Furthermore, $f(\mathbf{\Omega})$ is the orientational distribution function, which gives the probability of finding a spherocylinder with an orientation characterised by the solid angle $\mathbf{\Omega}$. To distinguish between the symbols for the grand potential and the solid angle, we use here the boldface $\mathbf{\Omega}$. The distribution function $f(\mathbf{\Omega})$ must be normalised:

$$\int f(\mathbf{\Omega}) d\mathbf{\Omega} = 1. \qquad (7.14)$$

In the isotropic phase, all orientations are equally probable, which implies that

$$f(\mathbf{\Omega}) = \frac{1}{4\pi}. \tag{7.15}$$

For large values of the scaling parameters λ and ν, the work required to insert an additional particle is just the work to create the volume of the scaled particle against the pressure exerted by the fluid of spherocylinders:

$$W(\lambda, \nu) = \left(\frac{1}{6}\pi\nu^3 D^3 + \frac{1}{4}\pi\nu^2 D^2\lambda L \right) P^R \quad (\lambda, \nu \gg 1). \tag{7.16}$$

For intermediate values of the scaling parameters, it is again assumed that the work $W(\lambda, \nu)$ can be found from a Taylor expansion of Eq. (7.12) around $\lambda = \nu = 0$, with the terms up to order of the terms given by expression Eq. (7.16), giving

$$W(\lambda, \nu) = \sum_{m=0}^{1} \sum_{n=0}^{1} \frac{1}{m!n!} \frac{\partial^{m+n} W}{\partial\lambda^m \partial\nu^n} \lambda^m \nu^n + \frac{1}{2} \frac{\partial^2 W}{\partial\nu^2} \nu^2 \tag{7.17}$$
$$+ \left(\frac{1}{6}\pi\nu^3 D^3 + \frac{1}{4}\pi\nu^2 D^2\lambda L \right) P^R.$$

The excess chemical potential of a spherocylinder with length L and diameter D is obtained by setting $\lambda = \nu = 1$ in the above expression and integrating over all possible orientations with the orientation distribution function $f(\mathbf{\Omega})$,

$$\mu_2^{ex} = \int f(\mathbf{\Omega})W(\mathbf{\Omega}, 1, 1)d\mathbf{\Omega}. \tag{7.18}$$

Equation (7.15) holds in the isotropic phase. In that case, the average value of the (absolute) sine of the angle between the axes of the spherocylinders is $\pi/4$. This leads to

$$\frac{\mu_2^{ex}}{kT} = -\ln(1 - \phi_2^R) + \frac{2n^R(\pi D^2 + \pi DL)(\frac{1}{2}D + \frac{1}{4}L)}{1 - \phi_2^R} \tag{7.19}$$
$$+ \frac{(n^R)^2(\pi D^2 + \pi DL)^3}{8\pi(1 - \phi_2^R)^2} + \frac{P^R(\frac{\pi}{6}D^3 + \frac{\pi}{4}D^2 L)}{kT}.$$

By applying the Gibbs–Duhem relation (see Appendix A.2)

$$\frac{1}{kT}\left(\frac{\partial P^R}{\partial n_2^R} \right)_T = 1 + n_2^R\left(\frac{\partial\mu_2^{ex}/kT}{\partial n_2^R} \right), \tag{7.20}$$

Fig. 7.5 Pressure of a dispersion of spherocylinders with $\Gamma = 6$ as a function of the rod volume fraction ϕ_2. Data points are Monte Carlo simulations from McGrother et al. [25] and the solid curve is the SPT result Eq. (7.22)

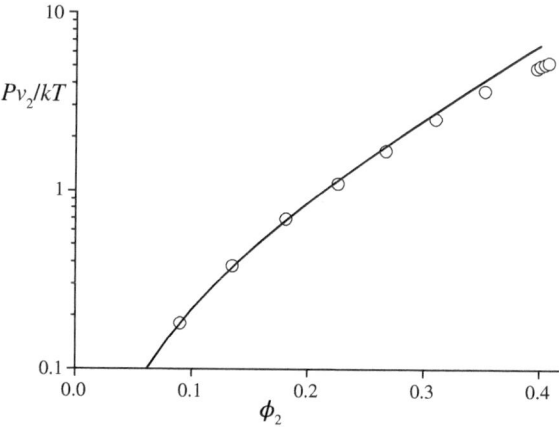

we find

$$\frac{P^{\mathrm{R}}}{n_2^{\mathrm{R}} kT} = \frac{1}{1 - \phi_2^{\mathrm{R}}} + \frac{n_2^{\mathrm{R}}(\frac{1}{2}D + \frac{1}{4}L)(\pi D^2 + \pi DL)}{(1 - \phi_2^{\mathrm{R}})^2}$$
$$+ \frac{(n_2^{\mathrm{R}})^2(\pi D^2 + \pi DL)^3}{12\pi (1 - \phi_2^{\mathrm{R}})^3}. \tag{7.21}$$

Note that for $L = 0$, where the spherocylinder reduces to a sphere, the above expression reduces to the pressure of hard spheres (Eq. 3.37). The dimensionless pressure $P^{\mathrm{R}} v_2 / kT$, where v_2 is the volume of the spherocylinder, can be written as

$$\frac{P^{\mathrm{R}} v_2}{kT} = \frac{\phi_2^{\mathrm{R}}}{1 - \phi_2^{\mathrm{R}}} + A \left(\frac{\phi_2^{\mathrm{R}}}{1 - \phi_2^{\mathrm{R}}}\right)^2 + B \left(\frac{\phi_2^{\mathrm{R}}}{1 - \phi_2^{\mathrm{R}}}\right)^3. \tag{7.22}$$

Here,

$$A = \frac{3\Gamma(\Gamma + 1)}{3\Gamma - 1},$$
$$B = \frac{12\Gamma^3}{(3\Gamma - 1)^2},$$

with

$$\Gamma = \frac{L + D}{D}.$$

Figure 7.5 presents a comparison of the SPT result for the pressure of spherocylinders (Eq. 7.22) with computer simulation results of McGrother et al. [25] for $\Gamma = 6$, showing that there is (except for high volume fractions) close agreement.

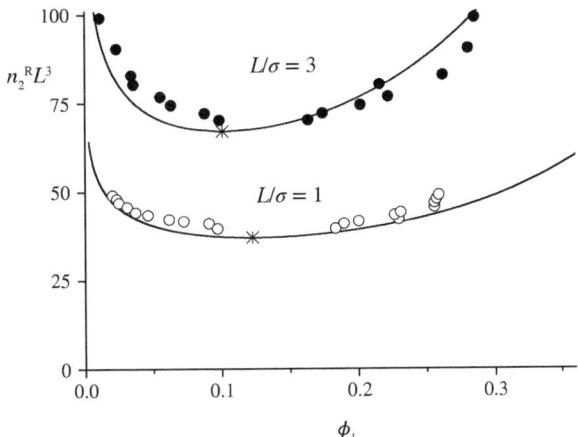

Fig. 7.6 Colloidal gas–liquid coexistence of mixtures of rods and hard spheres for $L/d = 2$ and 3. Data points are Monte Carlo simulation results [22]. Solid curve is the FVT prediction

7.2.3   Phase Behaviour Predictions of Simple FVT Theory

We now have all the ingredients for analysing the properties of the semi-grand potential (Eq. 7.1) of a colloidal sphere–rod mixture. From it, we obtain the pressure P of the system and the chemical potential μ_1 of the large hard spheres using the standard thermodynamic relations

$$P = -\left(\frac{\partial \Omega}{\partial V}\right)_{N_1, T, \mu_2} = P^0 + P^R\left(\alpha - n_1 \frac{\partial \alpha}{\partial n_1}\right), \qquad (7.23)$$

and

$$\mu_1 = \left(\frac{\partial \Omega}{\partial N_1}\right)_{V, T, \mu_2} = \mu_1^0 - P^R \frac{\partial \alpha}{\partial n_1}, \qquad (7.24)$$

where P^0 and μ_1^0 are the pressure and chemical potential of the pure (big) hard-sphere system (for which we use the expressions derived in Chap. 3). The dimensionless forms of Eqs. (7.23) and (7.24) are given in Eqs. (3.45) and (3.46). We can then calculate the phase behaviour of the colloidal sphere–rod mixture by solving the coexistence relations, see Appendix A. As a test of the quality of the FVT for the phase behaviour of colloidal sphere–rod mixtures, we present in Fig. 7.6 a comparison between FVT and simulation results for infinitely thin rods with $\mathcal{L} = L/d = 2$ and $L/d = 3$ taken from the work of Bolhuis and Frenkel [22]. The close agreement is, given the approximations made in FVT, remarkable.

In order to compare FVT with experiments, described in the next section, we need results for rods with a finite thickness. In Fig. 7.7, we give results [21] for spherocylinders with $L/D = 20$ and $\mathcal{L} = 0.2, 0.5$ and 1, both in the $\phi_2^R - \phi_1$ representation as well as in the experimentally relevant $\phi_2 - \phi_1$ representation by using Eq. (7.3): $\phi_2 = \alpha \phi_2^R$. Given that the depletion interaction between two spheres in a sea of thin rods scales as $\mathcal{L}/q^2 = Ld/D^2$ (see Eq. (2.124)), we have scaled the volume fractions of rods by multiplication with Ld/D^2; $\tilde{\phi}_2^R = \phi_2^R Ld/D^2$.

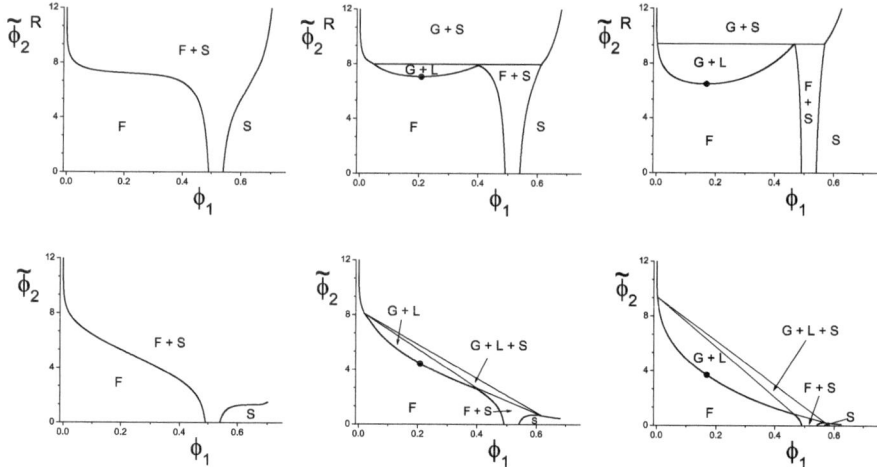

Fig. 7.7 FVT phase diagrams of the volume fractions of hard spheres (ϕ_1) mixed with sphero-cylinders (ϕ_2) [21] for $L/D = 20$ and $q = 0.01$ (*left*), $q = 0.025$ (*middle*), $q = 0.05$ (*right*). Upper curves are in the ϕ_2^R–ϕ_1-representation and lower curves are in the ϕ_2–ϕ_1-plane

For $L/d = 0.2$, only fluid–solid coexistence is found. For $L/d > 0.3$, a region of three-phase coexistence (colloidal gas–liquid–crystal) bounded by three distinct two-phase regions (gas–liquid, liquid–crystal and gas–crystal) is found.

The topology of these phase diagrams exhibits the same global features as those for hard spheres mixed with penetrable hard spheres in Chap. 3 and colloid–polymer mixtures in Chap. 4. Note that, for $L/D = 20$, FVT predicts that the rods can cause a fluid–crystal transition in dilute suspensions of spheres for rod volume fractions as low as 0.003. This is confirmed experimentally as we shall see in the next section.

7.3 Phase Behaviour of Colloidal Sphere–Rod Mixtures. Experiment

Koenderink et al. [3] and Vliegenthart et al. [2] studied depletion-induced crys-tallisation in a mixture of silica spheres (diameter of $d = 740$ nm) and boehmite ($\gamma - AlOOH$) rods (length $L = 230$ nm and diameter $D = 9$ nm) dispersed in DMF with 0.001 M LiCl added to screen electrostatic interactions. The rods are coated with a thin layer of silica to make them compatible with the silica spheres. The silica spheres are labelled with fluorescein isothiocyanate (FITC) to make them visible with fluorescence confocal microscopy. Transmission electron micrographs of the boehmite rods and a mixture of the boehmite rods and the silica spheres are presented in Fig. 7.8.

In Fig. 7.9a, a time series of confocal microscopy images is presented of a sam-ple containing silica spheres ($\phi_1 = 0.025$) and boehmite rods ($\phi_2 = 0.0025$). In the images, it is seen that the silica number density increases with time by sedimenta-

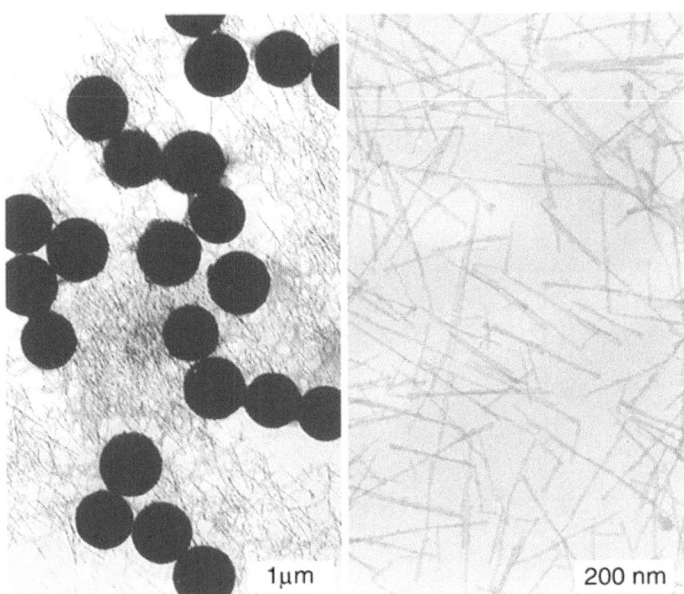

Fig. 7.8 Transmission electron micrographs of silica-coated boehmite rods with (*left*) and without (*right*) silica spheres. Reprinted with permission from Ref. [3]. Copyright 1999 American Chemical Society

tion. Locally ordered structures are formed but no signs of depletion-induced phase transitions are found.

Figure 7.9b presents a time series of confocal microscopy images of a sample of silica spheres with the same volume fraction as that in Fig. 7.9a but instead for a larger volume fraction of boehmite rods ($\phi_2 = 0.005$). In this case, the morphology of the system is totally different. Clusters are rapidly formed (within minutes) and those aggregates rapidly transform into crystallites while they grow and coalesce [2]. The initial clusters contain typically 10^3 particles. In the final stage, re-orientation of different crystalline patches and annealing of defect lines is seen. This results in large crystalline areas. The entire process does not take more than 8 min, much faster than the formation of locally ordered structures under the influence of sedimentation in the system with a 0.0025 volume fraction of rods. Apparently, in the case of a 0.005 volume fraction of rods, we are in the biphasic region (fluid–solid) of the phase diagram. Let us compare this to theory. Figure 7.10 shows the theoretical FVT phase diagram for a mixture of hard spheres (diameter $d = 740$ nm) and hard rods (length $L = 230$ nm and diameter $D = 10$ nm). The experimentally investigated systems discussed above are indicated by dots. The experimental observations of no phase separation ($\phi_1 = 0.025$ and $\phi_2 = 0.0025$) and phase separation ($\phi_1 = 0.025$ and $\phi_2 = 0.005$) are in agreement with theory. The experiments clearly indicate that rods are very efficient depletion agents.

Bakker et al. [26] performed an experimental study on the phase behaviour of charged silica rods ($L = 3.6$ μm, $D \approx 0.6$ μm) mixed with charged silica spheres

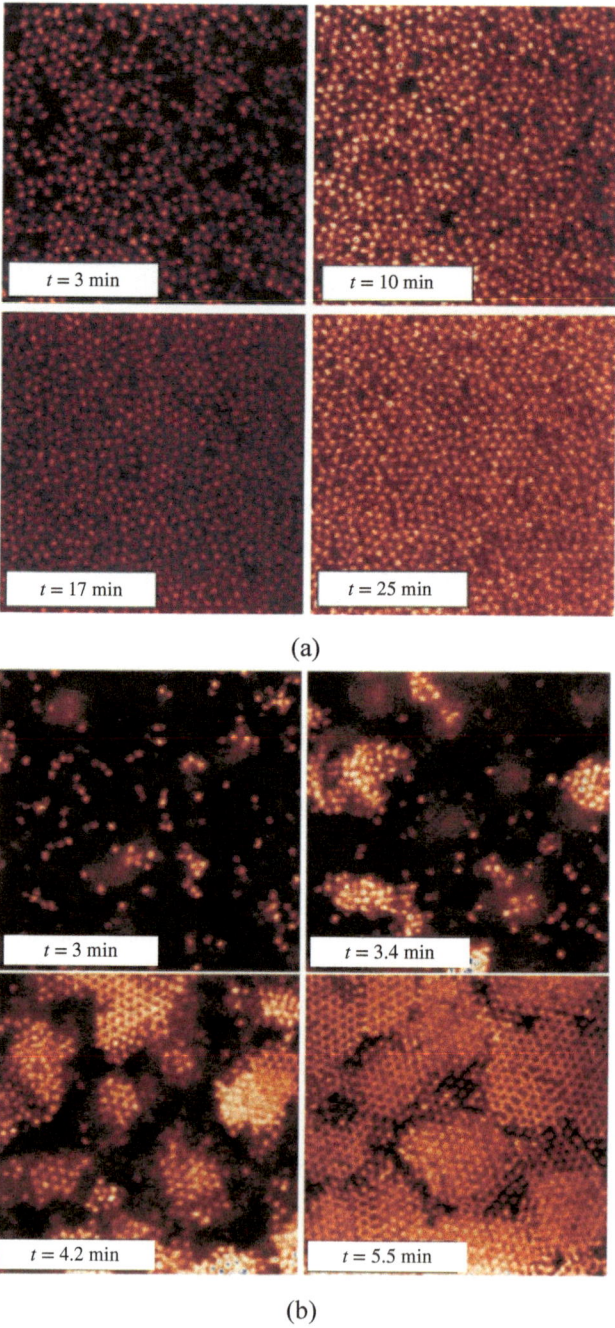

Fig. 7.9 Confocal microscopy images of fluorescently labelled silica spheres ($d = 740$ nm) mixed with boehmite rods ($L = 230$ nm, $D = 10$ nm) various times t after mixing. The samples are composed of 2.5 wt% spheres and **a** 0.25 wt% rods and **b** 0.5 wt % rods. Images are 50 μm × 50 μm. Reprinted with permission from Ref. [2]. Copyright 1999 Royal Society of Chemistry

Fig. 7.10 FVT phase diagram for a dispersion of spheres mixed with rods for $L/D = 23$ and $q = 1/74$. The open diamonds indicate the mixtures studied in Fig. 7.9a (0.25 wt%) and Fig. 7.9b (0.5 wt%)

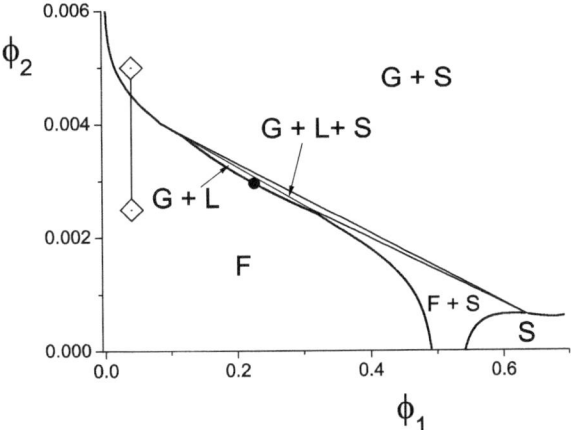

($d \approx 0.4 \ \mu m$). They also determined sedimentation–diffusion equilibria. For relatively large amounts of spheres, the mixtures form a stable isotropic phase for low rod concentrations, but phase-separated isotropic–smectic A coexistence regions at elevated rod concentrations. In Chap. 8, the rich phase behaviour of colloidal rods mixed with nonadsorbing polymers including higher order phase states of rods are discussed.

As mentioned in the introduction of this chapter, Guu et al. [14] studied the phase behaviour of mixtures of fd-virus and polystyrene (PS) spheres. In Fig. 7.11a, the data points refer to the fd-concentration above which the authors [27] experimentally observed a transition from a single isotropic phase to an isostructural fluid/fluid coexistence for various values of $\mathcal{L} = L/d$. The radii of the spheres used in the experiments are 248, 300, 354 and 497 nm and the length of fd-virus is 880 nm. Note the low rod volume fractions at which the phase transitions take place. The dashed curves are the predictions using the simple extension FVT theory.

Figure 7.11b shows the effect of rod thickness for $\mathcal{L} = 1.75$ Also here, the experimental data points lie significantly above the FVT predictions. In this case, the rods are long, and not infinitely thin. For such a system, excluded volume interactions between the rods are expected to be significant as fd-viruses themselves already exhibit an I–N phase transition at low concentrations (as will be shown and discussed in Chap. 8). In summary, the FVT description of the previous section works well for thin rods but starts to deviate for thicker rods. For such mixtures of hard spheres and hard spherocylinders with sufficient thickness of the rods, one needs to account for excluded volume interactions between the rods at the reservoir level [27]. This is the subject of Sect. 7.4.

7.4 Free Volume Theory for Sphere–Rod Mixtures: A Rigorous Approach

We follow a similar procedure to that in Sect. 6.2.2 but apply this to the case of dispersions of hard spheres (component 1) mixed with hard rods (component 2) as depletants. We again use Eq. (6.4) and write this for the fluid phase as

$$\phi_2 = \phi_2^R \frac{\alpha(\phi_1, \phi_2)}{\alpha^R(\phi_2^R)}. \tag{7.25}$$

As in the previous chapter, we use the definitions $\phi_2 = N_2 v_2 / V$, $\phi_2^R = N_2^R v_2 / V^R$, $\alpha(\phi_1, \phi_2) = \langle V_{\text{free}}(N_1, N_2) \rangle / V$ and $\alpha^R(\phi_2^R) = \langle V_{\text{free}}(N_2^R) \rangle / V^R$.

The volume fraction of rods in the system ϕ_2 can be found numerically by solving Eq. (7.25) at a given reservoir rod volume fraction ϕ_2^R. Substituting Eq. (7.25) into the definition of the semi-grand potential given by Eq. (6.1) and applying the Gibbs–Duhem relation (see Appendix A.2) yield an expression for the semi-grand potential of a mixture of hard spheres and hard rods for the fluid phase:

$$\widetilde{\Omega} = \widetilde{F}_0 - \int_0^{\phi_2^R} \frac{\alpha(q, \phi_1, \phi_2')}{\alpha^R(\phi_2^{R'})} \left(\frac{\partial \widetilde{P}^R}{\partial \phi_2^{R'}} \right) d\phi_2^{R'}, \tag{7.26}$$

(as in the previous chapter); and in a similar fashion, the grand potential of the solid phase can be obtained.

We use Eq. (7.22), the scaled particle theory (SPT) result for the osmotic pressure of a fluid dispersion of hard spherocylinders [28], for $\widetilde{P}^R$ in the reservoir, which is recast into

$$\widetilde{P}^R = \frac{v_1}{v_2} \left[\frac{\phi_2^R}{1 - \phi_2^R} + \frac{3\Gamma(\Gamma + 1)}{3\Gamma - 1} \left(\frac{\phi_2^R}{1 - \phi_2^R} \right)^2 + \frac{12\Gamma^3}{(3\Gamma - 1)^2} \left(\frac{\phi_2^R}{1 - \phi_2^R} \right)^3 \right], \tag{7.27}$$

with v_1 as the hard-sphere volume $(\pi/6)d^3$ and v_2 defined in Eq. 7.2.

Exercise 7.2. Show that Eqs. (7.22) and (7.27) are consistent.

Opdam et al. [27] also calculated the work of insertion W required to insert a hard spherocylinder into a binary mixture of hard spheres and hard spherocylinders. Using Eq. (7.3) yields the following expression for the free volume fraction α in the system:

$$\alpha = (1 - \phi_1 - \phi_2) \times$$
$$\exp \left[-a_2 \frac{\phi_1 + \phi_2}{1 - \phi_1 - \phi_2} - b_2 \left(\frac{\phi_1 + \phi_2}{1 - \phi_1 - \phi_2} \right)^2 - \frac{v_2}{v_1} \widetilde{P} \right], \tag{7.28}$$

with

$$a_1 = 6\frac{\phi_1}{\phi_1 + \phi_2} + \left[\frac{1}{q}\frac{6\Gamma}{3\Gamma - 1} + \frac{1}{q^2}\frac{3(\Gamma + 1)}{3\Gamma - 1}\right]\frac{\phi_2}{\phi_1 + \phi_2}, \tag{7.29}$$

$$a_2 = \left[\frac{3}{2}\mathcal{L}(1 + 2q) + 3q(1 + q)\right]\frac{\phi_1}{\phi_1 + \phi_2} + \left[6 + \frac{6(\Gamma - 1)^2}{3\Gamma - 1}\right]\frac{\phi_2}{\phi_1 + \phi_2}. \tag{7.30}$$

Predictions of Eq. (7.28) are plotted in Fig. 7.12a as solid curves for $\Gamma = 6$ and $\mathcal{L} = 1$ and are compared to Monte Carlo computer simulations (data points). Obviously, there is close agreement.

An expression for the osmotic pressure of the binary system $\tilde{P}$ can also be obtained using SPT [29], resulting in

$$\tilde{P} = \left(\phi_1 + \frac{v_1}{v_2}\phi_2\right)$$
$$\times \left[\frac{1}{1 - \phi_1 - \phi_2} + \frac{A_{1,2}}{2}\frac{(\phi_1 + \phi_2)}{(1 - \phi_1 - \phi_2)^2} + \frac{2B_{1,2}}{3}\frac{(\phi_1 + \phi_2)^2}{(1 - \phi_1 - \phi_2)^3}\right]. \tag{7.31}$$

In Eq. (7.31) the following coefficients are used:

$$A_{1,2} = a_1\frac{\phi_1}{\phi_1 + \phi_2\frac{v_1}{v_2}} + a_2\left(1 - \frac{\phi_1}{\phi_1 + \phi_2\frac{v_1}{v_2}}\right), \tag{7.32}$$

$$B_{1,2} = b_1\frac{\phi_1}{\phi_1 + \phi_2\frac{v_1}{v_2}} + b_2\left(1 - \frac{\phi_1}{\phi_1 + \phi_2\frac{v_1}{v_2}}\right), \tag{7.33}$$

$$b_1 = \left(\frac{3}{2}\frac{\phi_1}{\phi_1 + \phi_2} + \frac{1}{q}\frac{3\Gamma}{3\Gamma - 1}\frac{\phi_2}{\phi_1 + \phi_2}\right)^2, \tag{7.34}$$

$$b_2 = \left[\frac{3}{2}(\mathcal{L} + q)\frac{\phi_1}{\phi_1 + \phi_2} + \left(\frac{3(2\Gamma - 1)}{3\Gamma - 1} + \frac{3(\Gamma - 1)^2}{3\Gamma - 1}\right)\frac{\phi_2}{\phi_1 + \phi_2}\right]$$
$$\times \left(3q\frac{\phi_1}{\phi_1 + \phi_2} + \frac{6\Gamma}{3\Gamma - 1}\frac{\phi_2}{\phi_1 + \phi_2}\right). \tag{7.35}$$

As all ingredients are now available to compute the semi-grand potential of Eq. (7.26), we can predict the colloidal fluid–fluid (or gas–liquid) phase transition concentrations. Predictions of the phase diagram for $\Gamma = 6$ and $\mathcal{L} = 1$ are plotted in Fig. 7.12b as solid curve and are compared to Monte Carlo computer simulations [29] and classical FVT predictions (grey curves). The symbols in Fig. 7.12b depict the simulation data: open symbols denote state points at which a homogeneous binary fluid was obtained, and closed symbols indicate that phase separation was observed. There is reasonable agreement for the phase transition concentrations predicted by rigorous FVT and the simulations. The classical FVT prediction differs quantitatively and qualitatively.

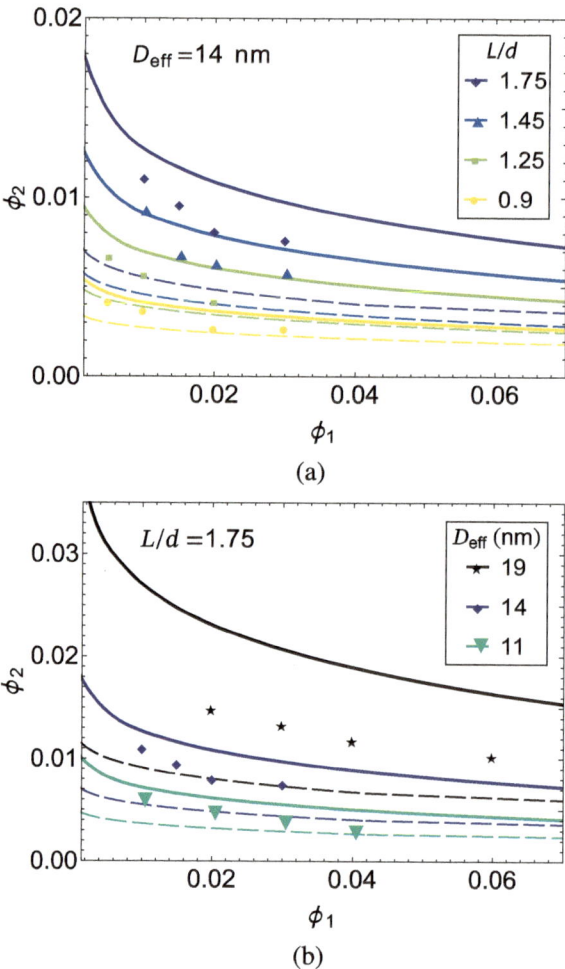

(a)

(b)

Fig. 7.11 Phase diagrams of polystyrene spheres mixed with *fd*-rods with a length of 880 nm. Symbols denote experimental data. Dashed curves represent original FVT from Vliegenthart and Lekkerkerker [21]; solid curves are binodals calculated using improved FVT, taking into account the excluded volume of the rods in the reservoir [27]. **a** Influence of size ratio $\mathcal{L} = L/d$ at fixed (effective) rod diameter $D_{\mathrm{eff}} = 14$ nm. The four sphere diameters used were $d = 496, 600, 708$ and 994 nm. **b** Influence of (effective) rod diameter D_{eff} at fixed $\mathcal{L}$ for a diameter of the spheres $d = 496$ nm. For $D_{\mathrm{eff}} \approx 11$ nm and 14 nm, experiments were performed at an ionic strength of 100 mM and 25 mM, respectively. For $D_{\mathrm{eff}} \approx 19$ nm, *fd*-virus sterically-stabilised with PEG was used

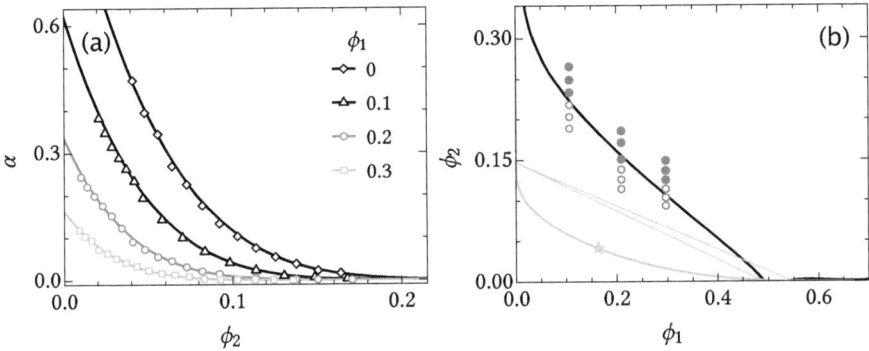

Fig. 7.12 Comparison of predictions (solid curves) from the improved FVT approach with Monte Carlo simulations (symbols) [29] for mixtures of hard spheres and hard rods for $\Gamma = 6$ and $\mathcal{L} = 1$ ($q = 1/5$). **a** The free volume fraction available for the hard rods in the system as a function of the rod volume fraction ϕ_2 for four hard-sphere volume fractions ϕ_1. **b** The phase behaviour. Data points from simulations show single phase ($\circ$) and demixing ($\bullet$). Black curve indicates improved FVT (showing colloidal fluid–solid coexistence); grey curves represent classical FVT [21] (fluid–fluid coexistence with ($\star$) critical point; triple gas–liquid–solid coexistence)

In Fig. 7.11a, the rigorous FVT predictions are given by the solid curves. It is clear the experimental data are now somewhat closer to the experimental data points, although FVT now predicts phase transitions at higher concentrations. It may be that charges actually blur the picture here since it is known [30] that charges can modify depletion effects [31], even under conditions of significant screening [32,33]. Figure 7.11b shows the comparison of rigorous FVT for various values of the effective rod thickness. The picture is the same: overall, a somewhat closer agreement with experiments (data) but overprediction of the phase transition concentrations, especially for $D_{\text{eff}} = 19$ nm.

References

1. Asakura, S., Oosawa, F.: J. Pol. Sci. **33**, 183 (1958)
2. Vliegenthart, G.A., van Blaaderen, A., Lekkerkerker, H.N.W.: Faraday Discuss. **112**, 173 (1999)
3. Koenderink, G.H., Vliegenthart, G.A., Kluijtmans, S.G.J.M., van Blaaderen, A., Philipse, A.P., Lekkerkerker, H.N.W: Langmuir **15**, 4693 (1999)
4. Oversteegen, S.M., Wijnhoven, J.G.E.J., Vonk, C., Lekkerkerker, H.N.W.: J. Phys. Chem. B **108**, 18158 (2004)
5. Yasarawan, N., van Duijneveldt, J.S.: Soft Matter **6**, 353 (2010)
6. Fraden, S., Maret, G., Casper, D.L.D., Meyer, R.B.: Phys. Rev. Lett. **63**, 2068 (1989)
7. Dogic, Z., Fraden, S.: Curr. Opin. Colloid Interface Sci. **11**, 47 (2006)
8. Dogic, Z., Fraden, S.: In: Soft Matter: Complex Colloidal Suspensions, Gompper, G., Schick, M., (Eds.), vol. 2, Chap. 1, pp. 1–86. Wiley Ltd. (2006)
9. Zimmermann, K., Hagedorn, H., Heuck, C., Hinrichsen, M., Ludwig, H.: J. Biol. Chem. **261**, 1653 (1986)
10. Lin, K.H., Crocker, J., Zeri, A.C., Yodh, A.G.: Phys. Rev. Lett. **87**, 888301 (2001)
11. Holmqvist, P., Kleshchanok, D., Lang, P.R.: Eur. Phys. J. E **26**, 177 (2008)
12. July, C., Lang, P.R.: Langmuir **26**, 18647 (2010)

13. Adams, M., Fraden, S.: Biophys. J. **74**, 669 (1998)
14. Guu, D., Dhont, J.K.G., Vliegenthart, G.A., Lettinga, M.P.: J. Phys.: Condens. Matter **24**, 464101 (2012)
15. Schmidt, M.: Phys. Rev. E **63**, 050201 (2001)
16. Brader, J.M., Esztermann, A., Schmidt, M.: Phys. Rev. E **66**, 031401 (2002)
17. Esztermann, A., Schmidt, M.: Phys. Rev. E **70**, 022501 (2004)
18. Harnau, L., Dietrich, S.: Phys. Rev. E **71**, 011504 (2005)
19. Wu, L., Malijevský, A., Jackson, G., Müller, E.A., Avendano, C.: J. Chem. Phys. **143**, 044906 (2015)
20. Wu, L., Malijevský, A., Avendano, C., Müller, E.A., Jackson, G.: J. Chem. Phys. **148**, 164701 (2018)
21. Vliegenthart, G.A., Lekkerkerker, H.N.W.: J. Chem. Phys. **111**, 4153 (1999)
22. Bolhuis, P.G., Frenkel, D.: J. Chem. Phys. **101**, 9869 (1994)
23. Cotter, M.A.: J. Chem. Phys. **66**, 1098 (1977)
24. Vroege, G.J., Lekkerkerker, H.N.W.: Rep. Progr. Phys. **55**, 1241 (1992)
25. McGrother, S.C., Williamson, D.C., Jackson, G.: J. Chem. Phys. **104**, 6755 (1996)
26. Bakker, H.E., Dussi, S., Droste, B.L., Besseling, T.H., Kennedy, C.L., Wiegant, E.I., Liu, B., Imhof, A., Dijkstra, M., van Blaaderen, A.: Soft Matter **12**, 9238 (2016)
27. Opdam, J., Guu, D., Schelling, M.P.M., Aarts, D.G.A.L., Tuinier, R., Lettinga, M.P.: J. Chem. Phys. **154**, 204906 (2021)
28. Cotter, M.A.: Phys. Rev. A **10**, 625 (1974)
29. Holovko, M.F., Hvozd, M.V.: Condens. Matter Phys. **20**, 1 (2017)
30. Fortini, A., Dijkstra, M., Tuinier, R.: J. Phys.: Condens. Matter **17**, 7783 (2005)
31. Tuinier, R., Rieger, J., de Kruif, C.G.: Adv. Colloid Interface Sci. **103**, 1 (2003)
32. Zhou, J., van Duijneveldt, J.S., Vincent, B.: Langmuir **26**, 9397 (2010)
33. van Gruijthuijsen, K., Tuinier, R., Brader, J.M., Stradner, A.: Soft Matter **9**, 9977 (2013)

Phase Behaviour of Colloidal Rods Mixed with Depletants

8

So far, we have considered the phase behaviour of colloidal spheres mixed with depletants. In Chap. 3, we considered the simplest type of depletant, the penetrable hard sphere (PHS). We then extended this treatment to ideal and excluded volume polymers in Chap. 4; and in Chap. 6, we considered small colloidal spheres (including micelles). Colloidal rods as depletants were addressed in Chap. 7; however, Chap. 7 only considered dilute dispersions of rods, in which the rods assume all configurations and are hence isotropic. In this chapter, we consider the phase behaviour of mixtures of colloidal rods and polymeric depletants, and we also account for higher rod concentrations and the corresponding phase states.

8.1 Experimental Observations with Rod-Like Particle Dispersions

Colloidal rods can be subdivided into synthetic inorganic rods, rod-like clay particles and biological rods (see also [1]). Examples are given in Fig. 8.1. Suspensions of rod-like particles exhibit interesting phase transitions and can assume various phase states. For a description of the various liquid crystalline phases, we refer the reader to the standard textbook on liquid crystals by de Gennes and Prost [2]. Lyotropic liquid crystalline phases were recognised a long time ago in suspensions of rod-like inorganic vanadium pentoxide (V2O5) colloids by Zocher [3], and later in solutions of biological particles comprising the tobacco mosaic virus (TMV) by Bawden et al. [4].

Upon concentrating a dilute rod suspension, a transition from an isotropic phase to an orientationally ordered phase occurs for rods with a length–diameter ratio $L/D > 3.5$. This is the so-called nematic liquid crystal phase, for which examples are given in Figs. 8.2 and 8.3. The first step of the isotropic–nematic phase transition

© The Author(s) 2024
H. N. W. Lekkerkerker, R. Tuinier, M. Vis, *Colloids and the Depletion Interaction*,
Lecture Notes in Physics 1026, https://doi.org/10.1007/978-3-031-52131-7_8

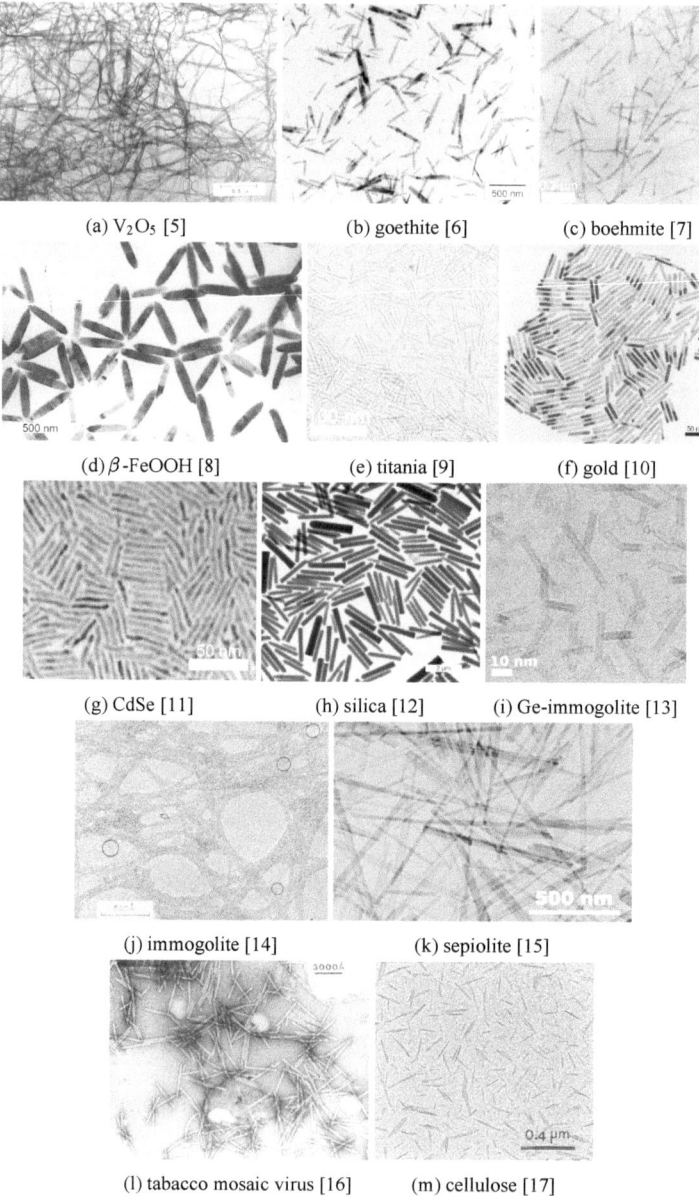

(a) V_2O_5 [5]   (b) goethite [6]   (c) boehmite [7]

(d) β-FeOOH [8]   (e) titania [9]   (f) gold [10]

(g) CdSe [11]   (h) silica [12]   (i) Ge-immogolite [13]

(j) immogolite [14]   (k) sepiolite [15]

(l) tabacco mosaic virus [16]   (m) cellulose [17]

Fig. 8.1 Transmission electron microscopy (TEM) micrographs of **a–i** synthetic inorganic, **j, k** clay and **l, m** biological rod-like colloids. Reprinted with permission from **a** Ref. [5], copyright 1991 American Chemical Society (ACS); **b** Ref. [6], copyright 2006 ACS; **c** Ref. [7], copyright 1994 ACS; **d** Ref. [8], copyright 1996 ACS; **e** Ref. [9], copyright 2020 Wiley; **f** Ref. [10], copyright 2009 Elsevier; **g** Ref. [11], copyright 2002 ACS; **h** Ref. [12], copyright 2011 ACS; **i** Ref. [13], copyright 2013 the Royal Society of Chemistry (RSC); **j** Ref. [14], copyright 1970 Cambridge University Press; **k** Z. Zhang, Soochow University, China; **l** Ref. [16], copyright 1985 EDP Sciences; **m** Ref. [17], copyright 1996 ACS

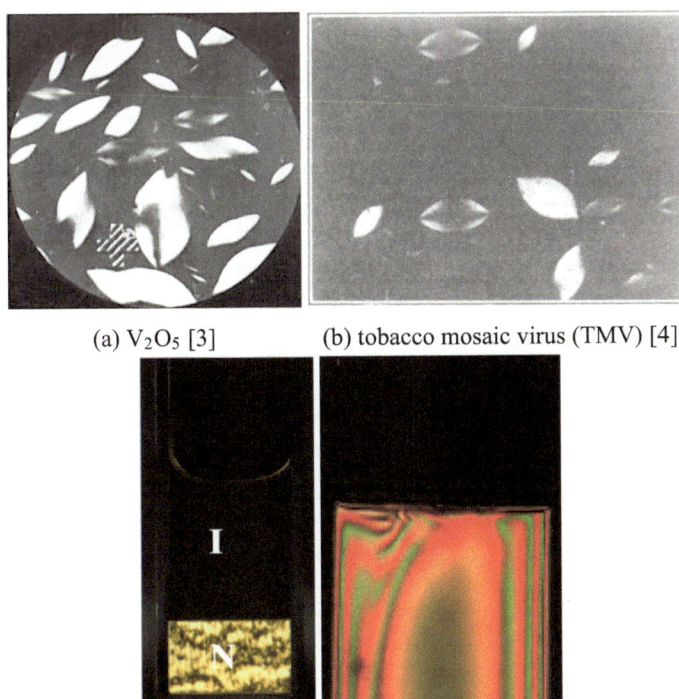

(a) V$_2$O$_5$ [3]   (b) tobacco mosaic virus (TMV) [4]

(c) boehmite [18]   (d) TMV [19]

Fig. 8.2 Examples of **a, b** nematic tactoids and **c, d** macroscopic isotropic–nematic phase coexistence observed between crossed polarisers. Reprinted with permission from: **a** Ref. [3], copyright 1925 Wiley; **b** Ref. [4], copyright 1936 Nature; **c** P. A. Buining, Utrecht University; **d** Ref. [19], copyright 2006 Wiley

is the formation of spindle-like droplets (so-called tactoids) of the nematic phase that float in the isotropic phase (Fig. 8.2a, b). Over time, these droplets coalesce to give rise to a macroscopic nematic bottom phase and an isotropic top phase. The rods have a different refractive index in parallel and perpendicular directions. As a result, the nematic phase displays typical interference colours under crossed polarisers due to a difference in retardation of light in different directions (Fig. 8.2c, d).

Oster [24] found an additional liquid crystal phase in suspensions of TMV in which the particles are ordered in periodic layers. On average, the axes of the rods are perpendicular to the layers (Fig. 8.3) and within the layers the rods behave like a two-dimensional fluid. This phase is known as the smectic A phase (SmA) [2] . For a long time, it was argued that attractive interactions between the rods were necessary for the occurrence of this phase. Frenkel, Stroobants and Lekkerkerker [25], however, showed by using Monte Carlo simulations that smectic ordering occurs in a fluid of hard rod-like particles, i.e., a smectic phase may appear solely driven by entropy.

Figure 8.3b–d gives examples of confocal microscopy images of silica rod suspensions with $L/D \approx 6$. Upon increasing the silica rod concentration, Kuijk et al. [22]

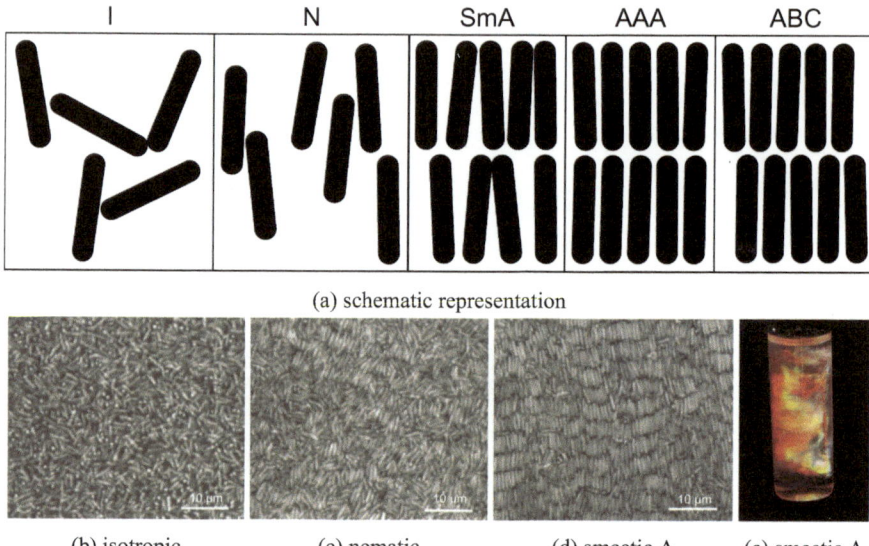

(a) schematic representation

(b) isotropic   (c) nematic   (d) smectic A   (e) smectic A

Fig. 8.3 a Sketches [20,21] and **b–d** experimental observations [22] of the structure of different phases of silica rods. **e** Iridescence observed in a smectic liquid crystal of TMV [23]. Reprinted with permission from **a** Peters et al. [20] under the terms of CC-BY-4.0; **b–d** Ref. [22], copyright 2012 RSC and **e** Ref. [23], copyright 1985 Wiley

found isotropic (b), nematic (c) and smectic ordering (d) also. The layered structure of the SmA with layer distances of the order of the wavelength of light gives rise to impressive iridescence (see Fig. 8.3e). Computer simulations have revealed that hard rods can also give rise to crystal phases AAA and ABC at high rod densities (see Fig. 8.3a) [26,27].

The effect of nonadsorbing polymer on the isotropic–nematic phase transition has been studied since the 1940s with a focus on the practical possibilities of isolation and separation of viruses [28,29] (see Sect. 1.3.2.4). It was observed that the addition of relatively small amounts of polymer to virus suspensions led to the 'precipitation' of the virus particles (i.e., the formation of a concentrated phase of the virus particles). Only much later in the 1990s were experiments initiated on model suspensions of rod-like colloids mixed with polymers to study the treatment given in the previous chapters to rod-like colloids mixed with depletant, and the results compared to theory and computer simulations.

To connect the experimental observations discussed here to theoretical predictions, we first discuss the Onsager theory (Sect. 8.2) to quantify the I–N phase transition of long hard rods and its extension to describe charged rods. Subsequently, scaled particle theory of rods is discussed to approximate finite size effects of rods in Sect. 8.3. In Sect. 8.4, theory for the phase behaviour of mixtures of hard rods and nonadsorbing polymers is presented, and experimental examples are provided in

Sect. 8.5. Finally, higher order phase states are included in the theoretical description in Sect. 8.6. All sections are supplemented by comparison with experiment and/or computer simulation results.

8.2   Onsager Theory of the Isotropic–Nematic Transition

8.2.1   Long Hard Rods

As we saw in the previous chapters, colloidal phase transitions of hard particles are governed by entropy. This was first revealed by Onsager, who showed that the isotropic–nematic (I–N) phase transition in assemblies of hard rods is driven solely by entropy. He realised that an attractive force is not necessary for the I–N transition by showing that an assembly of repelling rods exhibits a transition from an isotropic to a nematic state due to a gain of packing entropy that compensates the loss of orientational entropy. Onsager also demonstrated that the I–N transition may be treated within a virial expansion of the free energy. In fact, this is a unique example of a phase transition that can be treated using a virial expansion. For very thin, rigid, hard particles the transition occurs at a very low volume fraction and the virial expansion may even be truncated after the second virial term, leading to an exact theory for infinitely thin particles. In the following, we give a brief exposition on Onsager's theory. For more details, we refer the reader to [30, 31].

The Helmholtz free energy F for a dispersion of N hard rods (which we, as in Sect. 7, model as spherocylinders with length L and diameter D) in a volume V in the second virial approximation can be written as

$$\frac{F[f]}{NkT} = \text{constant} - 1 + \ln c + \mathfrak{s}[f] + c\rho[f]. \tag{8.1}$$

We have lumped in the constant quantities that do not affect the phase transition, i.e., have the same value in the coexisting phases. The quantity c is the dimensionless concentration

$$c = bn, \tag{8.2}$$

where $b = (\pi/4)L^2D$ is the excluded volume and $n = N/V$ is the number density of the rods. The orientational entropy is expressed through $\mathfrak{s}[f]$:

$$s_{\text{or}} = -k \int f(\mathbf{\Omega}) \ln[4\pi f(\mathbf{\Omega})] d\mathbf{\Omega} = -k\mathfrak{s}[f], \tag{8.3}$$

where $f(\mathbf{\Omega})$ is the orientational distribution function, which gives the probability of finding a spherocylinder with an orientation characterised by the solid angle $\mathbf{\Omega}$. Finally, $-kc\rho[f]$ is the packing entropy per particle, with

$$\rho[f] = \frac{4}{\pi} \int \int |\sin\gamma| f(\mathbf{\Omega}) f(\mathbf{\Omega}') d\mathbf{\Omega} d\mathbf{\Omega}', \tag{8.4}$$

where γ is the angle between the rods, which depends on their orientations $\mathbf{\Omega}$ and $\mathbf{\Omega}'$ (see Fig. 7.4).

As already remarked, the I–N transition originates from a competition between the orientational and packing entropy. For low concentrations the orientational entropy dominates and attains a maximum value for an isotropic distribution $f = (4\pi)^{-1}$; whereas for high concentrations the packing entropy becomes more important, favouring a nematic orientation distribution. The orientational distribution is determined by the fact that the free energy must be a minimum. Upon minimising Eq. (8.1), the integral equation

$$\ln[4\pi f(\theta)] = \lambda - \frac{8c}{\pi} \int |\sin \gamma(\mathbf{\Omega}, \mathbf{\Omega}')| f(\theta')d\mathbf{\Omega}' \tag{8.5}$$

is obtained. Here, we have taken into account that f does not depend on the azimuthal angle but only on θ, the polar angle between the rod vector and the nematic vector. Furthermore, the distribution function $f(\theta)$ must satisfy inversion symmetry, implying the angles θ and $\pi - \theta$ are equivalent. Note that we assume the nematic phase is apolar. The Lagrange multiplier λ is determined by requiring that $f(\theta)$ fulfils the normalisation condition

$$\int f(\mathbf{\Omega})d\mathbf{\Omega} = 1. \tag{8.6}$$

It is easily seen that the isotropic distribution function

$$f = \frac{1}{4\pi} \tag{8.7}$$

satisfies Eq. (8.5) for all concentrations (although it is only for low concentrations that this corresponds to a minimum of the free energy). For the isotropic phase, s and ρ attain the values

$$s_I = 0, \quad \rho_I = 1, \tag{8.8}$$

and hence

$$\frac{F_I}{NkT} = \text{constant} - 1 + \ln c + c. \tag{8.9}$$

Exercise 8.1. Derive Eq. (8.8) from Eqs. (8.3) and (8.4) using Eq. (8.7).

An exact solution to the non-linear integral equation Eq. (8.5) for higher concentrations, where a nematic distribution minimises the free energy, has not yet been found but ways to solve it numerically have appeared [32,33]. For a didactic account of how to solve Eq. (8.5) numerically, see Ref. [34]. This allows the determination of $s[f]$ and $\rho[f]$ and, from thereon, the free energy in the nematic phase. To be in mechanical and chemical equilibrium, both phases must have the same osmotic pressure and the same chemical potential (see Eqs. (A.13) and (A.14)),

$$P_{\mathrm{I}}(c_{\mathrm{I}}) = P_{\mathrm{N}}(c_{\mathrm{N}}), \tag{8.10a}$$

$$\mu_{\mathrm{I}}(c_{\mathrm{I}}) = \mu_{\mathrm{N}}(c_{\mathrm{N}}). \tag{8.10b}$$

These quantities can be obtained (see Appendix A) from the free energy using the standard thermodynamic relations:

$$P = - \left(\frac{\partial F}{\partial V} \right)_{N,T}, \tag{8.11a}$$

$$\mu = \left(\frac{\partial F}{\partial N} \right)_{V,T}. \tag{8.11b}$$

For the isotropic phase, we find from Eq. (8.9)

$$\frac{P_{\mathrm{I}} b}{kT} = c_{\mathrm{I}} + c_{\mathrm{I}}^2, \tag{8.12a}$$

$$\frac{\mu_{\mathrm{I}}}{kT} = \text{constant} + \ln c_{\mathrm{I}} + 2 c_{\mathrm{I}}. \tag{8.12b}$$

Exercise 8.2. Show that Eqs. (8.12a) and (8.12b) follow from Eqs. (8.9) and (8.11a) and (8.11b).

In the nematic phase, Eq. (8.1) gives

$$\frac{P_{\mathrm{N}} b}{kT} = c_{\mathrm{N}} + c_{\mathrm{N}}^2 \rho[f], \tag{8.13a}$$

$$\frac{\mu_{\mathrm{N}}}{kT} = \text{constant} + \ln c_{\mathrm{N}} + \mathfrak{s}[f] + 2 c_{\mathrm{N}} \rho[f]. \tag{8.13b}$$

where the distribution f must be obtained numerically for each concentration from Eq. (8.5). Solving the coexistence equations (8.10a) and (8.10b) with the above expressions for the osmotic pressure and chemical potential numerically yields the coexistence concentrations

$$c_{\mathrm{I}} = 3.290, \quad c_{\mathrm{N}} = 4.191. \tag{8.14}$$

The usual measure of the ordering in the nematic phase is given by the nematic order parameter S, which is defined as

$$S = \int \mathcal{P}_2(\cos\theta) f(\hat{\mathbf{u}}) d\hat{\mathbf{u}}, \tag{8.15}$$

where $\hat{u}$ is an orientational unit vector, and $\mathcal{P}_2$ is the second Legendre polynomial. For a dispersion of isotropic rods $S = 0$, whereas for ordered phases S can attain larger values, approaching $S \to 1$ for highly ordered rods. For the nematic phase of infinitely long rods considered here,

$$S = 4\pi \int_0^{\pi/2} f(\theta) \left[\frac{3}{2} \cos^2 \theta - \frac{1}{2} \right] \sin(\theta) d\theta \qquad (8.16)$$

has the value

$$S = 0.7922 \qquad (8.17)$$

for the coexisting nematic phase.

More convenient calculations of the phase transition can be performed by choosing a trial function for the orientational distribution function f with one or more variational parameters. The free energy as a function of these parameters can then be minimised with respect to these parameters. Onsager [30] chose the following function:

$$f_O(\theta) = \frac{\kappa \cosh(\kappa \cos \theta)}{4\pi \sinh \kappa}. \qquad (8.18)$$

This expression only has a single variational parameter (κ) and gives the following results for the coexisting concentrations and nematic order parameter at coexistence:

$$c_I = 3.340 , \quad c_N = 4.486 , \quad S = 0.848. \qquad (8.19)$$

Comparison of these results with the exact values in Eqs. (8.14) and (8.17) shows that the trial function chosen by Onsager works quite well.

Odijk [35, 36] realised that for large values of κ (and thus, for highly ordered nematics), Onsager's orientational distribution function can be approximated by a simple Gaussian distribution function:

$$f_G \sim \tilde{N}(\kappa) \exp\left(-\frac{1}{2}\kappa\theta^2\right) \qquad\qquad 0 \le \theta \le \frac{\pi}{2}, \qquad (8.20a)$$

$$\sim \tilde{N}(\kappa) \exp\left(-\frac{1}{2}\kappa(\pi - \theta)^2\right) \qquad \frac{\pi}{2} \le \theta \le \pi. \qquad (8.20b)$$

where $\tilde{N}(\kappa)$ is a normalisation constant. The advantage of this Gaussian distribution function is that for large values of κ the quantities $\mathfrak{s}[f]$ and $\rho[f]$ can be represented by the analytic expressions

$$\mathfrak{s}[f_G] \sim \ln \kappa - 1, \qquad (8.21)$$

and

$$\rho[f_G] \sim \frac{4}{\sqrt{\pi\kappa}}. \qquad (8.22)$$

This leads to the following expression for the free energy in the nematic phase:

$$\frac{F}{NkT} \sim \text{constant} - 1 + \ln c + \ln \kappa - 1 + \frac{4c}{\sqrt{\pi \kappa}}. \tag{8.23}$$

Minimising this expression with respect to κ,

$$\frac{\partial F}{\partial \kappa} = 0, \tag{8.24}$$

leads to

$$\kappa \sim \frac{4c^2}{\pi}. \tag{8.25}$$

Hence,

$$\frac{F}{NkT} \sim \text{constant} + \ln \frac{4}{\pi} + 3 \ln c. \tag{8.26}$$

Applying Eqs. (8.11a) and (8.11b) yields the following results for the (osmotic) pressure and chemical potential of the rods in the nematic phase:

$$\frac{P_N b}{kT} = 3c_N, \tag{8.27a}$$

$$\frac{\mu_N}{kT} = \text{constant} + \ln \frac{4}{\pi} + 3 + 3 \ln c_N. \tag{8.27b}$$

Combining these with the expressions given by Eqs. (8.12a) and (8.12b) for the pressure and chemical potential in the isotropic phase, the coexistence Eqs. (8.10a) and (8.10b) now take the simple forms:

$$c_I + c_I^2 = 3c_N \tag{8.28a}$$

$$\ln c_I + 2c_I = 3 \ln c_N + \ln \left(\frac{4}{\pi} \right) + 3 \tag{8.28b}$$

From this, we find the following coexisting concentrations:

$$c_I = 3.451, \quad c_N = 5.122, \tag{8.29}$$

implying, via Eq. (8.25), that $\kappa = 33.4$. Insertion of Eqs. (8.20a) and (8.20b) into Eq. (8.16) using this value for κ gives

$$S = 0.910 \tag{8.30}$$

for the nematic order parameter in the coexisting nematic phase.

While the results for the Gaussian distribution function differ more from the exact results than the Onsager trial function (although in both cases the values are too high for the coexisting concentrations and for the order parameter in the coexisting nematic phase), the calculations are substantially simpler [37] and provide a good estimate of the I–N transition in more complicated situations (that we will encounter in the next sections). Although the results of Onsager's theory are of great fundamental and methodological interest, they refer strictly to infinitely thin hard rods. Hence, the applicability of the theory to experimental results is limited. In real suspensions of rod-like particles, we have to take into account one or more of the following aspects:

- particles are not infinitely thin,
- particles may be polydisperse in size,
- particles are not hard but may show (long-range) repulsions, for instance, due to charges or anchored polymeric brushes,
- there may be attractions between the particles and
- particles may be semiflexible.

Onsager [30] already addressed the issues of additional particle repulsions and polydispersity. These and the other issues raised above have been considered extensively (for a review, see [31]). Some of these complex elements will be treated in the rest of this chapter.

8.2.2   Charged Rods

In experimental systems, the rod-like particles are often charged. This means that, besides the hard-core excluded volume interaction, there is a double layer repulsion between the rods that gives rise to a soft repulsive interaction. Double layer forces between charged colloids in a polar solvent are specified by the range and the strength of the repulsive interaction [38]. The density of surface charge groups, which is directly related to the electrostatic surface potential Ψ at the rod surface, determines the strength of the repulsion. The ionic strength of the medium dictates the Debye length, which mediates the range of the double layer repulsion. Onsager [30] proposed to describe charged rods as hard rods with an *effective* diameter $D_{\text{eff}} > D$.

Stroobants, Lekkerkerker and Odijk [39] used the pair interaction between two charged rods to compute the second virial coefficient. This revealed an effective rod diameter that is given by

$$\frac{D_{\text{eff}}}{D} = 1 + \frac{\ln A' + k_{\text{E}} + \ln 2 - \frac{1}{2}}{D/\lambda_{\text{D}}}, \tag{8.31}$$

where λ_{D} is the ionic strength-dependent Debye length (see Sect. 1.2.2), k_{E} is Euler's constant ≈ 0.577 and A' follows from the pair interaction as

$$A' = \frac{\pi \lambda_{\text{D}} \zeta^2 \exp\left[-D/\lambda_{\text{D}}\right]}{2\lambda_{\text{B}}}, \tag{8.32}$$

with λ_B the Bjerrum length (see Sect. 1.2.2). The parameter ζ is the proportionality constant of the outer part of the double layer electrostatic potential profile near a charged rod [40]:

$$\frac{e\Psi(r)}{k_B T} \sim \zeta K_0(r/\lambda_D), \tag{8.33}$$

where e denotes the elementary charge, r represents the distance from the centre line of the rod, and K_0 is the modified Bessel function of the second kind of order 0. For a weakly charged rod, the Debye–Hückel approximation provides [41]

$$\frac{e\Psi_{DH}(r)}{k_B T} = \frac{4Z\lambda_D\lambda_B K_0(r/\lambda_D)}{D K_1(D/2\lambda_D)}, \tag{8.34}$$

where Z is the linear charge density (per unit length) of the rod and K_1 is the modified Bessel function of the second kind of order 1. Comparison of Eqs. (8.33) and (8.34) enables ζ to be expressed in Eq. (8.32), giving

$$A'_{DH} = \frac{8\pi Z^2 \lambda_D^3 \lambda_B \exp[-D/\lambda_D]}{D^2 K_1^2(D/2\lambda_D)}. \tag{8.35}$$

For thick and thin double layers, this results in the following respective asymptotic analytical results [31]:

$$A'_{DH} \simeq \begin{cases} 2\pi Z^2 \lambda_D \lambda_B & D \ll \lambda_D, \\ \frac{8Z^2 \lambda_D^2 \lambda_B}{D} & D \gg \lambda_D. \end{cases} \tag{8.36}$$

Insertion of D_{eff} for D into the equations for hard rods then enables the physical properties of charged rods to be predicted. One may, for instance, insert the effective rod concentration given in Eq. (8.2) using D_{eff} to quantify the isotropic–nematic phase transition of long thin rods outlined in the previous section. It is noted that electrostatic twisting effects are not accounted for here. For those interested, see, for instance, Ref. [39].

8.3 Scaled Particle Theory of the Isotropic–Nematic Transition

When considering finite-sized rods (and later on, the effect of depletion attraction on the I–N transition in rod-like suspensions), we must take into account that the second virial term B_2 no longer strongly exceeds the higher virial terms. When there are attractions between the rods, nearly parallel configurations are of paramount importance and B_2 is no longer the dominating virial coefficient, as in the case of long, repulsive rods. It was shown that, for even slightly attractive rods, the third virial coefficient B_3 is almost as large as B_2 [42]. This means that we must start from a theory that takes into account higher virial coefficients. Here, we use scaled particle theory (SPT) [43], which will be treated in this section. SPT for rods mixed with polymers will subsequently be addressed in Sect. 8.4, following [44, 45].

SPT—a convenient and tractable way to incorporate higher virial coefficients in the treatment of the isotropic–nematic phase transition—was applied in Sect. 7.3 to obtain the osmotic pressure of an isotropic suspension of rods. The starting point of SPT is the calculation of the reversible work W to insert an additional spherocylinder into the system of spherocylinders to obtain the excess part of the chemical potential:

$$\mu^{\text{ex}} = \int f(\mathbf{\Omega}) W(\mathbf{\Omega}, 1, 1) d\mathbf{\Omega}, \tag{8.37}$$

where $W(\mathbf{\Omega}, 1, 1)$ is the reversible work to insert a spherocylinder with length L and diameter D and orientation $\mathbf{\Omega}$ in a system of hard spherocylinders. In Sect. 7.3, we considered an isotropic assembly of rods but Eq. (8.37) applies equally well to an orientationally ordered (nematic) system of rods, as long as we use an accurate expression for the orientation distribution function, $f(\mathbf{\Omega})$. After replacing the second virial contribution $2c\rho[f]$ in Eq. (8.13b) with the chemical potential μ^{ex}, we obtain

$$\frac{\mu}{kT} = \text{constant'} + \ln y + \mathfrak{s}[f],$$
$$+ (1 + 2 A[f]) y + \left(A[f] + \frac{3}{2} B[f] \right) y^2 + B[f] y^3. \tag{8.38}$$

Here, y has its usual meaning

$$y = \frac{\phi}{1 - \phi},$$

with ϕ the volume fraction of the rods (which equals $n v_0$), and v_0 as the spherocylinder volume given by

$$v_0 = \frac{\pi}{4} L D^2 + \frac{\pi}{6} D^3.$$

The quantities $A[f]$ and $B[f]$ are defined as

$$A[f] = 3 + \frac{3(\Gamma - 1)^2}{3\Gamma - 1} \rho[f] \tag{8.39}$$

$$B[f] = \frac{12\Gamma(2\Gamma - 1)}{(3\Gamma - 1)^2} + \frac{12\Gamma(\Gamma - 1)^2}{(3\Gamma - 1)^2} \rho[f], \tag{8.40}$$

where

$$\Gamma = \frac{L}{D} + 1 \tag{8.41}$$

is the overall length-to-diameter ratio. Using the Gibbs–Duhem equation (see Eq. (A.12)), one obtains for the pressure

$$\frac{P v_0}{kT} = y + A[f] y^2 + B[f] y^3 \quad . \tag{8.42}$$

Exercise 8.3. Show that Eq. (8.42) recovers the SPT expression for the pressure of a hard-sphere fluid (Eq. (3.37)) by setting $\Gamma = 1$ and imposing Eq. (8.7).

Finally, the Helmholtz energy can be obtained from the relation

$$F = N\mu - PV,$$

leading to

$$\frac{F[f]}{NkT} = \frac{\widetilde{F}[f]}{\phi} = \text{constant}' - 1 + \mathfrak{s}[f] + \ln y + A[f]y + \frac{1}{2}B[f]y^2, \quad (8.43)$$

with $\widetilde{F} = Fv_0/kTV$ (see Appendix A.2), which will be used in later sections.

Exercise 8.4. Show that in the limit $L/D \to \infty$ and low concentrations the above expression for the free energy reduces to the free energy in the second virial approximation Eq. (8.1) with constant$'$ = constant + $\ln(L/D)$.

As indicated earlier, the I–N phase equilibria can be found simultaneously:

- using Eq. (8.24) to minimise $F[f]$ numerically with respect to the orientational distribution function f,
- calculating the orientation distribution function of f,
- calculating the (osmotic) pressure and chemical potential and
- solving the coexistence equations.

Hence, there are three equations with three unknowns: the two coexistence concentrations and κ. The results for the coexisting concentrations, which now depend on L/D, are given in Fig. 8.4 (see also [45]). In this figure, we also present Monte Carlo simulation results [27] and the Onsager limit result ($L/D \to \infty$). Clearly, the agreement between numerically solving the SPT expressions (solid curves) and computer simulation results is quite good. In Fig. 8.5, we give the coexistence pressure at isotropic–nematic coexistence for hard spherocylinders as a function of the aspect ratio L/D.

It is interesting to compare the results obtained with the numerical orientational distribution function with the results obtained with the Gaussian orientational distribution function Eqs. (8.20a) and (8.20b). Minimising the free energy of Eq. (8.43) is possible by substituting the expressions for $\mathfrak{s}[f_G]$ and $\rho[f_G]$ that are given by Eqs. (8.21) and (8.22). This yields the following expression for κ:

$$\kappa = \frac{36}{\pi} \frac{(\Gamma - 1)^4}{(3\Gamma - 1)^2} \left(y + \frac{2\Gamma}{3\Gamma - 1} y^2 \right)^2. \quad (8.44)$$

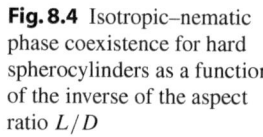

Fig. 8.4 Isotropic–nematic phase coexistence for hard spherocylinders as a function of the inverse of the aspect ratio L/D

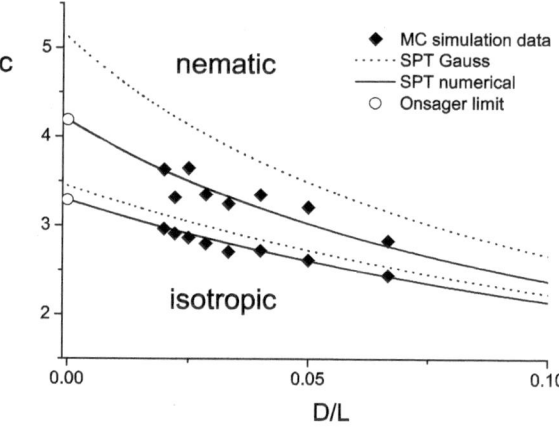

Fig. 8.5 Pressure Pb/kT at isotropic–nematic coexistence for hard spherocylinders as a function of D/L. Note $v_0 = b[D/L + (2/3)(D/L)^2]$

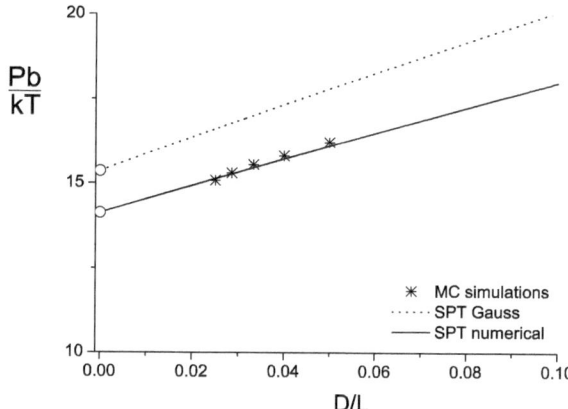

Exercise 8.5. Show that in the limit $L/D \to \infty$ and at low rod concentrations Eq. (8.44) reduces to Eq. (8.25).

Using Eq. (8.44) for κ in $\mathfrak{s}[f_G]$ and $\rho[f_G]$, and substituting these expressions in Eqs. (8.38) and (8.42), provides us with analytical expressions for the chemical potential and pressure in the nematic phase. The expressions for these quantities in the isotropic phase are obtained by setting $\mathfrak{s} = 0$ and $\rho = 1$ in equations (8.38) – (8.42). We can then solve the coexistence equations (8.10a) and (8.10b). The results for the coexisting concentrations and the coexistence pressure obtained using the Gaussian orientational distribution function are also given in Figs. 8.4 and 8.5 as the dashed curves.

As in the Onsager limit, the results lie somewhat above the numerical solution but are still quite reasonable. Given the fact that the Gaussian approximation is transparent and simple, its use provides an extremely valuable method to scan through a large parameter space as we shall see in the next sections.

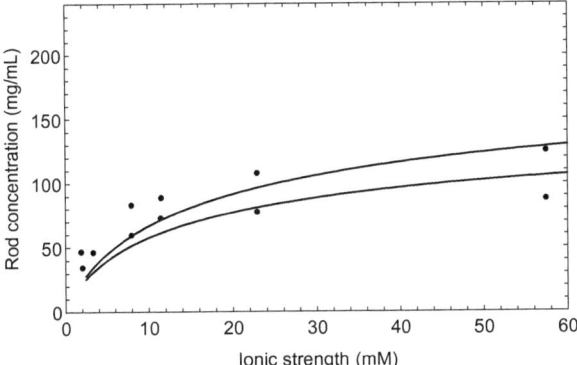

Fig. 8.6 Isotropic–nematic phase transition concentrations of TMV as a function of the ionic strength. The data points are experimental results redrawn from Ref. [46]. Curves represent SPT with D_{eff} calculated by using Eq. (8.31) and following [39]. Equation (8.35) was used to calculate A'. Parameters used: $L = 282$ nm, $D = 18$ nm, charge density $Z = -10$ e/nm and virus particle molar mass $M_{\text{p}} = 4 \cdot 10^7$ g/mol

This SPT description can be extended to also include the influence of a double layer surrounding the rods in a polar solvent due to charges at the rod surface. Basically, one can still apply Eq. (8.43) with Eq. (8.44) to describe charged rods, but D is instead replaced with D_{eff} given by Eq. (8.31). The I–N phase transition concentrations for TMV virus as a function of salt concentration are given in Fig. 8.6, as measured by Fraden et al. [46]. As the ionic strength increases the concentration of virus in the coexisting phases increases. Without added salt, an isotropic phase of 15 mg/mL TMV coexists with a nematic phase of 23 mg/mL, while, at an ionic strength of 60 mM, the coexisting concentrations are 90 mg/mL in the isotropic phase and 125 mg/mL in the nematic phase. Replacing the electrostatic potential between TMV particles with an appropriate effective diameter gives a reasonably good description of the experimentally observed phase boundaries [46]. This provides information on how the I–N phase coexistence varies with D_{eff}, which in turn depends on the ionic strength for charged rods.

8.4 Isotropic–Nematic Phase Behaviour of Rods Mixed with Penetrable Hard Spheres

We now consider the effect of added polymer on the phase behaviour of a system of hard rods. The simplest representation of a polymer is a penetrable hard sphere (PHS) with diameter $\sigma = 2\delta$ and radius δ equal to the depletion thickness. See Sect. 2.1 for details about the PHS model.

The starting point for the calculation of the phase behaviour is the semi-grand potential for the colloidal rods–PHSs system that is in osmotic equilibrium with a reservoir of PHSs, which sets the chemical potential of the PHSs. This system is depicted in Fig. 8.7. In the free volume approximation (see Sect. 3.3), we can write

Fig. 8.7 Osmotic equilibrium between a dispersion of hard rods and penetrable hard spheres (the system) and a reservoir containing a penetrable hard-sphere dispersion

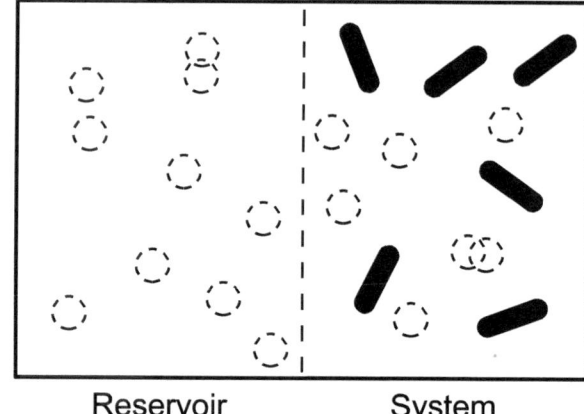

Reservoir   System

Fig. 8.8 Illustration of the available free volume (the unshaded volume) in a dispersion of hard spherocylinders

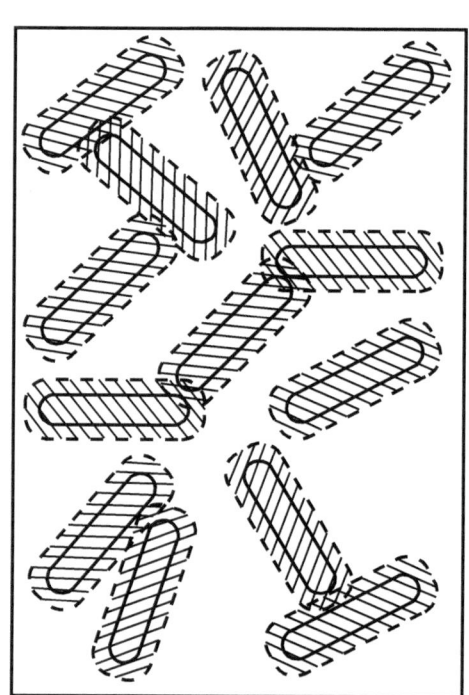

Eq. (3.26) as

$$\Omega(N_1, V, T, \mu_2) = F_0(N_1, V, T) - P^R \langle V_{\text{free}} \rangle_0, \tag{8.45}$$

where N_1 stands for the number of rods, μ_2 represents the chemical potential of the depletants (PHSs), P^R is the pressure in the reservoir and $\langle V_{\text{free}} \rangle_0$ is the free volume for PHSs in the system of rods, illustrated in Fig. 8.8.

For $F(N_1, V, T)$, we use the SPT expression (Eq. (8.43)) [43], and the osmotic pressure of the PHSs in the reservoir is given by

$$P^R = n_2^R kT,$$

where n_2^R is the number density of PHSs in the reservoir. The free volume is again calculated using the relation

$$\frac{\langle V_{\text{free}} \rangle_0}{V} = \alpha = e^{-W/kT}, \tag{8.46}$$

where W is the reversible work for inserting the PHSs into the hard rod suspension. An expression for the work of insertion W can again be conveniently obtained using SPT. The work W is calculated by expanding the PHS to be inserted from zero to its final size. By writing the size of the scaled PHS as $\lambda\sigma$ $(= 2\lambda\delta)$ in the limit that $\lambda \to 0$, the inserted sphere approaches a point particle. In this limit, it is very unlikely that excluded volumes of the hard rods and added scaled PHS overlap. So,

$$W(\lambda) = -kT \ln [1 - n_1 v_{\text{excl}}(\lambda)] \quad \text{for} \quad \lambda \ll 1, \tag{8.47}$$

where $v_{\text{excl}}(\lambda)$ is the excluded volume of the added scaled PHS and a hard spherocylinder with length L and diameter D:

$$v_{\text{excl}}(\lambda) = \frac{\pi}{4}(D + \lambda\sigma)^2 L + \frac{\pi}{6}(D + \lambda\sigma)^3. \tag{8.48}$$

The opposite limit $\lambda \gg 1$ corresponds to the case when the size of the inserted PHS is very large. Then W is, to a good approximation, equal to the volume work needed to create a cavity with volume

$$\frac{\pi}{6}(\lambda\sigma)^3$$

and is given by

$$W = \frac{\pi}{6}(\lambda\sigma)^3 P \quad \text{for} \quad \lambda \gg 1, \tag{8.49}$$

where P is the (osmotic) pressure of the hard rod system given by (Eq. (8.42)).

In SPT, the above two limiting cases are connected by expanding W in a series in λ:

$$W(\lambda) = W(0) + \left(\frac{\partial W}{\partial \lambda}\right)_{\lambda=0} \lambda + \frac{1}{2}\left(\frac{\partial^2 W}{\partial \lambda^2}\right)_{\lambda=0} \lambda^2 + \frac{\pi}{6}(\lambda\sigma)^3 P. \tag{8.50}$$

This yields

$$\frac{W(\lambda = 1)}{kT} = -\ln(1 - \phi_1) + \left[\frac{6\Gamma q}{3\Gamma - 1} + \frac{3(\Gamma + 1)q^2}{3\Gamma - 1}\right] y_1$$
$$+ \frac{1}{2}\left(\frac{6\Gamma}{3\Gamma - 1}\right)^2 q^2 y_1^2 + \frac{2q^3}{3\Gamma - 1}\frac{Pv_0}{kT}, \tag{8.51}$$

where

$$y_1 = \frac{\phi_1}{1 - \phi_1}$$

$$q = \frac{\sigma}{D} = \frac{2\delta}{D}$$

Inserting (Eq. (8.42)) for the pressure P of spherocylinders leads to the following expression for the free volume fraction:

$$\alpha = (1 - \phi_1) \exp[-Q(\phi_1)], \tag{8.52}$$

where

$$Q(\phi_1) = ay_1 + by_1^2 + cy_1^3 \tag{8.53}$$

with

$$a = \frac{6\Gamma}{3\Gamma - 1}q + \frac{3(\Gamma + 1)}{3\Gamma - 1}q^2 + \frac{2}{3\Gamma - 1}q^3, \tag{8.54a}$$

$$b = \frac{1}{2}\left(\frac{6\Gamma}{3\Gamma - 1}\right)^2 q^2 + \left(\frac{6}{3\Gamma - 1} + \frac{6(\Gamma - 1)^2}{(3\Gamma - 1)^2}\rho[f]\right)q^3, \tag{8.54b}$$

$$c = \frac{2}{3\Gamma - 1}\left(\frac{12\Gamma(2\Gamma - 1)}{(3\Gamma - 1)^2} + \frac{12\Gamma(\Gamma - 1)^2}{(3\Gamma - 1)^2}\rho[f]\right)q^3. \tag{8.54c}$$

Exercise 8.6. Check that, in the appropriate limit, Eq. (8.52) with Eqs. (8.53) and (8.54), α reduces to Eq. (3.38) with a, b and c given by Eq. (3.38).

We now have all the contributions to construct the semi-grand potential Ω given in (Eq. (8.45)). In order to obtain the phase behaviour, we proceed along the same lines as for the system of pure rods involving the following steps:

- minimise Ω with respect to the orientation distribution function f (compute the value of κ at which $\partial\Omega/\partial\kappa = 0$). Note that, in Eq. (8.45), both the free energy of the pure rod system F_0 and the free volume $\langle V_{\text{free}}\rangle_0$ depend on the orientation distribution function f
- evaluate the orientation distribution function f
- calculate the (osmotic) pressure and chemical potential of the rods, which are given by

$$P = -\left(\frac{\partial\Omega}{\partial V}\right)_{N_1,\mu_2} = P^0 + P^R\left(\alpha - n_1\frac{d\alpha}{dn_1}\right), \qquad (8.55)$$

$$\mu_1 = \left(\frac{\partial\Omega}{\partial N_1}\right)_{V,\mu_2} = \mu_1^0 - P^R\frac{d\alpha}{dn_1}, \qquad (8.56)$$

where P^0 and μ^0 are the pressure and chemical potential of the pure rod system, and

- solve the coexistence relations (Eqs. (A.13) and (A.14))

$$\mu_1^I\left(n_1^I, \mu_2\right) = \mu_1^{II}\left(n_1^{II}, \mu_2\right), \qquad (8.57)$$

$$P^I\left(n_1^I, \mu_2\right) = P^{II}\left(n_1^{II}, \mu_2\right). \qquad (8.58)$$

Instead of formal minimisation of the free energy leading to an integral equation for the orientational distribution function f, we will first use the Gaussian distribution function, which simplifies the calculations considerably while leading to reasonably accurate results. This is illustrated in Fig. 8.9, where we plot the isotropic–nematic phase coexistence curve for $L/D = 10$ and $q = 1$. On the ordinate, the relative reservoir concentration of PHSs in the reservoir ϕ_2^R is plotted versus the volume fraction of hard spherocylinders on the abscissa. The solid curves are the results for the binodals using the Gaussian distribution function, while the dashed curves were obtained using formal minimisation (see Ref. [45]). In Fig. 8.4, it was demonstrated that the Gaussian overestimates the I–N concentrations somewhat for the pure hard

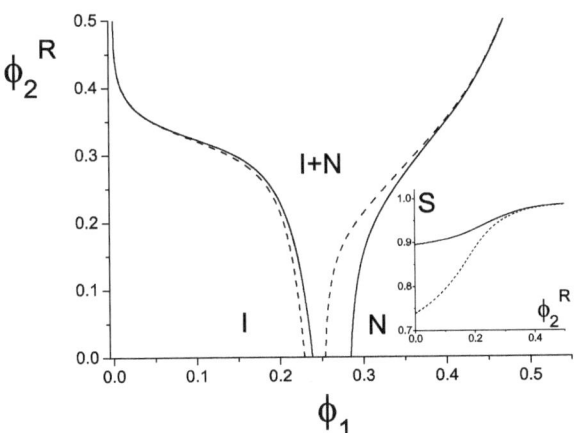

Fig. 8.9 Isotropic–nematic phase coexistence for $L/D = 10$ and $q = 1$ in the reservoir representation. The Gaussian orientational distribution function result (solid curves) is compared to the coexistence computed using formal minimisation of the oriental distribution function (dashed curves). *Inset*: Plot of the nematic order parameter S as a function of ϕ_2^R of the nematic phase (numerical approach: dotted curve, Gaussian approximation: solid curve) that coexists with the isotropic phase

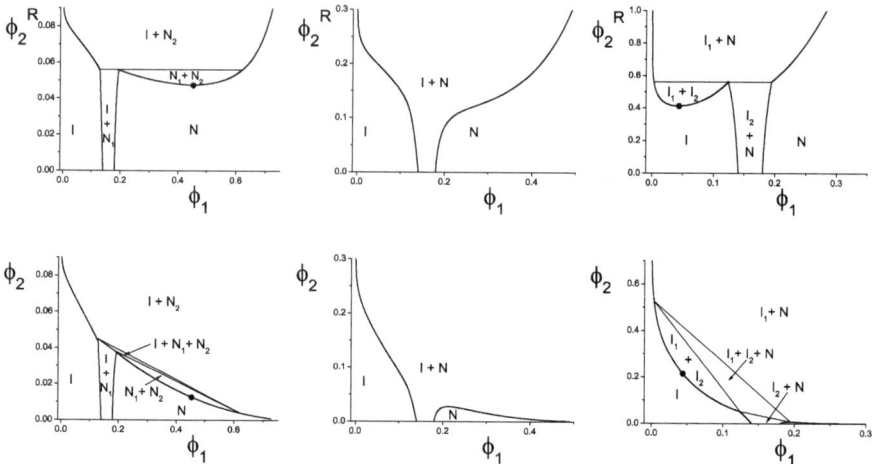

Fig. 8.10 Phase diagrams calculated using free volume theory for spherocylinders ($L/D = 20$) with added PHSs at three size ratios: $q = 0.3$ (*left*), $q = 1$ (*middle*) and $q = 2.5$ (*right*). The upper three curves are in the reservoir representation and the lower curves are the system results. The Gaussian form for the ODF was used to minimise the semi-grand potential and compute the coexistence concentrations

spherocylinder dispersion, which is also shown here. For a pure rod dispersion the Gaussian approximation provides a too sharply peaked orientation distribution function f, reflected by a too large value for the nematic order parameter S. Hence, the loss of orientational entropy is overestimated for the pure rod dispersion.

As the depletant concentration becomes significant and attractions play a dominating role f becomes sharply peaked. This is reflected in a strong increase of the nematic order parameter S (see the inset in Fig. 8.9). Hence, the Gaussian orientational distribution function becomes increasingly accurate at larger depletant concentrations.

Phase diagrams for $L/D = 20$, computed using the Gaussian f, are plotted in Fig. 8.10 for $q = 0.3, q = 1$ and $q = 2.5$. The upper plots are the reservoir depletant–rod representations, while the lower plots are the system representations. These three size ratios reflect different scenarios that are found in mixtures of spherocylinders and depletants when accounting for rods in the isotropic and/or nematic phase states. Depending on the length-to-width ratio of the rod-like particles and the ratio of the depletant diameter over the rod diameter, we find the following types of phase equilibria:

- coexistence between two isotropic phases (dilute and concentrated are the equivalent of vapour and liquid) and a nematic phase. This phase behaviour is predicted to occur for mixtures of relatively short rods and large depletants, so long-ranged attractions.
- coexistence between an isotropic and a nematic phase.
- equilibria between two coexisting nematic phases for rods mixed with small depletants, so short-ranged attractions.

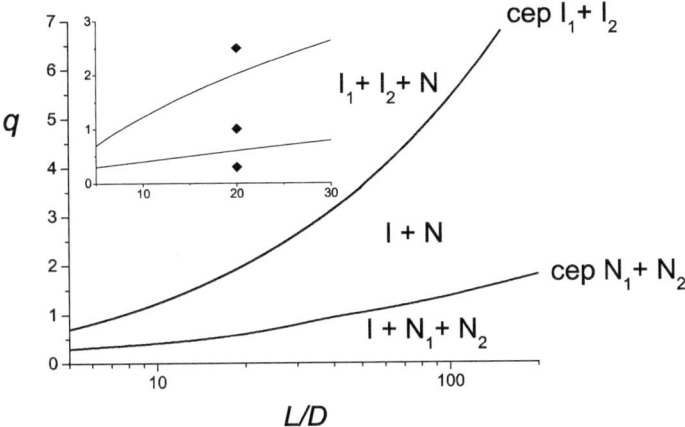

Fig. 8.11 Critical end point (cep) curves for isotropic–isotropic (I_1–I_2) and nematic–nematic (N_1–N_2) coexistence in dispersions of hard spherocylinders and PHSs as a function of the aspect ratio L/D. Computed using the Gaussian approximation for the orientational distribution function. *Inset*: magnified region for relatively small L/D

- coexistence between one isotropic phase and two nematic phases differing in concentration. This phase behaviour is predicted to occur for long rod-like particles and relatively small depletants.

A critical end point (CEP) exists for both the critical isotropic–isotropic and nematic–nematic points at given L/D. This CEP identifies the conditions for which a certain phase transition ceases to exist. The occurrence of the three different regimes as a function of the geometrical parameters L/D and q is shown in Fig. 8.11, as marked by the isotropic–isotropic and nematic–nematic critical end points. As a function of L/D the CEP values provide critical end curves. In the inset of Fig. 8.11, we have marked the conditions for which we plotted the phase diagrams in Fig. 8.10. The three types of phase behaviour are illustrated in Fig. 8.10 in a representation showing colloid volume fraction ϕ_1 against depletant concentration ϕ_2^R. Experimentally, one controls the depletant (for instance, nonadsorbing polymer) concentration in the system:

$$n_2 = -\frac{1}{V}\left(\frac{\partial \Omega}{\partial \mu_2}\right)_{N_2, V},\qquad(8.59)$$

rather than the polymer concentration (chemical potential) in the reservoir. Using the relation

$$\alpha = \frac{n_2}{n_2^R} = \frac{\phi_2}{\phi_2^R},\qquad(8.60)$$

phase diagrams in the experimentally accessible (ϕ_1, ϕ_2) plane can be obtained from the results in the (ϕ_1, ϕ_2^R) plane. The resulting phase diagrams are presented in Fig. 8.10 (lower diagrams).

8.5   Experimental Phase Behaviour of Rod–Polymer Mixtures

In this section, experimental results on the isotropic–nematic transition in mixed suspensions of colloidal rods and polymer are discussed and compared to the theory presented in the previous sections. The experimental results refer to three types of rod-like colloidal particles, which in suspension give rise to isotropic–nematic phase separation above a critical concentration:

- stiff and semiflexible virus particles,
- cellulose nanocrystals and
- colloidal boehmite (γ-AlOOH) rods.

In several experimental examples in this chapter, the rods are semiflexible. These are usually described using the worm-like chain model, (see, for instance, Refs. [47, 48]). In this model, the rod-like object has some flexibility by assuming a gradual change of the direction of the chain, which is in between the random walk character of a Gaussian chain and a rigid rod. This gradual change is described by assuming fluctuations in bond lengths and bond angles. In Fig. 8.12, $\mathbf{u}(s)$ is the direction vector of the chain at position s along the contour of the chain, and Δs is the angle (θ_P) between two direction vectors that are a distance apart. The persistence length l_P follows from:

$$\langle \mathbf{u}(s) \cdot \mathbf{u}(s + \Delta s) \rangle = \langle \cos \theta_P(\Delta s) \rangle = \exp\left(-\frac{\Delta s}{l_P}\right). \tag{8.61}$$

It follows that l_P is the characteristic length scale, on which the direction vector $\mathbf{u}$ of the chain varies.

The polymers added to the rod-like particles range from polysaccharides (heparin, chondroitin sulfate, dextran) to polyethylene oxide (PEO) and polystyrene (PS).

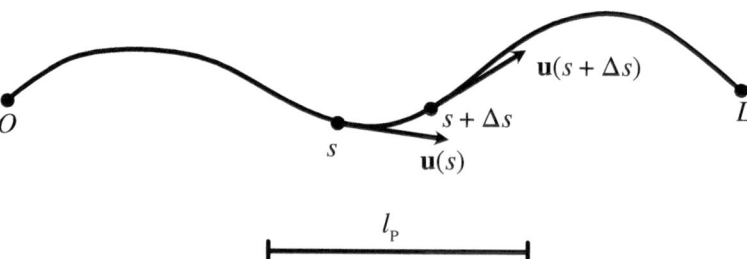

Fig. 8.12 An example of a configuration of a semiflexible rod described as a worm-like chain starting at position O, having a length L. The quantity $\mathbf{u}(s)$ is the direction vector of the chain at position s. For explanation of the other symbols, see the main text. Inspired by Fig. 8 in Ref. [31]

8.5.1 Stiff and Semiflexible Virus Particles Mixed with Polymer

Illustrative examples of rod-like colloidal particles are stiff and semiflexible virus particles, such as the plant virus tobacco mosaic virus (TMV) and the filamentous bacteriophages *fd*-virus, Pf1, Pf4. Table 8.1 summarises the characteristics of the virus particles discussed in this section.

Suspensions of TMV (see Fig. 8.13, taken from Ref. [56]) have long been recognised as an interesting system to study the I–N transition [4]. TMV is a rigid cylindrical particle consisting of a protein shell enclosing double-stranded RNA. Fraden et al. [46] measured the coexisting isotropic and nematic concentrations over a wide range of ionic strengths (Fig. 8.6).

It was as early as 1942 that Cohen [28] conducted a study directed at the isolation of TMV from infectious juice and observed that the addition of 5 mg/mL of the polysaccharide heparin to a dilute TMV suspension (2 mg/mL) in 0.1 M phosphate buffer (pH = 7.1) resulted in the production of needle-shaped paracrystals 5–20 μm in length (see Fig. 1.11 in Ref. [28]). These crystals may be considered as precursors of the I–N transition [4]. In the 1990s, Sano and co-workers [56–58] added the polysaccharide chondroitin sulfate (Chs) to dilute TMV suspensions with a view to establish the antiviral activity of these polysaccharides. With electron microscopy, Urakami et al. [56] observed that the addition of very low concentrations of Chs (1 mg/mL) to dilute TMV suspensions (1 mg/mL) caused the formation of large raft-

Table 8.1 Characteristics of TMV [46], (wild-type) *fd* [19,49–51], Pf1 [50,52,53] and Pf4 [54,55] virus particles. Acknowledgements to A. Tarafder, T. Bharat, P. Secor and P. Janmey for their help with compiling this table

Virus	M_p (kDa)	L (μm)	D (nm)	L/D	l_P (μm)
TMV	$4 \cdot 10^4$	0.3	18	17	$> 10L$
fd	$1.64 \cdot 10^4$	0.88	6.6	133	$2.5 \approx 2.8L$
Pf1	$3.5 \cdot 10^4$	2	6	333	$2 \approx L$
Pf4	$6.08 \cdot 10^4$	3.8	6	633	$2 \approx 0.5L$

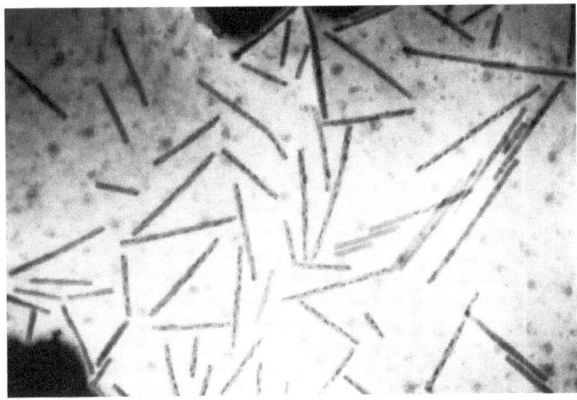

Fig. 8.13 TEM micrograph of TMV particles. Reprinted with permission from Ref. [56]. Copyright 1999 AIP Publishing

Fig. 8.14 TEM micrograph of ordered TMV particles as induced by added chondroitin sulfate. Reprinted with permission from Ref. [56]. Copyright 1999 AIP Publishing

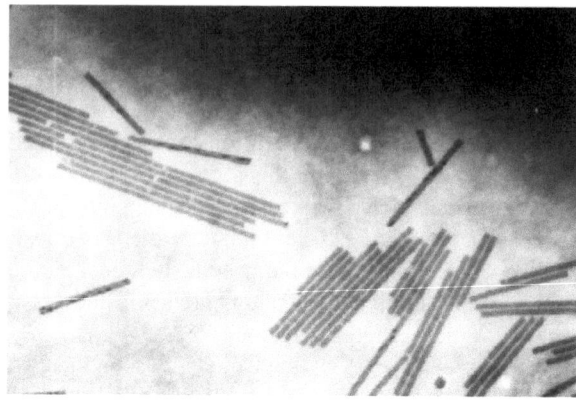

Fig. 8.15 I–N phase coexistence for hard spherocylinders ($L/D = 17$) mixed with PHSs, mimicking TMV mixed with polymers for size ratios $q = 0.4$ (solid curves) and $q = 1.1$ (dot-dashed curves)

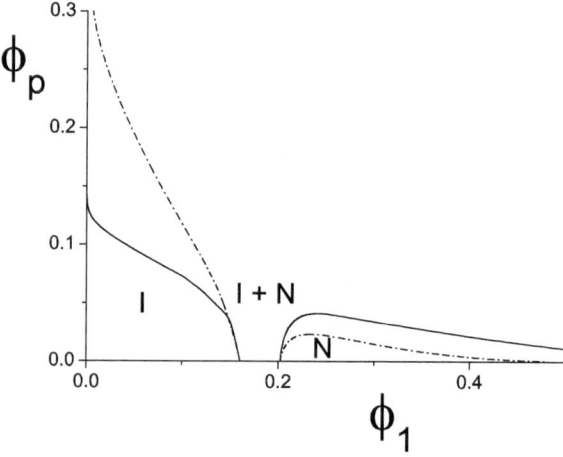

like aggregates (Fig. 8.14). The effect of Chs on infectivity may, according to Sano [57], be ascribed to these raft-like aggregates blocking the decapsulation process of TMV protein on the cell membrane surface. The fact that very low Chs concentrations lead to aggregation of TMV is attributed to its semirigidity [56,58].

Leberman [29] observed that addition of 6 mg/mL of the flexible polymer polyethylene oxide (PEO) ($M = 6$ kDa, $R_g = 3.6$ nm) to a dilute 1 mg/mL TMV suspension leads to precipitation of TMV, which may be considered as a sign of the I–N transition. Figure 8.15 presents a comparison of this experimental observation with the theoretical phase diagram for $L/D = 300/18 = 17$ and $q = 2R_g/D = 2 \cdot 3.6/18 = 0.4$, which are the relevant parameters for this mixed TMV–PEO suspension.

From this calculated phase diagram, we observe that at low TMV concentrations a relative polymer concentration $\phi_p = 0.125$ is required to cause I–N phase separation, which corresponds in this case to a mass concentration of $c_p = 3\phi_p M/4\pi N_{Av} R_g^3 = 6.4$ mg/mL. The agreement with the experiment should be considered with care since the electrostatic interactions have not been taken into account in the theoretical calculation.

More extensive measurements on the I–N transition in TMV suspensions with added PEO ($M = 100$ kDa, $R_g = 10$ nm) were carried out by Adams and Fraden [49]. At TMV concentrations of 20 mg/mL (where the pure rod system is in the isotropic phase), they observed the first signs of I–N phase separation at 5 mg/mL added PEO and a more definite phase transition for 10 mg/mL added PEO. For a direct comparison with the experimental observation, Fig. 8.15 also presents the theoretical phase diagram for $L/D = 17$ (TMV as before) but now with $q = 2R_g/D = 2 \cdot 10/18 = 1.1$, which are the relevant parameters for this mixed TMV–PEO suspension. From this calculated phase diagram, we observe that at low TMV concentration a relative polymer concentration $\phi_p = 0.25$ is required to cause I–N phase separation, which corresponds in this case to a mass concentration of $c_p = 3\phi_p M/4\pi N_{Av} R_g^3 = 10$ mg/mL. This is again in reasonable agreement with theory. As mentioned before, the electrostatic interactions that certainly play a role have not been taken into account, and therefore the comparison with experiment should be considered with care.

In addition to TMV, the liquid crystal phase behaviour of the semirigid cylindrical *fd*-virus has been investigated extensively. The *fd*-virus particle consists of a protein shell wound around a single ribbon of single-stranded DNA [59]. Figure 8.16 shows an AFM image of some *fd*-viruses. Near a neutral pH the linear charge density is -5 to -20 e/nm. Fraden and co-workers [50] measured the coexisting isotropic and (chiral) nematic concentrations over a wide range of ionic strengths of the wt *fd*-virus. The onset of the (chiral) nematic phase occurs from 10 to 20 mg/mL of *fd*-virus as the ionic strength is increased from 1 to 100 mM. Dogic et al. [60,61] studied the phase diagram of mixed suspensions of *fd* and dextran ($M_p = 500$ kDa, $R_g = 18$ nm) and an example is plotted in Fig. 8.17a.

A clear widening of the I–N transition of the *fd*-virus rod dispersion takes place upon increasing the dextran concentration. Although this finding is quite general for the I–N binodal, the corresponding spinodal points related to this phase transi-

Fig. 8.16 Image of *fd*-virus mutant type Y21M particles dried on a mica surface made using AFM. Scale and depth indicated. Image kindly provided by O. Deschaume, M.P. Lettinga and C. Bartic, KU Leuven, Belgium

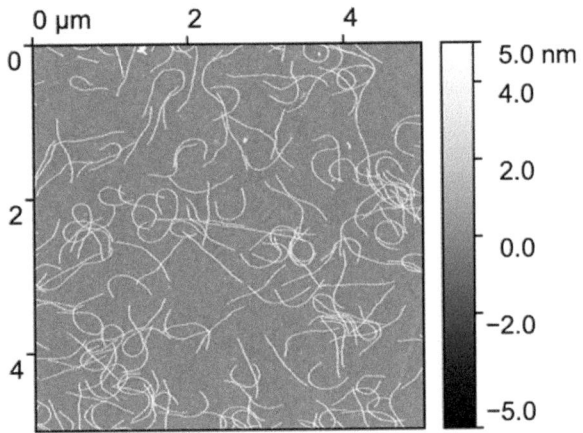

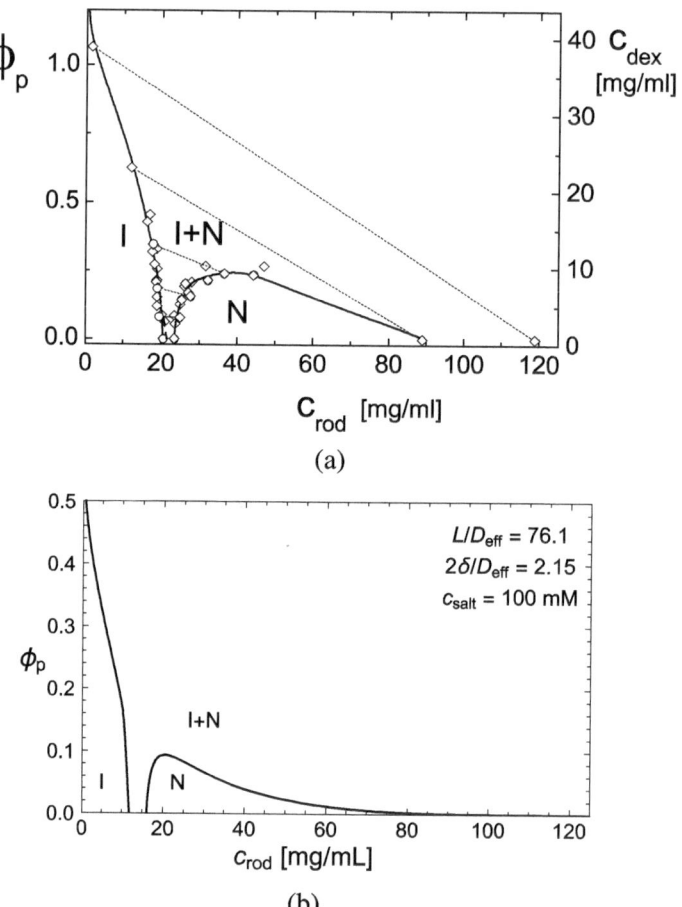

Fig. 8.17 **a** Phase diagrams of dispersions containing *fd*-virus and dextran. Data points are measured phase coexistences. The phase diagram was measured at an ionic strength of 100 mM at pH = 8.15 with added 500 kDa dextrans. Data replotted from Ref. [61]. **b** Predicted I–N phase coexistence for spherocylinders ($L/D = 133$, $D_{eff} = 11.57$ nm, $L/D_{eff} = 76.1$, ionic strength of 100 mM) mixed with PHSs ($2\delta/D_{eff} = 2.15$). The dextran polymer has $R_g = 17.6$ nm and the charge density is taken as -10 *e*/nm

tion seem to be much less affected by adding nonadsorbing polymers [62]. At low *fd*-virus concentrations, the I–N transition takes place upon adding large polymer concentrations.

For direct comparison with the experimental phase diagram, we present the theoretical phase diagram for $L/D = 880/6.6 = 133$ and $q = 2R_g/D = 2 \cdot 11/6.6 = 3.3$ in Fig. 8.17b. For detailed accounts of the ideal polymer chains, see Refs. [45,61]. The overall agreement between theory and experiment (compare Figs. 8.17a and 8.17b), while far from perfect, is satisfactory considering that *fd* is not completely rigid and that dextrans are branched polymers. The rod flexibility is known to have a significant effect on the I–N phase behaviour [31]. This is demonstrated by the inter-

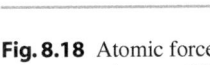

Fig. 8.18 Atomic force
microscopy image of Pf1
viruses in monovalent salt.
Image kindly provided by
P.A. Janmey from their study
reported in Ref. [64]

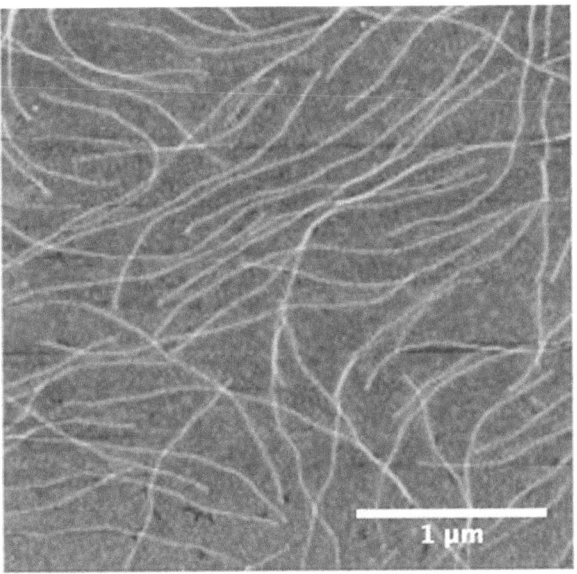

esting work of Barry, Beller and Dogic [63], who compared the phase behaviour of
a mutant filamentous virus, *fd* Y21M (Fig. 8.16), to that of a conventional *fd* wt.
The persistence length of *fd* wt is 2.8 ± 0.7 μm, whereas the persistence length of *fd*
Y21M is 9.9 ± 1.6 μm. Compared to *fd* wt, the location of the isotropic–cholesteric
phase transition for *fd* Y21M shifts to lower densities and approaches values that are
remarkably close to the Onsager prediction for rigid rods.

The filamentous bacteriophages Pf1 [50,53,64] and Pf4 [54,55] are structurally
similar to *fd*-virus. An atomic force microscopy image of the filamentous bacteriophage Pf1 [64] is presented in Fig. 8.18.

Booy and Fowler [65] observed small domains of smectic organisation (cybotactic
clusters [2]) in a nematic phase in suspensions of Pf1 at a concentration of 40 mg/mL.
Using optical microscopy Dogic and Fraden [50] observed coexisting regions of the
nematic and smectic phases in suspensions of Pf1 with abrupt boundaries between
the phases, which is evidence of a first-order phase transition.

So far, no liquid crystal phases in pure suspensions of Pf4 have been reported; but
Secor et al. [54] and Tarafder et al. [55] observed liquid crystal tactoids in suspensions
of Pf4 upon adding sodium alginate. In Fig. 8.19, we show the concentrations of Pf4
and sodium alginate where the tactoids appear [54].

Tarafder et al. [55] provided beautiful fluorescence microscopy images of Pf4
tactoids observed after mixing Pf4 with sodium alginate (Fig. 8.20a). The key role
of the tactoids is that they can encapsulate the pathogenic bacterium *Pseudomonas
aeruginosa*, shielding them from antibiotics [54,55,66]. Illustrative fluorescence
microscopy images of this encapsulation (Fig. 8.20b), can be found in [55].

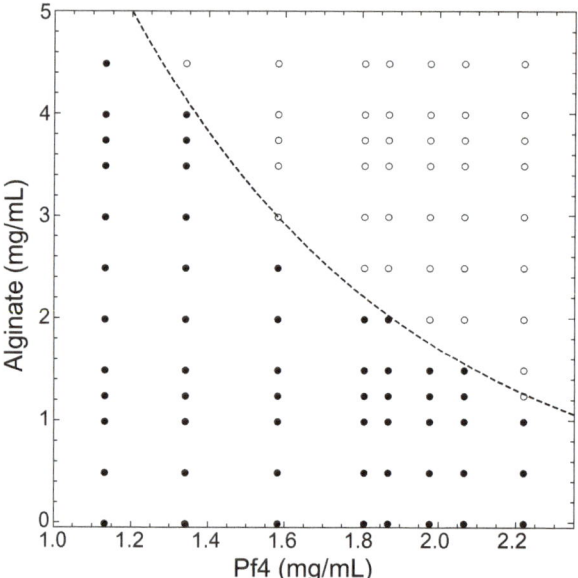

Fig. 8.19 Phase diagram indicating the stable (one phase) region (closed symbols) and unstable demixing region (open symbols) in terms of the polymer (alginate) concentration and virus (Pf4) concentration. The dotted curve that indicates the phase transition concentrations is drawn to guide the eye. Replotted from the data in Ref. [54]

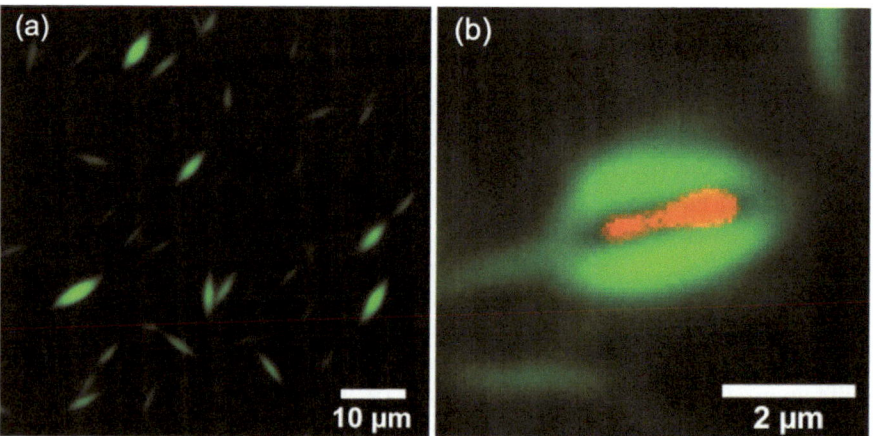

Fig. 8.20 Fluorescence microscopy images of Pf4 tactoids. **a** Tactoids observed in a mixture of 1 mg/mL Pf4 and 10 mg/mL sodium alginate 24 h after mixing and **b** tactoids of Pf4 encapsulate *Pseudomonas Aeruginosa* [55]. Figures were kindly provided by A.K. Tarafder and T.A.M. Bharat

8.5.2   Cellulose Nanocrystals Mixed with Nonadsorbing Polymers

In 1959, Marchessault, Morehead and Walter [67] reported on the formation of liquid crystals in suspensions of cellulose nanocrystals prepared from cellulose by acid hydrolysis in sulfuric acid (see Fig. 8.1m for a microscopy image of cellulose nanorods). The study of the isotropic–(chiral) nematic phase transition in suspensions of cellulose nanocrystals [68] has since developed into a blossoming and fruitful field of research (for an overview see Ref. [69]. See also [70,71] for more recent work). Edgar and Gray [72] studied the effect of 2000 kDa dextran ($R_g = 34$ nm) on the phase behaviour of cellulose nanocrystals (average length $L = 110$ nm, average diameter $D = 10$ nm), prepared by acid hydrolysis of cotton filter paper.

In Fig. 8.21, we redraw the I–N phase behaviour at low dextran concentrations. Above 7 wt % suspensions of these cellulose nanocrystals start to phase separate in an isotropic and chiral nematic liquid crystal phase. At 13.3 wt %, the relative volume of chiral nematic phase (compared to the total volume) is 79%. This wide biphasic range is a direct consequence of the polydispersity of the cellulose nanocrystals [35,73] and has been observed in other dispersions containing polydisperse rod-like colloids as well [15,74]. When dextran was added to the biphasic region, it led to a significant broadening of the coexistence region and the dextran preferentially partitions in the isotropic phase. These features are in agreement with the theory described in Sect. 8.4.

8.5.3   Sterically Stabilised Colloidal Boehmite Rods Mixed with Polymer

As mentioned in the introduction of this chapter, suspensions of rod-like inorganic colloids were the first systems in which the I–N transition was observed. In the early 1960s, Zocher and Török [75–77] and Bugosh [78] observed interesting liquid

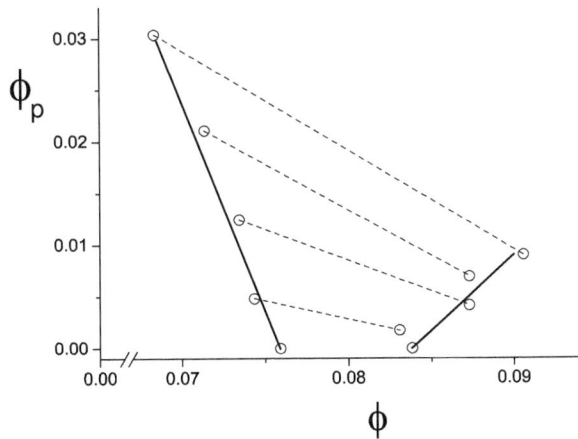

Fig. 8.21 Influence of blue dextran ($R_g/D = 3.4$) concentration (normalised to ϕ_p) on the isotropic–nematic phase coexistence in dispersions of rod-like cellulose nanorods ($L/D = 11$) with volume fraction ϕ. Replotted from Edgar and Gray [72]

Fig. 8.22 TEM micrograph of boehmite rods. Image kindly provided by J. Buitenhuis, Forschungszentrum Jülich, Germany

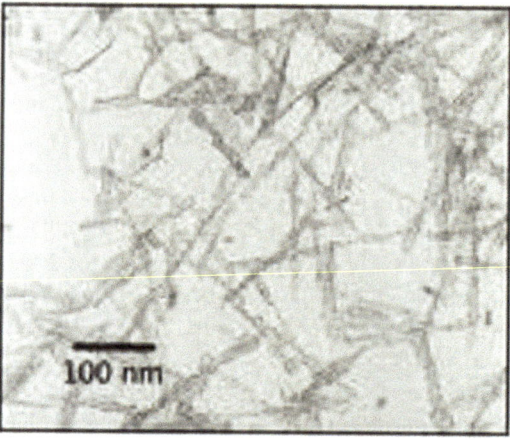

Fig. 8.23 Triphasic I_1–I_2–N equilibrium in dispersions of boehmite rods and polystyrene chains in *ortho*-dichlorobenzene [79]. Image kindly provided by J. Buitenhuis, Forschungszentrum Jülich, Germany

crystal phase behaviour in aqueous dispersions of colloidal boehmite rods, shown in Fig. 8.22.

Later, Buining and Lekkerkerker [74] observed isotropic–nematic phase separation in a dispersion of sterically stabilised boehmite rods, which approximate hard rods, in cyclohexane. Buitenhuis et al. [79] studied the effect of added 35 kDa polystyrene ($R_g = 5.9$ nm) on the liquid crystal phase behaviour of sterically stabilised boehmite rods with average $L = 71$ nm and average $D = 11.1$ nm in *ortho*-dichlorobenzene. Different phase equilibria were observed: two biphasic I–N equilibria (both dilute isotropic phase I_1 with nematic N and concentrated isotropic phase I_2 with nematic N) and a triphasic equilibrium I_1–I_2–N (Fig. 8.23). In this system, the boehmite rods are quite polydisperse. Therefore, comparison with theory should be done with an approach that includes polydisperse rods (see, for instance, Refs. [33, 80]). We further note no I_1–I_2 coexistence was observed experimentally but rather an I_1-gel at high polymer concentrations. The depletant-mediated appearance of a nonequilibrium long-lived metastable state such as a gel resembles the behaviour of colloidal sphere/polymer mixtures (see Chap. 4).

8.6   Phase Diagrams of Rod/polymer Mixtures Including Highly Ordered Phases

In the previous sections we focused on the isotropic–nematic phase transition in mixed suspensions of rod-like colloids and flexible polymers. In Sect. 8.3, it was shown that scaled particle theory (SPT) provided the pressure and chemical potential of the hard spherocylinder reference system. SPT also enabled a route for the calculation of the free volume fraction, the quantities required in the FVT calculation of the phase diagram (see Sect. 8.4). Depending on L/D and δ/D, three types of phase diagrams were obtained (presented in Fig. 8.10). For intermediate q values, a significant broadening of the I–N biphasic region was obtained. For large q values an isostructural I_1–I_2 transition arises in addition, while for small q values an additional N_1–N_2 transition arises. The broadening of the biphasic I–N region and also a triphasic I_1–I_2–N equilibrium have indeed been observed experimentally in mixed suspensions of rod-like colloids and flexible polymers [60,72,79]. We noted that the theoretical prediction of the N_1–N_2 transition (which, so far, has not been observed experimentally in mixed suspensions of rod-like colloids and flexible polymers) should be treated with reservation.

The N_1–N_2 phase transition is predicted to occur at quite high volume fractions of rods. At these high volume fractions, the N_1–N_2 transition may be superseded by more highly ordered (liquid) crystal phases such as the colloidal smectic phase. Experimentally, this colloidal smectic phase has been observed [24,50,81] in suspensions of monodisperse rods. Simulations confirmed that hard rods can form a thermodynamically stable smectic phase [25–27,82]. In this section, we outline how the more dense highly ordered phases can be accounted for in the phase diagram of mixtures of rod-like colloids and flexible polymers using FVT.

8.6.1   Full Phase Diagrams of Hard Spherocylinders

Computer simulation results of suspensions containing hard spherocylinders [25–27] revealed that, with increasing concentration, isotropic (I), nematic (N), smectic–A (SmA), AAA crystal and ABC crystal phases appear as preferred phase states (Fig. 8.3a).

As discussed in Sect. 8.2.1, the rods have a random orientation in phase I, while, in the other phases, they are aligned along a common nematic director. Both I and N phases are fluids and have no long-range positional order. While SPT [43] provides a reasonable equation of state for long rods, Parsons–Lee (PL) theory [83–85] is more accurate for short rods. PL theory is basically an extension of the Carnahan–Starling equation of state and is discussed in more detail in Sect. 9.2.2. See also [31]. To accurately cover the full range of aspect ratios, Peters et al. [20] combined SPT and PL using a sigmoidal interpolation procedure. In the SmA, AAA and ABC phases, the particles are confined in layers and the nematic director is perpendicular to the layers. For the SmA phase, there is, however, still no positional order within the layers, while, in the AAA and ABC phases, the particles are ordered hexagonally. In the AAA phase, the rods of adjacent layers are stacked on top of each other, while, in the ABC phase, they are stacked in between the rods of adjacent layers. For all three phase states, an extended cell theory was developed by Peters et al. [20], based upon an approach proposed by Graf and Löwen [86]. The final results were cast into algebraic equations [20], which are summarised below.

In general, for any of these phases, the free energy can be split into an ideal, orientational and packing contribution:

$$\widetilde{F} = \frac{F v_0}{kTV} = \widetilde{F}_{id} + \widetilde{F}_{or} + \widetilde{F}_{pack}, \tag{8.62}$$

with $\widetilde{F}_{id} = \phi \ln(\phi \Lambda^3 / v_c) - \phi$ being independent of the phase state and $\widetilde{F}_{or} = \mathfrak{s}[f]$ given by Eq. (8.3). For the SmA phase, the cell model includes a thermodynamic description of 2D discs with area fraction ϕ_{2D} that captures the in-plane fluidity of rods projected onto the smectic plane. The expression for the packing free energy $\widetilde{F}_{pack}$ of the smectic phase reads

$$\frac{\widetilde{F}_{pack}^{SmA}}{\phi} = -\ln\left(1 - \phi_{2D}\bar{D}_{eff}^2\right)$$
$$+ \frac{\phi_{2D}\bar{D}_{eff}^2}{1 - \phi_{2D}\bar{D}_{eff}^2} - \ln\left(1 - \frac{\Gamma}{\widetilde{\Delta}_\perp}\right). \tag{8.63}$$

The quantities $\bar{D}_{eff}$ and $\bar{\Delta}_\perp$ in Eq. (8.63), as well as $\widetilde{F}_{or}$, can be determined by simultaneously minimising the total free energy with respect to $f(\boldsymbol{\Omega})$ and $\bar{\Delta}_\perp$ (see Ref. [20] for details). The effective rod diameter $\bar{D}_{eff}$ can be calculated using

$$D_{eff} = 1 + A(\Gamma - 1) \int f(\boldsymbol{\Omega})|\sin(\theta)|d\boldsymbol{\Omega}, \tag{8.64}$$

where A was chosen such as to fit the resulting equations of state and nematic–smectic-A phase transitions to those obtained from computer simulations [26,27]. The quantity A varies depending on whether the equations of state for the nematic phase is based on SPT ($A = 0.41\phi$) or PL ($A = 0.28\phi$). For the sigmoidal interpretation approach $A = 0.41\phi h$, with $h = g + (1 - g)0.28/0.41$, with g defined by [20,21]

$$g = \frac{1}{1 + e^{\Gamma_t - \Gamma}}, \tag{8.65}$$

with $\Gamma_t = 6$, connecting the Onsager limit ($\Gamma \to \infty$) and the sphere limit ($\Gamma \to 1$).

For the AAA phase, Peters et al. [20] derived

$$\widetilde{F}_{pack}^{AAA} = \phi + \phi \ln \phi_{cp}^{AAA} - \phi \ln \left(\bar{\Delta}_\parallel^{AAA} - \bar{D}_{eff}\right)^2$$
$$- \phi \ln \left(\frac{\phi_{cp}^{AAA}/\phi}{(\bar{\Delta}_\parallel^{AAA})^2} - 1\right), \tag{8.66}$$

where for AAA the parameter $A = 0.225\phi h$ in the definition of $\bar{D}_{\text{eff}}$ was chosen based on comparison with simulation results for the equations of state and the AAA–ABC phase transition [26,27]. The resulting normalised layer spacing $\bar{\Delta}_{\parallel}$ reads

$$\bar{\Delta}_{\parallel}^{\text{AAA}} = \frac{6^{1/3}\left(\frac{\phi_{\text{cp}}^{\text{AAA}}}{\phi}\right) + \left(9\frac{\phi_{\text{cp}}^{\text{AAA}}}{\phi} + \frac{\phi_{\text{cp}}^{\text{AAA}}}{\phi}\sqrt{3\left(27 - 2\frac{\phi_{\text{cp}}^{\text{AAA}}}{\phi}\right)}\right)^{2/3}}{6^{2/3}\left(9\frac{\phi_{\text{cp}}^{\text{AAA}}}{\phi} + \frac{\phi_{\text{cp}}^{\text{AAA}}}{\phi}\sqrt{3\left(27 - 2\frac{\phi_{\text{cp}}^{\text{AAA}}}{\phi}\right)}\right)^{1/3}}. \tag{8.67}$$

The spacing parameter $\Delta_{\parallel}$ is normalised through $\bar{\Delta}_{\parallel} = \Delta_{\parallel}/D$.

For the ABC crystal phase, an extended cell theory approach leads to the following expression [20,86]:

$$\frac{\tilde{F}_{\text{pack}}^{\text{ABC}}}{\phi} = 1 + \ln\phi_{\text{ref}} - \ln\left(\bar{\Delta}_{\parallel}^{\text{ABC}} - \bar{D}_{\text{eff}}\right)^2 - \ln\left(\frac{\phi_{\text{ref}}/\phi}{(\bar{\Delta}_{\parallel}^{\text{ABC}})^2} - 1\right), \tag{8.68}$$

with

$$\bar{\Delta}_{\parallel}^{\text{ABC}} = \frac{6^{1/3}\left(\frac{\phi_{\text{ref}}}{\phi}\right) + \left(9\frac{\phi_{\text{ref}}}{\phi} + \frac{\phi_{\text{ref}}}{\phi}\sqrt{3\left(27 - 2\frac{\phi_{\text{ref}}}{\phi}\right)}\right)^{2/3}}{6^{2/3}\left(9\frac{\phi_{\text{ref}}}{\phi} + \frac{\phi_{\text{ref}}}{\phi}\sqrt{3\left(27 - 2\frac{\phi_{\text{ref}}}{\phi}\right)}\right)^{1/3}},$$

and

$$\phi_{\text{ref}} = \frac{\pi\left(3\Gamma - 1\right)}{6\left(\sqrt{3}\left(\Gamma - 1\right) + B\sqrt{2}\right)},$$

including the parameter $B = 1.16$. Next, the accuracy of these expressions is compared to computer simulation results.

For $L/D = 4$, the rod concentration dependence of the resulting theoretical osmotic pressures (curves) of the I, N, SmA and ABC phase states are plotted in Fig. 8.24. The predictions are compared to computer simulation data of McGrother et al. [26]. It is clear the equations of states correspond reasonably well to the computer simulation data. For comparisons at other aspect ratios, see Ref. [20].

Phase coexistence between two phases can be established by imposing mechanical and chemical equilibrium expressed by equality of osmotic pressure P and chemical potential μ (Appendix A). Algebraic expressions for P and μ were derived [20] for all phase states of the hard spherocylinders from the free energy expressions given above.

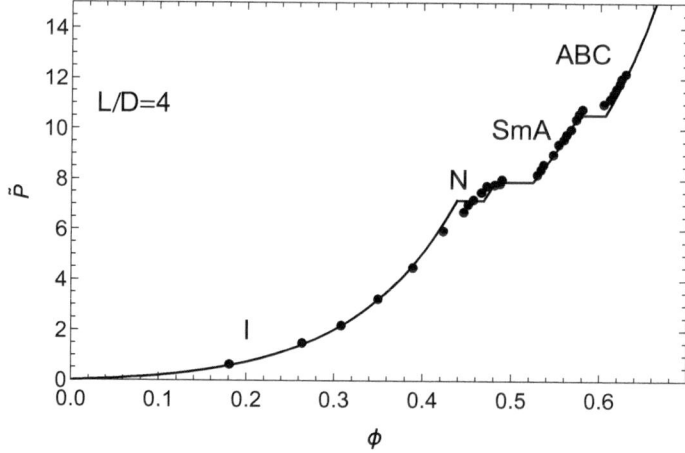

Fig. 8.24 Osmotic pressure of hard spherocylinders with $L/D = 4$ as a function of rod volume fraction ϕ from both theory (solid curves) [20] and computer simulation results (data points) [26]. The stable phases include the isotropic (I), nematic (N), smectic-A (SmA), AAA crystal and ABC crystal phases

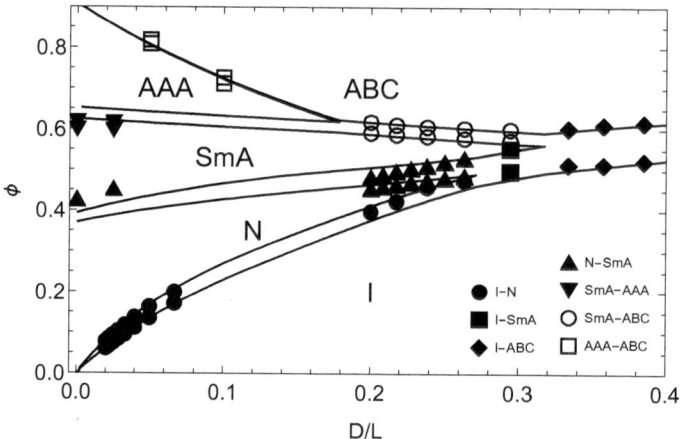

Fig. 8.25 Phase behaviour of hard spherocylinders as a function of rod volume fraction ϕ and aspect ratio D/L from both theory (solid curves) [20] and simulation (data points) [27]. The stable phases include the isotropic (I), nematic (N), smectic-A (SmA), AAA crystal and ABC crystal phases

The theoretical predictions [20] for the binodals using the analytical equations of states for all hard spherocylinder phase states discussed are shown in Fig. 8.25 (solid curves) as a function of volume fraction ϕ and aspect ratio D/L. Computer simulation data of Bolhuis and Frenkel [27] are plotted as data points. The phase diagram of rods without endcaps (see Ref. [82]) is very close to these results.

It is noted (not shown) that in the sphere limit of $L/D \to 0$ the equations of state for the isotropic and ABC phases become equivalent to those of a hard-sphere fluid and an FCC crystal, respectively (see Sect. 3.2.3). The agreement with the computer simulation results is quite reasonable for these phase equilibria.

8.6.2 Phase Behaviour of Rod–Polymer Mixtures Including Highly Ordered Phases

Next, the phase behaviour of mixtures of rod-like colloidal particles and polymers is considered, including the highly ordered phases discussed in the previous subsection. The rods are again described as hard spherocylinders and the nonadsorbing polymers are treated as ideal depletants by describing them as PHSs with radius δ (see Chaps. 2 and 3). It is possible to explicitly include polymer–polymer interactions (see Chap. 4), which demonstrates that the PHS approximation for the polymer works well for the relatively small polymers discussed here. Hence, phase diagrams are discussed for mixtures of hard spherocylinders and PHSs.

The mixture is again described using FVT. The hard spherocylinder–PHS system of interest is in contact with a PHS reservoir through a semi-permeable membrane that is impermeable to the colloids but fully permeable for the polymers. Solvent plays the role of continuum background again.

We use Eq. (8.43) for Ω, which we rewrite here as

$$\widetilde{\Omega} = \widetilde{F} - \alpha \widetilde{\Pi}^{R}, \tag{8.69}$$

with $\widetilde{\Omega} = \Omega v_c / kTV$ as the normalised semi-grand potential for the system of interest, and $\widetilde{F} = F v_c / kTV$ as the normalised Helmholtz free energy for a pure dispersion of hard spherocylinders. The free volume fraction $\alpha = \langle V_{\text{free}} \rangle / V$ is the average fraction of the system volume available to PHSs, and $\widetilde{\Pi}^{R}$ is the normalised osmotic pressure of the polymers in the reservoir, which is proportional to the reservoir polymer coil volume fraction ϕ_p^R by $\widetilde{\Pi}^{R} = \phi_p^R (3\Gamma - 1)/(2q^3)$.

A representative set of phase diagrams for colloid–polymer mixtures is shown in Fig. 8.26 in terms of the polymer reservoir concentration ϕ_p^R versus the rod volume fraction ϕ (see Refs. [21,87,88]). The rod aspect ratio is fixed at $L/D = 12$ and polymer–rod size ratio is set to $q = 0.4, 0.525$ and 0.57. In the plots, the binodals (solid curves) and three- and four-phase coexistences (dashed lines) are shown. In most cases, the miscibility gaps widen as the polymer concentration is increased. At the points where two binodals coincide, there is three-phase coexistence. For instance, at $q = 0.4$ (left panel) the miscibility gap of N–SmA and SmA–AAA coexistence widens as ϕ_p^R increases. At around $\phi = 0.4$–0.65 and $\phi_p^R \approx 0.05$, the binodals coincide and a triple N–SmA–AAA equilibrium emerges.

Increasing the polymer size qualitatively changes the phase diagram. The trends are similar to those reported by Savenko and Dijkstra [89] in their Monte Carlo simulation study. For example, at $q = 0.57$ (right panel) the N–SmA binodal coincides

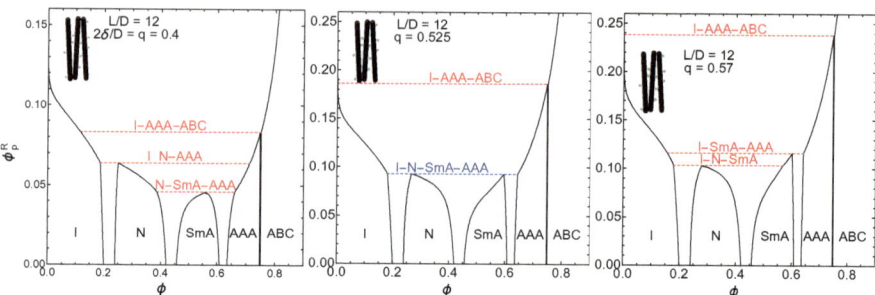

Fig. 8.26 Phase diagrams of colloid–polymer mixtures in terms of the colloid volume fraction ϕ and polymer reservoir concentration ϕ_p^R for colloidal rods of aspect ratio $L/D = 12$ and polymers of size $q = 2\delta/D = 0.4$ (*left*), 0.525 (*middle*), and 0.57 (*right*). Binodals are displayed as solid curves, while three- and four-phase coexistences are indicated as dashed lines. Reprinted with permission from Ref. [87] under the terms of CC-BY-4.0

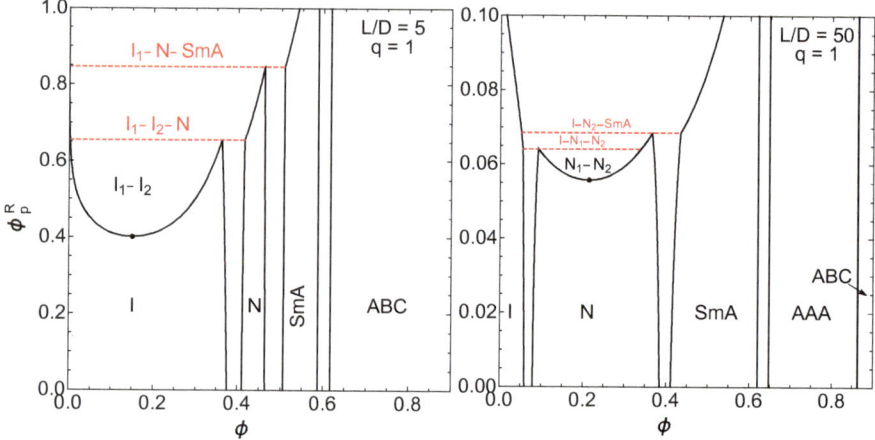

Fig. 8.27 Phase diagrams of colloid–polymer mixtures in a similar representation as Fig. 8.26, but for parameters exhibiting isotropic I_1–I_2 (*left*) and N_1–N_2 (*right*) phase coexistence. The colloidal rods have an aspect ratio $L/D = 5$ (*left*) and $L/D = 50$ (*right*) and the polymers have a fixed size $2\delta = D$, so $q = 1$. Binodals are displayed as solid curves, while three-phase coexistences are indicated as dashed lines

with the I–N binodal and instead leads to a triple I–N–SmA coexistence. Similarly, the N–AAA and I–N–AAA coexistences are only present at the smaller $q = 0.4$, while the I–SmA binodals and I–SmA–AAA triphasic coexistence are only stable at $q = 0.57$. The intermediate polymer size of $q = 0.525$ (middle panel) marks the exact size ratio where all three binodals coincide at the same polymer reservoir concentration. This leads to an I–N–SmA–AAA four-phase coexistence.

Next, the isostructural phase coexistences [88] discussed in Sect. 8.4 are re-evaluated. In Fig. 8.27, two examples are given for $L/D = 5$ and $L/D = 50$ and $q = 1$. As could be expected, the I_1–I_2 coexistence region (left panel) is large for sufficiently large polymer sizes compared to the rod length. For $L/D = 50$ and $q = 1$ (right panel), a region where N_1–N_2 phase equilibria are predicted appears.

This is not superseded by I_1–SmA_2 or I–crystalline phase equilibria. So, although these N_1–N_2 phase equilibria have not been observed they also appear as results from theoretical calculations when taking higher ordered phases of the rods into account. Moreover, the calculations reveal that (non-metastable) isostructural phase equilibria are possible for all phase states [21].

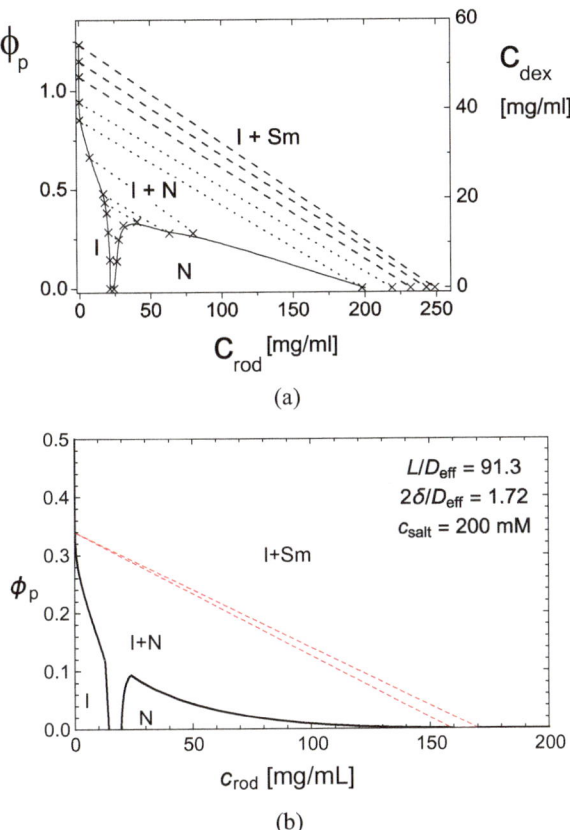

Fig. 8.28 Phase diagrams of dextran/fd-virus mixtures. Data points are measured coexistences at an ionic strength of 200 mM and pH = 8.10. In addition to the isotropic–nematic coexistence (dotted coexistence lines), isotropic–smectic coexistence (dashed lines) is found at high polymer and rod concentrations. Replotted from Ref. [90]. **b** Predicted phase diagram of a hard spherocylinder–penetrable hard-sphere dispersion. Spherocylinders were modelled with $L/D = 133$, $D_{eff} = 9.64\,nm$, $L/D_{eff} = 91.31$, an ionic strength of 200 mM and mixed with PHSs with $2\delta/D_{eff} = 1.72$. The dextran polymers have $R_g = 11\,nm$, the charge density is taken as $-10\,e$/nm to mimic the system of **a**. Solid curves represent I–N phase equilibria. The dashed lines represent I–N–Sm triple coexistence. Above the triple region, there is I–Sm phase coexistence

8.6.3   Comparison with Experiments

We now compare predictions including higher ordered phases with the limited number of experiments that have been reported in the literature. Dogic [90] extended the earlier work of Dogic and Fraden [60] on mixed suspensions of *fd* and dextran to higher dextran concentrations at a salt concentration of 200 mM. The phase diagram he observed is plotted in Fig. 8.28a. Above dextran concentrations of 55 mg/mL the I–N transition is superseded by the I–SmA transition. In Fig. 8.28b the predicted phase diagram is plotted for the relevant size parameters. The effective diameter of the rods was calculated using the theory presented in Sect. 8.2.2. The phase diagram computed corresponds to the experimental one. No observations were reported on the (narrow) triphasic I–N–SmA equilibrium that is expected between the biphasic I–N and I–SmA phase equilibria.

An interesting aspect of bacteria is that they can assume a wide range of shapes [91,93], examples of which are shown in Fig. 8.29. In 1954 Goldacre [92] showed that, like viruses, some bacteria can be crystallised. The bacterial cells form regular three-dimensional arrays, in which each cell corresponds to a molecule in a conventional crystal. In the case of rod-shaped bacteria, the rods align in a parallel fashion, as shown in Fig. 8.30 for *Amoeba proteus* [92].

Experimental work [94–96] demonstrated that, upon exceeding a certain concentration, suspensions of (non-motile) bacteria and nonadsorbing polymers exhibit phase separation, just as colloid–polymer mixtures. Guided by the ideas of Goldacre [92], we apply free volume theory to describe this phase separation. We highlight the work presented by Schwarz-Linek et al. [95], who focused on mixtures of *Escherichia coli* (*E. coli*) (Fig. 8.31) and sodium polystyrene sulfonate (NaPSS).

The added NaPSS polymers have a molar mass of 64.7 kDa and a radius of gyration of 14 nm. The mixtures were studied in aqueous solutions containing 0.18 M salt, at which the Debye screening length is 0.8 nm. In Fig. 8.32, we present results of the phase behaviour of a suspension of *E. coli* bacteria with a volume fraction of 12.5% with different polymer weight fractions ranging from 0 to 10%.

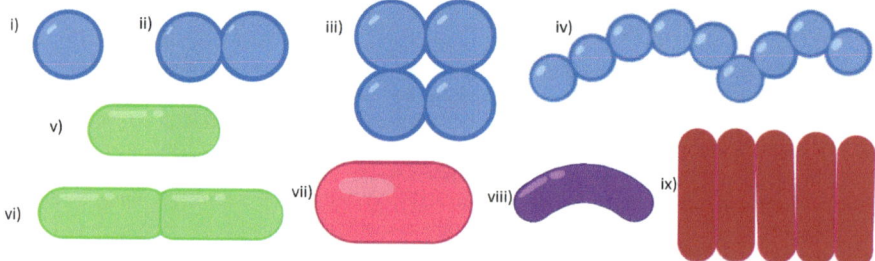

Fig. 8.29 Examples of bacteria and their shapes. Cocci have a spherical to ovoid shape and appear not only as single cells (i) but also as pairs (e.g., Diplococci) (ii), clusters (e.g., tetrad) (iii) or chains (e.g., Streptococci) (iv). Bacilli (v) and Diplobacilli (vi) are rod-shaped bacteria. Coccobacilli (vii) resemble cocci and bacilli. Spiral bacteria (viii) are slightly curved microbes with a comma shape. Pallisades are bacteria with a picket fence structure of connected rods (ix). Sketches made by Luuk Tuinier, inspired by Ref. [91]

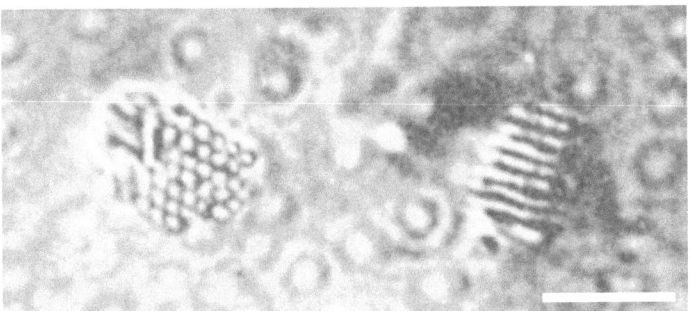

Fig. 8.30 Microscopy image of three-dimensional arrays of *Amoeba proteus* bacteria. Scale bar represents 10 μm. Reprinted with permission from Ref. [92]. Copyright 1954 Nature

Fig. 8.31 AFM image of short, rod-like, *E. coli* bacteria. The width of the image represents 10 μm. Reprinted with permission from Ref. [95]. Copyright 2010 RSC

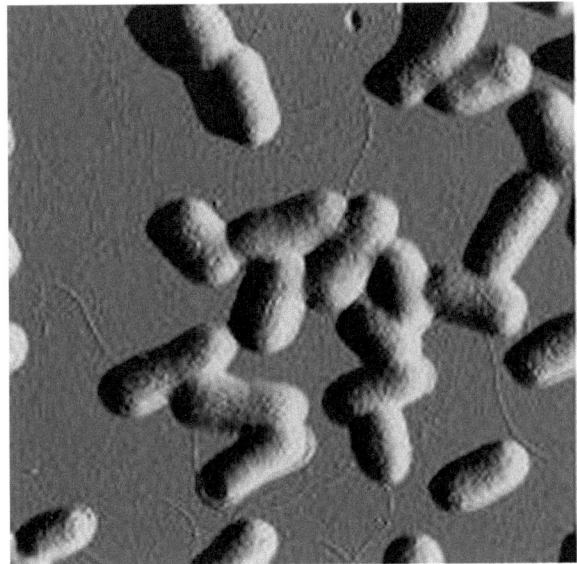

The pictures in Fig. 8.32 shows that the phase transition for a bacterial suspension of a volume fraction of 12.5% takes place at a polymer weight fraction of about 0.2% and that the bacterial volume fraction of the concentrated phase has a volume fraction of about 70%.

For the calculation of the phase behaviour, we assume that the *E. coli* bacteria (see Fig. 8.31) can be modelled as spherocylinders with $L = D \approx 1$ μm. We present calculations for the phase separation between the isotropic phase and the ABC crystal phase.

Schwarz-Linek et al. [95] present no experimental evidence that the concentrated phase is an ABC crystal, but it is known from the work of Goldacre [92] that such phases can occur in suspension of bacteria. By applying the theory presented in Sect. 8.6.2, we obtain the phase diagram plotted in Fig. 8.33b (see Ref. [97]).

Fig. 8.32 *E. coli* cell samples (cell density $\approx 9.6 \cdot 10^{10}$ cfu/mL ($\phi \approx 0.125$)) dispersed in phosphate buffer with NaPSS polyelectrolytes. The polymer concentration (in weight fraction) increases from left to right, with samples 1–11 containing 0%, 0.1%, 0.2%, 0.3%, 0.4%, 0.5%, 0.75%, 1%, 2%, 5% and 10% of NaPSS, respectively. Times: **a** $t=0$, **b** $t=30$ min, **c** $t=100$ min, **d** $t=24$ h. **e** Shows the bottom parts of samples 2—5 at 24 h at higher magnification. Reprinted with permission from Ref. [95]. Copyright 2010 RSC

In Fig. 8.33b, the drawn curves are the result of the theoretical calculations. The datapoints are experimental results, indicating that the system shows (●) a single-phase or (+) two-phase coexistence. The free volume calculations are in good agreement with the experiments on phase separation in mixed suspensions of non-motile bacteria and nonadsorbing polymers. For more details, see Ref. [97].

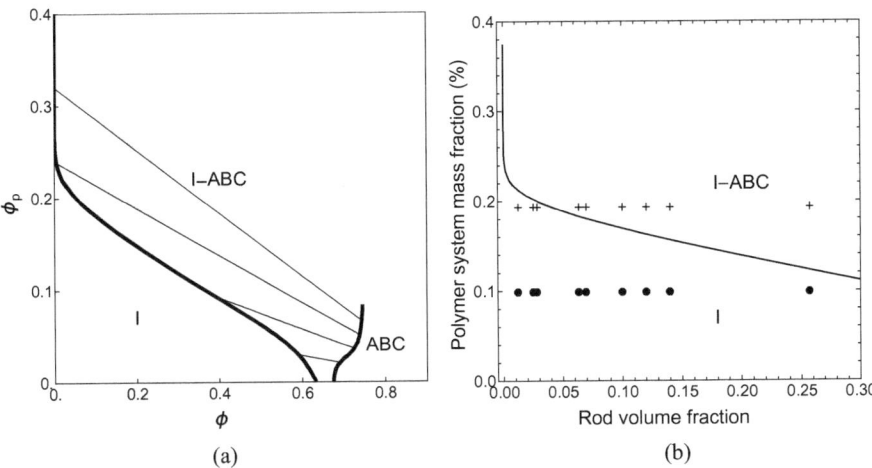

Fig. 8.33 Phase diagram of hard spherocylinders ($L/D = 1$) mixed with polymers ($q = 2R_g/D = 0.028$). **a** Predicted isotropic–ABC crystal phase coexistence (thick curves) with a few illustrative tie-lines (thin lines) [97]. **b** Comparison between predictions (curve) and experimental observations of single-phase (•) or two-phase (+) systems [95]

References

1. Solomon, M.J., Spicer, P.T.: Soft Matter **6**, 1391 (2010)
2. De Gennes, P.G., Prost, J.: The Physics of Liquid Crystals. Oxford University Press, Oxford (1974)
3. Zocher, H., Anorg, Z.: Chem. **147**, 91 (1925)
4. Bawden, F.C., Pirie, N.W., Bernal, J.D., Fankuchen, I.: Nature **138**, 1051 (1936)
5. Livage, J.: Chem. Mat. **3**, 578 (1991)
6. Vroege, G.J., Thies-Weesie, D.M.E., Petukhov, A.V., Lemaire, B.J., Davidson, P.: Adv. Mat. **18**, 2565 (2006)
7. Buining, P.A., Philipse, A.P., Lekkerkerker, H.N.W.: Langmuir **10**, 2106 (1994)
8. Maeda, H., Maeda, Y.: Langmuir **12**, 1446 (1996)
9. Hosseini, S.N., Grau-Carbonell, A., Nikolaenkova, A.G., Xie, X., Chen, X., Imhof, A., van Blaaderen, A., Baesjou, P.J.: Adv. Funct. Mater. **30**, 2005491 (2020)
10. V. Sharma, K. Park, M. Srinivasarao, Mater. Sci. Eng.: R: Rep. **65**, 1 (2009)
11. Li, L.S., Walda, J., Manna, L., Alivisatos, A.P.: Nano Lett. **2**, 557 (2002)
12. Kuijk, A., van Blaaderen, A., Imhof, A.: J. Am. Chem. Soc. **133**, 2346 (2011)
13. Amara, M.S., Paineau, E., Bacia-Verloop, M., Krapf, M.E.M., Davidson, P., Belloni, L., Levard, C., Rose, J., Launois, P., Thill, A.: Chem. Commun. **49**, 11284 (2013)
14. Wada, K., Yoshinaga, N., Yotsumoto, H., Ibe, K., Aida, S.: Clay Miner. **8**, 487 (1970)
15. Zhang, Z.X., van Duijneveldt, J.S.: J. Chem. Phys. **124**, 154910 (2006)
16. Fraden, S., Hurd, A.J., Meyer, R.B., Cahoon, M., Caspar, D.L.D.: J. Phys. Colloques **46**, C3 (1985)
17. Dong, X.M., Kimura, T., Revol, J.F., Gray, D.G.: Langmuir **12**, 2076 (1996)
18. P. Buining. personal communication
19. Z. Dogic, S. Fraden, in *Soft Matter: Complex Colloidal Suspensions*, vol. 2, ed. by G. Gompper, M. Schick (John Wiley & Sons, Ltd, 2006), chap. 1, pp. 1–86
20. Peters, V.F.D., Vis, M., Wensink, H.H., Tuinier, R.: Phys. Rev. E **101**, 062707 (2020)

21. V.F.D. Peters, Phase behaviour of mixtures of structurally complex colloids and polymers. Ph.D. thesis, Eindhoven University of Technology (2021)
22. Kuijk, A., Byelov, D.V., Petukhov, A.V., van Blaaderen, A., Imhof, A.: Faraday Discuss. **159**, 181 (2012)
23. Wetter, C.: Biol. unserer Zeit **15**, 81 (1985)
24. Oster, G.: J. Gen. Physiol. **33**, 445 (1950)
25. Frenkel, D., Lekkerkerker, H.N.W., Stroobants, A.: Nature **332**, 822 (1988)
26. McGrother, S.C., Williamson, D.C., Jackson, G.: J. Chem. Phys. **104**, 6755 (1996)
27. Bolhuis, P.G., Frenkel, D.: J. Chem. Phys. **106**, 666 (1997)
28. Cohen, S.S.: J. Biol. Chem. **144**, 353 (1942)
29. Leberman, R.: Virology **30**, 341 (1966)
30. Onsager, L.: Ann. NY Acad. Sci. **51**, 627 (1949)
31. Vroege, G.J., Lekkerkerker, H.N.W.: Rep. Progr. Phys. **55**, 1241 (1992)
32. Herzfeld, J., Berger, A.E., Wingate, J.W.: Macromolecules **17**, 1718 (1984)
33. Lekkerkerker, H.N.W., Coulon, P., van der Haegen, R., Deblieck, R.: J. Chem. Phys. **80**, 3427 (1984)
34. van Roij, R.: Eur. J. Phys. **26**, S57 (2005)
35. Odijk, T., Lekkerkerker, H.N.W.: J. Phys. Chem. **89**, 2090 (1985)
36. Odijk, T.: Macromolecules **19**, 2313 (1986)
37. Franco-Melgar, M., Haslam, A.J., Jackson, G.: Mol. Phys. **106**, 649 (2008)
38. Verwey, E.J.W., Overbeek, J.T.: Theory of the Stability of Lyophobic Colloids. Elsevier, Amsterdam (1948)
39. Stroobants, A., Lekkerkerker, H.N.W., Odijk, T.: Macromolecules **19**, 2232 (1986)
40. Philip, J., Wooding, R.: J. Chem. Phys. **52**, 953 (1970)
41. Hill, T.L.: Arch. Biochem. Biophys. **57**, 229 (1955)
42. van der Schoot, P., Odijk, T.: J. Chem. Phys. **97**, 515 (1992)
43. Cotter, M.A.: J. Chem. Phys. **66**, 1098 (1977)
44. Lekkerkerker, H.N.W., Stroobants, A.: Il Nuovo Cimento D **16**, 949 (1994)
45. Tuinier, R., Taniguchi, T., Wensink, H.H.: Eur. Phys. J. E **23**, 355 (2007)
46. Fraden, S., Maret, G., Casper, D.L.D., Meyer, R.B.: Phys. Rev. Lett. **63**, 2068 (1989)
47. Yamakawa, H.: Modern Theory of Polymer Solutions. Harper and Row, New York (1971)
48. Strobl, G.: The Physics of Polymers, 3rd edn. Springer, Berlin Heidelberg New York (2007)
49. Adams, M., Fraden, S.: Biophys. J . **74**, 669 (1998)
50. Dogic, Z., Fraden, S.: Phys. Rev. Lett. **78**, 2417 (1997)
51. Grelet, E.: Phys. Rev. X **4**, 021053 (2014)
52. Zimmermann, K., Hagedorn, H., Heuck, C., Hinrichsen, M., Ludwig, H.: J. Biol. Chem. **261**, 1653 (1986)
53. Janmey, P.A., Slochower, D.R., Wang, Y.H., Wen, Q., Cēbers, A.: Soft Matter **10**, 1439 (2014)
54. Secor, P.R., Sweere, J.M., Michaels, L.A., Malkovskiy, A.V., Lazzareschi, D., Katznelson, E., Rajadas, J., Birnbaum, M.E., Arrigoni, A., Braun, K.R., Evanko, S.P., Stevens, D.A., Kaminsky, W., Singh, P.K., Parks, W.C., Bollyky, P.L.: Cell Host Microbe **18**, 549–559 (2015)
55. Tarafder, A.K., von Kügelgen, A., Mellul, A.J., Schulze, U., Aarts, D.G.A.L., Bharat, T.A.M.: Proc. Natl. Acad. Sci. **117**, 4724 (2020)
56. Urakami, N., Imai, M., Sano, Y., Tasaku, M.: J. Chem. Phys. **111**, 2322 (1999)
57. Sano, Y.: Macromol. Symp. **99**, 239 (1995)
58. Imai, M., Urakami, N., Nakamura, A., Takada, R., Oikawa, R., Sano, Y.: Langmuir **18**, 9918 (2002)
59. Levy, J.A., Fraenkel-Conrat, H., Owens, R.A.: in: Virology. Prentice Hall, Englewood Cliffs, New Jersey (1994)
60. Dogic, Z., Fraden, S.: Phil. Trans. R. Soc. Lond. A **359**, 997 (2001)
61. Dogic, Z., Purdy, K.R., Grelet, E., Adams, M., Fraden, S.: Phys. Rev. E **69**, 051702 (2004)
62. Holmqvist, P., Ratajczyk, M., Meier, G., Wensink, H.H., Lettinga, M.P.: Phys. Rev. E **80**, 031402 (2009)
63. Barry, E., Beller, D., Dogic, Z.: Soft Matter **5**, 2563 (2009)

64. Huisman, E.M., Wen, Q., Wang, Y.H., Cruz, K., Kitenbergs, G., Erglis, K., Zeltins, A., Cebers, A., Janmey, P.A.: Soft Matter **7**, 7257 (2011)
65. Booy, F.P., Fowler, A.G.: Int. J. Biol. Macromol. **7**, 327 (1985)
66. Wettstadt, S.: Environ. Microbiol. **22**, 2461 (2020)
67. Marchessault, R.H., Morehead, F.F., Walters, N.M.: Nature **184**, 632 (1959)
68. Habibi, Y., Lucia, L.A., Rojas, O.J.: Chem. Rev. **110**(6), 3479 (2010)
69. M. Roman, W.T. Winter, TAPPI Techn. Papers **06NANO09**, 5 (2006)
70. Oguzlu, H., Danumah, C., Boluk, Y.: Curr. Opin. Colloid Interface Sci. **29**, 46–56 (2017)
71. Oguzlu, H., Boluk, Y.: Cellulose **24**, 131–146 (2017)
72. Edgar, C.D., Gray, D.G.: Macromolecules **35**, 7400 (2002)
73. Vroege, G.J., Lekkerkerker, H.N.W.: J. Phys. Chem. **97**, 3601 (1993)
74. Buining, P.A., Lekkerkerker, H.N.W.: J. Phys. Chem. **97**, 11510 (1993)
75. Zocher, H., Török, C.: Kolloid Z. **170**, 140 (1960)
76. Zocher, H., Török, C.: Kolloid Z. **173**, 1 (1960)
77. Zocher, H., Török, C.: Kolloid Z. **180**, 41 (1962)
78. Bugosh, J.: J. Phys. Chem. **65**, 1791 (1961)
79. Buitenhuis, J., Donselaar, L.N., Buining, P.A., Stroobants, A., Lekkerkerker, H.N.W.: J. Colloid Interface Sci. **175**, 46 (1995)
80. Wensink, H.H., Vroege, G.J.: J. Chem. Phys. **119**, 6868 (2003)
81. Kreibig, U., Wetter, C.: Z. Naturforsch. **35C**, 750 (1980)
82. Lopes, J.T., Romano, F., Grelet, E., Franco, L.F.M., Giacometti, A.: J. Chem. Phys. **154**, 104902 (2021)
83. Parsons, J.D.: Phys. Rev. A **19**, 1225 (1979)
84. Lee, S.D.: J. Chem. Phys. **87**, 4972 (1987)
85. Lee, S.D.: J. Chem. Phys. **89**, 7036 (1988)
86. Graf, H., Löwen, H.: Phys. Rev. E **59**, 1932 (1999)
87. V.F.D. Peters, M. Vis, Á. González García, H.H. Wensink, R. Tuinier, Phys. Rev. Lett. **125**, 127803 (2020)
88. V.F.D. Peters, Á. González García, H.H. Wensink, M. Vis, R. Tuinier, Langmuir **37**, 11582 (2021)
89. S.V. Savenko, M. Dijkstra, J. Chem. Phys. **124** (2006)
90. Dogic, Z.: Phys. Rev. Lett. **91**, 165701 (2003)
91. Yang, D.C., Blair, K.M., Salama, N.R.: Microbiol. Mol. Biol. Rev. **80**, 187 (2016)
92. Goldacre, R.J.: Nature **174**, 732 (1954)
93. Typas, A., Banzhaf, M., Gross, C., Vollmer, W.: Nat. Rev. Microbiol. **10**, 123–136 (2012)
94. Schwarz-Linek, J., Dorken, G., Winkler, A., Wilson, L.G., Pham, N.T., French, C.E., Schilling, T., Poon, W.C.K.: Europhys. Lett. **89**, 68003 (2010)
95. Schwarz-Linek, J., Winkler, A., Wilson, L.G., Pham, N.T., Schilling, T., Poon, W.C.K.: Soft Matter **6**, 4540 (2010)
96. Dorken, G., Ferguson, G.P., French, C.E., Poon, W.C.K., Soc, J.R.: Interface **9**, 3490 (2012)
97. Peters, V.F.D., Vis, M., Tuinier, R., Lekkerkerker, H.N.W.: J. Chem. Phys. **154**, 151101 (2021)

Phase Behaviour of Colloidal Platelet–Depletant Mixtures

9

9.1 Introduction to Colloidal Platelets

Colloidal platelets are encountered in a wide range of systems in nature and technology. Examples are hydroxides and layered double hydroxides, smectite clays and exfoliated inorganic nanosheets. Suspensions of these platelets have been found to exhibit liquid crystal ordering, including gibbsite [1–4], nickel hydroxide [5], layered double hydroxides [6,7], nontronite [8–10], beidellite [11,12], fluorohectorite [13,14], solid phosphatoantimonate acid [15,16], zirconium phosphate [17–19], niobate [20,21] and titanate [22]. TEM micrographs of dispersions made from some of these particles are displayed in Fig. 9.1, giving some insight into the morphology. Table 9.1 provides the chemical composition of the platelets.

Upon increasing the concentration of platelets a dispersion of initially isotropic platelets can become liquid crystalline, just as for rods, see Chap. 8. Concentrated dispersions of platelets may display nematic [1,6,8,14,16,18,20,22], columnar [5, 24] and smectic [4,7,15,16,19] phase states, illustrated in Fig. 9.2.

While each of these nematic (N), columnar (C) and smectic (Sm) phases exhibits long-range orientational order, they differ by the positional correlations between the particles. Long-range positional order is absent in the N phase. The C phase has a two-dimensional lattice of columns, which consist of liquid-like stacks of particles. The Sm phase is characterised by a one-dimensional periodic array of layers of particles.

Figure 9.3 depicts gibbsite dispersions which exhibit these liquid-crystalline phase states [2]. These images illustrate the phases and phase transitions that can be detected when using crossed polarisers for samples varying in platelet concentration. The Sm phase is only rarely observed [4,5,7,15,16,19,25]. The smectic phases observed in Refs. [7], [25] and [4], probably originated from the high surface charge of the platelets. The large polydispersity [19,26] of the diameter of the platelets probably

H. N. W. Lekkerkerker, R. Tuinier, M. Vis, *Colloids and the Depletion Interaction*,
Lecture Notes in Physics 1026, https://doi.org/10.1007/978-3-031-52131-7_9

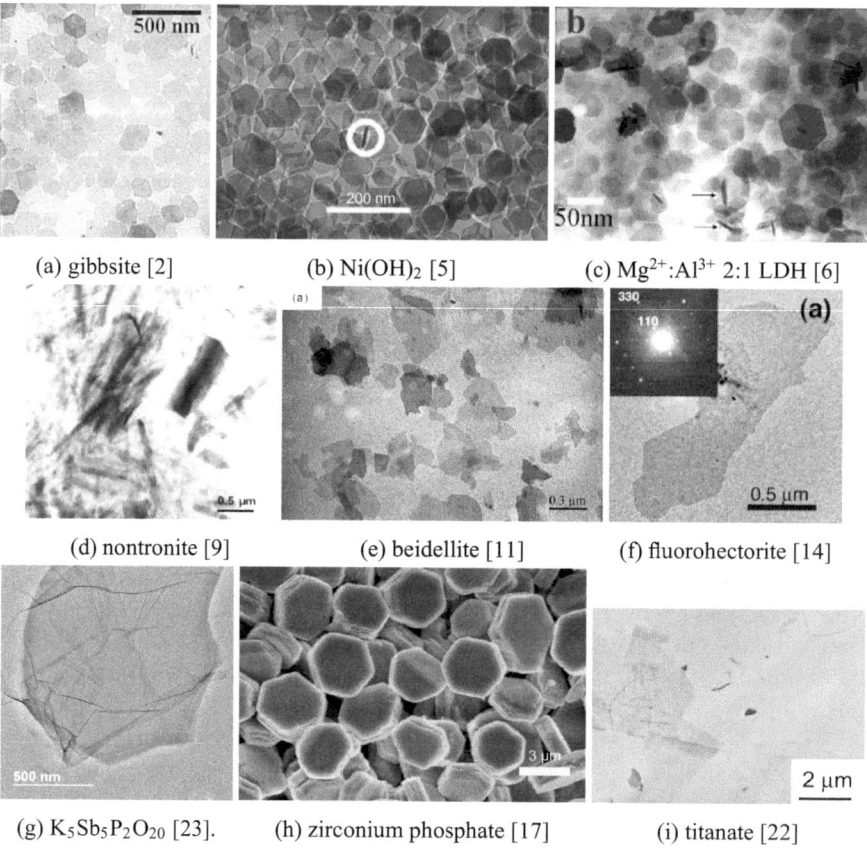

(a) gibbsite [2]   (b) Ni(OH)$_2$ [5]   (c) Mg^{2+}:Al^{3+} 2:1 LDH [6]

(d) nontronite [9]   (e) beidellite [11]   (f) fluorohectorite [14]

(g) K$_5$Sb$_5$P$_2$O$_{20}$ [23].   (h) zirconium phosphate [17]   (i) titanate [22]

Fig. 9.1 Examples of plate-like colloids: **a–c** synthetic platelets, **d–f** smectite clay particles, **g–i** exfoliated inorganic nanosheets. Scale bars are as indicated. Reprinted with permission from **a** Ref. [2], copyright 2000 Nature; **b** Ref. [5], copyright 1999 Springer; **c** Ref. [6], copyright 2003 the American Chemical Society (ACS); **d** Ref. [9], copyright 2008 ACS; **e** Ref. [12], copyright 2011 ACS; **f** Ref. [14], copyright 2010 the Royal Society of Chemistry (RSC); **g** P. Davidson; **h** Ref. [17], copyright 2012 the American Physical Society (APS); **i** Ref. [22], copyright 2014 RSC

explains the Sm phase observed in Ref. [19]. These smectic phases can show bright iridescence, see Fig. 9.4, depending on the spacing between the smectic layers [4].

Disc-like colloidal particles are also found in biological systems such as the red blood cells in blood (already discussed in Sect. 1.3.2) and plate-like proteins such as kinetochore [27]. Smectite clays, which are inorganic plate-like nanoparticle mixed suspensions, are ubiquitous on Earth. Mixed suspensions of colloidal platelets and inorganic nanoparticles display interesting rheological [28] and electronic properties [29].

In this chapter we consider the phase behaviour of mixtures of colloidal plates and depletants. The focus is on nonadsorbing polymers as depletants, although experimental examples of added small colloidal particles are also considered. First, a treatment of the phase behaviour of pure platelets is given.

Table 9.1 Chemical composition of colloidal plate-like particles

Platelet type	Chemical structure	Fig. 9.1
Gibbsite	$Al(OH)_3$	a
Nickel hydroxide	$Ni(OH)_2$	b
Mg_2Al layered double hydroxide	$[Mg_{2/3}Al_{1/3}(OH)_2]^{+1/3}$	c
Nontronite	$(Si_{7.55} Al_{0.16} Fe_{0.29})(Al_{0.34} Fe_{3.54} Mg_{0.05}) O_{20} (OH)_4 (Na)_{0.72}$	d
Beidellite	$(Si_{7.27} Al_{0.73})(Al_{3.77} Fe_{0.11} Mg_{0.21}) O_{20} (OH)_4 (Na)_{0.76}$	e
Fluorohectorite	$Mg_{2.60} Li_{0.46} Si_4 O_{10} F_2 Na_{0.46}$	f
Phosphoantimonate acid	$H_3Sb_3P_2O_{14}$	g
Zirconium phosphate	$Zr(HPO4)2H2_O$	h
Titanate	$H_{1.07}Ti_{1.73}O_4$	i

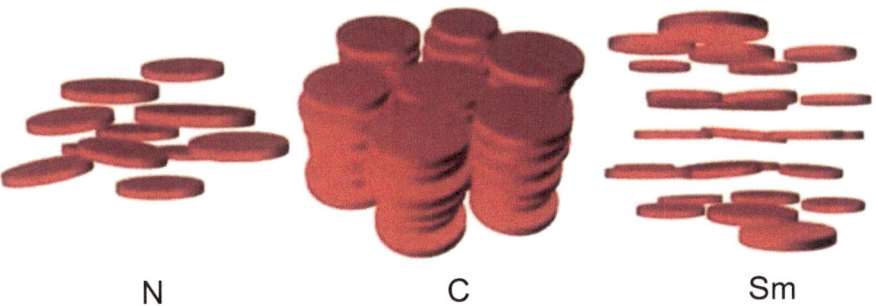

N   C   Sm

Fig. 9.2 Structure of the three main classes of liquid crystals made from disc-like particles: the nematic phase (N), the columnar phase (C) and the smectic phase (Sm). Reprinted with permission from Ref. [2]. Copyright 2000 Nature

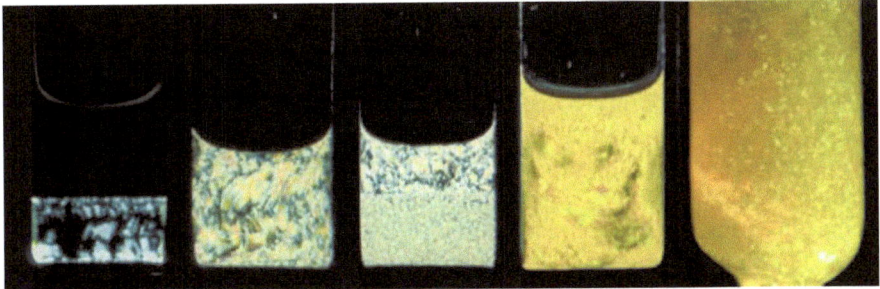

Fig. 9.3 Tubes containing gibbsite suspensions [2] in toluene at varying platelet concentrations photographed between crossed polarisers. Volume fractions from left to right: 0.19 (I–N), 0.28 (N), 0.28 (N–C) and 0.47 (C). The tube to the right depicts a monophasic columnar sample illuminated by white light. The colours of its Bragg reflections (visible as small bright spots) vary from yellow to green as the angle between the incident light and viewing direction is in the range of 50°–70°. Reprinted with permission from Ref. [2]. Copyright 2000 Nature

Fig. 9.4 Silica-coated gibbsite platelets suspended in DMF without added salt show iridescence upon illumination with white light in a 10 mm wide capillary, indicating a lamellar phase

9.2   Phase Diagram of Hard Colloidal Platelets

Hard colloidal platelets are theoretically described here in a simplified way as monodisperse discs with diameter D and thickness L, with a focus on $L \ll D$. The volume of a colloidal platelet is given by $v_0 = \pi D^2 L/4$ and the volume fraction of colloidal platelets $\phi = nv_0$, with n the number density N/V. It is noted that in experimental dispersions of these platelets there is often size dispersity in D and/or L. In this section theoretical and computer simulation results on the phase diagram of pure monodisperse hard platelets are reviewed.

9.2.1   Computer Simulations

In previous chapters it was identified that colloidal phase transitions of hard spheres and hard rods are governed by entropy. That is also the case for hard platelets. In Sect. 8.2 it was explained that, as was realised by Onsager, a theory based on the second virial coefficient suffices to accurately predict the thermodynamic properties of dilute long and thin rods. This includes the rod concentrations at the isotropic–nematic (I–N) phase transition. The reason for that is that the higher virial coefficients for long thin rods are very small compared to the second virial coefficient, as was first rationalised by Onsager [30] on the basis of geometric arguments. Using Monte Carlo simulations, Frenkel [31,32] showed that the longer and thinner the rods the smaller the higher virial coefficients become.

As was pointed out by Onsager [30], there are no geometrical arguments that the higher virial coefficients become small even in the limit of infinitely thin hard platelets. This can be illustrated by comparing the results of the second virial approach for the I–N transition of infinitely thin hard platelets with simulations.

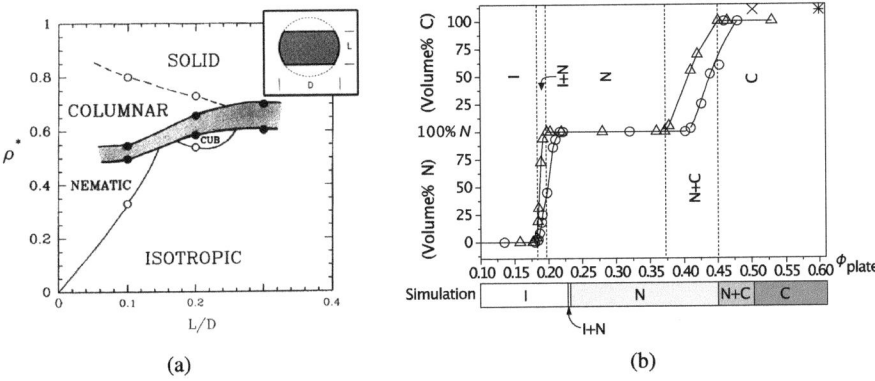

(a)   (b)

Fig. 9.5 a Phase diagram from computer simulations of dispersions of hard cut spheres to mimic platelets. *Inset*: side view of a cut sphere with diameter D and thickness L [35]. **b** Experimentally measured relative volume of the nematic and columnar phases as a function of the volume fraction of gibbsite platelets [2]. The platelets have $D/L = 13$ and a polydispersity in D of 17% ($\triangle$) and 25% ($\bigcirc$). The predicted phase states from panel **a** are indicated below the abscissa of (b). Reprinted with permission from **a** Ref. [35], copyright 1992 APS and **b** Ref. [2], copyright 2000 Nature

Forsyth et al. [33] used the second virial approach for the I–N transition and found $n_{\mathrm{iso}}D^3 = 5.3$ and $n_{\mathrm{nem}}D^3 = 6.8$ as coexisting platelet concentrations. Using Monte Carlo simulations on infinitely thin hard platelets, Eppenga and Frenkel [34] found $n_{\mathrm{iso}}D^3 = 4.04$ and $n_{\mathrm{nem}}D^3 = 4.12$. Clearly the second virial coefficient theory does not accurately predict the I–N phase transition of hard platelets.

Veerman and Frenkel [35] extended the simulations of Eppenga and Frenkel to hard platelets of finite thickness. The phase diagram that results from their computer simulations is reproduced in Fig. 9.5a. Note that the platelets were simulated as cut spheres (see the inset of Fig. 9.5a).

In Fig. 9.5a ρ^* is the density relative to the close packed density. For $L/D < 0.15$ (i.e. $D/L > 6.7$) the simulations of Veerman and Frenkel reveal an isotropic phase, an I–N phase transition, an N phase, a subsequent nematic–columnar (N–C) transition and finally a pure C phase upon increasing the platelet concentration. This simulation result was confirmed experimentally by using model systems of plate-like gibbsite particles with $D/L = 13$ with different polydispersities by Van der Kooij, Kassapidou and Lekkerkerker [24] (see the observed phases in Fig. 9.5b).

The computer simulations of Veerman and Frenkel also predicted that coexistence between isotropic and columnar phases (i.e. without forming a nematic phase) is possible for thicker platelets with $L/D > 0.15$ (Fig. 9.5a). Brown et al. [5] studied nickel hydroxide platelets with $D/L = 3.5$ and indeed found this direct I–C transition.

Veerman and Frenkel [35] also observed an unexpected region in the phase diagram, which they refer to as the cubatic phase (CUB). In this CUB phase the cut spheres are assembled in short columns. The columns themselves have a random orientation, and hence there is appreciable interaction between different columns. The interaction between the columns eventually becomes so severe that the column segments try to order in a manner that minimises the packing problems.

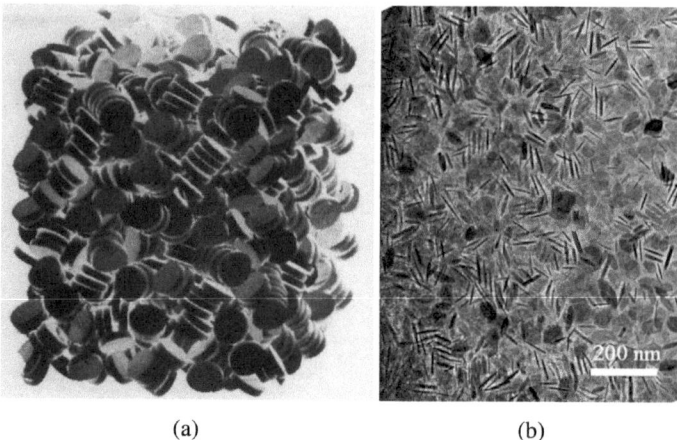

(a)　　　　　　　　　　　　　　　　(b)

Fig. 9.6 **a** Snapshot of the configuration of a system containing 1728 cut hard spheres with $D/L = 5$ at a reduced density of $\rho^* = 0.575$. The platelets spontaneously assemble in short stacks containing four or five particles. Neighbouring stacks tend to be approximately perpendicular [35]. Reprinted with permission from Ref. [35]. Copyright 1992 APS. **b** Cryo TEM image of a suspension of nickel hydroxide platelets at 20 wt%. Reprinted with permission from Ref. [36]. Copyright 1984 Elsevier

This is illustrated in the computer simulation snapshot in Fig. 9.6a. Qazi, Karlsson and Rennie [36] have presented experimental evidence for cubatic order in a dispersion of plate-like colloids (see Fig. 9.6b).

In the next subsection, an approximation is outlined that aims to develop a theoretical prediction for the phase diagram of hard platelets.

9.2.2　Theoretical Account

9.2.2.1 Onsager–Parsons–Lee Theory for the Isotropic and Nematic Phase States

In Sect. 8.3 we used scaled particle theory (SPT) to incorporate higher virial coefficients in the treatment of the isotropic–nematic phase transition of hard rods. An approach which is similar to SPT (in the fact that it may be considered a renormalised two-particle theory) has been given by Parsons [37] and was used by Lee [38] to calculate the isotropic–nematic transition in solutions of hard spherocylinders. This approximate theory [39,40] may be considered an extension of the Carnahan–Starling equation [41] for hard spheres (see Sect. 3.2.1).

The Helmholtz free energy F within the Onsager–Parsons–Lee approach [37,38] can be expressed as

$$\frac{F}{NkT} = \frac{\widetilde{F}}{\phi} = \ln\left(\frac{\Lambda^3}{v_0}\right) + \ln\phi - 1 + \mathfrak{s}[f] + \frac{2}{\pi}\frac{D}{L}\phi G_{\mathrm{P}}(\phi)\langle\langle\tilde{v}_{\mathrm{excl}}(\gamma)\rangle\rangle, \quad (9.1)$$

We further focus on the excess free energy $\widetilde{F}^{\text{ex}}$, defined through $\widetilde{F} = \widetilde{F}_{\text{id}} + \widetilde{F}_{\text{exc}}$, with $\widetilde{F}_{\text{id}} = \phi \ln(\phi \Lambda^3/v_0) - \phi$. As before, $\widetilde{F} = F v_0/kTV$. The orientational entropy of the platelets is (see Section 8.2.1) related to $\mathfrak{s}[f]$ and can be calculated using

$$\mathfrak{s}[f] = \int f(\mathbf{\Omega}) \ln[4\pi f(\mathbf{\Omega})] d\mathbf{\Omega}, \tag{9.2}$$

which includes the orientational distribution function $f(\mathbf{\Omega})$, which is normalised according to

$$\int f(\mathbf{\Omega}) d\mathbf{\Omega} = 1, \tag{9.3}$$

where $\mathbf{\Omega}$ is the solid angle (see Sect. 8.2.1). In Eq. 9.1, $\tilde{v}_{\text{excl}}(\gamma)$ is the excluded volume $v_{\text{excl}}(\gamma)$ between two hard platelets divided by D^3 at fixed interparticle angle γ:

$$\tilde{v}_{\text{excl}}(\gamma) = \frac{\pi}{2}|\sin(\gamma)| + \frac{L}{D}\left\{ \frac{\pi}{2} + 2E[\sin(\gamma)] + \frac{\pi}{2}\cos(\gamma)\right\}$$
$$+ 2\left(\frac{L}{D}\right)^2 |\sin(\gamma)|, \tag{9.4}$$

including the complete elliptic integral of the second kind $E[x]$. The average $\langle\langle \tilde{v}_{\text{excl}}(\gamma)\rangle\rangle$ is defined as

$$\langle\langle \tilde{v}_{\text{excl}}(\gamma)\rangle\rangle = \int \int d\mathbf{\Omega} d\mathbf{\Omega}' f(\mathbf{\Omega}) f(\mathbf{\Omega}') \tilde{v}_{\text{excl}}(\gamma). \tag{9.5}$$

The effects of higher order virial terms are incorporated via a Parsons–Lee scaling factor G_{P}:

$$G_{\text{P}}(\phi) = \frac{4 - 3\phi}{4(1 - \phi)^2}, \tag{9.6}$$

The factor G_{P} ensures that the ratios of the third and higher virial coefficients to the second virial coefficient are the same as for hard spheres.

At low concentrations the system is isotropic (I). In this isotropic phase, all platelets are oriented randomly and $f^{\text{I}} = (4\pi)^{-1}$ so that $\mathfrak{s}^{\text{I}} = 0$. Within the Parsons–Lee approximation, the isotropic excess free energy (Eq. (9.1)) becomes

$$\frac{F_{\text{I}}^{\text{exc}}}{NkT} = \frac{\widetilde{F}_{\text{I}}^{\text{exc}}}{\phi} = \frac{2}{\pi}\frac{D}{L}\phi G_{\text{P}}(\phi)\tilde{v}_{\text{excl}}^{\text{I}}, \tag{9.7}$$

with $\tilde{v}_{\text{excl}}^{\text{I}} = \langle\langle \tilde{v}_{\text{excl}}(\gamma)\rangle\rangle$, which becomes

$$\tilde{v}_{\text{excl}}^{\text{I}} \approx \frac{\pi^2}{8} + \left(\frac{3\pi}{4} + \frac{\pi^2}{4}\right)\frac{L}{D} + \frac{\pi}{2}\left(\frac{L}{D}\right)^2, \tag{9.8}$$

where the last term (of order $(L/D)^2$) is usually omitted because the focus is often on thin platelets ($L/D \lesssim 0.1$), for which its magnitude is negligible.

The Helmholtz energy (Eq. (9.7)) directly provides the (dimensionless) osmotic pressure ($\tilde{P} = Pv_0/kT$) and chemical potential ($\tilde{\mu} = \mu/kT$) of the platelets in suspension (see Appendix A):

$$\tilde{P}_I = \phi + \frac{2}{\pi}\phi^2 \frac{D}{L}\frac{1-\phi/2}{(1-\phi)^3}\tilde{v}^I_{excl}, \tag{9.9}$$

$$\tilde{\mu}_I = \ln\left(\frac{\Lambda^3}{v_0}\right) + \ln\phi + \frac{2}{\pi}\frac{D}{L}\frac{8\phi - 9\phi^2 + 3\phi^3}{4(1-\phi)^3}\tilde{v}^I_{excl}, \tag{9.10}$$

with the reference chemical potential $\ln(\Lambda^3/v_c)$.

Above a certain concentration, the platelets spontaneously assume a preferred orientation, the nematic state. One may then compute the orientational distribution function (ODF) at each concentration numerically by minimising the Helmholtz free energy expression (Eq. (9.1)), while using the condition of Eq. 8.6. Since the nematic phases we consider are uniaxial in symmetry the solid angle $\boldsymbol{\Omega}$ only depends on the polar angle θ between a nematic director and the orientation of the platelet.

As in Chap. 8, Odijk's Gaussian approximation f_G for the ODF $f(\theta)$ [42] is used:

$$f_G(\theta) = \frac{\kappa}{4\pi}\exp\left[-\frac{1}{2}\kappa\theta^2\right], \tag{9.11}$$

which applies to angles $-\pi/2 \leq \theta \leq \pi/2$. The prefactor of the Gaussian ODF follows from Eq. (8.6). Insertion of Eq. (9.11) into Eq. (9.2) gives

$$\mathfrak{s}^N \approx \ln\kappa - 1. \tag{9.12}$$

The normalised excluded volume in the nematic phase follows as [39]

$$\tilde{v}^N_{excl} = \langle\langle\tilde{v}_{excl}(\gamma)\rangle\rangle^N = 2\pi\frac{L}{D} + \frac{\pi}{2}\sqrt{\frac{\pi}{\kappa}}. \tag{9.13}$$

Using the Gaussian ODF in the free energy (Eq. (9.1)), the excess nematic state free energy can now be written as

$$\frac{F^{exc}_N}{NkT} = \frac{\tilde{F}^{exc}_N}{\phi} = \ln\kappa - 1 + \phi G_P(\phi)\left[\frac{D}{L}\sqrt{\frac{\pi}{\kappa}} + 4\right]. \tag{9.14}$$

The chemical potential and osmotic pressure in the nematic state can be easily obtained:

$$\tilde{\mu}_N = \ln\left(\frac{\Lambda^3}{v_0}\right) + \ln\phi + \ln\kappa - 1 + \frac{8\phi - 9\phi^2 + 3\phi^3}{4(1-\phi)^3}\left[\frac{D}{L}\sqrt{\frac{\pi}{\kappa}} + 4\right], \tag{9.15}$$

and

$$\tilde{P}_N = \phi + \phi^2\frac{1-\phi/2}{(1-\phi)^3}\left[\frac{D}{L}\sqrt{\frac{\pi}{\kappa}} + 4\right]. \tag{9.16}$$

Once the variational parameter κ is known, the free energy and various thermo-dynamic properties can be calculated explicitly. The κ parameter follows from the minimisation of the free energy w.r.t. κ (Eq. (8.24)). Applying this to Eq. (9.14) yields

$$\kappa = \frac{\pi}{4}\left(\frac{D}{L}\right)^2 \phi^2 G_P^2(\phi).\tag{9.17}$$

Insertion of this result into Eq. (9.14) gives

$$\frac{\widetilde{F}_N^{exc}}{\phi} = 2\ln\left(\frac{D}{L}\frac{\sqrt{\pi}}{2}\phi G_P(\phi)\right) + 4\phi G_P(\phi) + 1,\tag{9.18}$$

for the excess Helmholtz energy. The chemical potential of the hard platelets in the nematic phase follows as

$$\begin{aligned}\widetilde{\mu}_N = &\ln\left(\frac{\Lambda^3}{v_0}\right) + 1 + 2\ln\left(\frac{D}{L}\frac{\sqrt{\pi}}{2}\right) + \ln\phi + 2\ln[\phi G_P(\phi)]\\&+ 4\phi G_P(\phi) + \frac{2 - \phi - \phi^2 + \phi^3/2}{(1 - 3\phi/4)(1 - \phi)^3},\end{aligned}\tag{9.19}$$

and their osmotic pressure becomes

$$\widetilde{P}_N = \phi + \frac{2\phi - \phi^2 - \phi^3 + \phi^4/2}{(1 - 3\phi/4)(1 - \phi)^3}.\tag{9.20}$$

9.2.2.2 Isotropic–Nematic Phase Transition of Hard Platelets

It is now possible to compute the coexisting isotropic and nematic concentrations of hard platelets within the Parsons–Lee approximation using the Gaussian form for the ODF. In general, coexisting concentrations (the binodals) follow from solving the concentrations for which the chemical potentials μ and osmotic pressures P are equal (see also Appendix A).

Theoretical Parsons–Lee predictions (curves) for the I–N phase coexistence con-centrations are plotted in Fig. 9.7 as a function of L/D and are compared to Monte Carlo computer simulation results (data points). Two 'flavours' of the Parsons–Lee predictions are shown, using the Gaussian form of the ODF (dotted curves) and a numerical optimisation of the ODF (solid curves).

Compared to the numerical approach, the Gaussian approximation predicts a wider coexistence region and slightly higher coexisting platelet concentrations, espe-cially on the nematic side. This is similar to the situation for the I–N phase transition of hard spherocylinders as was discussed in Chap. 8 (see Fig. 8.4). Still, the rela-tively simple Gaussian ODF approach—combined with the Parsons–Lee—provides a reasonable description for the I–N phase transition of hard platelets. The Gaussian ODF approach deviates most near the limit $L/D \to 0$.

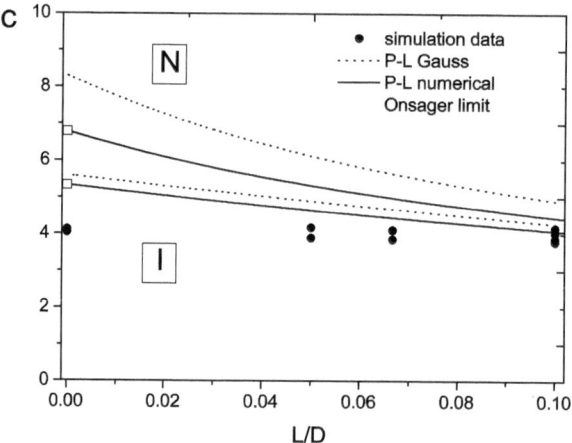

Fig. 9.7 I–N phase coexistence for pure hard platelets. Platelet concentration is given in terms of the quantity $c = ND^3/V = (4/\pi)\phi D/L$. Curves are theoretical predictions using the Parsons–Lee (P–L) approximation combined with a numerical minimisation (solid curves) and minimisation using the Gaussian approximation to the ODF (dotted curves). Data points are computer simulation results by Eppenga and Frenkel [34] for $L/D \to 0$ (Onsager limit) and Veerman and Fenkel [35] for other L/D values. Cut spheres were used to simulate the platelets, which slightly differs from the theoretical description of cylindrical platelets. Reprinted with permission from Ref. [43]. Copyright 2015 Taylor & Francis

9.2.2.3 Lennard-Jones–Devonshire Cell Theory for the Columnar Phase

To predict the thermodynamic properties of a columnar phase, an extended cell theory by Lennard-Jones and Devonshire (LJD) can be used (see Wensink [44]). Within this model the columnar phase is described as a superposition of a 1D liquid and a 2D solid. The configurational (excess) free energy associated with the LJD cell theory is given by

$$\frac{F_{LJD}^{exc}}{NkT} = \frac{\widetilde{F}_{LJD}^{exc}}{\phi} = 2\ln\left(\frac{\bar{\Delta}_C^{-1}}{1 - \bar{\Delta}_C^{-1}}\right), \tag{9.21}$$

where $\bar{\Delta}_C = \Delta_C/D$ is the (lateral) spacing, with Δ_C the nearest-neighbour distance (see Appendix C for details). Near close packing densities, Eq. (9.21) is expected to provide an accurate description of a 2D solid. Per column, it is assumed that the particles assume liquid-like configurations in one direction only. By applying the condition of single-occupancy, $\bar{\Delta}_C$ provides

$$\phi^* \bar{\Delta}_C^2 = \tilde{\rho}, \tag{9.22}$$

which relates the plate volume fraction $\phi = Nv_0/V$ (with $v_0 = (\pi/4)LD^2$ as the particle volume) to the reduced linear density $\tilde{\rho}$, where the reduced packing fraction $\phi^* = \phi/\phi_{cp}$ with $\phi_{cp} = \pi/2\sqrt{3} \approx 0.907$ for the area fraction of discs at close packing.

The excess free energy of the columnar state now follows from adding the fluid and LJD contributions:

$$\frac{F_{col}^{exc}}{NkT} = \frac{\widetilde{F}_{exc}^{col}}{\phi} = 2\ln\left\{\frac{3}{2}\frac{D}{L}\left(\frac{\phi^*\bar{\Delta}_C^2}{1-\phi^*\bar{\Delta}_C^2}\right)\right\} \tag{9.23}$$
$$-\ln(1-\phi^*\bar{\Delta}_C^2)(1-\bar{\Delta}_C^{-1})^2.$$

The final step is to minimise the total free energy with respect to $\bar{\Delta}_C$ using

$$\frac{\partial F}{\partial \bar{\Delta}_C} = 0, \tag{9.24}$$

which leads to

$$\bar{\Delta}_C = \frac{2^{1/3}\mathcal{K}^{2/3} - 3^{1/3}4\phi^*}{6^{2/3}\phi^*\,\mathcal{K}^{1/3}}, \tag{9.25}$$

in which $\mathcal{K}$ is defined by

$$\mathcal{K} = 27(\phi^*)^2 + [3(\phi^*)^3(32 + 243\phi^*)]^{1/2}. \tag{9.26}$$

With this, the free energy for the columnar state is fully specified. Unlike the nematic free energy, the columnar free energy is entirely algebraic and does not involve any implicit minimisation condition to be solved (see Eq. (C.6)). The pressure and chemical potential can be found in the usual way (see Appendix A). The nematic free energy can also be recast in closed algebraic form using a simple variational form for the ODF, similar to Eq. (C.7) (see Ref. [39]).

9.2.2.4 Theoretical Prediction of the Phase Diagram of Hard Platelets

Using the free energies for the different phase states for the isotropic, nematic and columnar phase states discussed above, standard thermodynamic relations can be applied to calculate the osmotic pressure and chemical potential of the pure platelet suspension for every phase state.

This enables the phase diagram for a system of hard platelets to be resolved, as is presented in Fig. 9.8. The relatively high excluded volume between thin platelets explains the I–N phase transition occurring at very low packing fractions for very small values of the aspect ratio ($L/D \to 0$). See also Fig. 9.7, which illustrates that $c \sim D/L$ at the I–N coexistence hardly varies. With increasing L/D, the I–N phase coexistence widens and its boundaries shift towards higher packing fractions. From Fig. 9.8 it also follows that the N–C phase coexistence concentrations barely depend on L/D. For sufficiently thick discs ($L/D \gtrsim 0.16$), transitions from an isotropic to a columnar phase occur without an intermediate nematic phase: thick discs are not sufficiently anisotropic to stabilise the occurrence of a nematic phase [39]. The grey vertical line in Fig. 9.8 at $L/D \approx 0.16$ indicates an I–N–C triple coexistence for hard colloidal platelets. Computer simulation data (symbols) from Refs. [35,45–47] have

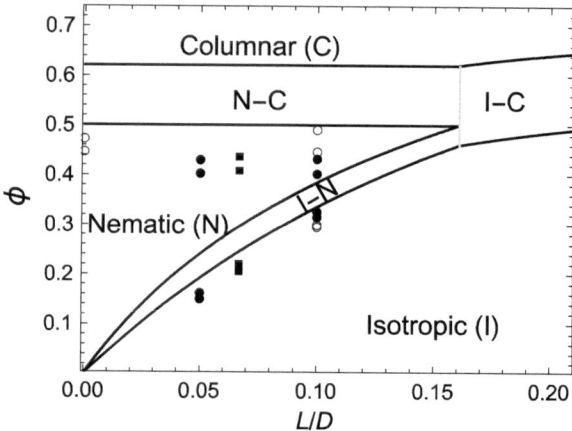

Fig. 9.8 Phase diagram of a monodisperse hard disc suspension. The grey triple line indicates the platelet aspect ratio of the I–N–C triple phase coexistence. For $L/D > 0.16$ there is only isotropic–columnar (I–C) phase coexistence. For $L/D < 0.16$ there are I–N and nematic–columnar (N–C) phase coexistences. Curves are computed using the Gaussian approximation for the nematic phase state; data points are Monte Carlo computer simulation results ($\circ$, [35,45]), ($\bullet$, [46]), ($\blacksquare$, [47]). Figure is based upon Refs. [39,48]

been added to have an idea of the accuracy of the equations of states used. Qualitative agreement is found but quantitatively the theoretical phase transitions occur at somewhat higher disc concentrations as predicted theoretically. The phase diagram presented in Fig. 9.8 constitutes the pure platelet reference point for calculating the thermodynamics of platelet–depletant mixtures.

Exercise 9.1. What are the fundamental differences between the theoretical description of hard plates in this chapter and hard rod-like particles in Chapter 8?

9.3 Phase Behaviour of Hard Platelet–Penetrable Hard Sphere Mixtures

To predict the phase behaviour of hard platelets mixed with penetrable hard spheres (PHSs), the same steps are followed as outlined in Sect. 3.3 for hard spheres mixed with PHSs and in Sect. 8.4 for hard rods mixed with PHSs (see also Appendix A). The system of interest contains hard platelets (modelled as cylinders with diameter D and thickness L) in osmotic equilibrium with a reservoir that only contains PHSs (with diameter σ). The size ratio $q = \sigma/D$ and the depletion thickness of the PHS is constant ($\delta = \sigma/2$). The FVT expression for the semi-grand potential in case of ideal depletants can be written as [49]

$$\tilde{\Omega} = \tilde{F}_0 - \frac{v_0}{v_d}\alpha \tilde{P}^R. \tag{9.27}$$

Here, F_0 is the free energy of the pure hard platelet dispersion, v_0 is the volume of the platelets, v_d is the volume of the depletants, α is the free volume fraction for the depletants, and

$$\tilde{P}^R = \frac{P^R v_d}{kT} = \phi_d^R,$$

is the dimensionless Van 't Hoff osmotic pressure of an ideal solution of depletants in reservoir R having a volume fraction ϕ_d^R. The depletant volume fraction in the system follows from

$$\phi_d = \alpha \phi_d^R.$$

The free volume fraction available for depletants in the system $\alpha = \langle V_{\text{free}} \rangle_0 / V$ and can be calculated using the relation $\alpha = e^{-W/(kT)}$, where W is the reversible work for inserting the PHSs in the hard platelet suspension [50]. Following scaled particle theory (SPT) [51,52], W is calculated by scaling the size of the PHSs as $\lambda \sigma$. In the case $\lambda \ll 1$, it is unlikely that the platelets and PHSs overlap. Hence,

$$W(\lambda) = -kT \ln[1 - n v_{\text{excl}}(\lambda)] \quad \text{for } \lambda \ll 1, \tag{9.28}$$

where v_{excl} is the excluded volume between a PHS and a hard platelet, given by

$$v_{\text{excl}}(\lambda) = \frac{\pi}{4}D^2\lambda\sigma + \frac{\pi}{4}L(D + \lambda\sigma)^2 + \frac{\pi^2}{8}D(\lambda\sigma)^2 + \frac{\pi}{6}(\lambda\sigma)^3. \tag{9.29}$$

In the opposite limit $\lambda \gg 1$, the inserted PHS is very large; in good approximation, W is then equal to the work required to create a cavity with volume $\frac{\pi}{6}(\lambda\sigma)^3$ against the pressure P of the hard platelets:

$$W = \frac{\pi}{6}(\lambda\sigma)^3 P \quad \text{for } \lambda \gg 1. \tag{9.30}$$

In SPT, these two limiting cases are connected by expanding W as a series in λ (Eqs. (3.32) and (8.50)), which yields an expression for $W(\lambda = 1)$ [48,53,54]:

$$\alpha = (1 - \phi) \exp[-Q], \tag{9.31}$$

with

$$Q = q\left(\frac{D}{L} + \frac{\pi q D}{2L} + q + 2\right)y$$
$$+ 2q^2\left[\frac{1}{4}\left(\frac{D}{L}\right)^2 + \frac{D}{L} + 1\right]y^2 + \frac{2}{3}\frac{D}{L}q^3\tilde{P}, \tag{9.32}$$

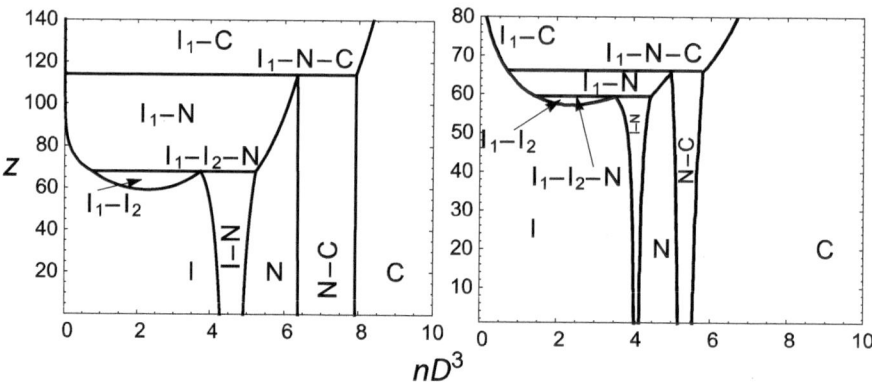

Fig. 9.9 Comparison of phase diagrams for mixtures of plate-like particles and PHSs for $L/D = 0.1$ and $q = 0.2$. *Left*: phase diagram for hard discs mixed with PHSs obtained from FVT with theoretical expressions for the thermodynamic properties of the pure plate system and SPT results for the free volume fractions [48]. *Right*: phase diagram for cut sphere–PHS mixtures by Zhang et al. [46] obtained from Monte Carlo simulations for the thermodynamic properties of pure plate systems combined with FVT with free volume fractions measured by a trial insertion method. Reprinted with permission from Ref. [48]. Copyright 2018 Taylor & Francis

where y is defined by Eq. (3.39e). It should be noted that Zhang, Reynolds and Van Duijneveldt [55] obtained α from SPT for a mixture of cut spheres and PHSs. For infinitely thin discs, the expressions for α obtained by Zhang et al. [55] and obtained here are identical. González García et al. [48] compared Eq. (9.31) (with Q given by Eq. (9.32)) with computer simulation results of Refs. [46,56] and found good agreement between theory and simulations.

With all the components required to calculate the grand potential at hand, determination of phase coexistence is straightforward in principle (see Appendix A). This theoretical approach is now compared to phase diagrams computed by Zhang et al. [46] for $L/D = 0.1$ and $q = 0.2$. They calculated the phase diagram also using FVT, but employed Monte Carlo simulations to obtain the thermodynamic properties of the pure platelet system and measured the free volume fraction in such simulations by a trial insertion method. In Fig. 9.9 the phase diagrams are plotted in terms of a dimensionless fugacity $z = n_d^R D^3$ of PHSs versus platelet concentration nD^3. The overall topology of the phase diagram from theory agrees with the one obtained from the hybrid simulation method.

The details of the phase diagrams depend on L/D and q. A few typical representative phase diagrams, calculated using the theory outlined above, are presented in Fig. 9.10a (in depletant reservoir concentrations along the ordinate) and Fig. 9.10b (ordinate plotted as system depletant concentrations). In Fig. 9.10a, the phase diagram for $L/D = 0.15$ and $q = 0.158$ shows that the I–N and N–C biphasic regions in the phase diagrams join in an I–N–C three-phase coexistence upon increasing the depletant concentration due to the widening of both I–N and N–C phase coexistences. By increasing q we encounter an I_1–I_2–N three-phase coexistence, in addition to a I_1–N–C three-phase coexistence for $L/D = 0.15$ and $q = 0.25$. Note that these triple lines lead to triple coexistences (see Fig. 9.10b).

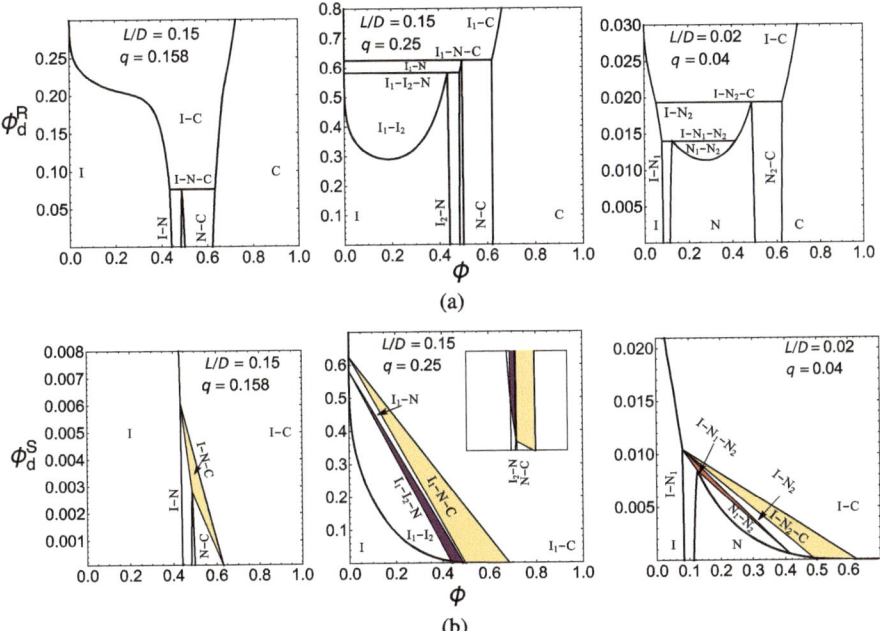

Fig. 9.10 Phase diagrams for platelet–polymer mixtures for various L/D and various q values as indicated [48]. **a** Diagrams in the $\{\phi, \phi_d^R\}$ polymer reservoir phase space. Horizontal lines mark multiple-phase coexistence. **b** As in **a**, but in the $\{\phi, \phi_d^S\}$ phase space. The coloured triangles indicate the system representation of the triple point lines on the top panels. The inset plot zooms into the low depletant concentration regime. Reprinted with permission from Ref. [48]. Copyright 2018 Taylor & Francis

For smaller L/D and q a nematic–nematic phase coexistence also becomes possible, for instance for $L/D = 0.02$ and $q = 0.04$. Now there are I–N$_1$–N$_2$ and I–N$_2$–C triple lines. As can be seen in the lower panels (b) the triple regions may be accessible experimentally because the concentrations are realistic and the regions are not too narrow. For a full overview of the possible phase diagrams, including a four-phase coexistence, see Refs. [48,53].

9.4   Experimentally Observed Phase Behaviour of Mixtures Containing Colloidal Platelets

In this section selected experimental examples of phase behaviour that includes colloidal platelets and depletants are discussed. The focus is on nonadsorbing polymers as depletants in Sects. 9.4.1 and 9.4.2, although we also illustrate some studies on colloidal mixtures of platelets and added colloidal spheres in Sects. 9.4.3–9.4.5, 9.4.4.

9.4.1 Sterically Stabilised Gibbsite Platelets Mixed with Polymers

As we have seen in Sects. 9.1 and 9.2 sterically-stabilised gibbsite platelets dispersed in toluene display a rich liquid crystal phase behaviour. With increasing concentration the isotropic phase, isotropic–nematic phase coexistence, the nematic phase, nematic–columnar phase coexistence and the columnar phase are observed [2]. Van der Kooij et al. [24] found that the depletion attraction, brought about by the addition of nonadsorbing polymer, enriches the phase behaviour of these platelet suspensions even further. They used sterically stabilised gibbsite platelets with an average diameter of 208 nm and thickness (including the thickness of the stabilising grafted polymer layer) of 14 nm, leading to an aspect ratio L/D of $1/15 \approx 0.067$.

The nonadsorbing polymer used by Van der Kooij et al. [24] is a trimethyl-siloxy terminated polydimethylsiloxane (PDMS) with a weight-averaged molar mass $M_w = 4.2 \cdot 10^5$ g/mol. The radius of gyration R_g of this polymer is estimated as 33 nm. Hence, the ratio of the polymer coil diameter over the plate diameter is about 0.3.

The observed phase behaviour is presented in Fig. 9.11. The overall topology of the plate–polymer phase diagram is characterised by a wealth of one-, two- and three-phase equilibria and even a four-phase equilibrium. Each of these phase regions can be rationalised, based on possible combinations of the I_1, I_2, N, and C phases.

In Fig. 9.12 examples of phase-separated plate–polymer mixtures are shown as observed between crossed polarisers. A calculated phase diagram for $L/D = 0.0673$ and $2\delta/D = 0.317$ using the theoretical approach outlined in Sect. 9.3 is shown in Fig. 9.13.

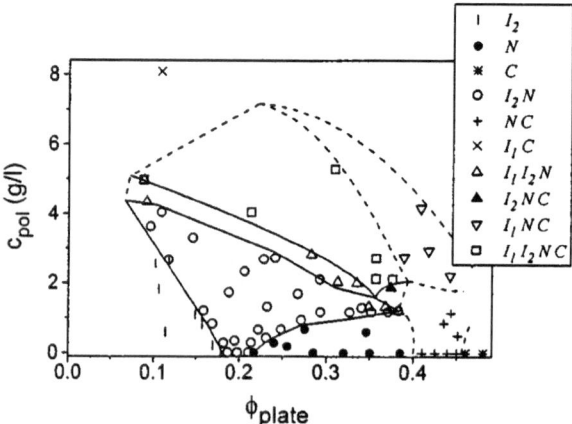

Fig. 9.11 Experimental phase diagram of gibbsite platelet–PDMS polymer mixtures [24] in toluene. Phase boundaries are indicated by solid curves, their shape and position being based on the data points they enclose, and on the consistency with surrounding phase regions. Curves are dashed in cases where the location of the phase boundary is not known precisely due to local scarcity of data points. Reprinted with permission from Ref. [24]. Copyright 2000 APS

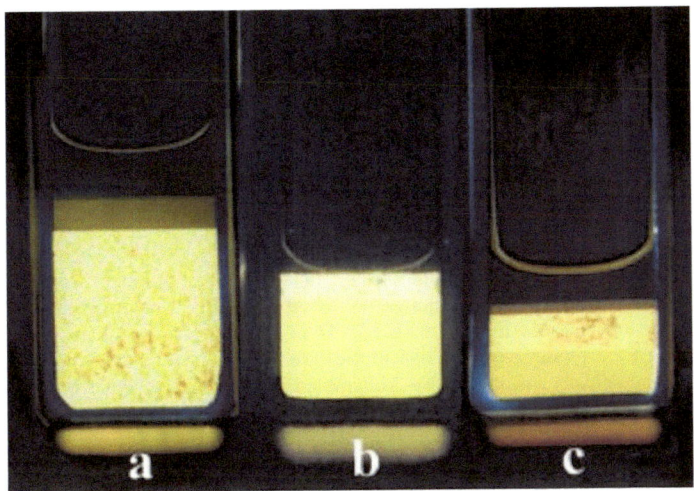

Fig. 9.12 Phase separated gibbsite platelet–PDMS polymer mixtures in toluene as observed between crossed polarisers [24]. Depicted are **a** triple phase coexistence I_1–I_2–N (*top to bottom*: dilute isotropic, concentrated isotropic I_2 and nematic N at the composition $\phi_{plate} = 0.31$ and $c_{pol} = 2.0$ g/L); **b** triple phase coexistence I_1-N-C (dilute isotropic, nematic, columnar for $\phi_{plate} = 0.44$ and $c_{pol} = 2.2$); and **c** four-phase coexistence I_1–I_2–N–C (dilute isotropic, concentrated isotropic, nematic and columnar phase for $\phi_{plate} = 0.31$ and $c_{pol} = 5.3$). Reprinted with permission from Ref. [24]. Copyright 2000 APS

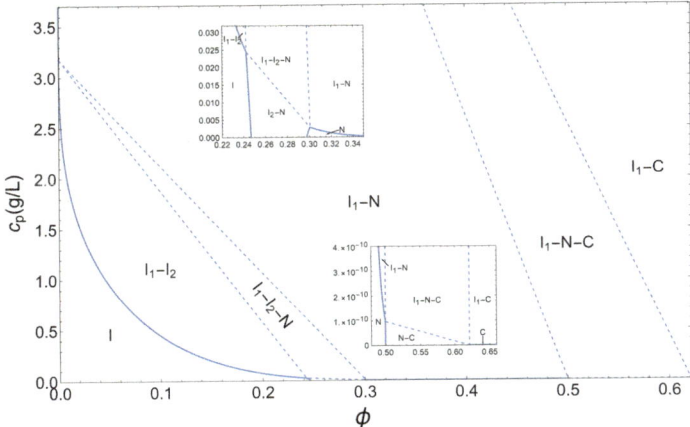

Fig. 9.13 Predicted phase diagram using free volume theory for $L/D = 0.0673$ and $2\delta/D = 0.317$, calculated using the approach outlined in Sect. 9.3, following Ref. [48]. The I_2–N and single-phase N regions are magnified in the upper inset, while the lower inset shows the N–C region

Fig. 9.14 Three-phase sedimentation equilibrium in a system of sterically stabilised gibbsite platelets [47]. **a** The complete sample between crossed polarisers, where the upper right part is digitally enhanced to visualise the I–N interface. The columnar phase contains a dark region at the upper right of the phase, probably due to the orientation of the platelets along the sample walls. Although not clearly visible, the N–C interface is horizontal and sharp. **b** Columnar phase illuminated with white light to capture the red Bragg reflections. Reprinted with permission from Ref. [47]. Copyright 2004, with permission from AIP Publishing

The richness of the observed phase diagram of colloidal plate dispersions with added nonadsorbing polymer chains (see Fig. 9.11) raises the question of how to explain the observed topology, including the four-phase region and the three-, two- and single-phase regions surrounding it. It contrasts to some degree with the theoretical prediction of Fig. 9.13. At first sight, such four-phase coexistence seems to conflict with the phase rule of Gibbs, which (at a given temperature) limits the maximum number of coexisting phases to three for a system of effectively two components (platelet–polymer). See Refs. [57,58] for further discussion.

An explanation for the observed phase diagram can be found by considering the effect of gravity [59]. The height distribution of colloidal particles in a dispersion is influenced by gravity, particularly when the sedimentation length l_{sed} (Eq. (1.1)) is much smaller than the sample height. At each height the system is thermodynamically different because locally, at each position, the external gravity field provides a different potential energy to the system. This implies that gravity has an influence on the system over the length scale of the sample, and therefore mediates the number of coexisting phases present, as well as their stacking. One can account for this using a so-called local density approximation (LDA), which assumes that at any height there is a local equilibrium.

Exercise 9.2. Argue why very rich apparent multi-phase coexistence is expected for a polydisperse colloidal dispersion in the field of gravity in case of a significant solvent-particle density difference.

We first consider the effect of gravity in a system of platelets without added polymer. Van der Beek et al. [47] observed that a suspension of sterically stabilised gibbsite platelets, which is initially an isotropic–nematic biphasic sample, develops a columnar phase on the bottom after prolonged standing (Fig. 9.14). By employing the theoretical approach of Wensink and Lekkerkerker [59], Van der Beek et al. [47] presented a simple calculation of the heights of the phases based on the sedimentation–diffusion equilibrium using the LDA. Consider a suspension of monodisperse, hard discs with number density $n(z)$ at position z and buoyant mass $m^* = v_0 \Delta \rho$, where v_0 is the colloid particle volume and $\Delta \rho$ is the difference in the mass densities of the colloidal particles and solvent. The condition for sedimentation–diffusion equilibrium reads as

$$-\left(\frac{\partial \Pi}{\partial n}\right)_{T,\mu_{\text{solvent}}} \frac{\partial n}{\partial z} = m^* g n. \tag{9.33}$$

It is convenient to use reduced quantities. The osmotic pressure of platelets is given by $\tilde{\Pi} = \Pi D^3 / kT$, the reduced concentration by $\tilde{n} = n D^3$, and positions can be scaled with the sedimentation length ℓ_{sed}. Substituting these expressions in Eq. (9.33) yields

$$-\frac{1}{\tilde{n}}\left(\frac{\partial \tilde{\Pi}}{\partial \tilde{n}}\right)_{T,\mu_{\text{solvent}}} d\tilde{n} = \frac{dz}{\ell_{\text{sed}}}. \tag{9.34}$$

The height $H = z_{\text{top}} - z_{\text{bottom}}$ can be found for a single-phase state by integrating (Eq. (9.34)) from the bottom to the top of that phase:

$$H = \int_{z_{\text{bottom}}}^{z_{\text{top}}} dz = -\ell_{\text{sed}} \int_{\tilde{n}_{\text{bottom}}}^{\tilde{n}_{\text{top}}} \frac{1}{\tilde{n}}\left(\frac{\partial \tilde{\Pi}}{\partial \tilde{n}}\right)_{T,\mu_{\text{solvent}}} d\tilde{n}. \tag{9.35}$$

The average concentration $\bar{n}$ of this phase now follows as

$$\bar{n} = \frac{\int_{z_{\text{bottom}}}^{z_{\text{top}}} \tilde{n}(z)dz}{\int_{z_{\text{bottom}}}^{z_{\text{top}}} dz} = \frac{1}{H} \int_{\tilde{n}_{\text{bottom}}}^{\tilde{n}_{\text{top}}} \tilde{n}(z)\left(\frac{\partial z}{\partial \tilde{n}}\right) d\tilde{n}. \tag{9.36}$$

Using (Eq. (9.34)), this yields

$$\bar{n} = \frac{\ell_{\text{sed}}}{H} \left[\tilde{\Pi}(\tilde{n}_{\text{bottom}}) - \tilde{\Pi}(\tilde{n}_{\text{top}})\right]. \tag{9.37}$$

For a sedimentation equilibrium that includes multi-phase coexistences, Eqs. (9.35) and (9.37) apply to every phase. The total sample height H_{sample} can be written as the sum of all individual phase heights H^i:

$$H_{\text{sample}} = \sum_i H^i. \tag{9.38}$$

The average overall sample concentration $\bar{n}_{\text{sample}}$ can now be written as

$$\bar{n}_{\text{sample}} = \frac{1}{H_{\text{sample}}} \sum_i H^i \bar{n}^i, \tag{9.39a}$$

$$= \frac{\ell_{\text{sed}}}{H_{\text{sample}}} \sum_i \left[\tilde{\Pi}(\tilde{n}_{\text{bottom}}^i) - \tilde{\Pi}(\tilde{n}_{\text{top}}^i)\right], \tag{9.39b}$$

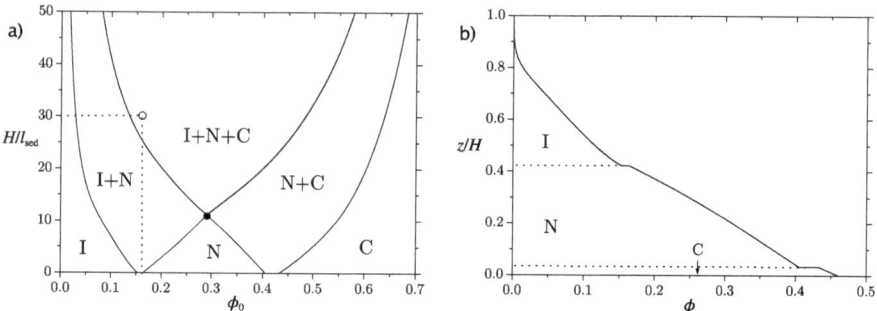

Fig. 9.15 **a** Phase diagram for colloidal platelets with $L/D = 0.05$ in a gravitational field. Plotted is the reduced sample height H/ℓ_{sed} versus the overall plate volume fraction ϕ_0. The three-phase region opens up at $H = 11.15\ell_{sed}$. **b** Concentration profile of a sample with overall volume fraction $\phi_0 = 0.157$ and vessel height $H = 30\ell_{sed}$ (corresponding to the open dot in **a**). Plotted is the relative height z/H as a function of ϕ. The I–N and N–C phase boundaries are indicated by the horizontal dotted lines. Reprinted with permission Ref. [59]. Copyright 2004 Institute of Physics (IOP)

where the average phase concentration of phase i is given by $\bar{n}^i$. For two coexisting phases A and B (where A is on top of B) the osmotic pressures are equal:

$$\widetilde{\Pi}(\tilde{n}^{A}_{bottom}) = \widetilde{\Pi}(\tilde{n}^{B}_{top}); \qquad (9.40)$$

so, Eqs. (9.39b) and (9.37) lead to

$$\bar{n}_{sample} = \frac{l_{sed}}{H_{sample}}\left[\widetilde{\Pi}(\tilde{n}^{sample}_{bottom}) - \widetilde{\Pi}(\tilde{n}^{sample}_{top})\right]. \qquad (9.41)$$

Using the above equations and computer simulation data for the equation of state for cut spheres ($L/D = 1/20$) from Zhang, Reynolds and Van Duijneveldt [46] the phase diagram for colloidal platelets with [59] can be calculated. Results are plotted in Fig. 9.15. The calculated heights of the I, N and C phases are in reasonable agreement with the experimental data for the sample shown in Fig. 9.14.

The role of gravity on the phase behaviour of mixtures of colloidal plates with nonadsorbing polymer is more complicated but follows the same lines. Wensink and Lekkerkerker [59] performed calculations for $L/D = 1/20$ and a ratio of the polymer coil diameter over the plate diameter of 0.355 to mimic the experimental system of Ref. [24]. They obtained the phase diagram shown in Fig. 9.16 for a sample with a height of 15 mm, a gravitational length of $\ell_{sed} = 0.9$ mm for the platelets, and $\ell_{sed} \rightarrow \infty$ for the polymers. All the experimentally observed multi-phase equilibria shown in Fig. 9.11 appear—even the four-phase equilibrium.

De las Heras and Schmidt [60] also used the local density approximation (LDA) but applied it to account for multiple sedimenting components. The LDA implies that at any z there is a chemical potential $\mu_i(z)$ for each component that can be expressed as [61,62]

$$\mu_i(z) = \mu_i^{bulk} - zgm_{i,eff}, \qquad (9.42)$$

Fig. 9.16 Phase diagram under gravity of a plate–polymer mixture with $L/D = 1/20$ and $q = 0.355$ in the representation of dimensionless polymer fugacity $z_p D^3 = n_d^R{}^3 D^3$ versus volume fraction representation for a vessel height of 15 mm ($H = 16.67\ell_{sed}$). Reprinted with permission from Ref. [59]. Copyright 2004 IOP

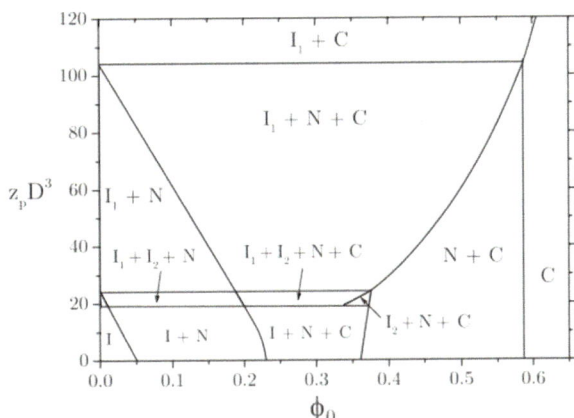

where $\widetilde{\mu}_i^{bulk}$ is the bulk chemical potential of each component i. In Eq. (9.42) an external potential due to gravity is given by $zgm_{i,eff}$, with $m_{i,eff}$ denoting the buoyant mass of component i. This means that along the sample height there is a spectrum of chemical potentials for every component, termed the sedimentation path. A possible scenario for a binary mixture composed of components 1 and 2 is sketched in Fig. 9.17. Using an appropriate model for the thermodynamics of the bulk (e.g. DFT, FVT, TPT), De las Heras and Schmidt related the bulk phase diagram to its phase stacking in the field of gravity [60,61,63]. This enables one to predict the phase states in the field of gravity.

Along the dashed line of Fig. 9.17 the chemical potentials of both components from top to bottom are now position-dependent due to the external field of gravity. In this hypothetical example, two-phase transitions occur along the sedimentation path. At the bottom there is a certain phase A of mixed components 1 and 2. However, for

Fig. 9.17 Hypothetical bulk phase diagram of a binary mixture of two components 1 and 2 in terms of the chemical potentials of these components μ_1 and μ_2. Two phases are possible in this diagram: A (left region) and B (right region). The solid curve is a binodal at which two phases A and B coexist. The dashed line is the sedimentation path along the sample with height H. Reprinted with permission from Ref. [61]. Copyright 2015 IOP

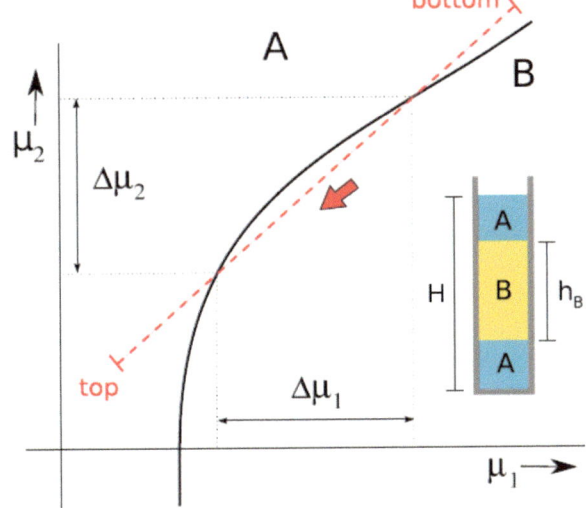

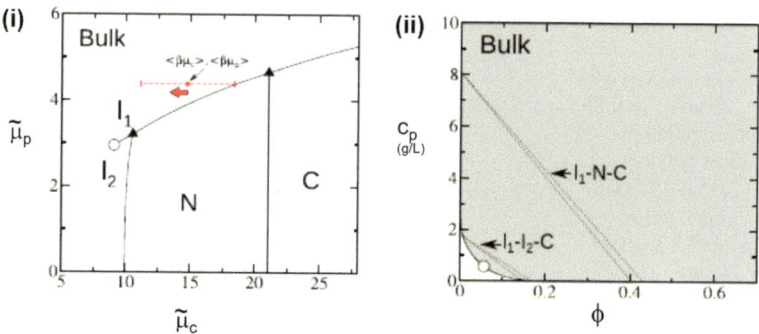

Fig. 9.18 Bulk phase diagrams for mixtures of colloidal platelets and nonadsorbing polymers in terms of (i) $(\widetilde{\mu}_p, \widetilde{\mu}_c)$ plots and (ii) concentrations $\{c_p, \phi\}$. Reprinted with permission from Ref. [61]. Copyright 2015 IOP

a certain range of chemical potentials, phase B is the preferred phase state. Close to the top, phase A is again the preferred phase state. This explains the possibility of a floating phase (B in this case) between two A phases [63,64]. The chemical potential differences $\Delta\mu_i$ are directly related to the height of phase B:

$$\Delta\mu_i = -h_B g m_{i,\text{eff}}. \tag{9.43}$$

In this case the stacking sequence is ABA.

De las Heras and Schmidt [61] applied the LDA approach to various colloidal mixtures, including mixtures of sterically stabilised gibbsite platelets ($D = 208$ nm, $L = 14$ nm; $L/D \approx 0.067$) with PDMS polymers ($R_g = 33$ nm; $M_w = 4.2 \cdot 10^5$ g/mol, so $q \approx 0.32$) [24] as were discussed earlier (see the experimental phase diagram in Fig. 9.11). The difference with the approach presented above is that De las Heras and Schmidt [61] took into account both sedimenting components explicitly, which is especially essential for the description of multi-component colloidal mixtures [65,66].

The bulk diagram for this mixture was computed by De las Heras and Schmidt [61] using the perturbation approach of Zhang, Reynolds and van Duijneveldt [46] for $L/D = 0.05$ and $q = 0.35$. It is presented in Fig. 9.18i in terms of a chemical potential plot $(\widetilde{\mu}_p, \widetilde{\mu}_c)$, where $\widetilde{\mu}_p$ is the (normalised) chemical potential of the nonadsorbing polymers and $\widetilde{\mu}_c$ is the chemical potential of the colloidal platelets. This phase diagram is relatively simple. There are two isotropic phases, a nematic phase and a columnar phase. Additionally, there are two triple points (I_1–I_2–N and I_1–N–C; ▲) and an isostructural isotropic critical point (○).

Each sedimentation path has an associated stacking sequence, for instance, CNI_1. The complete set of paths can be represented in a stacking diagram, e.g. in the plane of average chemical potential along the path. Each point in the stacking diagram corresponds to a sedimentation path in bulk. Boundaries in the stacking diagram between different stacking sequences are connected to paths that cross a binodal. A tiny change of such a path can alter the stacking sequence.

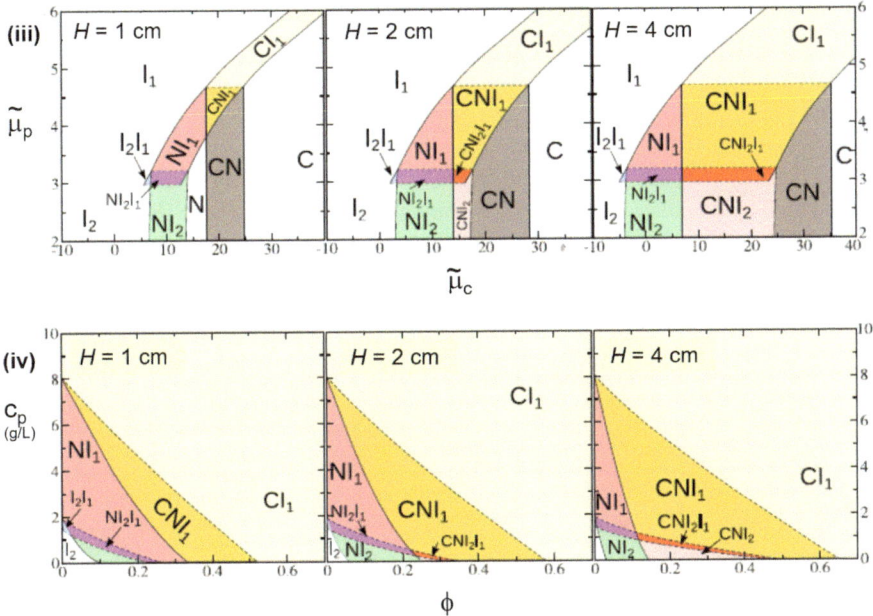

Fig. 9.19 Sample height-dependent phase diagrams for mixtures of (both) sedimenting colloidal platelets and nonadsorbing polymers in terms of (iii) $(\tilde{\mu}_p, \tilde{\mu}_c)$ plots and (iv) concentrations $\{c_p, \phi\}$. Reprinted with permission from Ref. [61]. Copyright 2015 IOP

In Fig. 9.18(ii) the bulk phase diagram is presented, which follows from (i) but is now given in terms of the polymer concentration (c_p (g/l)) and volume fraction of platelets ϕ. In terms of concentrations the triple points now become (small) regions and the coexistence lines are now wide regions.

In Fig. 9.19 the stacking diagrams are plotted at heights of $H = 1$, 2 and 4 cm. These phase diagrams of finite heights are much richer and also include a quadruple region (I_1–I_2–N–C). The phase diagrams for various heights are shown in Fig. 9.19(iv) and reveal a quadruple region (I_1–I_2–N–C) that is absent in (iii). Comparison with Fig. 9.11 shows that the phase diagrams in Fig. 9.19, and especially those for $H = 2$ and 4 cm, are much closer to what is observed experimentally. These phase diagrams of finite heights exhibit a quadruple region (I_1–I_2–N–C) that was observed in Ref. [24].

A full quantitative comparison requires exact knowledge of the height distributions of both particles in the field of gravity. This is a challenge since there are differences in sample preparation methods and solvent evaporation is possible. Further, polydispersity also affects the details of the experimental phase diagram.

9.4.2 Mixtures of Magnesium Aluminide Layered Double Hydroxide Platelets and Polymers

Liu et al. [6] observed isotropic–nematic phase coexistence in aqueous suspensions of Mg_2Al-layered double hydroxide platelets ($D = 120$ nm, $L = 3.2$ nm). The same research group [67] also studied the phase behaviour of mixtures of Mg_2Al-layered double hydroxide platelets and nonadsorbing polyvinylpyrrolidone (PVP, $M_p = 630$ kg/mol). The radius of gyration of the PVP used in water is 42 nm, hence, in the dilute polymer concentration regime, one estimates a depletion thickness (see Chaps. 2 and 4) $\delta \approx 1.13 R_g = 47$ nm. The observations of the phase behaviour are shown in Fig. 9.20.

In Fig. 9.21 the transient phase transition for a sample is shown as a function of time.

The predicted theoretical phase diagram for $L/D = 3.2/120$ and $q = 94/120 = 0.78$ is plotted in Fig. 9.22. Calculations were done using FVT as outlined in this chapter.

This predicted theoretical phase diagram reveals an expected I_1–I_2–N three-phase coexistence region, but not a four-phase I_1–I_2–N_1–N_2 equilibrium, as is observed (see Fig. 9.20). The difference can probably be explained by effects of gravity and polydispersity.

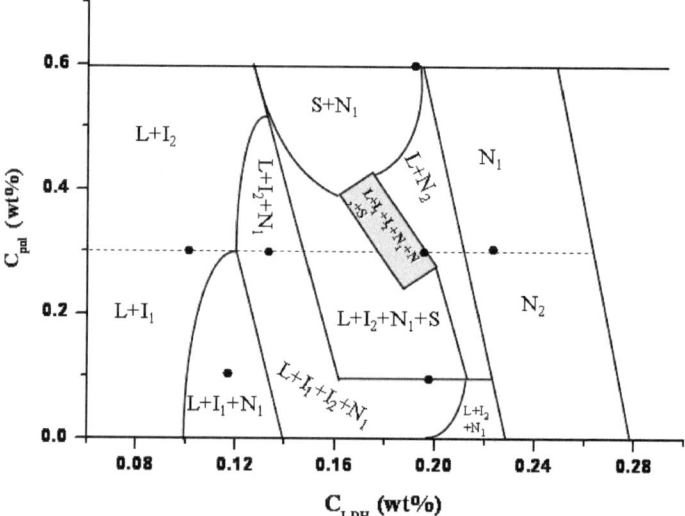

Fig. 9.20 Phase states and coexistence region observed in mixtures of Mg_2Al layered double hydroxide platelets and PVP. The solid curves represent estimated phase boundaries between different regions: liquid phase (L), dilute isotropic region (I_1), concentrated isotropic region (I_2), faint birefringent (dilute nematic) phase (N_1), concentrated nematic phase (N_2) and sediment phase (S). Data points are experimental observations. Reprinted with permission from Ref. [67]. Copyright 2009 ACS

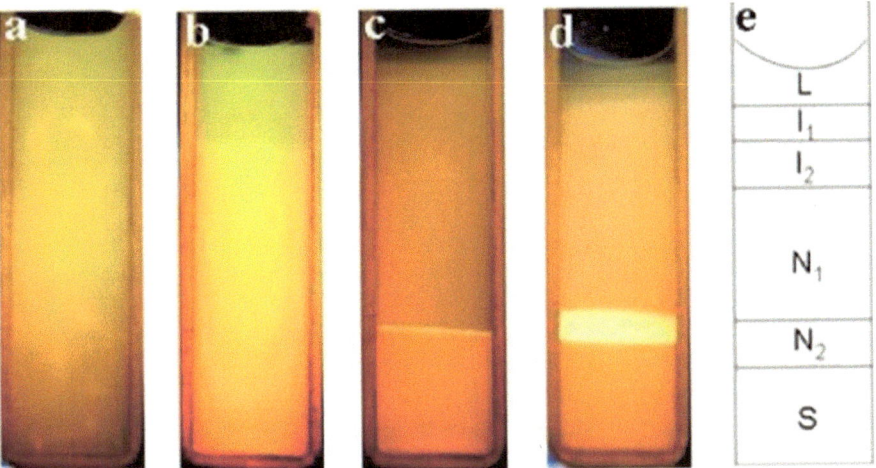

Fig. 9.21 Illustration of the evolution of the phase states of a mixture of Mg_2Al layered double hydroxide platelets (20 wt %) and PVP (0.3 wt %), as observed using crossed polarisers [67]. **a** Just after preparation, **b** after 2 days, **c** after 30 days and **d** after 55 days. **e** Schematic representation of the multi-phase coexistence regions in the sample. Reprinted with permission from Ref. [67]. Copyright 2009 ACS

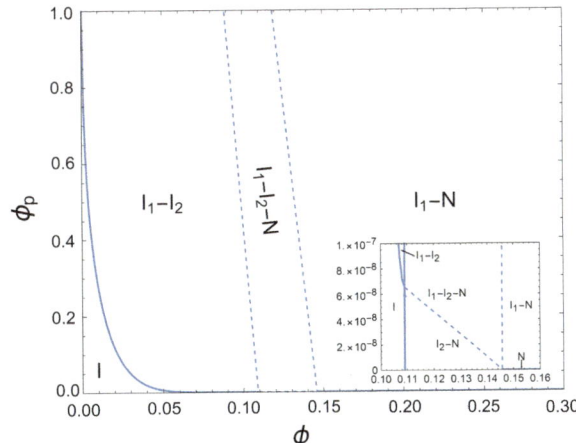

Fig. 9.22 Phase diagram of hard platelets $(L/D = 3.2/120 = 0.0267)$ mixed with nonadsorbing polymer chains modelled as PHSs; $q = 94/120 = 0.78$, calculated using the approach outlined in Sect. 9.3, following Refs. [48,54]. *Inset*: magnified I_2–N region

9.4.3   Gibbsite Plate—Silica Sphere Mixtures

Doshi et al. [63,68,69] studied the phase behaviour of aqueous suspensions containing gibbsite platelets mixed with alumina-coated silica spheres with diameter σ. See Fig. 9.23 for an illustration of the alumina coating, chemically bound to a silica sphere.

To inhibit double layer repulsions between the particles 5 mM NaCl was added, and commercially available stabilisers (Solplus D450 and Solsperse 41,000) were adsorbed on the particle surface to create near-hard particle interactions. The dimen-

Fig. 9.23 Sketch of the alumina coating at the silica surface. Reprinted from Ref. [69] with permission of the authors, copyright 2011

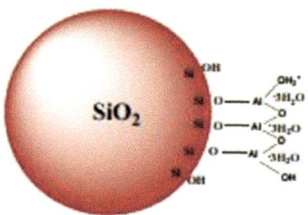

Table 9.2 Mean dimensions of the plates and spheres discussed in this section. Dimensions include size dispersity from TEM/AFM and average bare and effective dimensions of plates and spheres from scattering data

Particle	Dimension	Characterisation method			
		TEM/AFM		Scattering	
		Mean	Dispersity	Bare value	Effective value
		(nm)	(%)	(nm)	(nm)
gibbsite	$\langle D \rangle$	203	20	183	191
gibbsite	$\langle L \rangle$	5	21	7	15
Klebosol 30CAL25	$\langle \sigma \rangle$	30	15	30	40
Klebosol 30CAL50	$\langle \sigma \rangle$	74	21	74	90

sions of the particles studied are given in Table 9.2; bare and effective dimensions are quoted. The effective dimensions of the spheres and plates take into account the additional layer that is due to steric stabilisation or the Debye length of 4 nm at 5 mM of NaCl.

Figure 9.24a presents a TEM micrograph of the gibbsite plates and Klebosol 30CAL25 alumina-coated silica spheres, while panels (b) and (c) show phase diagrams of these gibbsite plate–silica sphere mixtures. The effective size ratios for a mixture of gibbsite platelets with Klebosol 30CAL25 silica spheres are $L_{eff}/D_{eff} = 0.08$ and $\sigma_{eff}/D_{eff} = 0.21$, so on theoretical grounds we only expect an I–N transition. This is confirmed experimentally, as can be seen in Fig. 9.24d, e.

For a similar mixture of gibbsite platelets and silica particles but with the larger Klebosol 30CAL50 silica spheres the phase behaviour is considerably richer. Now the ratio of $d_{eff}/D_{eff} = 0.47$ and, in addition to an isotropic–nematic phase, both an isotropic–isotropic and an isotropic–isotropic–nematic phase transitions are expected. The latter is indeed observed, as is shown in Fig. 9.24e.

De las Heras et al. [61,63] have shown that sedimentation–diffusion equilibria of binary colloidal mixtures can involve phase transitions, which can lead to complex phase stacks, such as the sandwich of a floating nematic layer between top and bottom isotropic phases. This may explain what is observed in mixtures of silica spheres and gibbsite platelets.

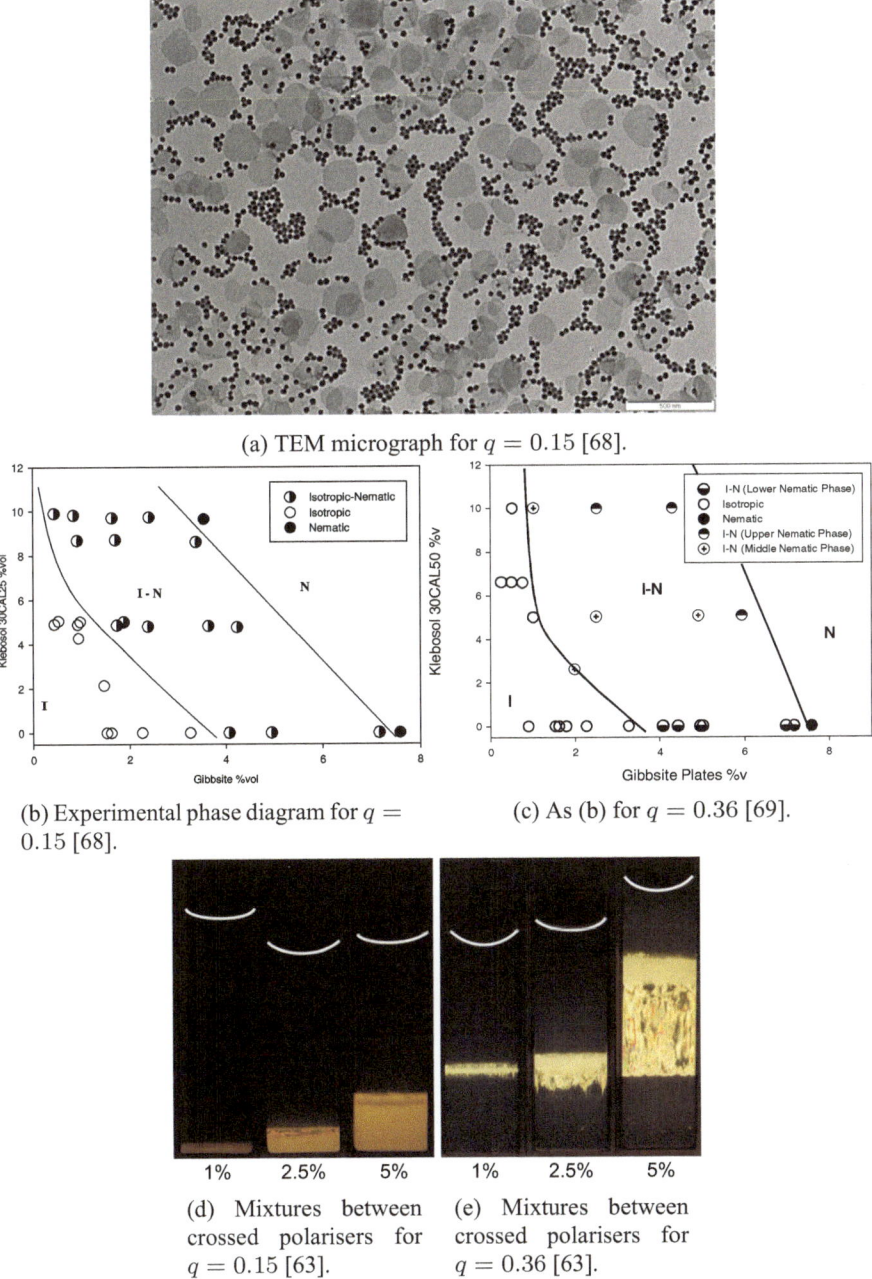

(a) TEM micrograph for $q = 0.15$ [68].

(b) Experimental phase diagram for $q = 0.15$ [68].

(c) As (b) for $q = 0.36$ [69].

(d) Mixtures between crossed polarisers for $q = 0.15$ [63].

(e) Mixtures between crossed polarisers for $q = 0.36$ [63].

Fig. 9.24 Plate–sphere mixtures composed of gibbsite platelets and Klebosol silica spheres. Particle dimensions: **a, b, d** silica $\sigma = 30$ nm, **c, e** silica $\sigma = 74$ nm. Platelet dimensions $\langle D \rangle = 203$ nm ± 20 %, $\langle L \rangle = 5$ nm ± 20 %. **d, e** $\phi_{\text{silica}} = 0.05$, gibbsite platelet volume fractions as indicated. Reprinted with permission from **a, b** Ref. [68], copyright 2012 IOP; **c** Ref. [69], copyright 2011; **d, e** Ref. [63], copyright 2012 Springer Nature

9.4.4   Mixtures of Zirconium Phosphate Platelets and Silica Spheres

Chen et al. [70,71] studied the phase behaviour of aqueous mixtures of zirconium phosphate (ZrP) platelets with silica spheres. In Ref. [71] they used ZrP $(D = 704.3\,\text{nm}, L = 2.68\,\text{nm})$ and added silica spheres $(\sigma = 162.7\,\text{nm})$.

Exercise 9.3. In aqueous dispersion platelets are typically charged. How would the I–N phase transition be affected upon adding salt?

The size ratio $q = \sigma/D$ of the diameters of the silica spheres and ZrP plates is 0.23; so we expect—in addition to the biphasic equilibria I_1–I_2, I_2–N and I_1–N—a triphasic phase triangle region I_1–I_2–N. The latter is indeed observed (see the left panel of Fig. 9.25).

Again, a nematic phase is observed floating between two isotropic phases. Note that the amounts of the phases I_1, I_2, and N reflect the positions of the samples A, B and C denoted in the three-phase triangle denoted by points O (I_1), P (I_2) and Q (N). From applying the lever rule [72] to the triangle region of the phase diagram (see left panel of Fig. 9.25) it follows that state point A, which is close to the I_1 vertex, will have a relatively large amount of phase I_1. Similarly, state point C, which is close to the I_2 vertex, has a relatively large amount of phase I_2. This is indeed seen experimentally (Fig. 9.25, right panel). Chen et al. [70] also studied the phase behaviour of a mixture of ZrP platelets and silica spheres with a size ratio $q = \sigma/D = 0.013$. As expected, an I–N_1–N_2 phase equilibrium is now observed (Fig. 9.26).

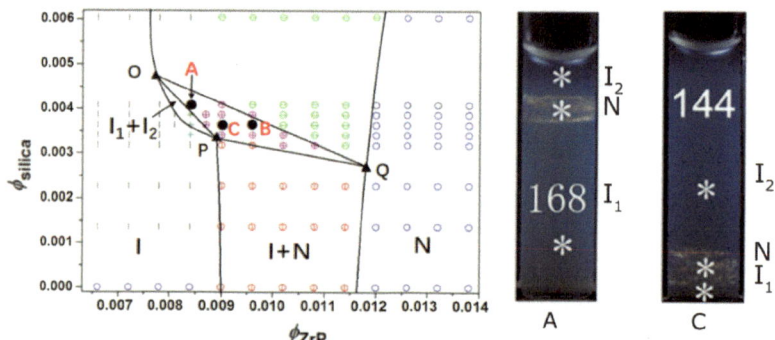

Fig. 9.25 *Left*: experimental phase diagram of aqueous mixtures of ZrP platelets and silica spheres [71]. *Right*: photographs of the (red) state points A and C in the phase diagram. The numbers on the tubes indicate the number of hours lapsed after preparing the samples. Reprinted with permission from Ref. [71]. Copyright 2017 RSC

Fig. 9.26 Triphasic
$I–N_1–N_2$ equilibrium
observed for an aqueous
mixture of ZrP platelets and
silica spheres with a
sphere–platelet size ratio
$q = \sigma/D$ of 0.013. The ZrP
volume fraction is 0.0063
and the silica concentration
is 2 wt % (corresponding to a
volume fraction of 0.015)
Reprinted with permission
from Ref. [70]. Copyright
2015 RSC

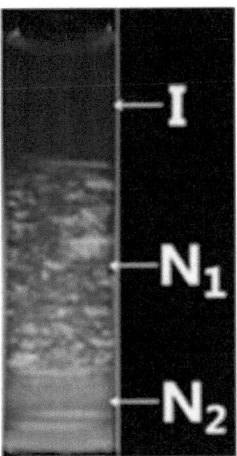

9.4.5 Effect of Added Silica Nanoparticles on the Nematic Liquid Crystal Phase Formation in Beidellite Suspensions

While virtually all smectite clays dispersed in water form gels at very low concentrations, aqueous suspensions of beidellite [11] as well as nontronite [8] exhibit a unique behaviour with a first order isotropic–nematic phase transition before gel formation. Landman et al. [73] studied the modification of the phase behaviour of beidellite suspensions upon addition of colloidal silica spheres. TEM images of their beidellite platelets are reproduced in Fig. 9.27a. Figure 9.27b shows a TEM micrograph of the silica spheres. Images giving some indications of the phase behaviour of the pure beidellite suspensions are presented in Fig. 9.28.

Note that sample (a) ($\phi = 0.27\%$) is still in the isotropic phase sample, (e) ($\phi = 0.40\%$) is completely nematic, and sample (f) ($\phi = 0.41\%$) is a nematic gel. Adding silica nanoparticles to the nematic gel sample (f) leads to an isotropic–nematic phase equilibrium as is displayed in Fig. 9.29.

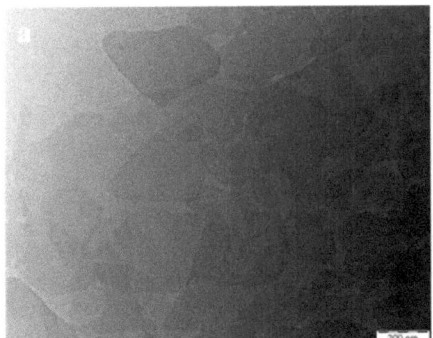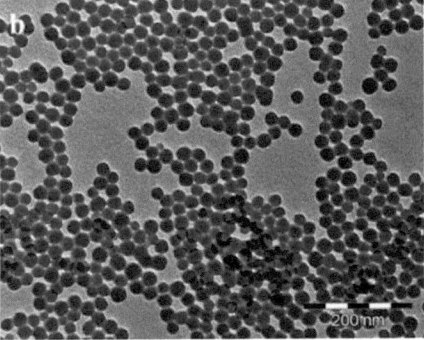

Fig. 9.27 TEM micrographs of **a** beidellite platelets and **b** silica Ludox AS-40 spheres used. Reprinted with permission from Ref. [73]. Copyright 2014 ACS

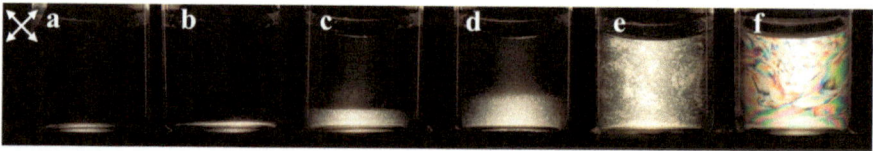

Fig. 9.28 Aqueous beidellite suspensions observed between crossed polarisers one month after preparation. Volume fractions of clay (ϕ_{clay}) are as follows: **a** 0.27%; **b** 0.32%; **c** 0.35%; **d** 0.37%; **e** 0.40%; **f** 0.41%. Note that the tiny bright layer just at the glass bottom of the vial in **a** is not due to a nematic phase but is a light reflection artefact. Reprinted with permission from Ref. [73]. Copyright 2014 ACS

Fig. 9.29 Mixed beidellite/silica suspensions observed between crossed polarisers one month after preparation: $\phi_{clay} = 0.41\%$ and (from left to right) $\phi_{silica} = 0, 0.034$ and 0.138%. Reprinted with permission from Ref. [73]. Copyright 2014 ACS

Fig. 9.30 Experimental phase diagram of aqueous beidellite/silica suspensions. (○) Isotropic, (◑) biphasic and (⋆) gelled states are indicated. The boundary between the isotropic and the biphasic samples was obtained from naked-eye observations of the test-tubes, while the sol–gel transition line was determined by rheological measurements. Reprinted with permission from Ref. [73]. Copyright 2014 ACS

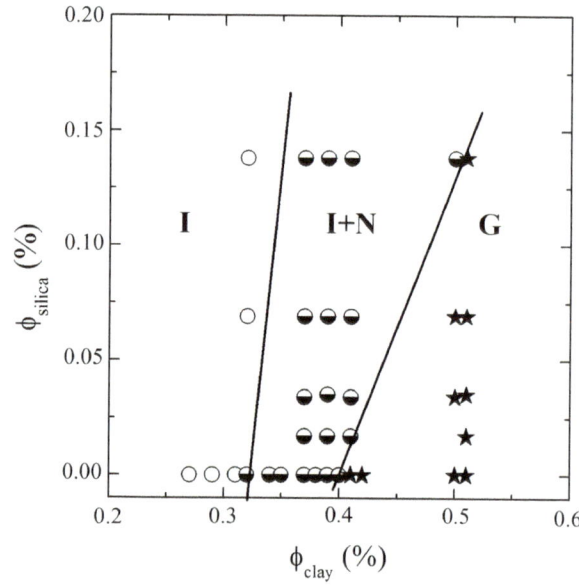

The modification of the beidellite phase diagram due to the addition of silica nanoparticles is indicated in the images of Fig. 9.30. Note that the addition of a tiny amount of silica nanoparticles (volume fraction of 10^{-3}) has a significant effect on the phase diagram.

References

1. Van der Kooij, F.M., Lekkerkerker, H.N.W.: J. Phys. Chem. B **102**, 7829 (1998)
2. Van der Kooij, F.M., Kassapidou, E., Lekkerkerker, H.N.W.: Nature **406**, 868 (2000)
3. Van der Beek, D., Lekkerkerker, H.N.W.: Langmuir **20**, 8582 (2004)
4. Vis, M., Wensink, H.H., Lekkerkerker, H.N.W., Kleshchanok, D.: Mol. Phys. **113**, 1053 (2015)
5. Brown, A.B.D., Ferrero, C., Narayanan, T., Rennie, A.R.: Eur. Phys. J. B **11**, 481 (1999)
6. Liu, S., Zhang, J., Wang, N., Liu, W., Zhang, C., Sun, D.: Chem. Mater. **15**, 3240 (2003)
7. Wang, N., Liu, S., Zhang, J., Wu, Z., Chen, J., Sun, D.: Soft Matter **1**, 428 (2005)
8. Michot, L.J., Bihannic, I., Maddi, S., Funari, S.S., Baravian, C., Levitz, P., Davidson, P.: Proc. Natl. Acad. Sci. **103**, 16101 (2006)
9. Michot, L.J., Bihannic, I., Maddi, S., Baravian, C., Levitz, P., Davidson, P.: Langmuir **24**, 3127 (2008)
10. Michot, L.J., Paineau, E., Bihannic, I., Maddi, S., Duval, J.F.L., Baravian, C., Davidson, P., Levitz, P.: Clay Miner. **48**, 663 (2013)
11. Paineau, E., Antonova, K., Baravian, C., Bihannic, I., Davidson, P., Dozov, I., Impéror-Clerc, M., Levitz, P., Madsen, A., Meneau, F., Michot, L.J.: J. Phys. Chem. B **113**, 15858 (2009)
12. Paineau, E., Bihannic, I., Baravian, C., Philippe, A.M., Davidson, P., Levitz, P., S.S. F., Rochas, C., Michot, L.J.: Langmuir **27**, 5562 (2011)
13. Fonseca, D.M., Méheust, Y., Fossum, J.O., Knudsen, K.D., Parmar, K.P.S.: Phys. Rev. E **79**, 021402 (2009)
14. Miyamoto, N., Iijima, H., Ohkubo, H., Yamauchi, Y.: Chem. Commun. **46**, 4166 (2010)
15. Gabriel, J.C., Camerel, F., Lemaire, B., Desvaux, H., Davidson, P., Batail, P.: Nature **413**, 504–508 (2001)
16. Davidson, P., Penisson, C., Constantin, D., Gabriel, J.C.: Proc. Natl. Acad. Sci. **115**, 6662 (2018)
17. Mejia, A.F., Chang, Y.W., Ng, R., Shuai, M., Mannan, M.S., Cheng, Z.: Phys. Rev. E **85**, 061708 (2012)
18. Chen, F., Chen, M., Chang, Y.W., Lin, P., Chen, Y., Cheng, Z.: Soft Matter **13**, 3789 (2017)
19. Sun, D., Sue, H.J., Cheng, Z., Martínez-Ratón, Y., Velasco, E.: Phys. Rev. E **80**, 041704 (2009)
20. Miyamoto, N., Nakato, T.: Adv. Mat. **14**, 1267 (2002)
21. Miyamoto, N., Nakato, T.: J. Phys. Chem. B **108**, 6152 (2004)
22. Nakato, T., Yamashita, Y., Mouri, E., Kuroda, K.: Soft Matter **10**, 3161 (2014)
23. Davidson, P.: Personal communication
24. Van der Kooij, F.M., Vogel, M., Lekkerkerker, H.N.W.: Phys. Rev. E **62**, 5397 (2000)
25. Kleshchanok, D., Holmqvist, P., Meijer, J.M., Lekkerkerker, H.N.W.: J. Am. Chem. Soc. **134**(13), 5985 (2012)
26. Martínez-Ratón, Y., Velasco, E.: J. Chem. Phys. **134**, 124904 (2011)
27. Santaguida, S., Musacchio, A.: EMBO J. **28**, 2511 (2009)
28. Bailey, L., Lekkerkerker, H.N.W., Maitland, G.C.: Soft Matter **11**, 222 (2015)
29. Nakato, T., Yamada, Y., Miyamoto, N.: J. Phys. Chem. B **113**, 1323 (2009)
30. Onsager, L.: Ann. NY Acad. Sci. **51**, 627 (1949)
31. Frenkel, D.: J. Phys. Chem. **91**, 4912 (1987)
32. Frenkel, D.: J. Phys. Chem. **92**, 3280 (1988)
33. Forsyth, P.A., Marcelja, S., Mitchell, D.J., Ninham, B.W.: J. Chem. Soc., Faraday Trans. 2 **73**, 84 (1977)
34. Eppenga, R., Frenkel, D.: Mol. Phys. **52**, 1303 (1984)

35. Veerman, J.A.C., Frenkel, D.: Phys. Rev. A **45**, 5632 (1992)
36. Qazi, S.J.S., Karlsson, G., Rennie, A.R.: J. Colloid Interface Sci. **348**, 80 (2010)
37. Parsons, J.D.: Phys. Rev. A **19**, 1225 (1979)
38. Lee, S.D.: J. Chem. Phys. **87**, 4972 (1987)
39. Wensink, H.H., Lekkerkerker, H.N.W.: Mol. Phys. **107**, 2111 (2009)
40. Wu, L., Wensink, H.H., Jackson, G., Müller, E.A.: Mol. Phys. **110**, 1296 (2012)
41. Carnahan, N.F., Starling, K.E.: J. Chem. Phys. **51**, 635 (1969)
42. Odijk, T.: Macromolecules **19**, 2313 (1986)
43. Lekkerkerker, H.N.W., Tuinier, R., Wensink, H.H.: Mol. Phys. **113**, 2666 (2015)
44. Wensink, H.H.: Phys. Rev. Lett. **93**, 157801 (2004)
45. Bates, M.A., Frenkel, D.: Phys. Rev. E **57**, 4824 (1998)
46. Zhang, S.D., Reynolds, P.A., van Duijneveldt, J.S.: J. Chem. Phys. **117**, 9947 (2002)
47. Van der Beek, D., Schilling, T., Lekkerkerker, H.N.W.: J. Chem. Phys. **121**, 5423 (2004)
48. González García, Á., Tuinier, R., Maring, J.V., Opdam, J., Wensink, H.H., Lekkerkerker, H.N.W.: Mol. Phys. **116**, 2757 (2018)
49. Lekkerkerker, H.N.W., Poon, W.C.K., Pusey, P.N., Stroobants, A., Warren, P.B.: Europhys. Lett. **20**, 559 (1992)
50. Widom, B.: J. Chem. Phys. **39**, 2808 (1963)
51. Helfand, E., Reiss, H., Frisch, H., Lebowitz, J.: J. Chem. Phys. **33**, 1379 (1960)
52. Lebowitz, J.L., Helfand, E., Praestgaard, E.: J. Chem. Phys. **43**, 774 (1965)
53. González García, Á., Wensink, H.H., Lekkerkerker, H.N.W., Tuinier, R.: Sci. Rep. **7**, 17058 (2017)
54. González García, Á.: Polymer-Mediated Phase Stability of Colloids. Springer, Berlin, Heidelberg (2019)
55. Zhang, S.D., Reynolds, P.A., van Duijneveldt, J.: Mol. Phys. **100**, 3041 (2002)
56. Bates, M.A., Frenkel, D.: Phys. Rev. E **62**, 5225 (2000)
57. Peters, V.F.D., Vis, M., González García, Á., Wensink, H.H., Tuinier, R.: Phys. Rev. Lett. **125**, 127803 (2020)
58. Opdam, J., Peters, V.F.D., Wensink, H.H., Tuinier, R.: J. Phys. Chem. Lett. **14**, 199 (2023)
59. Wensink, H.H., Lekkerkerker, H.N.W.: Europhys. Lett. **66**, 125 (2004)
60. de las Heras, D., Schmidt, M.: Soft Matter **9**, 8636 (2013)
61. de las Heras, D., Schmidt, M.: J. Phys.: Condens. Matter **27**, 194115 (2015)
62. Geigenfeind, T., de las Heras, D.: J. Phys.: Condens. Matter **29**, 064006 (2017)
63. de las Heras, D., Doshi, N., Cosgrove, T., Phipps, J., Gittins, D.I., Van Duijneveldt, J.S., Schmidt, M.: Sci. Rep. **2**, 789 (2012)
64. Schmidt, M., Dijkstra, M., Hansen, J.P.: Phys. Rev. Lett. **93**, 088303 (2004)
65. Eckert, T., Schmidt, M., de las Heras, D.: Commun. Phys. **4**, 202 (2021)
66. Eckert, T., Schmidt, M., de las Heras, D.: Phys. Rev. Res. **4**, 013189 (2022)
67. Luan, S., Li, W., Liu, S., Sun, D.: Langmuir **25**, 6394 (2009)
68. Doshi, N., Cinacchi, G., Van Duijneveldt, J., Cosgrove, T., Prescott, S., Grillo, I., Phipps, J., Gittins, D.: J. Phys.: Condens. Matter **23**, 194109 (2011)
69. Doshi, N.: Colloidal mixtures of spheres and plates. University of Bristol, Ph.D. thesis (2011)
70. Chen, M., Li, H., Chen, Y., Mejia, A.F., Wang, X., Cheng, Z.: Soft Matter **11**, 5775 (2015)
71. Chen, M., He, M., Lin, P., Chen, Y., Cheng, Z.: Soft Matter **13**, 4457 (2017)
72. DeHoff, R.T.: Thermodynamics in Materials Science. McGraw-Hill, Singapore (1993)
73. Landman, J., Paineau, E., Davidson, P., Bihannic, I., Michot, L.J., Philippe, A.M., Petukhov, A.V., Lekkerkerker, H.N.W.: J. Phys. Chem. B **118**, 4913 (2014)

Phase Behaviour of Colloidal Cubes Mixed with Depletants

10

In Chaps. 8 and 9 it was shown that the phase behaviour of anisotropic hard particles is considerably richer than that of hard spheres (see Sect. 3.2). Recent breakthroughs in colloidal synthesis allow the control of particle shapes and properties with high precision. This provides us with a constantly expanding library of new anisotropic building blocks, thus opening new avenues to explore colloidal self-assembly at a higher level of complexity [1,2]. One of these intriguing novel systems are cube-like colloids. In this chapter, a selective overview is given on the current knowledge of the phase behaviour of cube-like colloids with and without added depletants.

This chapter commences with an introduction to some experimental cube-like systems that have been developed, followed by an outline of the basic thermodynamics of dispersions comprising hard superballs, which are often used to model cubic colloids. Subsequently, an explanation is provided about how free volume theory can be applied to predict the phase stability of cube–polymer mixtures. We conclude by discussing the experimental work available on the phase stability of dispersions containing colloidal cubes mixed with nonadsorbing polymers.

10.1 Introduction to Colloidal Cubes

Recent improvements in the synthesis of a wide range of different types of colloidal particles have led to the realisation of well-defined building blocks that enable the structuring of matter (see also Sects. 8.1 and 9.1). The anisotropic shape of inorganic particles can be explained by the variation in growth rates of different crystal facets that develop during colloid synthesis [3,4]. The shape can be tuned by the adsorption of additives, which inhibit or accelerate the growth of certain crystal facets [5]. The inhibition or acceleration of crystal facet growth also evolved into the development of various types of cube-like colloidal particles. The ability to tune size, shape and

© The Author(s) 2024
H. N. W. Lekkerkerker, R. Tuinier, M. Vis, *Colloids and the Depletion Interaction*,
Lecture Notes in Physics 1026, https://doi.org/10.1007/978-3-031-52131-7_10

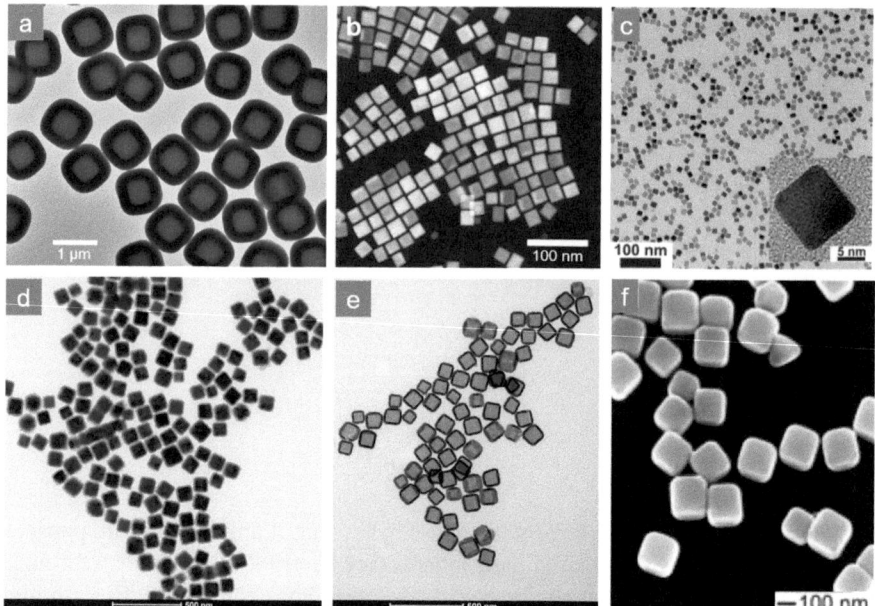

Fig. 10.1 Microscopy images of examples of cube-like colloidal particles. **a** Micrometre-sized hollow silica cubes prepared by growing a shell around iron oxide particles according to the method of Rossi et al. [20]; image kindly provided by L. Rossi. **b** Sharp Fe_3O_4 nanocubes from Ref. [32], synthesised according to the method by Park et al. [13]. **c** Pd nanocubes; *inset*: magnification [19]. **d, e** SiO_2-coated Cu_2O nanocubes [27,33], images kindly supplied by F. Dekker. The CuO_2 core is removed in **e**. **f** Ag cubes [11]. Images **b**, **c** and **f** reprinted with permission from: **b** Ref. [32] (CC-BY); **c** Ref. [19], copyright 2011 the American Physical Society [19]; **f** Ref. [11], copyright 2005 Wiley

chemistry of cube-like colloids [6–31] extends the range of possible applications. Some examples are shown in Fig. 10.1.

Many of the synthesised cube-like colloidal particles have a shape that lies in between that of a cube and a sphere. To a certain degree the shape is also tunable within that range [27,34]. A common way to quantify the shape of a cube-like particle with rounded edges is by describing it as a superball. Formally, superballs are a subset of a family of geometric shapes called superellipsoids [35]. The following expression describes the shape of a superball in Cartesian coordinates $\{x, y, z\}$ [36]:

$$\mathfrak{f}(x, y, z) = \left|\frac{x}{R}\right|^m + \left|\frac{y}{R}\right|^m + \left|\frac{z}{R}\right|^m \leq 1, \tag{10.1}$$

where R is the radius of the superball (the shortest distance from the centre of the superball to its surface), which is related to the edge length $R_{el} \equiv 2R$. The quantity m is the shape parameter. The surface of the superball is described for $\mathfrak{f}(x, y, z) = 1$, whereas the location of the material inside the superball is given by $\mathfrak{f}(x, y, z) < 1$. For $m = 2$ a sphere is recovered and $m = \infty$ corresponds to a cube. To describe cube-like colloidal particles the focus here is on $m \geq 2$. In Fig. 10.2 a collection of superballs is depicted for several m-values.

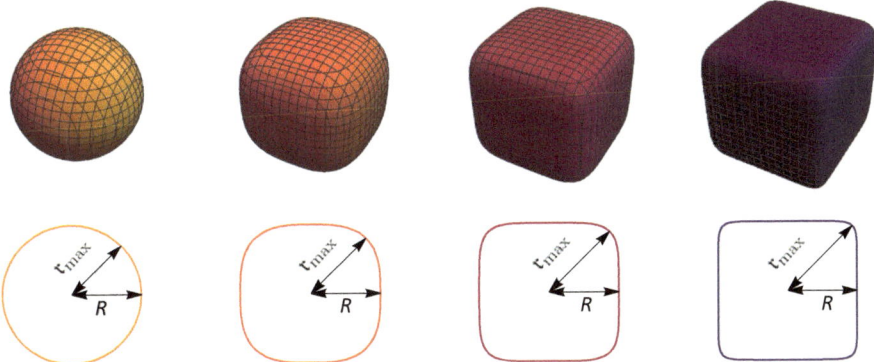

Fig. 10.2 Three-dimensional (*top panel*) and two-dimensional (*bottom panel*) representations of a superball for various values of the shape parameter (left to right): $m = 2, 3, 5,$ and 10. The superball radius (R) and the maximum distance of the superball surface to the centre (τ_{max}) are indicated. Adapted from Ref. [38] under the terms of CC-BY-4.0

Exercise 10.1. Compute the *maximum* distance τ_{max} between the particle centre and its surface in terms of R for the shapes in Fig. 10.2. *Hint*: see the Appendix of Ref. [37].

It is noted that experimentally prepared cubes may be even more accurately described as a sphube [39], or by using the Minkowski sum of a cube and a sphere [32]. Also the Minkowski particle and sphube shapes interpolate smoothly from perfect sharp cubes to perfect spheres. The advantage of the superball shape however is that it provides approximate analytic expressions for various thermodynamic properties of the particle dispersions, enabling the theoretical prediction of phase diagrams.

10.2   Equations of State of Hard Colloidal Superballs

In this section, equations of state are presented for a fluid and two solid phase states composed of hard colloidal superball dispersions. These approximate results will be used in subsequent sections to predict phase diagrams of hard superballs and superballs mixed with depletants.

10.2.1 Second (Osmotic) Virial Coefficient of Superballs

The first step is to quantify the second (osmotic) virial coefficient for superballs. To quantify the second (osmotic) virial coefficient (B_2) of hard superballs, the orientationally averaged excluded volume between two particles [40] should be calculated

Fig. 10.3 Normalised second virial coefficient B_2^* as a function of the shape parameter m. Numerical solutions are given by the grey dots (see Ref. [38] for details). The black curve shows a fit through the data points following Eq. (10.3). Adapted from Ref. [38] under the terms of CC-BY-4.0

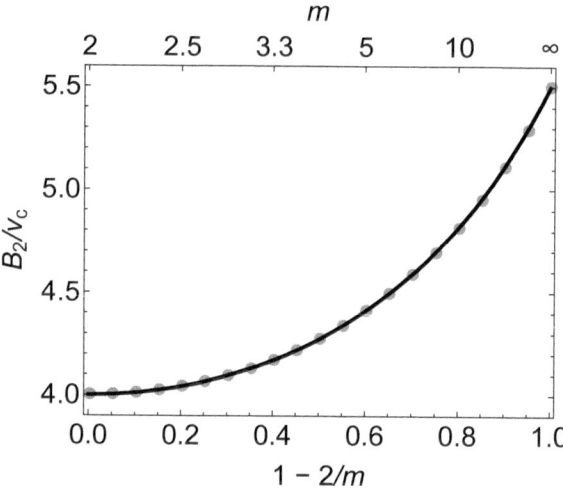

because the particles are anisotropic for $m > 2$. For convex particles (hence, also for superballs with $m \geq 2$), a general expression for B_2 reads [41,42]

$$\frac{B_2}{v_c} \equiv B_2^* = 3\varsigma + 1,$$ (10.2)

with the so-called asphericity ς, defined through

$$\varsigma = \frac{s_c c_c}{3 v_c},$$

which contains the geometrical superball characteristics volume v_c, surface area s_c and surface integrated mean curvature c_c. In [38,43] it is explained in detail how ς can be calculated as a function of the shape parameter m from numerical computation of v_c, s_c and c_c. This yields B_2^* via Eq. (10.2). The numerically obtained B_2^* values for hard superballs [38] are plotted in Fig. 10.3. It follows that B_2^* smoothly increases with m from the hard sphere (hs) limit ($m = 2$, $B_2^* = 4$) to the cube limit ($m = \infty$, $B_2^* = 5.5$). The numerical data can be described using the following closed expression for B_2 (solid curve in Fig. 10.3):

$$B_2^* \approx \frac{1}{0.42\sqrt{1 - \left(\frac{1-2/m}{1.83}\right)^2} - 0.17}.$$ (10.3)

10.2.2 Fluid Phase State of Superballs

To find an expression for the fluid state of superballs, consider a collection of N_c hard superballs in a volume V, so the volume fraction of superballs $\phi_c = N_c v_c / V$. An

accurate equation of state (EOS) for a fluid of hard convex particles was proposed by
Boublík [44–46] in terms of the reduced osmotic pressure of the pure hard superball
dispersion $\widetilde{P}_f$:

$$\widetilde{P}_f = \frac{P v_c}{kT} = \frac{\phi_c + \mathfrak{Q}\phi_c^2 + \mathfrak{R}\phi_c^3 - \mathfrak{T}\phi_c^4}{(1 - \phi_c)^3}, \tag{10.4}$$

with

$$\begin{aligned}
\mathfrak{Q} &= 3\varsigma - 2, \\
\mathfrak{R} &= 1 - 3\varsigma + 3\varsigma^2, \\
\mathfrak{T} &= 6\varsigma^2 - 5\varsigma.
\end{aligned} \tag{10.5}$$

It is noted that Gibbons [47] derived an earlier, less accurate, EOS using scaled
particle theory.

Exercise 10.2. Show that Eq. (10.4) equals the Carnahan–Starling prediction
Eq. (3.1) in the limit of hard spheres [48]. *Hint*: use Eq. (10.2) to find ς for
hard spheres.

Using the relation between B_2, ς and $\widetilde{P}_f$ of Eqs. (10.2)–(10.4), the EOS for a fluid
of hard superballs is completely defined for a given value of m. Computer simulations
have shown the accuracy of the Boublík EOS for a wide range of m-values [49,50].
In Fig. 10.4 computer simulation results for the limits of hard cubes ($m = \infty$) and
hard spheres ($m = 2$) are compared to predictions using Eq. (10.4).

The chemical potential of the superballs is related to the osmotic pressure through
the Gibbs–Duhem relation Eq. (A.12) for a single-component system at constant
temperature, which can also be written as

$$d\widetilde{\mu}_f = \frac{1}{\phi_c} \frac{d\widetilde{P}_f}{d\phi_c} d\phi_c; \tag{10.6}$$

Fig. 10.4 Volume fraction
dependence of the osmotic
pressure of hard cubes [50]
(■) and of hard spheres [51]
(●). Curves: Eq. (10.4)

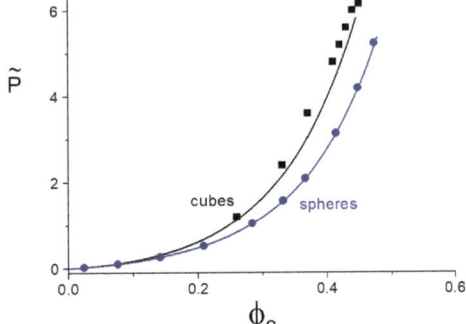

so, the chemical potential follows from combining Eqs. (10.4) and (10.6) (see also Eq. (3.5)):

$$
\begin{aligned}
\tilde{\mu}_f = \ln\left(\frac{\Lambda^3}{v_c}\right) &+ (\mathfrak{T} - 1)\ln(1 - \phi_c) + \ln\phi_c \\
&+ \frac{(10 + 4\mathfrak{Q} + 2\mathfrak{T})\phi_c - (13 + 3\mathfrak{Q} - 3\mathfrak{R} + 5\mathfrak{T})\phi_c^2}{2(1 - \phi_c)^3} \\
&+ \frac{(5 + \mathfrak{Q} - \mathfrak{R} + \mathfrak{T})\phi_c^3}{2(1 - \phi_c)^3},
\end{aligned}
\tag{10.7}
$$

with $\ln(\Lambda^3/v_c)$ the reference chemical potential of a superball fluid. The free energy follows from the chemical potential and the osmotic pressure through the thermodynamic relation $\tilde{F}_f = \phi_c\tilde{\mu}_f - \tilde{P}_f$ (see Eq. (A.11)).

10.2.3 Solid Phase States of Superballs

To approximate the free energy of the solid phases of hard superballs, the cell theory (see also Chaps. 3 and 9) proposed by Lennard-Jones and Devonshire (LJD) for hard spheres [52] was modified. Each particle is considered to be contained in a closed region whose shape is determined by neighbouring particles fixed at their lattice positions [53], as illustrated in Fig. 10.5. The free energy of the solid is computed from the number of configurations determined by the volume v^* that the *centre* of the particle explores, provided it does not overlap with its nearest neighbours. This leads to the following normalised free energy for a solid:

$$
\tilde{F}_s = \phi_c \ln\left(\frac{\Lambda^3}{v^*}\right).
\tag{10.8}
$$

The free volume v^* depends on the shape parameter m and the volume fraction ϕ_c of the superballs. It is, however, also dependent on the structure of the solid phase state, because this structure affects the relative position of the nearest neighbours. Solids that appear in dispersions of hard superballs are the face-centred cubic (FCC), the simple cubic (SC) structure and two families of other more complex lattice packings: the $\mathbb{C}_0$-lattice and $\mathbb{C}_1$-lattice [36,50,54].

These packings possess twofold ($\mathbb{C}_0$) and threefold ($\mathbb{C}_1$) rotational symmetries. Both packings can be considered as a continuous deformation of the FCC lattice for a sphere ($m = 2$) to a SC lattice for a perfect cube ($m \to \infty$). Since these solids have a distorted structure the corners of the superballs can be closer to one another. Hence, the voids between the particles are smaller, which increases the maximum packing density. The $\mathbb{C}_0$-lattice provides the densest packing for small m, while for $m \gtrsim 2.308$ the $\mathbb{C}_1$-lattice is the most efficient. The FCC and SC structures are the thermodynamically preferred structures in the limits of hard spheres and hard cubes, respectively [50]. Since there are no analytic expressions available for the $\mathbb{C}_0$- and

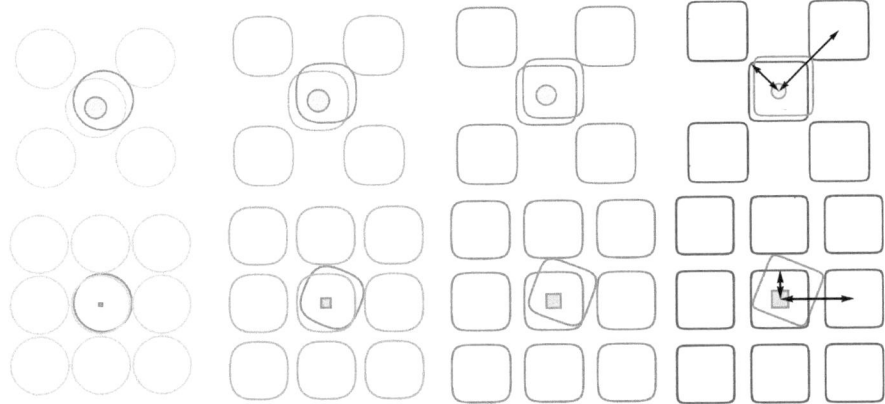

Fig. 10.5 Representations of the face-centred cubic (FCC) crystal lattice (*top panel*) and simple cubic (SC) lattice (*bottom panel*) at $\phi_c = 0.45$ for (left to right) $m = 2, 3, 5$, and 10. The free volume is illustrated as the shaded regions. As an illustration, a particle that just touches a nearest neighbour is also depicted. The small arrows indicate the superball radius R and the large arrows specify the nearest-neighbour distance r. Adapted from Ref. [38] under the terms of CC-BY-4.0

$\mathbb{C}_1$-lattice structures we consider only the FCC and SC phase states. A schematic view of the FCC and SC structures of superballs for several m-values is shown in Fig. 10.5.

For the FCC crystal the free volume depends on the shape of the Wigner–Seitz cell [53], which for an FCC crystal has a rather complicated geometry [55,56]. Usually it is approximated as a sphere (see Chap. 3). If one considers that this is still reasonable for superballs for small values of $m - 2$, the free volume $v^{*,\mathrm{FCC}}$ is then given by

$$v^{*,\mathrm{FCC}} = \frac{4\pi}{3} \left(r - r_{\mathrm{cp}}\right)^3, \tag{10.9}$$

where r is the distance between the centres of a superball and its nearest neighbours, and r_{cp} is r at close packing. For the FCC crystal, a 'frozen' crystal can be considered in which the particles are perfectly aligned. For the FCC lattice r_{cp} then is two times the distance between the edges of the superballs, $2\tau_{\mathrm{max}}^{\mathrm{2D}}$, in the two-dimensional representation. With ϕ_{cp} as the close packing fraction, the distance r at a certain volume fraction can be determined from r_{cp}:

$$r = r_{\mathrm{cp}} \left(\frac{\phi_{\mathrm{cp}}}{\phi_c}\right)^{1/3}. \tag{10.10}$$

Combining $r_{\mathrm{cp}} = 2\tau_{\mathrm{max}}^{\mathrm{2D}} = 2R\sqrt{2}\,(1/2)^{1/m}$ with Eqs. (10.8)–(10.10) provides the free energy for the FCC phase. Taylor expansion of $(\phi_{\mathrm{cp}}^{\mathrm{FCC}}/\phi_c)^{1/3} - 1$ was used in the original LJD approach for hard spheres [52]. The chemical potential and (osmotic) pressure are calculated from the free energy via the thermodynamic relations given in Appendix A, leading to the following closed, m-dependent, expressions for the

FCC phase:

$$\tilde{F}_{\text{FCC}} = \phi_{\text{c}} \ln \left(\frac{\Lambda^3}{v_{\text{c}}} \right) + \phi_{\text{c}} \ln \left[\frac{3^4 2^{3/m} f(m)}{\pi 2^{7/2}} \right] - 3\phi_{\text{c}} \ln \left(\frac{\phi_{\text{cp}}^{\text{FCC}}}{\phi_{\text{c}}} - 1 \right), \quad (10.11a)$$

$$\tilde{\mu}_{\text{FCC}} = \tilde{\mu}_0 + \ln \left[\frac{3^4 2^{3/m} f(m)}{\pi 2^{7/2}} \right] - 3 \ln \left(\frac{\phi_{\text{cp}}^{\text{FCC}}}{\phi_{\text{c}}} - 1 \right) + \frac{3}{1 - \phi_{\text{c}}/\phi_{\text{cp}}^{\text{FCC}}}, \quad (10.11b)$$

$$\tilde{P}_{\text{FCC}} = \frac{3\phi_{\text{c}}}{1 - \phi_{\text{c}}/\phi_{\text{cp}}^{\text{FCC}}}, \quad (10.11c)$$

with $\tilde{\mu}_0 = \ln \Lambda^3/v_{\text{c}}$. The m-dependency of the close packing volume fraction in an FCC crystal is given by [38,43]

$$\phi_{\text{cp}}^{\text{FCC}} = \frac{1}{2} f(m) 2^{3/m}, \quad (10.12)$$

with

$$f(m) = \frac{[\Gamma_{\text{E}}(1 + 1/m)]^3}{\Gamma_{\text{E}}(1 + 3/m)}. \quad (10.13)$$

Here, $\Gamma_{\text{E}}(x)$ is the Euler Gamma function of x. For $m = 2$, Eqs. (10.11)–(10.13) recover the result for hard spheres in the FCC phase [52], given in Sect. 3.2.2.

Following a similar procedure as for the FCC phase state, the thermodynamic functions of the SC phase read

$$\tilde{F}_{\text{SC}} = \phi_{\text{c}} \ln \left(\frac{\Lambda^3}{v_{\text{c}}} \right) + \phi_{\text{c}} \ln f(m) - 3\phi_{\text{c}} \ln \left[\left(\frac{\phi_{\text{cp}}^{\text{SC}}}{\phi_{\text{c}}} \right)^{1/3} - 1 \right], \quad (10.14a)$$

$$\tilde{\mu}_{\text{SC}} = \tilde{\mu}_0 + \ln f(m) - 3 \ln \left[\left(\frac{\phi_{\text{cp}}^{\text{SC}}}{\phi_{\text{c}}} \right)^{1/3} - 1 \right] + \frac{1}{1 - (\phi_{\text{c}}/\phi_{\text{cp}}^{\text{SC}})^{1/3}}, \quad (10.14b)$$

$$\tilde{P}_{\text{SC}} = \frac{\phi_{\text{c}}(\phi_{\text{cp}}^{\text{SC}}/\phi_{\text{c}})^{1/3}}{(\phi_{\text{cp}}^{\text{SC}}/\phi_{\text{c}})^{1/3} - 1} + \frac{\phi_{\text{c}}}{1 - (\phi_{\text{c}}/\phi_{\text{cp}}^{\text{SC}})^{1/3}}, \quad (10.14c)$$

with the close packing fraction in the SC phase given by

$$\phi_{\text{cp}}^{\text{SC}} = \frac{[\Gamma_{\text{E}}(1 + 1/m)]^3}{\Gamma_{\text{E}}(1 + 3/m)}. \quad (10.15)$$

In this simple approach [38] of estimating v^*, effects of particle rotations are only accounted for approximately. See also [57] for a comparison of cell theory of cubes and other methods. Algebraic expressions for nonaxisymmetric hard particles are

Table 10.1 Close packing volume fractions for superballs with $m = 2$ (spheres), $m = 3$ and $m = \infty$ (cubes). At $m = 3$ both crystals have the same close packing fraction

	$m = 2$	$m = 3$	$m = \infty$
FCC	$\pi/3\sqrt{2} \approx 0.74$	$[\Gamma_E (4/3)]^3 \approx 0.71$	0.5
SC	$\pi/6 \approx 0.52$	$[\Gamma_E (4/3)]^3 \approx 0.71$	1

scarce. No analytic expressions could even be derived for biaxial hard particles, and one must rely on computational approaches [58]. Table 10.1 provides close packing volume fractions for perfect spheres ($m = 2$) and perfect cubes ($m = \infty$) and for a limiting intermediate case. SC arrangements can pack closer for large m and FCC packings are more efficient for small m. It follows from Eqs. (10.12) and (10.15) that ϕ_{cp} attains the same value for both the FCC and the SC phase states at $m = 3$ (see Table 10.1).

Cell theory is known to give accurate results for FCC and SC crystals of hard *spheres* [55,56], but extending cell theory to other crystal structures is not straightforward. For a body-centred-cubic crystal of hard spheres, cell theory deviates from computer simulation results [56]. Still, the approach outlined above gives some semiquantitative insight into the effects of m on the free energy of FCC and SC solid phase states, of which the latter has been found for micrometre-sized superball-like colloids upon adding nonadsorbing polymer chains [20]. Interestingly, experiments reveal that superballs will form FCC (rotator phase) and dense $\mathbb{C}_1$-lattices upon sedimentation for $m \lesssim 3$ [59].

10.3   Phase Behaviour of Hard Colloidal Superballs

The complete phase diagram of hard superballs can now be calculated using the expressions of the chemical potential and osmotic pressure of the fluid, FCC and SC phase states. The resulting theoretical phase diagram for a suspension of pure hard superballs is presented in Fig. 10.6 (left panel). Results obtained using Monte Carlo computer simulations are plotted in the right panel of Fig. 10.6. The fluid–FCC coexistence for hard spheres [60] (discussed in Chap. 3) is recovered for $m = 2$. The theoretically predicted fluid–FCC equilibrium gradually shifts to a higher volume fraction upon increasing m. A similar shift is found using computer simulations [54]. The forbidden region (grey) identifies volume fractions beyond close packing. A discontinuity at $m = 3$ along the border of the forbidden region (left panel) corresponds to the transition between FCC to SC phase states. The preferred solid phase is related to the largest close packing volume fraction.

A triple F–FCC–SC point is found at $m \approx 3.71$ (left panel). Simulations also indicate a triple point at a somewhat higher m value [54] (right panel of Fig. 10.6), to be discussed later. Between $m = 3$ and $m \approx 3.71$, theory predicts SC–FCC coexistence (left panel). Above $m \approx 3.71$, only F–SC coexistence is found theoretically, which shifts towards lower packing fractions with increasing m, also in qualitative agreement with simulations [54]. In the cube limit ($m = \infty$), the simple theory presented

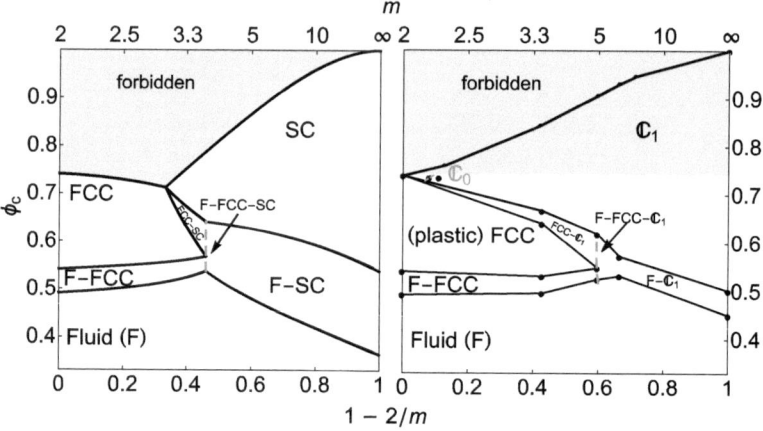

Fig. 10.6 *Left*: phase diagram for a suspension of superballs presented in terms of volume fraction and shape parameter. Two-phase coexistences take place in the regions bounded by two single-phase regions as indicated. The vertical dashed grey lines hold for the F–FCC–SC coexistence. *Right*: theoretical predictions based upon Monte Carlo computer simulations by Ni et al. [54]. Adapted from Ref. [38] under the terms of CC-BY-4.0

here predicts $\phi_c^F \approx 0.36$ and $\phi_c^{SC} \approx 0.54$. Computer simulation studies (not shown here) by Agarwal and Escobedo [50] indicate that phase coexistence between a fluid and a cubatic liquid crystal takes place at $\phi_c^F \approx 0.47$ and $\phi_c^{SC} \approx 0.58$ for perfect cubes.

Exercise 10.3. What is the theoretical maximum number of coexisting phases in a dispersion of hard superballs?

The overall topology of the theoretical phase diagram (left panel) corresponds roughly to computer simulations results (see Refs. [49,50,54]; right panel of Fig. 10.6). Differences can be justified because theory does not account for the same solid phases for superballs as in simulations, as mentioned earlier. The solid $\mathbb{C}_0$ and $\mathbb{C}_1$ phases can be accounted for in computer simulation studies [49,54]. Not surprisingly, the triple point from simulations is a fluid–plastic FCC–$\mathbb{C}_1$ [54] and lies at a larger m value than the theoretical triple point. Due to the limitations inherent to the simple theory used here, the $\mathbb{C}_0$ and $\mathbb{C}_1$ phases are not accounted for. The FCC phase features (and their coexistences) roughly match those of the plastic FCC. The role played in simulations by the $\mathbb{C}_1$ phase state is mimicked by the SC phase in the simpler model applied in the theoretical description.

10.4   Theory for the Phase Behaviour of Colloidal Superballs Mixed with Polymers

The presence of the nonadsorbing polymers leads to depletion zones around the cube-like colloidal particles. In Fig. 10.7, overlap volumes of depletion layers (indicated by the dashed zones) are illustrated for superballs with shape parameters $m = 2, 4$, and ∞. The volume of overlapping depletion zones V_{ov} (given by Eq. 1.19 for $m = 2$) increases with m when the superballs align their flat faces, under the condition that the superball radius R and polymer size (and thus, the depletion thickness) are fixed. At those configurations a maximum depletion attraction is achieved [61].

Upon adding nonadsorbing polymer (or colloidal particles) to a colloidal dispersion, one expects that the entropic patchiness effect, discussed in Sect. 1.3.6, leads to an enhanced depletion attraction for cube-like particles as compared to spheres.

When considering colloidal superball–polymer mixtures, the phase diagrams would only enrich upon refinements of the method. The liquid-crystalline and crystalline coexistence regions are found in simulations in a broader range of m-values [54].

10.4.1   Free Volume Theory

Adding depletants to dispersions of hard superballs is accounted for here in a semi-grand canonical fashion via free volume theory (FVT), as in the previous chapters. Within FVT, the superball–polymer system (S) is considered to be in equilibrium with a reservoir (R) of polymers. In R and S the solvent is treated as background

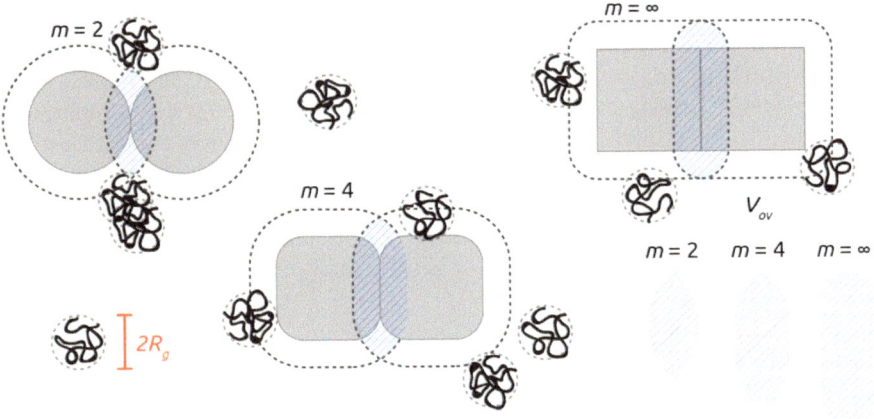

Fig. 10.7 Sketches of the maximum overlap of depletion zones for three types of hard superballs in nonadsorbing polymer solutions for fixed polymer size and constant particle radius R. The hatched areas reflect overlap volumes of depletion zones. The examples are given for superballs with shape parameters $m = 2, 4$ and ∞. These overlap volumes are drawn next to each other for comparison in the lower right section of the figure. Reprinted from Ref. [62] under the terms of CC-BY-4.0

as before (see Sect. 3.3.4), and system and reservoir are connected through a membrane permeable for the polymers and the common solvent but impermeable for the superball or cube-like particles. The relative volume available for depletants in the system, the free volume fraction α, relates the polymer concentrations in R and S. Following original FVT, it is assumed that this key quantity α is independent of the chemical potential of the depletants in R.

The simplest description is used here for polymeric depletants, namely the penetrable hard sphere (PHS) model: depletants are treated as ghost-like spheres (with radius δ) that can freely interpenetrate each other but do not overlap with the superballs. Further, the approximation is made again that the ensemble-averaged free volume $\langle V_{\text{free}} \rangle$ is independent of the concentration of depletants. This implies that, similar to what was discussed in the previous chapters, $\alpha = \langle V_{\text{free}} \rangle / V \approx \langle V_{\text{free}} \rangle_0 / V$, resulting in the following (normalised) expression for the grand potential of the system:

$$\widetilde{\Omega} = \frac{\Omega v_c}{kTV} = \widetilde{F} - \widetilde{P}_d^R \alpha \frac{v_c}{v_d}, \tag{10.16}$$

with V as the volume of the system, and v_d as the volume of the depletant ($v_d = 4\pi\delta^3/3$). Since the depletants are considered to behave ideally, the osmotic pressure in the reservoir R is simply given by Van't Hoff's law:

$$\widetilde{P}_d^R = \frac{P_d^R v_d}{kT} = \phi_d^R.$$

The depletant concentration in the system ϕ_d is again given by $\phi_d = \alpha \phi_d^R$.

From the grand potential, the chemical potential and osmotic pressure of superball–PHS mixtures are obtained through the standard thermodynamic relations given in Appendix A, where it is also indicated how to calculate binodals, critical points, spinodals and multi-phase coexistences. If coexistence between three phases takes place a triple point (TP) arises, and a four-phase coexistence is denoted as a quadruple point (QP) (see also the previous two chapters). Colloidal systems may exhibit isostructural phase coexistence (such as gas–liquid equilibrium) when attractive interactions between particles are present. In such a case, the low density phase will be entropically favourable and the high-density phase will be stabilised by attractive interactions between the particles. The limit of isostructural phase coexistence is defined via the critical point (CP).

The conditions of Eq. (A.20) enable one to determine the topology of the phase diagrams as a function of the system parameters, which are the colloidal shape (through m) and the relative depletant size trough:

$$q = \frac{\delta}{R}, \tag{10.17}$$

where δ is the radius of the PHSs.

10.4.2 Free Volume Fraction

The only unknown parameter in Eq. (10.16) is the free volume fraction for depletants in the system, α. Widom's insertion theorem [63] relates the free volume fraction α to the work (W) required to bring a depletant from R to S via

$$\alpha = \frac{\langle V_{\text{free}} \rangle_0}{V} = e^{-W/kT}, \tag{10.18}$$

where $\langle V_{\text{free}} \rangle_0$ is the free volume for depletants in the undistorted (depletant-free) system. This work (W) is obtained via Scaled Particle Theory (SPT) [64,65], by connecting the limits of inserting a very small depletant (up to second order) and a very big depletant in the system of interest, followed by scaling back to the actual size of the depletant. As the depletants are considered to be spherical here, a single scaling factor (λ) enables this work of insertion to be expressed as

$$W = \lim_{\lambda \to 1} W(\lambda),$$

$$W(\lambda) = \underbrace{W(0) + \left.\frac{\partial W}{\partial \lambda}\right|_{\lambda=0} \lambda + \frac{1}{2} \left.\frac{\partial^2 W}{\partial \lambda^2}\right|_{\lambda=0} \lambda^2}_{\lambda \ll 1} + \underbrace{v_{\text{d}} P}_{\lambda \gg 1}, \tag{10.19}$$

where P is the osmotic pressure of the pure hard superball dispersion to which the depletants are added (see Sect. 10.2).

In the limit of depletants with a vanishing size ($\lambda \to 0$) there is no overlap of depletion zones. Hence, the free volume fraction can then be written as a function of the excluded volume between a superball and a depletant (v_{exc}):

$$\alpha(\lambda \to 0) = 1 - \phi_{\text{c}} \left(\frac{v_{\text{exc}}(\lambda)}{v_{\text{c}}} \right); \tag{10.20}$$

so, W becomes

$$W(\lambda \to 0) = -kT \ln \left[1 - \phi_{\text{c}} \left(\frac{v_{\text{exc}}(\lambda)}{v_{\text{c}}} \right) \right]. \tag{10.21}$$

In the limit of big depletants, one finds

$$\frac{W(\lambda \gg 1)}{kT} = \frac{\pi}{6 f(m)} (\lambda q)^3 \widetilde{P}. \tag{10.22}$$

By combining Eqs. (10.19) and (10.22), a general expression for W in terms of $\widetilde{v}_{\text{exc}}(\lambda) = v_{\text{exc}}(\lambda)/v_{\text{c}}$ can be derived:

$$\frac{W}{kT} = -\ln(1 - \phi_c) + y \left. \frac{\partial \widetilde{v}_{exc}(\lambda)}{\partial \lambda} \right|_{\lambda=0}$$

$$+ \frac{1}{2}y^2 \left[\left. \frac{\partial \widetilde{v}_{exc}(\lambda)}{\partial \lambda} \right|_{\lambda=0} \right]^2 \qquad (10.23)$$

$$+ \frac{1}{2}y \left. \frac{\partial^2 \widetilde{v}_{exc}(\lambda)}{\partial \lambda^2} \right|_{\lambda=0} + \frac{\pi q^3}{6f(m)} \widetilde{P}.$$

In Eq. (10.23), we have used

$$y = \frac{\phi_c}{1 - \phi_c},$$

which is similar to Eq. (3.39e).

A simple expression is available for the (normalised) excluded volume between general, convex bodies and spheres [66]. For the excluded volume superball–sphere it reads:

$$\widetilde{v}_{exc} = 1 + \frac{3\widetilde{s}_c q + 6\pi \widetilde{c}_c q^2 + \pi q^3}{6f(m)}, \qquad (10.24)$$

where $\widetilde{s}_c = s_c/R_{el}^2$, $\widetilde{c}_c = c_c/R_{el}$ and $f(m)$ is defined in Eq. 10.13. Due to the linear relationship between δ and q ($\delta = qR$), $\widetilde{v}_{exc}(\lambda)$ is simply obtained from Eq. (10.24) by making the substitution $q \to \lambda q$.

It is not possible to solve Eq. (10.24) analytically (see Ref. [38] for details on the calculation of $\widetilde{s}_c$ and $\widetilde{c}_c$). Via interpolation of $\widetilde{s}_c$ and $\widetilde{c}_c$ it is possible to obtain an expression for Eq. (10.23). González García et al. [38] found that the depletion zone is accurately described by the following approximate, but accurate, algebraic expression:

$$\widetilde{v}_{exc} f(m) = 454.337 + 216.356(1 - 2/m) + 308.593q$$
$$+ \exp[-0.005(1 - 2/m) + 8.178] \qquad (10.25)$$
$$\times \sin[0.0604(1 - 2/m) + 0.087q - 3.014].$$

This enables straightforward calculation of α by inserting Eq. (10.25) into Eq. (10.23). Now the grand potential $\widetilde{\Omega}$ (Eq. (10.16)) is completely defined.

10.5   Phase Diagrams of Mixtures of Hard Superballs and Polymers: Theoretical Predictions

Firstly, superball–polymer mixtures with (hard) superballs are considered, whose shape is still close to a sphere. Phase diagrams of superballs with $m = 2.5$ and added depletants are presented in Fig. 10.8 for three relative size ratios q. For pure superballs ($\phi_d^R = 0$) the fluid–FCC coexistence corresponds to the densities shown in Fig. 10.6 (left panel) and hardly differs from the fluid–solid phase coexistence

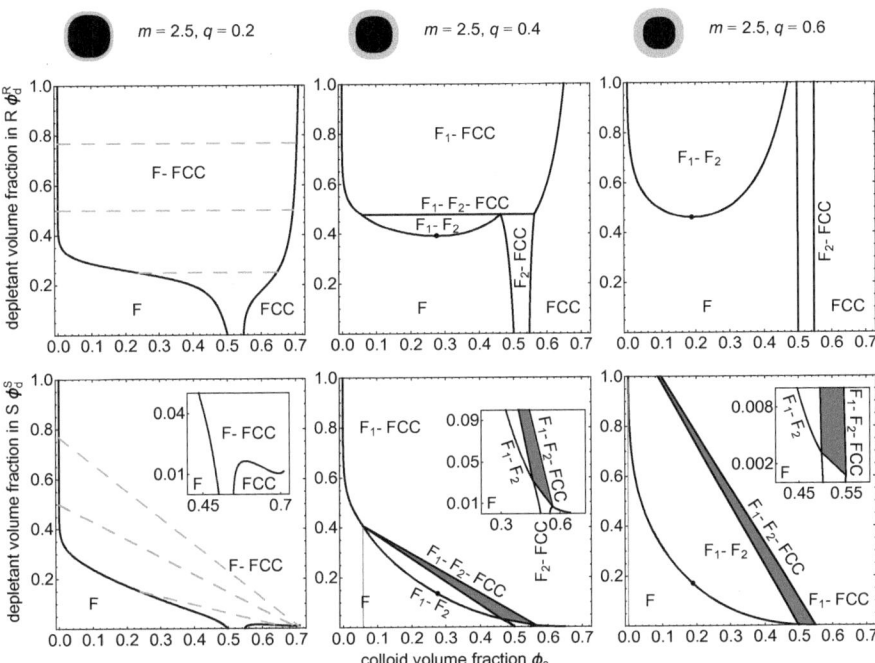

Fig. 10.8 Phase diagrams for a mixture of colloidal superballs for $m = 2.5$ and penetrable hard sphere depletants for several relative depletant sizes q. The top panels are in the reservoir representation and the bottom panels are in the system representation. The F_1–F_2 critical point is indicated by a black dot. Triple-phase coexistences are shown as a horizontal black line in the reservoir representation, and as a black area bounded by black lines in the system representation. All coexisting phases present are indicated in the reservoir representation. Insets in the system representation zoom in on the low depletant concentration region, and some of the coexistence regions are indicated. A few illustrative tie-lines are shown as dashed grey lines for $q = 0.2$. Above each phase diagram, a 2D illustration of the superball (black) and its depletion zone (grey) are shown, and the m and q-values are indicated. Reprinted from Ref. [38] under the terms of CC-BY-4.0

concentrations of hard spheres (see Sect. 3.2). Upon addition of depletants the FCC phase at coexistence gets denser, and the coexisting fluid phase becomes more dilute in order to maximise the total free volume available for the depletants in the system. For sufficiently large q-values ($q = 0.4$ and $q = 0.6$ in Fig. 10.8), an isostructural colloidal F_1–F_2 (also termed *gas–liquid*) coexistence appears (metastable for low q-values). For $q = 0.4$ a triple line is found (upper panel), which becomes a region in the system representation (lower panels of Fig. 10.8). For larger q, the F–S coexistence narrows and the F_1–F_2 critical point shifts to higher depletant volume fractions.

The coexistence regions in the system representation (bottom panels in Fig. 10.8) show that the fluid phase with a low concentration of superballs has a high concentration of depletants, whereas the FCC phase has a high concentration of superballs but a low concentration of depletants. The system representation also shows that for a superball–depletant mixture a single solid phase (*without* a coexisting fluid phase) only occurs at quite small depletant concentrations. Figure 10.8 reveals no special

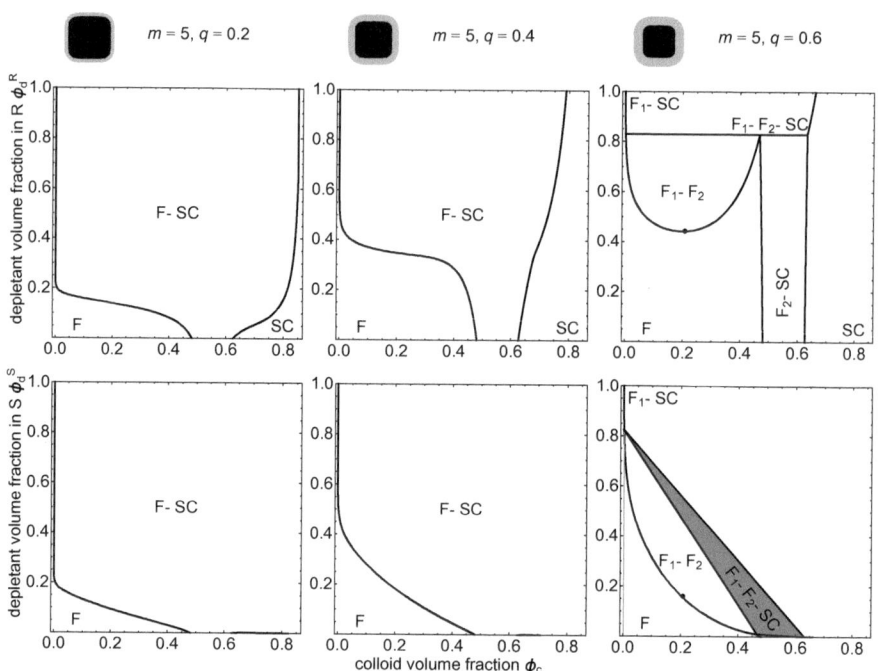

Fig. 10.9 Phase diagrams of superball–polymer mixtures as in Fig. 10.8, but for $m = 5$. Reprinted from Ref. [38] under the terms of CC-BY-4.0

features compared to FVT for hard spheres mixed with PHSs (see Sect. 3.3), even though the colloidal shape considered is not perfectly spherical.

Phase diagrams for more cubic particles that have $m = 5$ (see Fig. 10.2) are presented in Fig. 10.9, for which only a SC state is present in the pure hard superballs system (see left panel of Fig. 10.6) at high particle concentrations. Similar qualitative trends as in Fig. 10.8 are observed, but with the F–SC coexistence instead of the F–FCC equilibrium. For small q-values the broadening of the coexistence lines occurs at lower depletant concentrations with respect to the superball–polymer mixture with $m = 2.5$ in Fig. 10.8: the overlap of depletion zones is larger for particles with an increased cubicity (see Fig. 10.7), which results in a stronger depletion attraction. For $m = 5$ and $q = 0.4$, there is no F_1–F_2 equilibrium phase coexistence, whereas F_1–F_2 coexistence was found at this q-value for spheres (superballs with $m = 2$, see Sect. 3.3.4 or [67]), and for $m = 2.5$. Due to the tendency of flat faces to align upon addition of depletant into the system, stable F_1–F_2 coexistence shifts to higher q-values: for larger m-values, longer ranges of attraction are required to induce a (stable) F_1–F_2 coexistence. The trends in Fig. 10.9 hold for larger m and F_1–F_2 occurs at even larger q-values. Quantitatively, comparison of Figs. 10.8 and 10.9 reveals that the stable single-phase fluid region for colloid volume fractions of, say, smaller than 0.4, is larger for smaller m. This means that dispersions of particles which have a more cube-like shape, are expected to undergo phase transitions at smaller depletant concentrations. The increased overlap volume of the depletion

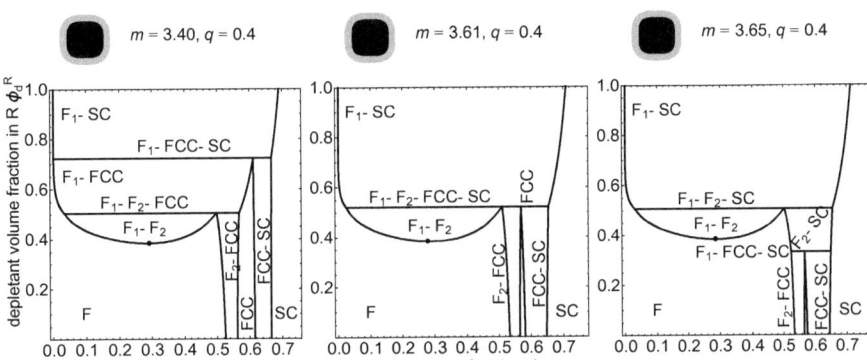

Fig. 10.10 Phase diagrams of mixtures of superball–polymer mixtures in the reservoir depletant concentration representation for $q = 0.4$ and various m-values as indicated. Reprinted from Ref. [38] under the terms of CC-BY-4.0

zones with increasing m, as illustrated in Fig. 10.7, drives this earlier onset of the phase transitions.

Based on Fig. 10.6, a physically interesting region is expected near $m = 3.65$, since here fluid, FCC *and* SC phase states are predicted upon increasing the colloid volume fraction. In particular, one may wonder what the effect of adding depletants is near the transitions between these phases. A few phase diagrams for $q = 0.4$ and a few selected m-values in that interesting region are plotted in Fig. 10.10. For $m = 3.4$, a F_1–FCC–SC triple coexistence occurs at a higher ϕ_d^R than the F_1–F_2–FCC triple line (F_1–F_2–SC coexistence is metastable). For $m = 3.65$, however, the F_1–F_2–FCC triple point becomes metastable and an F_1–FCC–SC triple point arises at lower ϕ_d^R than the F_1–F_2–SC isostructural coexistence. This is explained by the fact that the stability of the FCC decreases as m increases.

The condition at which the F_1–F_2–FCC and the F_1–F_2–SC coexistences merge results in a *quadruple* coexistence (F_1–F_2–FCC–SC). This four-phase coexistence is present for a range of m-values (at different q-values). As a consequence of the enhanced alignment of the flat faces upon the addition of depletants, F–SC coexistence takes place at m-values below those of the depletant-free system. Hence, depletion-mediated entropic patchiness promotes the appearance of the SC phase. As a F–FCC–$\mathbb{C}_1$ triple point has been detected experimentally [59] and in a Monte Carlo computer simulation study [54] for the depletant-free superball system, the corresponding quadruple phase coexistence may be found from simulations or experiments with the $\mathbb{C}_1$ phase instead of the SC phase used here.

At low q and high m-values an isostructural SC_1–SC_2 coexistence appears. A few illustrative phase diagrams are depicted in Fig. 10.11, where small isostructural SC_1–SC_2 coexistence regions appear. The single fluid phase and simple cubic regions get smaller upon decreasing q. For $m = 10$ the binodals shift towards lower ϕ_d^R-values with decreasing q. As can be observed in the rightmost panel of Fig. 10.11, the m-value tunes the depletant concentration at which SC_1–SC_2 coexistence is found: SC_1–SC_2 equilibria are driven by the alignment of the flat faces, and thus for more

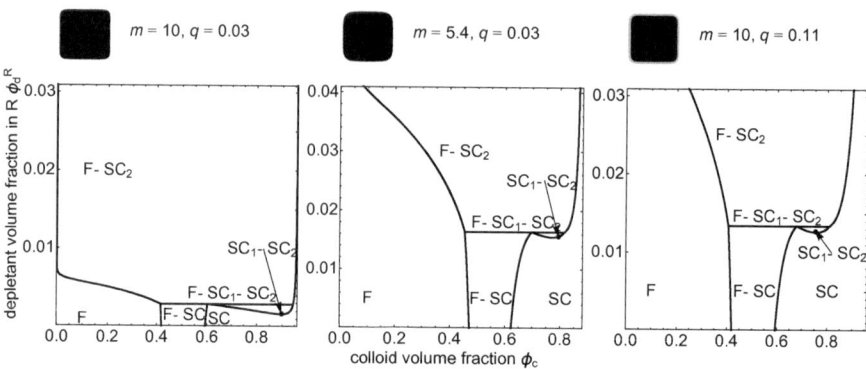

Fig. 10.11 Illustrative phase diagrams of superball–polymer mixtures, in which isostructural SC phase coexistences appear. Reprinted from Ref. [38] under the terms of CC-BY-4.0

curved particles (decreasing m) the SC_1–SC_2 coexistence requires a higher depletant concentration. This leads to demixing of the crystal state into two coexisting solids, as is expected for short-ranged attractions in colloidal systems [68,69]. With more accurate models or in experimental systems, these SC_1–SC_2 coexistences may be replaced, for example, by a $\mathbb{C}_1$ coexisting with a SC phase. The absence of a stable FCC_1–FCC_2 coexistence can be rationalised by the non-optimal overlap of depletion zones between the flat faces of the superballs in an FCC state.

10.6 Phase Stability of Cubes Mixed with Polymers: Experiments

In Sects. 10.4 and 10.5 theoretical predictions [38] were outlined for the rich phase behaviour of colloidal cubes mixed with nonadsorbing polymers. A thorough verification of this phase behaviour is still underway; experimental studies on the bulk phase behaviour of mixtures containing cubes and nonadsorbing polymers (or other nonadsorbing components) are scarce. Depletion effects in dispersions containing cubes were studied by Park et al. [70]. They mixed gold rods and cubes, and added nonadsorbing polymers to separate them. For more details, see Sect. 11.3.

Another early demonstration of the effects of adding nonadsorbing polymers to dispersions of cube-like particles was performed by Rossi et al. [20,34], who studied aqueous mixtures of polymers and micrometre-sized hollow silica superball-shaped particles (typically with m between 3 and 4). The experiments were performed in 10 mM NaCl, so the Debye length $\lambda_D \approx 3$ nm. Images of the cube-like particles were mapped onto the so-called superball shape to determine m. These particles are very suitable for experiments: their size enables them to study their shape and configurations using an optical microscope. However, the combination of their size and silica shells makes them susceptible to gravity, so that equilibrium studies of the bulk properties are challenging. Hence, it is noted that these experimental observations correspond to colloid–polymer mixtures confined at a surface, whereas equilibrium theoretical results presented earlier hold for bulk systems.

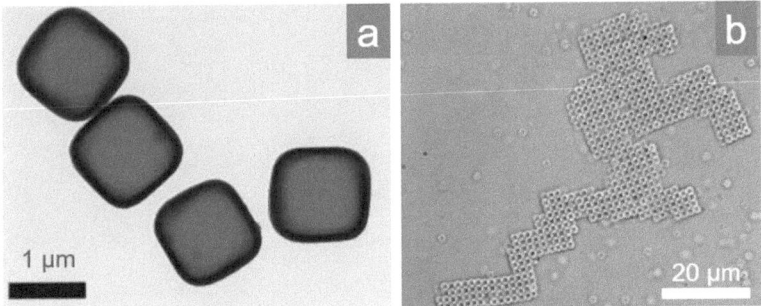

Fig. 10.12 a Hollow silica cubes with a total edge length $2R = 1.3$ μm, silica shell thickness of 100 nm and shape parameter $m = 3.5$. **b** Adding nonadsorbing PEO (molar mass 600 kDa, $R_g \approx 57$ nm) drives depletion-mediated self-assembly of the cubes to a simple cubic symmetry. Size ratio $q = R_g/R \approx 0.085$. Reprinted with permission from Ref. [20]. Copyright 2011 Royal Society of Chemistry

Still, by studying their properties at a glass plate after sedimentation, Rossi et al. [20,21,34] could investigate the influence of nonadsorbing polymers on the structures that appeared. Their experimental observations [20,34] revealed that the depletion attraction mediated by the nonadsorbing polymers leads to a phase transition towards a preferred simple cubic phase state, see Fig. 10.12. The appearance of such a solid phase state can be understood by the fact that the volume of depletion zones that is overlapping is then maximised (leaving space for the depletants to fit in the voids of the respective lattices). The size of the nonadsorbing polymers was shown to play a crucial role [20]. The cubic structures appear both in plane as well as in 3D.

The authors also studied adding nonadsorbing poly(N-isopropylacrylamide) (pNIPAM) particles with a radius of 65 nm. The size of these particles is highly temperature-responsive between 20 and 45 °C. Upon increasing the temperature the pNIPAM particles shrink in water as the solvency changes from good to poor. This enabled Rossi et al. [20] to reversibly induce thermoresponsive cube crystallisation.

The experiments on dispersions of these hollow micrometre-sized silica particles with added nonadsorbing polymers revealed a rich phase behaviour. It was demonstrated that the obtained phase states of the sediment depend on the colloid–depletant size ratio and the details of the shape of the cube-like colloidal particles [34] (Fig. 10.13).

Recently, it was shown that hollow silica nanocubes display effective hard-core interactions [29], which makes them promising particles for studying the effect of nonadsorbing polymers on the bulk phase behaviour of model anisotropic particles. Based on the estimated phase-transition points from light scattering, Dekker et al. [62] constructed an experimental phase diagram for mixtures of hollow silica (nano)shells ($R_{el} = 129$ nm and $m = 4.1$) and polystyrene polymers (PS) in DMF (40 mM LiCl). Three different regions can be discerned in the phase diagram depicted in Fig. 10.14: (1) a concentration range where the mixture is stable (○), (2) a concentration range where the mixture clearly phase separates (■) and (3) an intermediate transition region where no clear phase separation occurs, but where the scattering studies indicated that significant attraction is present (▲).

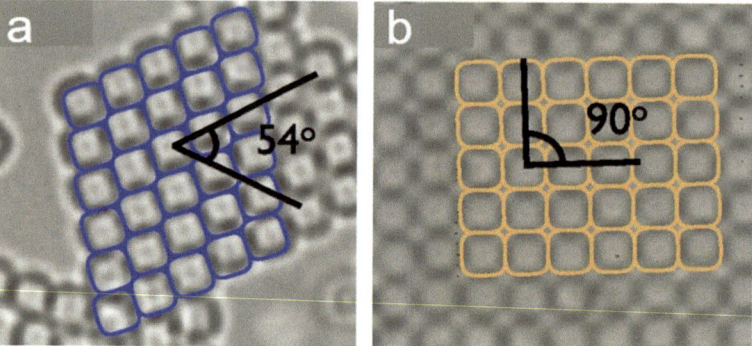

Fig. 10.13 Solid phases prepared from hollow silica cubes similar to those in Fig. 10.12 but with $m = 3.9$. **a** Solid structures obtained upon adding depletants larger than the pockets available in an SC arrangement. **b** SC arrangement that appears upon using smaller depletants than for. Reprinted with permission from Ref. [34]. Copyright 2015 the National Academy of Sciences

Theoretical predictions for the phase stability threshold are also plotted in Fig. 10.14 to compare with the experimental data. Particle volume fractions were calculated using the specific volumes obtained earlier [29] and the overlap concentration of polystyrene ($M_w = 600$ kg/mol, $R_g = 21.5 \pm 1.1$ nm) was 24 g/L. The curves correspond to the fluid branch of the fluid–solid coexistence binodal for superballs with $m = 2$, $m = 4.1$ and $m = 10^4$ added nonadsorbing polymers with size ratio $q = 0.32$. The dashed curves are phase coexistence lines for superballs with $m = 4.1$ and size ratios $q = 0.29$ and 0.35, representing the lower and upper limit of the polymer polydispersity. The experimental data are in remarkable agreement with the theoretical predictions, indicating that the theory is able to predict the depletion effects in experimental model systems and that dispersions of hollow silica cubes in DMF with 40 mM LiCl and polystyrene is such a model system.

Saez Cabezas et al. [71] compared the influence of nonadsorbing polymers on the phase stability of tiny spherical and cubic nanocrystals. The prepared spherical particles were composed of magnetite, iron oxide Fe_3O_4, following the method of Yu et al. [72]. The cubic (F,Sn:In_2O_3) cubes were made using the procedure of Cho et al. [30]. As followed from SAXS measurements, both particles were rather monodisperse in size and the diameters were similar: 8–9 nm. PEG nonadsorbing polymers with $M = 1$ kg/mol, $R_g = 1.0$ nm were added to the aqueous dispersions containing the colloidal particles and the observations are indicated in Fig. 10.15. The open symbols refer to a single-phase mixture, whereas the closed symbols refer to instability (the authors observed gelation). The curves are calculated spinodal points, computed from Eq. (A.15). It is clear that adding nonadsorbing polymer to the cubic particles leads to instability at much lower polymer concentrations, as is the case for spherical particles. This is confirmed by the predicted spinodal curves, which were computed using the theory outlined in Sect. 10.4. It is noted that both particles have similar charge densities at the surface in aqueous solution, so double layer interactions also play a role here.

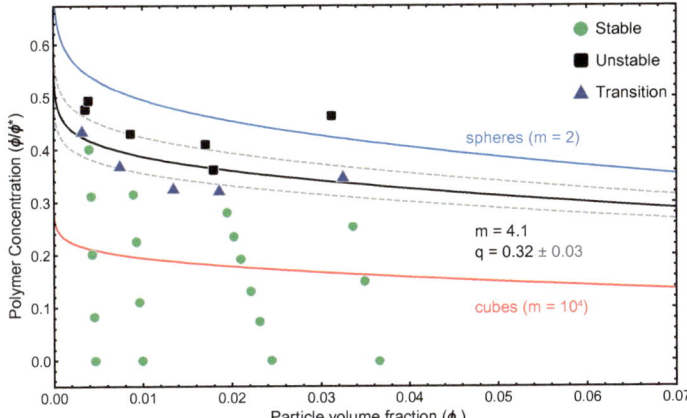

Fig. 10.14 Experimental phase diagram of hollow silica (nano)shells and polystyrene in DMF with 40 mM LiCl compared to theoretical predictions (curves) of hard superballs with nonadsorbing polymers. The curves are predicted fluid–solid coexistence binodal curves for superballs with m-values equal to 2, 4.1 and 10^4 in solution containing nonadsorbing polymers with size ratio $q = 2R_g/R_{el} = 0.32$. The dashed curves are binodal curves for superballs with $m = 4.1$ and size ratios $q = 2R_g/R_{el} = 0.29$ (bottom) and $q = 2R_g/R_{el} = 0.35$ (top). Reprinted from Ref. [62] under the terms of CC-BY-4.0

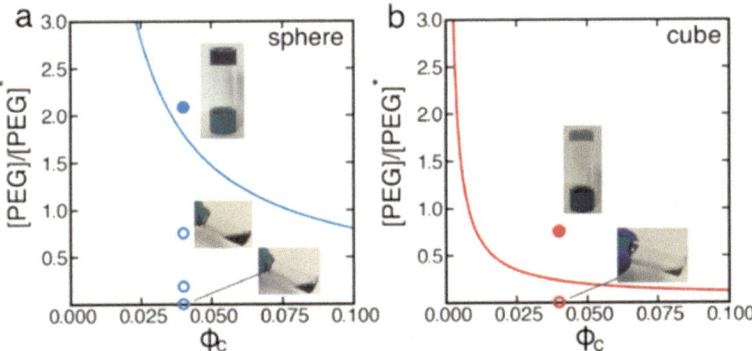

Fig. 10.15 Phase diagrams of **a** Sn:In$_2$O$_3$ spheres and **b** F,Sn:In$_2$O$_3$ cubes mixed with nonadsorbing polymers. Comparison of theoretical spinodal boundaries (curves) and experimental observations (data points). The polymer concentration is normalised by the overlap concentration. (○) Single phase mixtures and (●) gelled samples are indicated. Reprinted with permission from Ref. [71]. Copyright 2020 American Chemical Society

References

1. Glotzer, S.C., Solomon, M.J.: Nat. Mat. **6**, 557–562 (2007)
2. Sacanna, S., Pine, D.J.: Current Opin. Colloid. Interface Sci. **16**, 96 (2011)
3. Cozzoli, P.D., Manna, L.: Nat. Mater. **4**, 801 (2005)
4. Philipse, A.P.: In: Lyklema, J. (ed.) Fundamentals in Colloid and Interface Science, vol. 4, Chap. 2. Elsevier, Amsterdam (2005).
5. Sugimoto, T.: Monodispersed Particles, 2nd edn. Elsevier, Amsterdam (2019)
6. Sugimoto, T., Sakata, T.: J. Colloid Interface Sci. **152**, 587 (1992)
7. Sugimoto, T., Khan, M., Muramatsu, A., Itoh, H.: Colloids Surf. A **79**, 233 (1993)
8. Sugimoto, T., Wang, Y., Itoh, H., Muramatsu, A.: Colloids Surf. A **134**, 265–279 (1998)
9. Sugimoto, T., Wang, Y.: J. Colloid Interface Sci. **207**, 137 (1998)
10. Sun, Y., Xia, Y.: Science **298**, 2176–2179 (2002)
11. Wiley, B., Sun, Y., Mayers, B., Xia, Y.: Chem. Eur. J. **11**, 454 (2005)
12. Tao, A., Sinsermsuksakul, P., Yang, P.: Angew. Chem. **45**, 4597–4601 (2006)
13. Park, J., An, K., Hwang, Y., Park, J.G., Noh, H.J., Kim, J.Y., Park, J.H., Hwang, N.M., Hyeon, T.: Nature Mater. **3**, 891–895 (2004)
14. Bratlie, K.M., Lee, H., Komvopoulos, K., Yang, P., Somorjai, G.A.: Nano Lett. **7**, 3097–3101 (2007)
15. Sevonkaev, I., Goia, D.V., Matijević, E.: J. Colloid Interface Sci. **317**, 130 (2008)
16. Kim, D., Lee, N., Park, M., Kim, B.H., An, K., Hyeon, T.: J. Am. Chem. Soc. **131**, 454–455 (2009)
17. Park, J.C., Kim, J., Kwon, H., Song, H.: Adv. Mater. **21**, 803 (2009)
18. Adireddy, S., Lin, C.K., Cao, B.B., Zhou, W.L., Caruntu, G.: Chem. Mater. **22**, 1946–194 (2010)
19. Zhang, Y., Lu, F., van der Lelie, D., Gang, O.: Phys. Rev. Lett. **107**, 135701 (2011)
20. Rossi, L., Sacanna, S., Irvine, W.T.M., Chaikin, P.M., Pine, D.J., Philipse, A.P.: Soft Matter **7**, 4139 (2011)
21. Rossi, L.: Colloidal superballs. Ph.D. thesis, Utrecht University (2012)
22. Xie, K., Wu, P., Zhou, Y.Y., Ye, Y., Wang, H., Tang, Y., Zhou, Y.Y., Lu, T., Appl, A.C.S.: Mater. Interfaces **6**, 10602–10607 (2014)
23. Park, J., Porter, M.D., Granger, M.C.: Langmuir **31**, 3537–3545 (2015)
24. Meijer, J.M.: Colloidal Crystals of Spheres and Cubes in Real and Reciprocal Spaces. Springer, Berlin, Heidelberg (2015)
25. Ramasamy, P., Lim, D.H., Kim, B., Lee, S.H., Lee, M.S., Lee, J.S.: Chem. Commun. **52**, 2067–2070 (2016)
26. Jeon, S.J., Yazdi, S., Thevamaran, R., Thomas, E.L.: Cryst. Growth & Des. **17**, 284 (2017)
27. Dekker, F., Tuinier, R., Philipse, A.P.: Colloids Interfaces **2**, 44 (2018)
28. Dekker, F., Kuipers, B.W.M., Petukhov, A.V., Tuinier, R., Philipse, A.P.: J. Colloid Interface Sci. **571**, 419 (2020)
29. Dekker, F., Kuipers, B.W.M., González García, Á., Tuinier, R., Philipse, A.P.: J. Colloid Interface Sci. **571**, 267 (2020)
30. Cho, S.H., Roccapriore, K.M., Dass, C.K., Ghosh, S., Choi, J., Noh, J., Reimnitz, L.C., Heo, S., Kim, K., Xie, K., Korgel, B.A., Li, X., Hendrickson, J.R., Hachtel, J.A., Milliron, D.J.: J. Chem. Phys. **152**, 014709 (2020)
31. Meijer, J.M., Rossi, L.: Soft Matter **17**, 2354 (2021)
32. Wang, D., Hermes, M., Kotni, R., Wu, Y., Tasios, Y., Liu, Y., de Nijs, B., van der Wee, E.B., Murray, C.B., Dijkstra, M., van Blaaderen, A.: Nat. Commun. **9**, 2228 (2018)
33. Dekker, F.: Colloidal cubes : Preparation, optical properties and polymer-mediated interactions. Ph.D. thesis, Utrecht University (2020)
34. Rossi, L., Soni, V., Ashton, D.J., Pine, D.J., Philipse, A., Chaikin, P.M., Dijkstra, M., Sacanna, S., Irvine, W.T.M.: Proc. Natl. Acad. Sci. **112**, 5286 (2015)

35. Barr, A.H., Comput, I.E.E.E.: Graph. Appl. **1**, 11 (1981)
36. Jiao, Y., Stillinger, F.H., Torquato, S.: Phys. Rev. E **79**, 041309 (2009)
37. Linse, P.: Soft Matter **11**, 3900 (2015)
38. González García, Á., Opdam, J., Tuinier, R.: Eur. Phys. J. E **41** (2018)
39. Fong, C.: (2016). https://doi.org/10.48550/arXiv.1604.02174
40. Herold, E., Hellmann, R., Wagner, J.: J. Chem. Phys. **147**, 204102 (2017)
41. Isihara, A., Hayashida, T.: J. Phys. Soc. Jpn. **6**, 40 (1951)
42. Hadwiger, H.: Experientia **7**, 395 (1951)
43. González García, Á.: Polymer-Mediated Phase Stability of Colloids. Springer, Berlin, Heidelberg (2019)
44. Boublík, T.: Mol. Phys. **27**, 1415 (1974)
45. Boublík, T.: J. Chem. Phys. **63**, 4084 (1975)
46. Boublík, T.: Mol. Phys. **42**, 209 (1981)
47. Gibbons, R.: Mol. Phys. **17**, 81 (1969)
48. Carnahan, N.F., Starling, K.E.: J. Chem. Phys. **51**, 635 (1969)
49. Batten, R.D., Stillinger, F.H., Torquato, S.: Phys. Rev. E **81**, 061105 (2010)
50. Agarwal, U., Escobedo, F.A.: Nat. Mater. **10**, 230 (2011)
51. Fortini, A., Dijkstra, M., Tuinier, R.: J. Phys.: Condens. Matter **17**, 7783 (2005)
52. Lennard-Jones, J.E., Devonshire, A.F.: Proc. Roy. Soc. **163A**, 53 (1937)
53. Baus, M., Tejero, C.F.: Equilibrium Statistical Physics. Springer, Berlin, Heidelberg (2008)
54. Ni, R., Gantapara, A.P., de Graaf, J., van Roij, R., Dijkstra, M.: Soft Matter **8**, 8826 (2012)
55. Velasco, E., Mederos, L., Navascués, G.: Langmuir **14**, 5652 (1998)
56. Kwak, S.K., Park, T., Yoon, Y.J., Lee, J.M.: Mol. Sim. **38**, 16 (2012)
57. Groh, B., Mulder, B.: J. Chem. Phys. **114**, 3653 (2001)
58. Cuetos, A., Dennison, M., Masters, A., Patti, A.: Soft Matter **13**, 4720 (2017)
59. Meijer, J.M., Pal, A., Ouhajji, S., Lekkerkerker, H.N.W., Philipse, A.P., Petukhov, A.V.: Nat. Commun. **8**, 14352 (2017)
60. Hoover, W.G., Ree, F.H.: J. Chem. Phys. **49**, 3609 (1968)
61. Petukhov, A.V., Tuinier, R., Vroege, G.J.: Current Opin. Colloid. Interface Sci. **30**, 54 (2017)
62. Dekker, F., González García, Á., Philipse, A.P., Tuinier, R.: Eur. Phys. J. E **43**, 38 (2020)
63. Widom, B.: J. Chem. Phys. **39**, 2808 (1963)
64. Helfand, E., Reiss, H., Frisch, H., Lebowitz, J.: J. Chem. Phys. **33**, 1379 (1960)
65. Lebowitz, J.L., Helfand, E., Praestgaard, E.: J. Chem. Phys. **43**, 774 (1965)
66. Oversteegen, S.M., Roth, R.: J. Chem. Phys. **122**, 214502 (2005)
67. Lekkerkerker, H.N.W., Poon, W.C.K., Pusey, P.N., Stroobants, A., Warren, P.B.: Europhys. Lett. **20**, 559 (1992)
68. Bolhuis, P.G., Frenkel, D.: J. Chem. Phys. **101**, 9869 (1994)
69. Tejero, C.F., Daanoun, A., Lekkerkerker, H.N.W., Baus, M.: Phys. Rev. Lett. **73**, 752 (1994)
70. Park, K., Koerner, H., Vaia, R.A.: Nano Lett. **10**, 1433 (2010)
71. Saez Cabezas, C.A., Sherman, Z.M., Howard, M.P., Dominguez, M.N., Cho, S.H., Ong, G.K., Green, A.M., Truskett, T.M., Milliron, D.J.: Nano Lett. **20**, 4007 (2020)
72. Yu, W.W., Falkner, J.C., Yavuz, C.T., Colvin, V.L.: Chem. Commun. **20**, 2306 (2004)

Further Manifestations of Depletion Effects

In this chapter we provide examples of the manifestations of depletion effects in areas such as biology and technology. The addition of nonadsorbing polymers to colloidal suspensions can cause phase separation of the mixture into a colloid-rich and a polymer-rich phase. The understanding of this polymer-induced phase separation is very important, not only for colloid science but also for industrial systems, such as food dispersions [1–4] and paint [5–8]. Colloids and polymers (or surfactants) are both present in these systems and influence the stability and subsequent processing issues. This holds similarly for binary or multi-component colloidal mixtures.

It has been realised that procedures employing the depletion interaction have the potential to enable the fabrication of materials based on self-organised colloidal structures [9]. Adding depletants can for instance enable the formation of a Penrose quasi-crystal of mobile colloidal tiles [10].

Also, the importance of depletion effects in biological systems is recognised [11–15]. Nonadsorbing polymer chains promote the adhesion of cells to surfaces [16] and enhance adsorption of lung surfactants at the air–water interface in lungs so as to help patients suffering from acute respiratory syndrome [17]. The physical properties of actin networks are affected by nonadsorbing polymers [18], which also modify phase transitions in virus dispersions [19]. It has been shown that depletion forces can deform epithelial cells [20]. Rod-like depletants are even able to induce a plethora of shape transitions of red blood cells. To further illustrate this, we discuss a few examples of depletion effects in systems of biological and technological interest in this chapter.

© The Author(s) 2024
H. N. W. Lekkerkerker, R. Tuinier, M. Vis, *Colloids and the Depletion Interaction*,
Lecture Notes in Physics 1026, https://doi.org/10.1007/978-3-031-52131-7_11

11.1   Macromolecular Crowding

A longstanding question in molecular biology is the extent to which the behaviour of macromolecules observed in vitro accurately reflects their behaviour *in vivo* [21]. A characteristic of the cytoplasm of living cells is the high concentration of macro-molecules (including proteins and nucleic acids) that they contain (up to 400 g/L) [22,23]. Since the 1980s [22,24,25] it has been increasingly appreciated that the large volume fraction occupied by these macromolecules influences several intra-cellular processes [26–28], ranging from the bundling of biopolymers like DNA and actin, to the phase separation in a bacterial cell. These effects are known amongst biochemists and biophysicists as *macromolecular crowding* (see for instance Refs. [11,15,29–32]). The term 'crowding' is used rather than 'concentrated' because, in general, no single macromolecular species occurs at high concentration but, taken together, the macromolecules occupy a significant fraction (typically 10–30%) of the total volume [33].

The biological relevance of crowding such as chemical equilibria and rates, association reactions and enzyme kinetics has been studied extensively. For reviews, see Refs. [26,33]. Another important characteristic where macromolecular crowding plays an important role is phase separation in the cytoplasm. Walter and Brooks [34] put forward the hypothesis that macromolecular crowding is the basis for micro-compartmentalisation.

Phase separation between a nucleoid and cytoplasm in bacterial cells is a striking example of macromolecular crowding [35–37]. Chromosomes in bacterial cells do not occur in dispersed form but are organised in the nucleoid as a separate phase. Depletion forces that originate from the presence of proteins can explain the phase separation [36]. As a result, the proteins partition over the cytoplasm and nucleoid phases. Their concentration in the cytoplasm is about two times larger than their concentration in the nucleoid phase [37] (see Fig. 11.1).

Another example where macromolecular crowding plays a key role is fluid–fluid phase separation in the cell [15,38–46]. Here, phase transitions give rise to dense droplets in the cell such as nucleoli, germ granules and speckles [43,45] that have been collectively described as membraneless organelles. This is schematically illustrated in Fig. 11.2 [15].

There has been significant interest in the role that the depletion interaction plays in driving cellular organisation [11–13,47,48]. However, while the depletion interaction promotes fluid–fluid phase separation in the cell, Groen et al. [48] have argued that crowded macromolecular solutions are very prone to non-specific associative interactions that can potentially counteract depletion. It gradually becomes clear that excluded volume interactions can explain the assembly of, and liquid–liquid phase separation in, a wide range of cellular structures. These range from the cytoskeleton to chromatin loops and entire chromosomes [11,15].

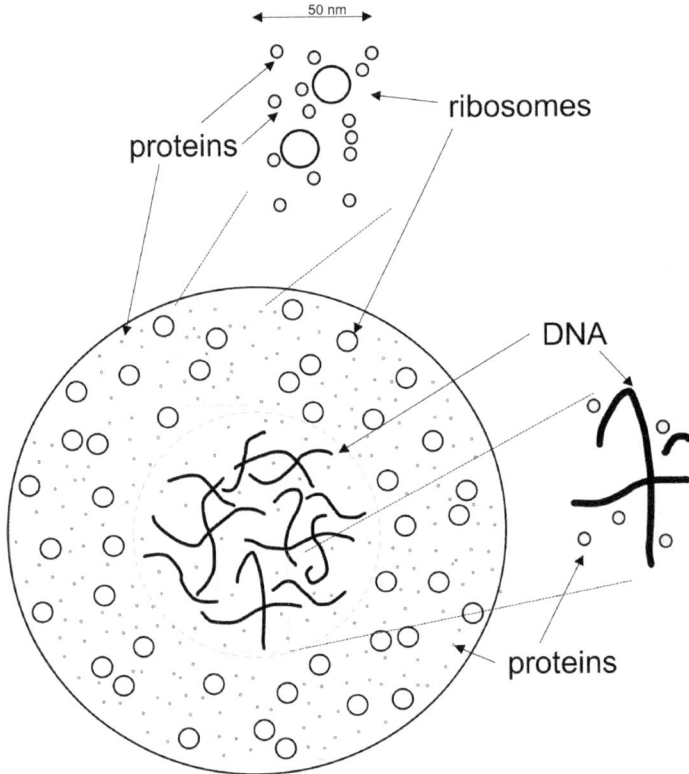

Fig. 11.1 Representation of a bacterial cell containing phase-separated nucleoid and cytoplasm. DNA is concentrated in the nucleoid, and ribosomes and proteins are concentrated in the cytoplasm. Inspired by a drawing in Ref. [37]

11.2   Depletion Interactions and Protein Crystallisation

In 1934, Desmond Bernal and Dorothy Crowfoot (later Hodgkin) discovered that crystals of the digestive enzyme pepsin give a well-resolved X-ray diffraction pattern [49]. It took 25 years before the first atomic structures of proteins using X-ray crystallography were determined. In 1958 Kendrew et al. published the structure of the protein myoglobulin [50], which stores oxygen in muscle cells; and in 1960, Perutz et al. [51] reported the structure of the protein haemoglobin, which transports oxygen in blood.

The first requirement for protein structure determination with X-ray diffraction is to grow suitable crystals [52]. While great strides have been made in the determination of protein structures (more than 200,000 protein structures have been resolved [53]), protein crystallisation (notwithstanding a history spanning more than 150 years [54, 55]) remains somewhat elusive. This actually holds for crystallisation in general [56].

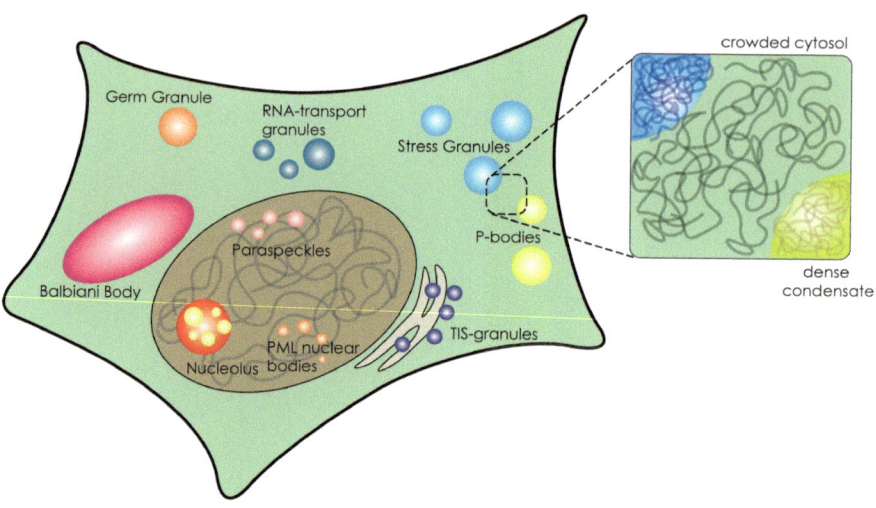

Fig. 11.2 A eukaryotic cell containing several types of organelles. *Inset*: the crowded state of the cytosol. Reprinted from Ref. [15] under the terms of CC-BY-4.0

Figure 11.3 characterises the state of the art in protein crystallisation in 1988 [57]. In recent years, significant progress has been made in understanding protein crystallisation on the basis of the phase diagram of protein solutions. The key observation that lies at the basis of this development was made by Benedek and co-workers [58,59]. In the course of their investigations of proteins involved in maintaining the transparency of the eye lens, they discovered that in aqueous solutions of several bovine lens proteins the solid-liquid phase boundary lies higher in temperature than the liquid–liquid coexistence curves. Thus, over a range of concentrations and temperatures for which liquid–liquid phase separation occurs, the coexistence of a protein crystal phase with a protein liquid solution phase is thermodynamically stable relative to the metastable separated liquid phases [60] (Fig. 11.4). Note also the metastable critical fluid–fluid point [58,59,61–65].

It was shown that this remarkable phase behaviour could be understood on the basis of the sensitivity to the form of the pair potential of the phase diagram of small attractive colloidal particles [66–69]. Moreover, it was soon realised that successful protein crystallisation depends on the location (protein concentration and temperature) in the phase diagram [65,70–74]. Control of protein crystal nucleation around the metastable 'liquid–liquid' phase boundary [74] appears key to the development of systematic crystallisation strategies (for a concise review, see Ref. [75]). This phase boundary can be manipulated by depletion interactions through the addition of nonadsorbing polymers such as polyethylene glycol [76–78].

Fig. 11.3 View upon the 'art' of protein crystallisation by J. Drenth. Reprinted with permission from Ref. [57]. Copyright 1988 Elsevier

Fig. 11.4 Typical phase diagram of a globular protein solution. The critical point is marked by the asterisk

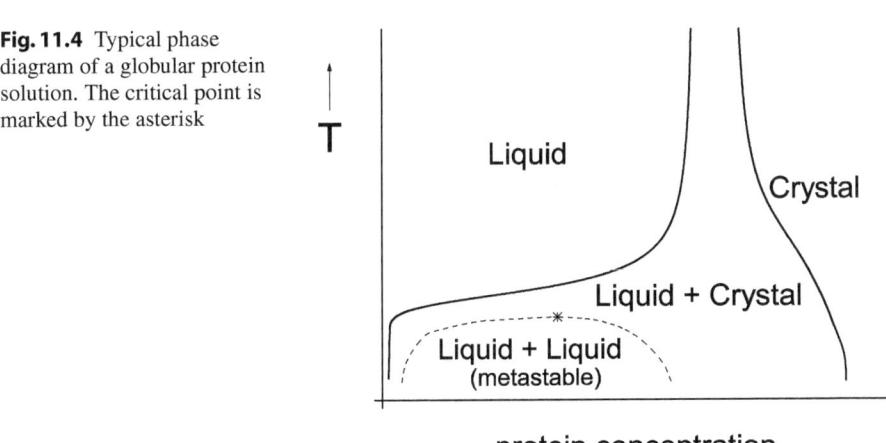

Exercise 11.1. Argue how Fig. 11.4 is modified upon increasingly adding nonadsorbing polymers for $q > 0.3$.

To illustrate the role of nonadsorbing polymer chains on the protein solution phase behaviour, we discuss the results of adding PEG to a solution with the protein apoferritin by Tanaka and Ataka [79]. Apoferritin is an iron storage protein consisting of 24 subunits. The effective radius is about 8 nm and the molar mass of apoferritin is 440 kg/mol. It is not easy to crystallise a solution of apoferritins by adding salt ions. Traditionally, a well-defined scale known as the Hofmeister series [80] is used as a measure for the efficiency of precipitating proteins. A solution of apoferritin cannot be crystallised in the common manner with the usual salt ions as precipitating agents. Adding PEG, however, does make it possible to induce crystallisation. Fig. 11.5 shows the experimental data obtained from visual and microscopic inspection of PEG–apoferritin in aqueous 0.6 M NaCl solutions. The concentration of apoferritin was fixed at 54 g/L. The PEG concentration and molar mass were varied. Four molar masses M_{PEG} (and radii of gyrations R_g) of the PEGs were used: 1.5, 4.0, 8.0 and 20 kg/mol (1.4, 2.5, 3.7 and 6.2 nm, respectively), corresponding to $q = 0.18, 0.31, 0.46,$ and 0.78.

Various situations were observed after mixing PEG with apoferritins [79]. For sufficiently small concentrations (depending on M_{PEG}, hence q) the mixture was stable ($\triangle$, Fig. 11.5), while further increasing the PEG concentration leads to a phase transition. At $q = 0.18$ random aggregates ($\bullet$) were found, which is typical for a protein solution undergoing a fluid-to-solid transition and does not give the proper conditions for obtaining good-quality crystals. The same happens for the highest concentrations at $q = 0.31$ and 0.46. For $q = 0.31, 0.46,$ and 0.78 there was a region where liquid domains (+) were formed, indicative of a gas–liquid phase transition, usually referred to as liquid–liquid phase separation. For $q = 0.78$, liquid domains were found in the entire unstable regime. Finally, good-quality crystals ($\circ$), in coexistence with liquid domains, were formed at $q = 0.31$ and 0.46 for intermediate PEG concentrations. For these q-values the critical point is close to the fluid–crystal coexistence line, in agreement with the findings of Ten Wolde and Frenkel [71]. Thus, it follows that adding PEG indeed provides the conditions for good crystallisation within a specific range of protein–polymer size ratios and polymer concentrations. The different states are illustrated with the micrographs in the right panel of Fig. 11.5.

Aided crystallisation is, of course, not limited to proteins. For instance, Kirner and Sturm [81] used depletant-mediated crystallisation to separate mixtures of nanocrystals of different sizes and shapes.

11.3 Shape and Size Selection

The depletion interaction, as argued in Sect. 1.2.5, depends on the concentration of the depletion agent and the overlap volume of the depletion zones. For a given concentration of depletant the only variable is the overlap volume, which in turn depends on the size (see Chap. 2) and shape of the colloidal particles. Tuning the strength of the depletion interaction therefore allows particles of different size and shape to be separated. For example, the separation of rod-like particles and spheres under the influence of polymers is schematically indicated in Fig. 11.6.

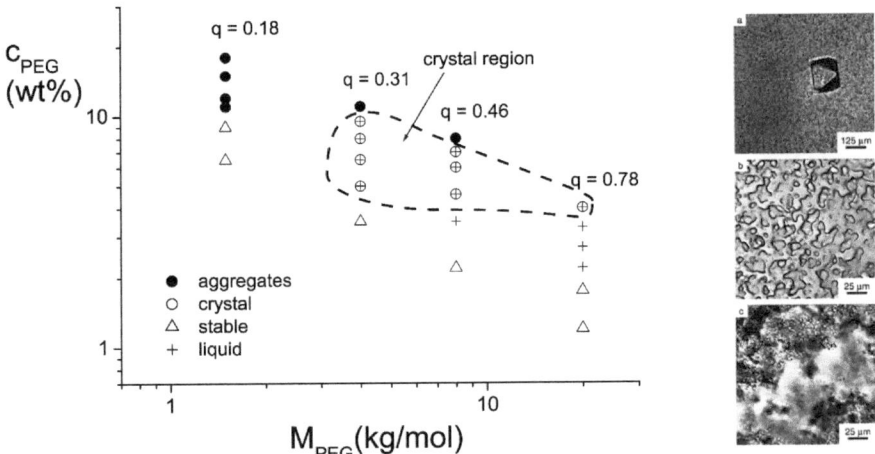

Fig. 11.5 Phase behaviour of apoferritin with PEG. *Left*: state diagram of apoferritin mixed with PEG of various molar masses. The apoferritin concentration was kept constant at 54 g/L and the molar mass and concentration of PEG was varied as indicated in the diagram. Results are redrawn from Ref. [79]. *Right*: Micrographs representing the various kinds of unstable solutions that were found in aqueous apoferritin–PEG mixtures: **a** crystals, **b** liquid domains and **c** random aggregates. Right panel: reprinted with permission from Ref. [79]. Copyright 2002, American Institute of Physics

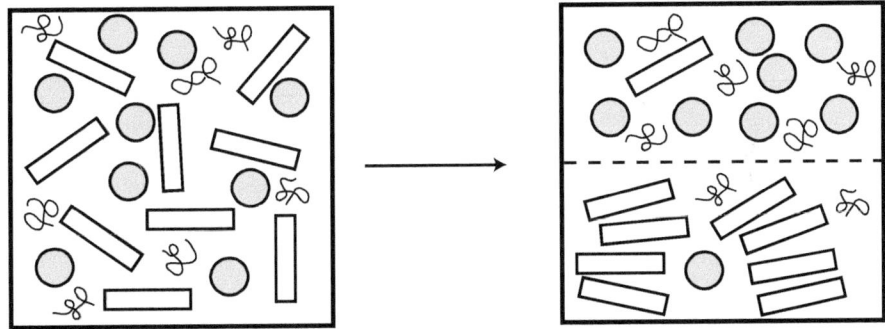

Fig. 11.6 Shape-selective separation induced by depletion forces

Unaware of the underlying principle, this had already been used by Cohen in 1941 [82] to separate two viruses, Tobacco Mosaic Virus and Tobacco Necrosis Virus. Tobacco Mosaic Virus is a rod-like virus with a length of 300 nm and diameter of 18 nm, and Tobacco Necrosis Virus a spherical virus with a diameter of about 26 nm. Cohen used the polysaccharide heparin as depletant to separate these viruses. This method to separate colloids of different sizes and shapes has recently gained new impetus. Obtaining particles of a specific size and shape is critical for optimising the nanostructure-dependent optical, electrical and magnetic properties in nano-based technologies.

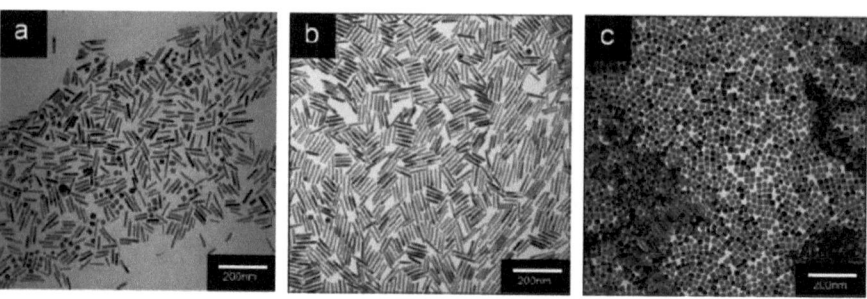

Fig. 11.7 TEM images of a dispersion of rod-like and cube-like gold colloids. **a** synthesised mixture, **b** sediment, **c** supernatant. Reprinted with permission from Ref. [86]. Copyright 2010, American Chemical Society (ACS)

While the self-organisation of nearly monodisperse spherical colloidal particles has been studied for a long time, the full potential of the self-assembly of aniso-metric colloidal particles (rods and plates) is far from being achieved. Nevertheless, important advances have been made. For example, CdSe semiconductor nanorods have been shown to form nematic liquid crystals [83] that can potentially be used as functional components in electro-optical devices. Kim et al. [84] succeeded in generating long-range assembly of colloidal anisotropic nanocrystals into thin films with orientational and positional order by adding depletants. Hence, the depletion interaction has the potential to enable the effective separation of anisometric colloids from a mixture of particles of different sizes and shapes.

Depletion-induced shape and size selection of colloidal particles could be a pow-erful tool to achieve the separation of different components. For instance, efficient purification of gold platelets in complex multi-component colloidal mixtures was realised by Zhao et al. [85] using surfactant micelles as depletants. Park et al. [86] reported the depletion-induced shape and size selection of gold rods and cubes. In Fig. 11.7 we show their transmission electron microscopy (TEM) images of gold rods ($L = 77$ nm, $D = 11$ nm) and cubes (20 nm), which could be separated by adding nonadsorbing polymers.

Baranov et al. [9] showed that the depletion attraction forces were effective in the shape selective separation of CdSe/CdS-rods from a mixture of rods and CdSe spheres. Mason [87] showed that the depletion interaction between plate-like parti-cles is much stronger than between spheres, leading to a separation between a phase enriched with plates and a phase mainly concentrated with spheres. The dependence of the depletion interaction on size can also be used to fractionate a bidisperse pop-ulation of colloidal spheres [88], or to obtain a monodisperse population of spheres from a collection of polydisperse spheres [89]. Bidisperse colloidal particle mix-tures have the potential to self-organise into colloidal crystals (see Chap. 6 for more details).

Ye et al. [90] presented an experimental-computational investigation of mixtures of rods and spheres showing that the mixture can co-assemble into a binary super-lattice. The formation of two-dimensional colloidal membranes from a suspension of rod-like viruses mixed with nonadsorbing polymer chains was studied by Kang et al. [91]. It is clear that depletion forces can be exploited in the design of a wide range of reconfigurable colloidal structures.

These procedures, based on the depletion interaction, have the potential to enable powerful fabrication procedures of materials based on self-organised colloidal structures. A computational study by Bevan et al. [92] showed that actuating colloidal assembly in a practical process is feasible, and they provided insights into how to optimise the process conditions.

11.4   Directing Colloidal Self-assembly Using Surface Microstructures

As indicated in Sect. 11.3, the depletion interaction depends on the overlap volume for a given depletant concentration. This dependence leads to a difference in depletion interaction between particles of different sizes and shapes and offers a powerful and cost-effective way to separate them.

The use of surface microstructures provides a promising route for creating colloidal assemblies via depletion forces. Dinsmore, Yodh and Pine [93] studied the interaction of large polystyrene spheres ($R = 203$ nm, $\phi = 10^{-5}$) in a sea of small polystyrene spheres ($R = 41$ nm, $\phi = 0.30$) with a wall with a step edge, see Fig. 11.8.

Clearly, the overlap volume depends on the position of the big sphere with respect to the step edge. Since the depletion interaction can be seen as the product of overlap volume and osmotic pressure of the depletants ($W_{\mathrm{dep}} \approx -PV_{\mathrm{ov}}$, where P is now the osmotic pressure of the small spheres (see Eq. 1.18)), a difference in overlap volume affects the depletion interaction accordingly. It is, therefore, also expected that confinement effects can mediate phase transitions [94,95]. Spannuth and Conrad

Fig. 11.8 A large colloidal sphere near a step edge in a sea of small spheres. The presence of the small spheres leads to depletion zones (light grey regions) near the walls of the container and around the big sphere. Overlap of depletion zones is indicated by the hatched area. This overlap volume increases the volume accessible to the small spheres, thereby increasing their entropy

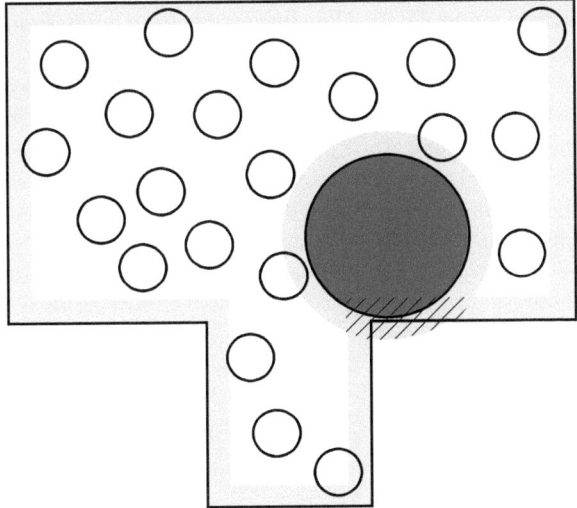

showed that confinement of a colloid–polymer mixture can induce solidification [96]. For an overview of theoretical accounts on confinement and depletion effects, see Ref. [97]. Confinement effects are also relevant for the microchannel flow of colloidal or colloid–polymer mixtures [94,98].

The depletion interaction can be derived by measuring the probability of the various positions (h) of the big particles on the terrace with the step edge using optical microscopy, and relating this probability $p(h)$ with the Boltzmann relation

$$p(h) \sim e^{-W_{dep}(h)/kT}, \tag{11.1}$$

the depletion interaction can be measured. For the system, the differences in the overlap volume amount to a difference in the depletion potential of about twice the thermal energy of the particles. This indicates that surface structures can create localised force fields that can trap particles.

Exercise 11.2. Rationalise what are more favourable positions inside the box for the big sphere in Fig. 11.8.

An interesting application of this concept can be found in the work of Sacanna et al. [99]. By clever colloid synthesis, they created 5 μm (diameter) polymerised silicon oil droplets with a well-defined spherical cavity. To these 'lock' particles they added appropriately sized spherical 'key' particles (silica, poly(methyl methacrylate) or polystyrene colloids) that can fit into the cavity. Nanometer sized nonadsorbing polymers were added to provide a depletion interaction. The depletion interaction, being proportional to the overlap volume of the depletion zones, attains a maximum when the key particle fits precisely into the spherical cavity of a lock particle (Fig. 11.9). The depletion-driven self-assembly of lock-and-key particles is demonstrated in Fig. 11.10. This time series (from left to right) illustrates the site-specificity of the attraction.

By developing colloids with well-defined multicavities, this concept has been extended to make lock particles with multiple key holes [101]. Adding appropriately sized depletants to dispersions of colloidal golf-balls [102] induces the formation of controlled self-assembled structures. Computer simulations have shown how the binding tendency in a dispersion of lock-and-key colloids can be controlled by adjusting the characteristics of polymeric depletants [103]. Theoretical predictions revealed interesting phase behaviour of such mixtures of lock-and-key particles with depletants [104].

Another way to manipulate the overlap volume of the depletion zones is to vary the roughness of the surface [105] of the colloidal particles (Fig. 11.11). The left drawings show that surface roughness does not affect the overlap volume for intermediate overlap of depletion zones. When the particles are in close contact surface roughness prevents overlap of certain zones that would normally overlap for smooth surfaces (see the sketches on the right).

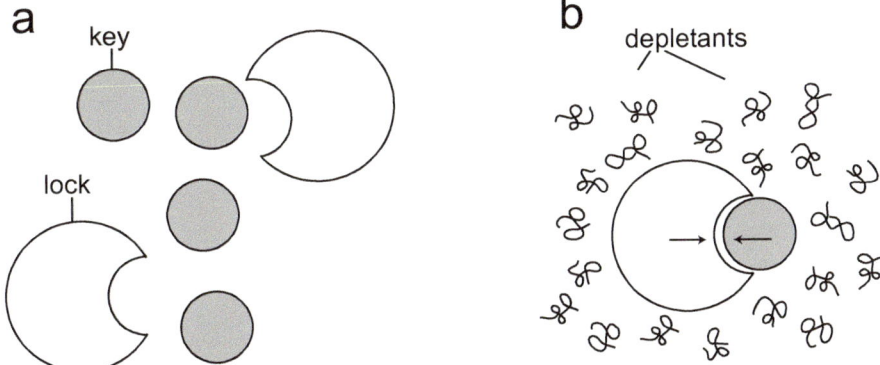

Fig. 11.9 a Colloidal 'lock' particles can be synthesised [99] to contain a dimple into which 'key' particles, spherical colloids with appropriate size, can fit. **b** By adding depletants (polymer chains) a key can be pushed into a lock using the depletion force. Inspired by Solomon [100]

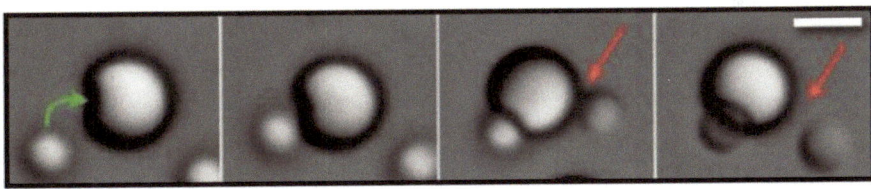

Fig. 11.10 Series of images demonstrating a colloidal sphere entering the lock of larger colloid. The curved arrow in the first micrograph indicates a successful lock–key binding. Scale bar is 2 μm. Reprinted with permission from Ref. [99]. Copyright 2010, Springer Nature

Hence, selecting the strength of the attraction is possible by introducing colloidal surface roughness [106–108]. This makes it possible to direct the self-assembly of particles by selectively controlling the roughness of different sides of colloidal particles. Badaire et al. [109,110] demonstrated the potential of this method in the assembly of lithographically designed colloidal particles. In Fig. 11.12 (left panel) we show the particles used by Badaire et al. [109,110] that consist of roughened, rounded side walls and flat ends. Upon adding surfactant micelles, these particles will attract one another due to the depletion force. Since the attraction is stronger between the flat sides of the particles, rod-like equilibrium structures are formed at a certain depletant concentration. An example of the work of Badaire et al. is depicted in Fig. 11.12 (right panel). Zhao and Mason [106] demonstrated the same principle on plate-like particles with manipulated roughness.

Kraft et al. [111] prepared patchy particles with smooth and rough parts. This made it possible to employ the depletion interaction to make 'colloidal micelles': the patchy particles assemble into clusters that resemble surfactant micelles with the smooth and attractive sides of the colloids located at the interior. Anzini and Parola [108] developed a simple model to describe the effects of surface roughness on the depletion interaction, yielding explicit expressions for a wide range of interesting conditions. The theoretical predictions compare well with the numerical simulations

Fig. 11.11 Representation of the overlap zones between two colloidal hard spheres with flat surfaces (*upper*) and two particles with roughened surfaces (*lower*) for small (*left*) and large (*right*) overlap. Drawn by C.M. Martens

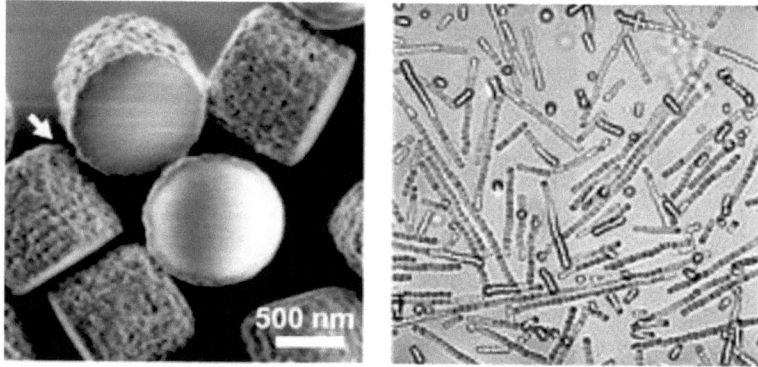

Fig. 11.12 *Left*: Scanning Electron Microscopy (SEM) image of colloidal particles that have sides with surface roughness and smooth sides. *Right*: Aggregated state of these particles under the influence of depletion forces. Image size 50 μm × 50 μm. Reprinted with permission from Refs. [109,110]. Copyright 2007 and 2008 ACS

of Kamp et al. [107]. This theory enables the onset of colloidal aggregation to be predicted in suspensions of rough particles.

In materials science, depletion effects were used in various ways to self-organise colloidal systems. Okabe et al. [112] used it to assemble artificially manufactured components larger than micrometres. By making use of shape complementarity, simple immersion of the microcomponents in polymer solutions enabled assembly.

11.5   Dynamic Depletion Effects

Macromolecular crowding also has consequences for transport properties [113]. One may wonder how protein transport occurs through a cell composed of a highly concentrated dispersion. The viscosity of the cytoplasm will be significantly larger than that of a physiological salt solution. The question arises of what friction a protein experiences as it moves through a cell. This relates to a fundamental problem in colloid physics: the dynamics of a colloidal sphere translating and rotating through a polymer solution.

Dzubiella, Löwen, and Likos [114] considered the flow of a dispersion containing non-interacting Brownian small particles around bigger hard spheres (Fig. 11.13). They found that the effective forces are highly anisotropic. The density profiles are obviously non-trivial. A detailed analysis of the nonequilibrium forces under dynamic circumstances was later performed by Dolata and Zia [115]. A theoretical framework for the non-Newtonian viscosity of a colloidal dispersion with short-ranged depletion attraction was developed by Huang and Zia [116].

As a colloidal particle diffuses or sediments through a solution containing non-adsorbing polymer chains, one may naively expect that the friction experienced by the particle is set by the bulk viscosity. In practice, it is smaller. An analysis of the velocity profile of a nonadsorbing polymer solution near a flat surface shows that depletion leads to effective slip [117]. The depletion layer implies a non-uniform viscosity profile near the surface, which explains this slip. Such effective slip effects also appear when considering the flow of colloidal particles at a wall [118]. Even in the case of simulating colloids in a solvent, the solvent molecules induce deple-

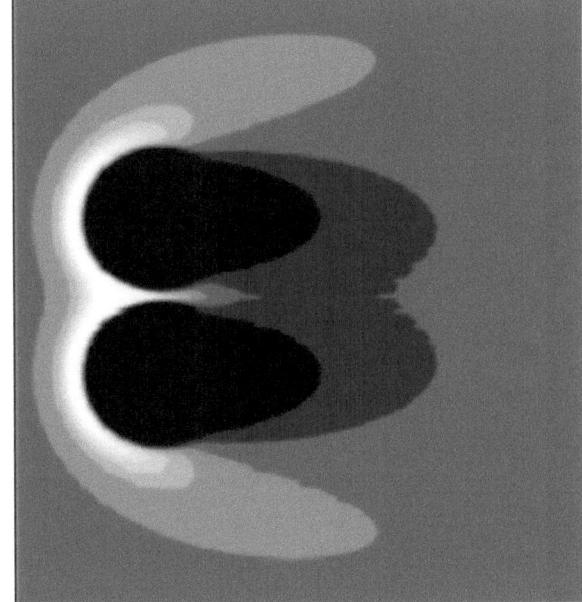

Fig. 11.13 Steady state contour density field (flow from left to right) of Brownian non-interacting spheres around two hard spheres (black). The brighter the region the higher the Brownian particle concentration, which is grey in the bulk (average concentration). Reprinted with permission from Ref. [114]. Copyright 2003 American Physical Society

tion effects [119, 120], and it is challenging to properly account for such effects in mesoscale simulation methods.

Phillies and co-workers [121, 122] studied the translational self-diffusion of well-defined colloidal spheres through polymer solutions, and showed that the interpretation of the measured friction coefficient of the particles is fairly complicated. For a spherical particle that moves through a medium containing small solvent molecules, the friction coefficient is proportional to the solvent viscosity. When the solvent is replaced by a polymer solution, one may naively expect that the friction coefficient is proportional to the viscosity of the polymer solution. Measurements indicate that this is only true when the chains are very small compared to the size of the particle.

Exercise 11.3. What viscosity is experienced by a tiny sphere in a dilute solution with very long polymer chains?

For polymer chains that are roughly as big as the particle, the apparent or effective viscosity experienced by a sphere is in between the viscosities of solvent and polymer solution. A similar finding was also reported for the rotational diffusion of colloidal particles [123] and for the sedimentation of colloids through a polymer solution [124]. The influence of depletion forces on sedimentation in itself is a rich and challenging topic [125, 126].

The fact that the effective viscosity is intermediate between that of solvent and polymer solution can be rationalised as follows. Within the depletion layer, the viscosity is expected to follow the polymer density distribution [117], and it gradually increases from the solvent viscosity at the solid surface to the bulk viscosity far from the particle. Therefore, as a particle diffuses, the hydrodynamic resistance force is also in between the two limits. Fan et al. [127–129] derived analytical expressions for the friction felt by a sphere when it moves through a macromolecular medium and showed that the friction is strongly reduced compared to Stokes' law. This means that depletion-induced slip effects facilitate protein transport through crowded media. This work has been extended to (i) understand the effect of shear flow on the segment density profile of a nonadsorbing polymer solution in a narrow slit [130] and (ii) the mimicking of colloid dynamics of interacting hard spheres mediated by depletants [131, 132]. For the diffusion of a slender object through a polymer solution, see Ref. [133].

Krüger and Rauscher [134] calculated the short-time and the long-time diffusion coefficients of a colloidal sphere in a polymer solution and took hydrodynamic interactions into account. It follows that the long-time diffusion coefficient can be described using a generalised Stokes-Einstein relation, whereas it deviates for the short-time coefficient. Ochab and Holyst [135] proposed a model of confined diffusion to describe the diffusion of a sphere through a polymer solution. Their model explains the anomalous diffusion that is observed experimentally [136].

Anomalous diffusive motion of particles in crowded environments such as the interior of cells and in cellular membranes was also evaluated by Höfling and Franosch [137]. For the analysis of anomalous transport, they reviewed the theory that underlies commonly applied techniques such as single-particle tracking, fluorescence correlation spectroscopy and fluorescence recovery after photobleaching. They show experimental evidence for anomalous transport in crowded biological media. Zöttl and Yeomans [138] investigated the transport of driven nano- and micro-particles in complex fluids. They measured the fluid flow fields and local polymer density and polymer conformation around the particles. Schuler et al. [139] performed extensive single-molecule experiments to investigate the interaction between two intrinsically disordered proteins. They studied the influence of crowding on the association and dissociation kinetics of the proteins and the translational diffusion. Theory by Fan et al. [127, 128] can accurately quantify the measured diffusion of proteins through crowded macromolecular media [139, 140].

References

1. Grinberg, V.Y., Tolstoguzov, V.B.: Food Hydrocolloids **11**, 145 (1997)
2. Syrbe, A., Bauer, W.K., Kostermeyer, H.: Int. Dairy J. **8**, 179 (1998)
3. Doublier, J.L., Garnier, C., Renard, C., Sanchez, C.: Curr. Opin. Colloid Interface Sci. **5**, 184 (2000)
4. de Kruif, C.G., Tuinier, R.: Food Hydrocolloids **15**, 555 (2001)
5. Overbeek, A., Bückmann, F., Martin, E., Steenwinkel, P., Annable, T.: Progr. Org. Coat. **48**, 125 (2003)
6. Tadros, T.: Colloids in Paints. Wiley (2011)
7. de With, G.: Polymer Coatings. Wiley, New York (2018)
8. Schulz, M., Keddie, J.L.: Soft Matter **14**, 6181 (2018)
9. Baranov, D., Fiore, A., van Huis, M., Giannini, C., Falqui, A., Lafont, U., Zandbergen, H., Zanella, M., Cingolani, R., Manna, L.: Nano Lett. **10**, 743 (2010)
10. Wang, P.Y., Mason, T.G.: Nature **561**, 94 (2018)
11. Marenduzzo, D., Finan, K., Cook, P.R.: J. Cell Biol. **175**, 681–686 (2006)
12. Ping, G., Yang, G., Yuan, J.M.: Polymer **47**, 2564 (2006)
13. Sapir, L., Harries, D.: Curr. Opin. Colloid Interface Sci. **20**, 3 (2015)
14. Secor, P.R., Sweere, J.M., Michaels, L.A., Malkovskiy, A.V., Lazzareschi, D., Katznelson, E., Rajadas, J., Birnbaum, M.E., Arrigoni, A., Braun, K.R., Evanko, S.P., Stevens, D.A., Kaminsky, W., Singh, P.K., Parks, W.C., Bollyky, P.L.: Cell Host Microbe **18**, 549–559 (2015)
15. André, A.A.M., Spruijt, E.: Int. J. Mol. Sci. **21** (2020)
16. Neu, B., Meiselman, H.J.: Biochim. et Biophys. Acta **1760**, 1772 (2006)
17. Stenger, P.C., Zasadzinski, J.A.: Biophys. J. **92**, 3 (2007)
18. Tharmann, R., Claessens, M.M.A.E., Bausch, A.R.: Biophys. J. **90**, 2622 (2006)
19. Dogic, Z., Purdy, K.R., Grelet, E., Adams, M., Fraden, S.: Phys. Rev. E. **69**, 051702 (2004)
20. Hashimoto, S., Yoshida, A., Ohta, T., Taniguchi, H., Sadakane, K., Yoshikawa, K.: Chem. Phys. Lett. **655–656**, 11 (2016)
21. McGuffee, S.R., Elcock, A.H.: PLoS Comput. Biol. **6**, e1000694 (2010)
22. Fulton, A.B.: Cell **30**, 345 (1982)
23. Goodsell, D.S.: Trends Biochem. Sci. **16**, 203 (1991)
24. Ralston, G.B.: J. Chem. Edu. **67**, 857 (1990)
25. Goodsell, D.S.: The Machinery of Life. Springer, New York (1998)
26. Minton, A.P.: Curr. Opin. Struct. Biol. **10**, 34 (2000)
27. Snoussi, K., Halle, B.: Biophys. J. **88**, 2855 (2005)
28. Cheung, M.S., Klimov, D., Thirumalai, D.: Proc. Natl. Acad. Sci. **102**, 4753 (2005)
29. Zimmerman, S.B., Minton, A.P.: Ann. Rev. Biophys. Biomol. Struct. **22**, 27 (1993)
30. Herzfeld, J.: Acc. Chem. Res. **29**, 31 (1996)

31. Ellis, R.J.: Curr. Opin. Struct. Biol. **11**, 114 (2001)
32. Ellis, R.J., Minton, A.P.: Nature **425**, 27 (2003)
33. Ellis, R.J.: Trends Biochem. Sci. **26**, 597 (2001)
34. Walter, H., Brooks, D.E.: FEBS Lett. **361**, 135 (1995)
35. Valkenburg, J.A.C., Woldringh, C.L.: J. Bacteriol. **160**, 1151 (1984)
36. Odijk, T.: Biophys. Chem. **73**, 23 (1998)
37. Woldringh, C.L., Odijk, T.: In: Charlebois, R.L. (Ed.) Organization of the Prokaryotic Genome, Chap. 10. ASM Press, Amsterdam (1999)
38. Iborra, F.J.: Theor. Biol. Med. Mod. **4**, 1 (2007)
39. Hyman, A.A., Brangwynne, C.P.: Dev. Cell **21**, 14 (2011)
40. Hyman, A.A., Weber, C.A., Julicher, F.: Annu. Rev. Cell. Dev. Biol. **30**, 39 (2014)
41. Brangwynne, C.P.: J. Cell Biol. **11**, 899 (2013)
42. Brangwynne, C.P., Tompa, P., Pappu, R.V.: Nat. Phys. **11**, 899 (2015)
43. Mitrea, D.M., Kriwacki, R.W.: Cell Commun. Signal **14**, 1 (2016)
44. Y. Shin, C.P. Brangwynne, Science **357**, eaaf4382 (2017)
45. Banani, H.O., Lee, A.A., Rosen, M.K.: Nat. Rev. Mol. Cell Biol. **18**, 285 (2017)
46. Berry, J., Brangwynne, C.P., Haataja, M.: Rep. Prog. Phys. **81**, 046601 (2018)
47. Sapir, L., Harries, D.: Bunsen-Mag. **19**, 152 (2017)
48. Groen, J., Foschepoth, D., te Brinke, E., Boersma, A.J., Imamura, H., Rivas, G., Heus, H.A., Huck, W.T.S.: J. Am. Chem. Soc. **137**, 13041 (2015)
49. Bernal, J.D., Crowfoot, D.: Nature **133**, 794 (1934)
50. Kendrew, J.C., Bodo, G., Dintzis, H.M., Parrish, R.G., Wyckoff, H., Phillips, D.C.: Nature **181**, 662 (1958)
51. Perutz, M.F., Rossmann, M.G., Cullis, A.F., Muirhead, H., Will, G.: Nature **185**, 416 (1960)
52. Drenth, J.: Principles of Protein X-Ray Crystallography. Springer (2007)
53. Word wide protein data bank: https://www.wwpdb.org/stats/deposition. Accessed 04-08-2023
54. McPherson, A.: J. Cryst. Growth **110**, 1 (1991)
55. McPherson, A.: Methods **34**, 254 (2004)
56. Gilman, J.J. (ed.): The Art and Science of Growing Crystals. Wiley, New York (1963)
57. Drenth, J.: J. Cryst. Growth **90**, 368 (1988)
58. Broide, M.L., Berland, C.R., Pande, J., Ogun, O., Benedek, G.B.: Proc. Natl. Acad. Sci. **88**, 5660 (1991)
59. Berland, C.R., Thurston, G.M., Kondo, M., Broide, M.L., Pande, J., Ogun, O., Benedek, G.B.: Proc. Natl. Acad. Sci. **89**, 1214 (1992)
60. Stradner, A., Schurtenberger, P.: Soft Matter **16**, 307 (2020)
61. Tanaka, T., Ishimoto, C., Chylack, L.T.: Science **197**, 1010 (1977)
62. Ishimoto, C., Tanaka, T.: Phys. Rev. Lett. **39**, 474 (1977)
63. Phillies, G.D.J.: Phys. Rev. Lett. **55**, 1341 (1985)
64. Thomson, J.A., Schurtenberger, P., Thurston, G.M., Benedek, G.B.: Proc. Natl. Acad. Sci. **84**, 7079 (1987)
65. Muschol, M., Rosenberger, F.: J. Chem. Phys. **107**, 1953 (1997)
66. Asherie, N., Lomakin, A., Benedek, G.B.: Phys. Rev. Lett. **77**, 4832 (1996)
67. Rosenbaum, D., Zamora, P.C., Zukoski, C.F.: Phys. Rev. Lett. **76**, 150 (1996)
68. This could be considered as a contribution to the 'Wiedergutmachung' between colloid science in the colloid/macromolecule controversy with protein chemistry [69]
69. Tanford, C., Reynolds, J.: Nature's Robots. Oxford University Press, New York (2001)
70. Poon, W.C.K.: Phys. Rev. E **55**, 3762 (1997)
71. Ten Wolde, P.R., Frenkel, D.: Science **277**, 1975 (1997)
72. Haas, C., Drenth, J.: J. Cryst. Growth **196**, 388 (1999)
73. Galkin, O., Vekilov, P.G.: Proc. Natl. Acad. Sci. **97**, 6277 (2000)
74. Dumetz, A.C., Chockla, A.M., Kaler, E.W., Lenhoff, A.M.: Biophys. J. **94**, 570–583 (2008)
75. Piazza, R.: Curr. Opinion Colloid Interface Sci. **5**, 38 (2000)
76. Kulkarni, A.M., Chatterjee, A.P., Schweizer, K.S., Zukoski, C.F.: Phys. Rev. Lett. **83**, 4554 (1999)

77. Annunziata, O., Asherie, N., Lomakin, A., Pande, J., Ogun, O., Benedek, G.B.: Proc. Natl. Acad. Sci. **99**, 14165 (2002)
78. Annunziata, O., Ogun, O., Benedek, G.B.: Proc. Natl. Acad. Sci. **100**, 970 (2003)
79. Tanaka, S., Ataka, M.: J. Chem. Phys. **117**, 3504 (2002)
80. "see various papers in", Curr. Opinion Colloid Interface Sci. **9**, 1 (2004)
81. Kirner, F., Sturm, E.V.: Cryst. Growth Des. **21**, 5192 (2021)
82. Cohen, S.S.: Proc. Soc. Exp. Biol. Med. **48**, 163 (1941)
83. Li, L.S., Walda, J., Manna, L., Alivisatos, A.P.: Nano Lett. **2**, 557 (2002)
84. Kim, D., Bae, W.K., Kim, S.H., Lee, D.C.: Nano Lett. **19**, 963 (2019)
85. Zhao, C., Wang, G., Takarada, T., Liang, X., Komiyama, M., Maeda, M.: Colloids Surf. A **568**, 216 (2019)
86. Park, K., Koerner, H., Vaia, R.A.: Nano Lett. **10**, 1433 (2010)
87. Mason, T.G.: Phys. Rev. E **66**, 60402 (2002)
88. Piazza, R., Iacopini, S., Pierno, M., Vignati, E.: J. Phys.: Condens Matter **14**, 7563 (2002)
89. Bibette, J.: J. Colloid Interface Sci. **147**, 474 (1992)
90. Ye, X., Millan, J.A., Engel, M., Chen, J., Diroll, B.T., Glotzer, S.C., Murray, C.B.: Nano Lett. **13**, 4980 (2013)
91. Kang, L., Gibaud, T., Dogic, Z., Lubensky, T.C.: Soft Matter **12**, 386 (2016)
92. Bevan, M.A., Ford, D.M., Grover, M.A., Shapiro, B., Maroudas, D., Yang, Y., Thyagarajan, R., Tang, X., Sehgal, R.M.: J. Proc. Contr. **27**, 64 (2015)
93. Dinsmore, A.D., Yodh, A.G., Pine, D.J.: Nature **383**, 239 (1996)
94. Nikoubashman, A., Mahynski, N., Pirayandeh, A., Panagiotopoulos, A.: J. Chem. Phys. **140**, 094903 (2014)
95. Moncho-Jordá, A., Odriozola, G.: Curr. Opin. Colloid Interface Sci. **20**, 24 (2015)
96. Spannuth, M., Conrad, J.C.: Phys. Rev. Lett. **109**, 028301 (2012)
97. Trokhymchuk, A., Henderson, D.: Curr. Opin. Colloid Interface Sci. **20**, 32 (2015)
98. Pandey, R., Conrad, J.C.: Soft Matter **8**, 10695 (2012)
99. Sacanna, S., Irvine, W.T.M., Chaikin, P.M., Pine, D.J.: Nature **464**, 575 (2010)
100. Solomon, M.J., Spicer, P.T.: Soft Matter **6**, 1391 (2010)
101. Wang, Y., Wang, Y., Zheng, X., Yi, G.R., Sacanna, S., Pine, D.J., Weck, M.: J. Am. Chem. Soc. **136**, 6866 (2014)
102. Watanabe, K., Tajima, Y., Shimura, T., Ishii, H., Nagao, D.: J. Colloid Interface Sci. **534**, 81 (2019)
103. Chang, H.Y., Huang, C.W., Chen, Y.F., Chen, S.Y., Sheng, Y.J., Tsao, H.K.: Langmuir **31**, 13085 (2015)
104. Ashton, D.J., Jack, R.L., Wilding, N.B.: Phys. Rev. Lett. **114**, 237801 (2015)
105. Bryk, P., Sokolowski, S.: Appl. Surf. Sci. **253**(13), 5802 (2007)
106. Zhao, K., Mason, T.G.: Phys. Rev. Lett. **99**, 268301 (2007)
107. Kamp, M., Hermes, M., van Kats, C.M., Kraft, D.J., Kegel, W.K., Dijkstra, M., van Blaaderen, A.: Langmuir **32**, 1233 (2016)
108. Anzini, P., Parola, A.: Soft Matter **13**, 5150 (2017)
109. Badaire, S., Cottin-Bizonne, C., Woody, J.W., Yang, A., Stroock, A.D.: J. Am. Chem. Soc. **129**, 40 (2007)
110. Badaire, S., Cottin-Bizonne, C., Stroock, A.D.: Langmuir **24**, 11451 (2008)
111. Kraft, D.J., Ni, R., Smallenburg, F., Hermes, M., Yoon, K., Weitz, D.A., van Blaaderen, A., Groenewold, J., Dijkstra, M., Kegel, W.K.: Proc. Natl. Acad. Sci. **109**, 10787 (2012)
112. Okabe, U., Okano, T., Suzuki, H.: Sens. Actuat. A: Phys. **254**, 43 (2017)
113. Tabaka, M., Kalwarczyk, T., Szymanski, J., Hou, S., Holyst, R.: Front. Phys. **2** (2014)
114. Dzubiella, J., Löwen, H., Likos, C.N.: Phys. Rev. Lett. **91**, 248301 (2003)
115. Dolata, B.E., Zia, R.N.: J. Fluid Mech. **836**, 694–739 (2018)
116. Huang, D.E., Zia, R.N.: J. Colloid Interface Sci. **562**, 293 (2020)
117. Tuinier, R., Taniguchi, T., Phys, J.: Condens. Matter **17**, L9 (2005)
118. Ghosh, S., van den Ende, D.T.M., Mugele, F., Duits, M.H.G.: Colloids Surf. A **491**, 50 (2016)
119. Wagner, M., Ripoll, M.: Int. J. Mod. Phys. C **30**, 1941008 (2019)

120. Barcelos, E.I., Khani, S., Boromand, A., Vieira, L.F., Lee, J.A., Peet, J., Naccache, M.F., Maia, J.: Comput. Phys. Commun. **258**, 107618 (2021)
121. Lin, T.H., Phillies, G.D.J.: J. Phys. Chem. **86**, 4073 (1982)
122. Ullmann, G.S., Ullmann, K., Lindner, R.M., Phillies, G.D.J.: J. Phys. Chem. **89**, 692 (1985)
123. Koenderink, G.H., Sacanna, S., Aarts, D.G.A.L., Philipse, A.P.: Phys. Rev. E **69**, 021804 (2004)
124. Ye, X., Tong, P., Fetters, L.J.: Macromolecules **31**, 5785 (1998)
125. Lattuada, E., Buzzaccaro, S., Piazza, R.: Phys. Rev. Lett. **116**, 038301 (2016)
126. Fiore, A.M., Wang, G., Swan, J.W.: Phys. Rev. Fluids **3**, 063302 (2018)
127. Tuinier, R., Dhont, J.K.G., Fan, T.H.: Europhys. Lett. **75**, 929 (2006)
128. Fan, T.H., Dhont, J.K.G., Tuinier, R.: Phys. Rev. E. **75**, 018803 (2007)
129. Fan, T.H., Xie, B., Tuinier, R.: Phys. Rev. E. **76**, 051405 (2007)
130. Taniguchi, T., Arai, Y., Tuinier, R., Fan, T.H.: Eur. Phys. J. E **35**, 88 (2012)
131. Karzar-Jeddi, M., Tuinier, R., Taniguchi, T., Fan, T.H.: J. Chem. Phys. **140**, 214906 (2014)
132. He, Y., Li, L., Taniguchi, T., Tuinier, R., Fan, T.H.: Phys. Rev. Fluids **5**, 013302 (2020)
133. Morozov, K.I., Leshansky, A.M.: Macromolecules **55**, 3116–3128 (2022)
134. Krüger, M., Rauscher, M.: J. Chem. Phys. **131**, 094902 (2009)
135. Ochab-Marcinek, A., Hołyst, R.: Soft Matter **7**, 7366 (2011)
136. Kalwarczyk, T., Sozanski, K., Ochab-Marcinek, A., Szymanski, J., Tabaka, M., Hou, S., Holyst, R.: Adv. Colloid Interface Sci. **223**, 55 (2015)
137. Höfling, F., Franosch, T.: Rep. Progr. Phys. **76**, 046602 (2013)
138. Zöttl, A., Yeomans, J.M.: J. Phys.: Condens. Matter **31**, 234001 (2019)
139. Zosel, F., Soranno, A., Buholzer, K.J., Nettels, D., Schuler, B.: Proc. Natl. Acad. Sci. **117**, 13480 (2020)
140. Galvanetto, N., Ivanovic, M., Chowdhury, A., Sottini, A., Nüesch, M., Nettels, D., Best, R., Schuler, B.: Nature **619**, 876 (2023)

Epilogue

12

This book mainly focuses on basic concepts and *model* systems; but in reality, soft materials are complex and have a practical impact on our daily lives. These materials make up common products such as pharmaceutical formulations, paints, dairy products and cosmetics [1]. To connect the insights into depletion effects to practical applications, we highlight some of the unresolved questions and future directions that could be pursued.

The basic concept of the depletion interaction can explain many phenomena in practical systems (Chap. 1). It also quantifies several properties of model colloid–polymer mixtures and can qualitatively describe phenomena in applications. This also holds for depletion *forces*, which are well understood in simple model systems (Chap. 2); but the challenge ahead is to understand the interactions in mixtures in which the direct interactions between colloids and/or depletants are more realistic than pure hard-core interactions. Additionally, depletion forces are typically not pair-wise additive—certainly not in the case of relatively large depletants; hence, it is important to account for multi-body interactions. Measuring these multi-body forces, as well as interactions in more complex mixtures (such as those including charged colloids and/or polyelectrolytes), is still a major challenge. The establishment of an increasing number of advanced techniques is helpful here.

Phase diagrams summarise a material's thermodynamic stability [2], quantifying the stable phase state(s) upon varying conditions [3]. For that reason, phase diagrams are crucial for materials design and/or process optimisation and constitute a major part of this book. It is clear that the size of the nonadsorbing polymers relative to the colloidal spheres plays a crucial role; this determines the phase diagram topology and the region over which such a mixture is stable. Theoretical approaches can describe the main equilibrium phase diagram of colloid–polymer mixtures [4] both qualitatively and semi-quantitatively [5] (Chaps. 3, 4). Nonequilibrium phenomena (e.g.

© The Author(s) 2024
H. N. W. Lekkerkerker, R. Tuinier, M. Vis, *Colloids and the Depletion Interaction*,
Lecture Notes in Physics 1026, https://doi.org/10.1007/978-3-031-52131-7_12

aggregation, gelation and glass formation) also play an important role in dictating whether certain phase states are experimentally accessible (Chap. 4).

Colloidal gas–liquid interfaces have unique characteristics, including ultra-low interfacial tension and observable thermal capillary waves (Chap. 5). Although theory and experiment show reasonable agreement for the interfacial tension and thickness, model systems have so far been the main focus. The interfacial properties of mixtures with more complex interactions and/or shapes remain an open field for exploration. Fundamentally, this is of great interest as these parameters are tunable for colloidal suspensions, which is in contrast to molecular systems. From a practical point of view, these interfacial properties may be relevant for water-in-water emulsions, which can be composed of phase separating aqueous protein–polysaccharide mixtures. Stabilising the fluid–fluid interface of these emulsion droplets against coalescence requires intricate knowledge of the details of the interface and could be a promising method to develop fat/oil-free food emulsions and other compartmentalised aqueous structures.

In some cases, interactions *between* depletants are of importance, such as in binary colloidal systems (e.g. mixtures of small and large spheres, mixtures of spheres and rods). The presence of these depletant–depletant interactions significantly influences the phase behaviour (Chaps. 6, 7). For binary mixtures of hard spheres, the colloidal gas–liquid phase transition is absent, while solid–solid phase equilibria appear. Rods turn out to be highly efficient depletants; free volume theory predicts that they induce phase transitions at very low volume fractions, in line with computer simulations and experiments of well-defined systems.

Nematic, smectic and columnar liquid-crystalline phases can be induced by the addition of polymers to rods or platelets (Chaps. 8,9), and their phase behaviour turns out to be remarkably rich: a zoo of three-, four- and five-phase coexistence is found, although this may appear to be at odds with the Gibbs phase rule.

The addition of colloidal spheres to rod-like particles leads to interesting phase behaviour, such as a smectic phase consisting of alternating two-dimensional liquidlike layers of rods and spheres [6–8]. It not only demonstrates the possibility for control of colloidal self-assembly using depletion phenomena, but also highlights the clear need for the use of models to guide such efforts. Anisotropic mixtures display remarkable nonequilibrium phenomena, e.g. the formation of gels and glasses, and unconventional responses to shear forces [9,10]; yet, these remain under-explored. The structure and dynamics of their phases have been studied using a range of experimental techniques, including X-ray and neutron scattering, microscopy and rheology; but further understanding of their properties is needed in order to capitalise on their potential applications.

The emergence of experimental model systems comprising cube-like colloids allows a new range of colloidal solids to be prepared. They show surprising structures depending on the exact shape and size of the cubes (Chap. 10). The parameter space (i.e. cube and polymer concentrations, cube shape and cube–polymer size ratio) of these mixtures is, however, almost impossible to fully explore experimentally. Therefore, the availability of a complete theoretical framework that successfully predicts the phase behaviour of cubes mixed with depletants is paramount for making scientific and technological progress. It must take the more complex solid phase states (e.g. $\mathbb{C}_0$-lattice and $\mathbb{C}_1$-lattice) in the theoretical descriptions into account. Further, the nonequilibrium behaviour of cubes and polymers remains yet unexplored.

Depletion interactions have also become relevant and/or recognised in fields beyond classic colloid science, such as biology and technology (Chap. 11). Depletion-induced phase separation can be used to concentrate or purify colloidal suspensions, and exploiting depletion insights in various separation and purification technology applications is still an open field.

Accurate prediction of the depletion forces between colloidal particles in crowded or confined spaces is another unresolved issue. Additionally, crowding phenomena in dense systems affect the dynamics. This is of relevance for understanding, for instance, the formation of structures [11] and the dynamics of proteins [12] in cells. Crowders and the related depletion effects can induce hierarchical assembly and mediate specific biomolecular interactions [13]. These are challenging and promising topics where chemistry, physics and biology and chemistry meet.

Photovoltaics [14], energy storage materials [15], emerging battery technologies, fuel cells and novel products often consist of multi-component colloids and/or colloid–polymer hybrid systems; and consequently, depletion phenomena play an important role as they provide structure, affect dynamics and modify the phase stability. The colloidal systems that underpin real-world examples are, however, much more complex than the relatively well-defined ones described in this book. It is crucial to extend this knowledge towards these complex systems.

Besides complexity due to shape, charge and crowding, the colloids used in a range of application areas are often soft (e.g. polymer brushes or surfactants), or attract one another due to Van der Waals and/or hydrophobic forces. Association colloids, such as surfactant or copolymeric micelles and vesicles, may drive depletion interactions but have hardly been explored. In practice, dispersity in size and surface chemistry is an issue with colloidal systems; and polymers may feature additional complexity by being, e.g. branched, multi-armed, comb-like, copolymeric, or even responsive to external stimuli, as is the case for some microgels and supramolecular polymers [16]. Predicting how these characteristics affect the physical properties of colloid–polymer mixtures is still difficult.

Another topic that has largely been neglected is the influence of depletion forces on the dynamic properties of multi-component colloidal mixtures. Without a doubt, the rheological properties have implications for the practical applications of these systems. The viscosity, for instance, is not just the result of the combined contributions of colloids and polymers of a colloid–polymer mixture; there must be a complex interplay [17] that also affects, for instance, the fluid-to-gel transition [18].

Despite the significant progress made in understanding the depletion interaction and resulting phase behaviour of colloidal systems, there is still a long way to go. With so many factors influencing the depletion interaction, it is almost impossible to explore them all experimentally; yet mastering their impact on the behaviour and properties is instrumental for the design of the next generation of soft materials. The application of theory and computer simulation to this challenge have significantly advanced our understanding; but there is still a need for more accurate and predictive models and advanced experimental and computer simulation tools that can take into account the various, more complex, factors that influence the depletion force and related properties in practice. Here, we also see opportunities for artificial intelligence and machine learning tools to be applied, which will undoubtedly accelerate the testing of concepts. This powerful combination of tools, and further interdisciplinary endeavours, will provide essential design rules for colloidal systems.

In summary, the depletion interaction is often employed as a tool to induce well-defined attractions in colloidal systems; but it is far more than that, profoundly impacting the phase behaviour and many other properties of colloidal dispersions. This makes it a fascinating and relevant field of research in its own right, with many fundamental questions still unanswered. Future research will, no doubt, lead to valuable insights into the behaviour of colloids, and lead to the development of new materials and technologies.

References

1. Piazza, R.: Soft Matter - the Stuff That Dreams Are Made of. Springer, Heidelberg (2010)
2. Chew, P.Y., Reinhardt, A.: J. Chem. Phys. **158**, 030902 (2023)
3. Schmid-Fetzer, R.: J. Phase Equil. Diff. **35**, 735–760 (2014)
4. Poon, W.C.K.: Mater. Res. Bull. **96** (2004)
5. Fleer, G.J., Tuinier, R.: Adv. Colloid Interface Sci. **143**, 1 (2008)
6. Koda, T., Numajiri, M., Ikeda, S.: J. Phys. Soc. Japan **65**, 3551 (1996)
7. Adams, M., Dogic, Z., Keller, S.L., Fraden, S.: Nature **393**, 349 (1998)
8. Dogic, Z., Frenkel, D., Fraden, S.: Phys. Rev. E. **62**, 3925 (2000)
9. Solomon, M.J., Spicer, P.T.: Soft Matter **6**, 1391 (2010)
10. Lettinga, M.P., Dhont, J.K.G.: J. Phys.: Condens. Matter **16**, S3929 (2004)
11. Hancock, R.: Front. Phys. **2**, 53 (2014)
12. König, I., Soranno, A., Nettels, D., Schuler, B.: Angew. Chem. **60**, 10724 (2021)
13. Xiu, F., Knezevic, A., Huskens, J., Kudernac, T.: Small **19**, 2207098 (2023)
14. Saunders, B.R., Turner, M.L.: Adv. Colloid Interface Sci. **138**, 1 (2008)
15. Parant, H., Muller, G., Le Mercier, T., Tarascon, J., Poulin, P., Colin, A.: Carbon **119**, 10 (2017)
16. Peters, V.F.D.: Phase behaviour of mixtures of structurally complex colloids and polymers. Ph.D. thesis, Eindhoven University of Technology (2021)
17. Miyajima, A., Inoue, R., Onishi, E., Miyake, M., Hyodo, R.: J. Oleo Sci. **68**, 837 (2019)
18. Parisi, D., Truzzolillo, D., Slim, A.H., Dieudonné-George, P., Narayanan, S., Conrad, J.C., Deepak, V.D., Gauthier, M., Vlassopoulos, D.: Macromolecules **56**, 1818–1827 (2023)

Thermodynamic Quantification of Phase Transitions and Equilibria

A

A.1 Relation Between Helmholtz Energy, (Osmotic) Pressure and Chemical Potential

For a multi-component system containing N_i particles of type i, classical thermodynamics relate the change of the Helmholtz (or free) energy F to the chemical potential μ, entropy S, interfacial tension γ and pressure P [1–3]:

$$dF = \sum_i \mu_i dN_i - SdT - PdV + \gamma dA_s. \tag{A.1}$$

For a colloidal suspension containing only one type of colloidal particles and considering the solvent as background (effective one-component system) at constant temperature T and constant surface area A_s, this expression simplifies to

$$dF = \mu_c dN_c - PdV, \tag{A.2}$$

where μ_c is now the chemical potential of the colloidal particles and the pressure becomes the osmotic pressure of the colloidal suspension.

The chemical potential and (osmotic) pressure now follow from F:

$$\mu_c = \left(\frac{\partial F}{\partial N_c}\right)_{V,T,A_s} \tag{A.3}$$

and

$$P = -\left(\frac{\partial F}{\partial V}\right)_{N_c,T,A_s} \tag{A.4}$$

© The Editor(s) (if applicable) and The Author(s) 2024
H. N. W. Lekkerkerker, R. Tuinier, M. Vis, *Colloids and the Depletion Interaction*,
Lecture Notes in Physics 1026, https://doi.org/10.1007/978-3-031-52131-7

A.2   Thermodynamics of Phase Transitions of Colloidal Dispersions

In this book we consider both canonical systems (for which we use the Helmholtz energy F to describe the thermodynamic properties) and grand canonical systems, for which the grand potential Ω is used. Here, we illustrate how to obtain the relevant thermodynamic quantities from either one of them by writing the quantity $\mathcal{F}$. This then becomes either F or Ω for the canonical and grand canonical descriptions of the system of interest, respectively. Dimensionless units are used for this generalised Helmholtz energy:

$$\widetilde{\mathcal{F}}_j = \frac{\mathcal{F}_j v_0}{kTV}, \tag{A.5}$$

where the subscript j refers to a certain phase state (fluid, solid, nematic, smectic, columnar, etc.). A dimensionless colloid concentration (volume fraction) ϕ, chemical potential and (osmotic) pressure are used:

$$\phi = \frac{N_c v_0}{V}, \tag{A.6}$$

$$\widetilde{\mu}_{c,j} = \frac{\mu_{c,j}}{kT}, \tag{A.7}$$

$$\widetilde{P}_j = \frac{P_j v_0}{kT}. \tag{A.8}$$

This normalisation simplifies numerical computations. The standard thermodynamic relations enable one to derive the osmotic pressure and chemical potential from the ϕ-dependence of $\widetilde{\mathcal{F}}_j$. For the chemical potential of a component, applying Appendix A.3 gives

$$\widetilde{\mu}_{c,j} = \left(\frac{\partial \widetilde{\mathcal{F}}_j}{\partial \phi} \right)_{T,V(,n_d^R)}, \tag{A.9}$$

where n_d^R is the depletant concentration in case of an external reservoir. For the osmotic pressure Appendix A.4 gives

$$\widetilde{P}_j = -\left(\frac{\partial (\widetilde{\mathcal{F}}_j/\phi)}{\partial (1/\phi)} \right)_{T,V(,n_d^R)} = \phi^2 \left(\frac{\partial (\widetilde{\mathcal{F}}_j/\phi)}{\partial \phi} \right)_{T,V(,n_d^R)}, \tag{A.10}$$

so that

$$\widetilde{P}_j = \phi\widetilde{\mu}_j - \widetilde{\mathcal{F}}_j. \tag{A.11}$$

The pressure $\tilde{P}_j$ and chemical potential $\tilde{\mu}_j$ can be directly connected via the Gibbs-Duhem relation:

$$\phi \left(\frac{\partial \tilde{\mu}_j}{\partial \phi} \right)_{T,V(,n_d^R)} = \left(\frac{\partial \tilde{P}_j}{\partial \phi} \right)_{T,V(,n_d^R)}, \tag{A.12}$$

which follows from differentiation of Eq. (A.11) with respect to ϕ. Binodal points are found by solving for the coexistence between the different phases:

$$\tilde{P}_I = \tilde{P}_{II}, \tag{A.13}$$

and

$$\tilde{\mu}_I = \tilde{\mu}_{II}. \tag{A.14}$$

The subscripts I and II denote the coexisting phases of the system. Spinodals can be found from

$$\left(\frac{\partial \tilde{\mu}_j}{\partial \phi} \right)_{T,V(,n_d^R)} = \left(\frac{\partial \tilde{P}_j}{\partial \phi} \right)_{T,V(,n_d^R)} = 0. \tag{A.15}$$

An isostructural phase coexistence region is marked by a critical point, which can be calculated from the conditions

$$\left(\frac{\partial \tilde{\mu}_j}{\partial \phi} \right)_{T,V(,n_d^R)} = \left(\frac{\partial^2 \tilde{\mu}_j}{\partial \phi^2} \right)_{T,V(,n_d^R)} = 0, \tag{A.16}$$

or from

$$\left(\frac{\partial \tilde{P}_j}{\partial \phi} \right)_{T,V(,n_d^R)} = \left(\frac{\partial^2 \tilde{P}_j}{\partial \phi^2} \right)_{T,V(,n_d^R)} = 0. \tag{A.17}$$

If there is a critical point one can often also find three coexisting phases, for which the conditions

$$\tilde{P}_I = \tilde{P}_{II} = \tilde{P}_{III}, \tag{A.18}$$

and

$$\tilde{\mu}_I = \tilde{\mu}_{II} = \tilde{\mu}_{III} \tag{A.19}$$

hold, which enables one to determine a triple point or region. Extrapolation to quadruple (Chaps. 9 and 10) or quintuple ([4,5]) phase coexistence is straightforward.

Whenever a phase state has a stable or metastable critical point (CP), isostructural phase coexistence takes place. The transition from a stable to a metastable isostructural phase coexistence is defined by the critical end point (CEP), where the CP and the TP (or in general, multi-phase coexistences) of the corresponding isostructural coexistences merge:

$$\widetilde{\mu}_I = \widetilde{\mu}_{II} = \cdots \quad \text{and} \quad \widetilde{P}_I = \widetilde{P}_{II} = \cdots \tag{A.20}$$

and

$$\frac{\partial \widetilde{\mu}_I}{\partial \phi} = \frac{\partial^2 \widetilde{\mu}_I}{\partial^2 \phi} = 0 \quad \text{and} \quad \frac{\partial \widetilde{P}_I}{\partial \phi} = \frac{\partial^2 \widetilde{P}_I}{\partial^2 \phi} = 0. \tag{A.21}$$

Statistical Mechanical Derivation of the Free Volume Theory

Here, we present a statistical mechanical derivation of the grand potential for a mixture of hard spheres mixed with penetrable hard spheres (PHSs), inspired by the work of Meijer and Frenkel [6] and Dijkstra et al. [7]. Following statistical mechanics [1] the grand potential is defined by

$$\Omega(N_c, V, T, \mu_d) = -kT \ln \Xi(N_c, V, T, \mu_d), \tag{B.1}$$

where Ξ is the grand canonical partition function

$$\Xi = \sum_{N_d=0}^{\infty} \exp(\mu_d N_d / kT) Q(N_c, V, T, N_d). \tag{B.2}$$

Here, Q is the canonical partition function

$$Q = \frac{1}{\Lambda_c^{3N_c} \Lambda_d^{3N_d} N_c! N_d!} \int \exp[-(U_c + U_{cd})/kT] d\mathbf{R}^{N_c} d\mathbf{r}^{N_d}, \tag{B.3}$$

where U_c is the interaction between the N_c hard spheres and U_{cd} the interaction between the N_c hard spheres and the N_d (depletants). The latter interaction term limits the integration over the position of the PHSs to the free volume V_{free}, which is a function of the positions $\mathbf{R}^{N_c}$ of the N_c hard spheres. This leads to

$$Q = \frac{1}{\Lambda_c^{3N_c} \Lambda_d^{3N_d} N_c! N_d!} \int \exp[-U_c/kT] V_{\text{free}}^{N_d} d\mathbf{R}^{N_c}. \tag{B.4}$$

© The Editor(s) (if applicable) and The Author(s) 2024
H. N. W. Lekkerkerker, R. Tuinier, M. Vis, *Colloids and the Depletion Interaction*,
Lecture Notes in Physics 1026, https://doi.org/10.1007/978-3-031-52131-7

We next take into account that

$$\sum_{N_{\mathrm{d}}=0}^{\infty} \frac{\exp(\mu_{\mathrm{d}} N_{\mathrm{d}}/kT) V_{\mathrm{free}}^{N_{\mathrm{d}}}}{\Lambda_{\mathrm{d}}^{3N_{\mathrm{d}}} N_{\mathrm{d}}!} = \exp[P^{\mathrm{R}} V_{\mathrm{free}}/kT], \tag{B.5}$$

where we have used the left-hand side of Eq. (B.5) as the grand canonical partition function of the PHSs with chemical potential μ_{d} in a volume V_{free}. Combining Eqs. (B.2), (B.4) and (B.5) yields [6]

$$\Xi = \frac{1}{\Lambda_{\mathrm{c}}^{3N_{\mathrm{c}}} N_{\mathrm{c}}!} \int \exp[-(U_{\mathrm{c}} - P^{\mathrm{R}} V_{\mathrm{free}})/kT] \mathrm{d}\mathbf{R}^{N_{\mathrm{c}}} \tag{B.6}$$

$$= Q(N_{\mathrm{c}}, V, T) \left\langle \exp\left(\frac{P^{\mathrm{R}} V_{\mathrm{free}}}{kT}\right)\right\rangle_0. \tag{B.7}$$

The quantity $Q(N_{\mathrm{c}}, V, T)$ is the canonical partition function of the N_{c} hard spheres, and the pointed brackets with subscript 0 indicate an average over the unperturbed configurations of the hard spheres. Using Eqs. (B.7), (B.1) can be written as

$$\Omega = -kT \ln Q(N_{\mathrm{c}}, V, T) - kT \ln\left\langle \exp\left(\frac{P^{\mathrm{R}} V_{\mathrm{free}}}{kT}\right)\right\rangle_0 \tag{B.8}$$

$$= F_0(N_{\mathrm{c}}, V, T) - kT \ln\left\langle \exp\left(\frac{P^{\mathrm{R}} V_{\mathrm{free}}}{kT}\right)\right\rangle_0. \tag{B.9}$$

This expression for Ω is exact but, from a computational point of view, difficult to handle. To make progress, we replace the average of the exponent with the exponent of the average and obtain the following approximate expression for the grand potential:

$$\overline{\Omega} = F_0(N_{\mathrm{c}}, V, T) - P^{\mathrm{R}} \langle V_{\mathrm{free}} \rangle_0 \tag{B.10}$$

This is precisely expression Eq. (3.26), obtained from the thermodynamic integration route using the approximation Eq. (3.24).

Configurational Integrals for a Columnar Phase of Colloidal Hard Platelets

In Wensink's approach [8], the structure of a columnar phase is envisioned in terms of columns ordered along a perfect lattice in two lateral dimensions with a strictly one-dimensional fluid behaviour of the constituents in the remaining direction along the columns. For the configurational integral of a system of N parallel platelets with thicknesses L and diameter D with their centre of mass moving along the plate normal on a line of length ℓ, Tonks [9] derived

$$Q_{\text{fluid}}(N, \ell, T) = \frac{1}{\Lambda^N N!} [\ell - NL]^N. \tag{C.1}$$

The columns are assumed to be strictly linear and rigid. At high packing fractions, the rotational freedom of each platelet is assumed to be asymptotically small, and Equation (C.1) may be written as

$$Q_{\text{fluid}}(N, \ell, T) \approx \frac{Q_{\text{or}}}{\mathcal{V}_1^N N!} [\ell - N \langle L_{\text{eff}} \rangle]^N \tag{C.2}$$

where $\mathcal{V}_1$ represents the total 1D thermal volume including contributions arising from the 3D rotational momenta of the platelet. Furthermore,

$$Q_{\text{or}} = \exp[-N \langle \ln 4\pi f \rangle], \tag{C.3}$$

is an orientational partition integral depending on the orientational probability distribution f. In the *mean-field* description implied by Eq. (C.2) there is no coupling between the orientational degrees of freedom of the platelets. Up to leading order in the polar angle θ, the rotational freedom of the platelets is expressed in an *effective entropic thickness*:

$$\langle L_{\text{eff}} \rangle = L \left\{ 1 + \frac{1}{2} \frac{D}{L} \int d(\cos\theta) |\theta| f(\theta) + \cdots \right\}. \tag{C.4}$$

© The Editor(s) (if applicable) and The Author(s) 2024
H. N. W. Lekkerkerker, R. Tuinier, M. Vis, *Colloids and the Depletion Interaction*,
Lecture Notes in Physics 1026, https://doi.org/10.1007/978-3-031-52131-7

The prefactor '1/2' in Eq. (C.4) (partly) corrects for the azimuthal rotational freedom, and accounts for the fact that the excluded length between two platelets at fixed polar angles minimises when the azimuthal orientations are the same. From $F = -kT \ln Q$, the 1D fluid Helmholtz energy then follows as

$$\frac{F_{\text{fluid}}}{kTN} = \ln \tilde{\mathcal{V}}_1 \rho - 1 + \langle \ln 4\pi f \rangle - \ln \left[1 - \rho \langle \tilde{L}_{\text{eff}} \rangle \right], \tag{C.5}$$

where $\tilde{\mathcal{V}}_1 = \mathcal{V}_1/L$, $\rho = NL/\ell$ is the reduced *linear density* and $\tilde{L}_{\text{eff}} = L_{\text{eff}}/L$ is the effective thickness.

The equilibrium form $f(\theta)$ can be found by a formal minimisation of the Helmholtz energy:

$$\frac{\delta}{\delta f} \left(\frac{F}{kTN} - \lambda \int d\hat{u} \, f(\hat{u}) \right) = 0 \tag{C.6}$$

under the normalisation constraint (see also Sect. 8.2). The Lagrange multiplier λ ensures the normalisation of f. The Helmholtz energy given by Eq. (C.5) only depends on single-particle orientational averages. This enabled Wensink [8] to obtain the following analytic form for the equilibrium ODF:

$$f(\theta) = \frac{\iota^2}{4\pi} \exp[-\iota|\theta|] \tag{C.7}$$

with

$$\iota = \left(\frac{3}{2} \frac{D}{L} \right) \frac{\rho}{1 - \rho}. \tag{C.8}$$

The resulting expressions for the orientational entropy and entropic thickness are

$$\langle \ln 4\pi f \rangle \sim 2 \ln \iota - 2 \tag{C.9}$$

$$\langle \tilde{L}_{\text{eff}} \rangle \sim 1 + \left(\frac{D}{L} \right) \frac{1}{\iota}. \tag{C.10}$$

Applying Eq. (8.15) for a column of discs enables the relation of the nematic order parameter S to ι:

$$S \equiv \langle \mathcal{P}_2(\cos\theta) \rangle \sim 1 - \frac{3}{2} \langle \theta^2 \rangle \sim 1 - \frac{9}{\iota^2}. \tag{C.11}$$

The Helmholtz energy associated with the positional order along the lateral directions of the columnar liquid crystal can be approximated as follows: one can map the system onto an ensemble of N disks ordered into a 2D lattice. To a good approximation, the configurational integral of the system near the close packing density can be computed using the Lennard-Jones–Devonshire (LJD) cell theory [10]. Within the framework of the cell model, particles are considered to be localised in 'cells' centred on the sites of a fully occupied lattice (of some prescribed symmetry). Each particle experiences a potential energy $u_{\text{cell}}^{\text{nn}}(\mathbf{r})$ generated by its nearest neighbours. In the simplest version, the theory presupposes each cell to contain one particle moving *independently* from its neighbours. The N-particle canonical partition function can then be written as

$$Q_{\text{LJD}}(N) = \frac{1}{\Lambda^{2N}} \int d\mathbf{r}^N \exp\left[-\frac{U(\mathbf{r}^N)}{kT}\right]$$

$$\approx \left(\frac{1}{\Lambda^2} \int d^2\mathbf{r} \exp\left[-\frac{u_{\text{cell}}^{\text{nn}}(\mathbf{r})}{2kT}\right]\right)^N. \tag{C.12}$$

For hard interactions, the second phase space integral is simply the cell *free area* available to each particle. If the nearest neighbours form a perfect hexagonal cage, the free area is given by $\sqrt{3}(\Delta_C - D)^2/2$. The configurational integral then becomes

$$Q_{\text{LJD}}(N) = \left(\frac{\frac{1}{2}\sqrt{3}\Delta_C^2}{\Lambda^2}\right)^N \left(1 - \bar{\Delta}_C^{-1}\right)^{2N} \tag{C.13}$$

where $\bar{\Delta}_C$ is a measure for the translational freedom that each particle experiences within the cage. This leads to Eq. (9.21).

References

1. Hill, T.L.: An Introduction to Statistical Thermodynamics. Addison-Wesley, New York (1962)
2. McQuarrie, D.A.: Statistical Mechanics. University Science Books, Sausalito (2000)
3. Atkins, P., De Paula, J., Keeler, J.: Physical Chemistry, 12th edn. Oxford University Press, Oxford (2022)
4. Peters, V.F.D., Vis, M., González García, Á., Wensink, H.H., Tuinier, R.: Phys. Rev. Lett. **125**, 127803 (2020)
5. Opdam, J., Peters, V.F.D., Wensink, H.H., Tuinier, R.: J. Phys. Chem. Lett. **14**, 199 (2023)
6. Meijer, E.J., Frenkel, D.: J. Chem. Phys. **100**, 6873 (1994)
7. Dijkstra, M., Brader, J.M., Evans, R., Phys, J.: Condens. Matter **11**, 10079 (1999)
8. Wensink, H.H.: Phys. Rev. Lett. **93**, 157801 (2004)
9. Tonks, L.: Phys. Rev. **50**, 955 (1936)
10. Lennard-Jones, J.E., Devonshire, A.F.: Proc. Roy. Soc. **163A**, 53 (1937)

Index

A

AAA crystal phase; phase state of hard rods, 244, 271–273

Aarts, 152, 168, 169, 188, 191, 193, 199, 200, 202, 203

ABC crystal phase; phase state of hard rods, 244, 271, 273, 274, 279

Adhesive hard sphere interaction, 11

Adhesive hard spheres, 11

Adsorbed polymer, 9, 143, 145

$Al(OH)_3$ gibbsite, 285

Aggregation, 3, 8, 11, 18–20, 22, 46, 49, 99, 172, 173, 220, 264, 353, 354

Anchored polymer chains: see brush, 10

Anchored polymer chains: see grafted polymer chains, 24

Anisotropic colloidal particles, 49, 50

Anisotropic particles, 50

Arrested state, 170

Asakura and Oosawa, 12, 13, 17, 23, 40, 42, 67, 68, 72, 80, 99, 101, 223

Asphericity, 322

Atomic force microscope, 110

Attached polymer: see adsorbed polymer, 11

Avogadro's number, 15, 121

B

Bacteria, 19, 22, 49, 267, 278–280, 344

Bacterial cells, 278

Bacteriophage, 263, 267

Beijerinck, 19

Binodal, 27, 28, 36, 38, 40, 41, 44, 47, 136–138, 149, 150, 161, 165, 210, 217–219, 275, 276, 305, 338, 339

Binodal: definition, 27

Bjerrum length, 7, 251

Boehmite rods, 269, 270

Bolhuis, 37, 39–41, 129, 162, 165, 227, 231, 244, 253, 271–275, 336

Boltzmann constant, 1

Boublík equation of state, 323

Boublík–Mansoori–Carnahan–Starling–Leland equation of state, 214

Bragg, 3, 302

Bridging flocculation, 46, 143

Bristol, 24

Brownian Dynamics simulation, 148

Brownian motion, 1, 20, 78, 113, 121, 148, 171, 202, 355

Brush, 3, 9, 10, 15, 24, 34, 36, 46, 128, 164, 250

Brush: see also grafted polymer chains, 46

B_2: see second osmotic virial coefficient, 11

C

Capillary length, 188, 190, 199

Capillary velocity, 199, 200

Capillary waves, 15, 43, 198, 199, 201

Carnahan–Starling, 123, 124, 213, 290, 323

Carnahan–Starling equation of state, 123, 125, 134, 290, 323

Casein micelles, 3, 34, 178

Casimir forces, 46

Cell model, 123, 125, 272, 294

Cells, 343

Cell: see living cell or bacteria, 344

Cellular organisation, 49, 344

CEP: critical end point, 138, 149, 162, 261, 368

Charged colloids, 4, 6–8, 36, 41, 47, 48, 250

© The Editor(s) (if applicable) and The Author(s) 2024
H. N. W. Lekkerkerker, R. Tuinier, M. Vis, *Colloids and the Depletion Interaction*,
Lecture Notes in Physics 1026, https://doi.org/10.1007/978-3-031-52131-7